Chemistry Dept
Copy #2

preface

to the second edition

In the years since the first edition of this book was published there has been an enormous influx into organic chemistry courses of students preparing for medical, dental, and other health-oriented careers. We believe that these students, no less than those majoring in chemistry or chemical engineering, deserve and, in fact, require a modern organic laboratory experience. To this end we have revised the book with the addition of experiments that should be of special interest and value to pre-meds, pre-dents, pharmacy majors, and other biologically oriented students.

We believe that the wide acceptance of the first edition can be attributed to its reasonable combination of theory and practice, of classical and modern techniques. These features have been retained in the second edition, with significant additions and improvements based on our own experience and on suggestions from other users of the text.

The new material includes an expanded carbohydrate chapter (Chapter 22) and new chapters on amino acids and peptides (Chapter 23), the isolation of natural products (Chapter 24), and the synthesis of medicinally related heterocyclic compounds (Chapter 18). In addition to the new bio-organic material, there are three other new chapters. Chapter 7 presents a general outline of isomerism and provides suggestions for the study of stereoisomers by means of molecular models. In Chapter 21 the resolution of racemic α-phenylethylamine with tartaric acid is described, and there are suggestions for optional projects of a related nature. Chapter 26 is an introduction to the literature of organic chemistry. It includes a detailed example of how to look up an organic compound in *Beilstein* and *Chemical Abstracts* and a set of Exercises that presents a challenge to the students to make their own introduction to the literature.

New experiments and additional material have been added to chapters retained from the first edition. These include a Grignard synthesis of a secondary alcohol which utilizes starting materials prepared in other experiments; procedures for vacuum distillations and dry-column chromatography; new appendices giving more complete infrared and nmr data; and detailed illustrations of solutions of "paper unknowns" on the basis of spectroscopic data combined with classical ("wet") identification procedures.

We would, of course, welcome criticisms from users of the text. We are indebted to many users of the first edition whose criticisms and suggestions have been helpful to us in the preparation of this edition, and we would acknowledge them individually if it were not for the large number and the fear that in naming some we would overlook others. However, we do wish to express special thanks to three of our colleagues for suggestions and advice on some of the new chapters: G. Barry Kitto and Eddie C. Smith (Chapters 23 and 24) and Aubrey Skinner (Chapter 26).

R.M.R. / J.C.G. / L.B.R. / A.S.W.

preface
to the first edition

Since organic chemistry is fundamentally an experimental science, this book is written with the view that laboratory experience should be an integral part of the first course in organic chemistry. In recent years the trend in teaching the elementary course has been to incorporate increasingly more theory into the lecture material, particularly in the first semester, thereby increasing the difficulty in correlating laboratory experiments with the lecture part of the course. In some schools this problem has been "solved" by postponing organic laboratory until after one semester of lecture, or even later. However, this solution is unsatisfactory, not only because of the inherent scheduling difficulties, but also because it begs the question.

A better approach is to design an up-to-date laboratory course that can be correlated with the most modern lecture texts. This book, having been designed to provide the basis for such a laboratory course, emphasizes both the theoretical and practical aspects of a variety of organic reactions. The experiments on the halogenation of hydrocarbons, for example, are illustrative of the mechanistic and the potential synthetic utility of free-radical substitution. The experiments involving electrophilic aromatic substitution are also designed to promote an understanding of the mechanistic and synthetic aspects of this important class of reactions.

Another factor that must be considered in planning a modern laboratory text is the application of recently developed spectroscopic (nmr and ir) and chromatographic (tlc and glpc) techniques to experimental organic chemistry. Since it is important that the student become acquainted with these techniques as soon as possible, we have discussed the fundamentals of the methods in early chapters, and have applied them in experiments throughout the book.

It is possible to give students experience with the modern instrumental methods suggested in the experiments even though access to the necessary instruments may be limited. For example, the instructor can demonstrate the use of the instrument(s) to the students and then obtain typical analyses to post in the laboratories. Even if no instrumentation is available, students can still gain valuable experience by studying the illustrative spectra and chromatograms presented in the text.

A third important factor is to provide the student with useful experimental challenges. Toward this end, we have included procedures involving the *in situ* generation of volatile or highly reactive reagents, the handling of substances that are sensitive to moisture or air, and the safe removal of noxious gases formed in the course of a reaction. Several multi-step syntheses are described to afford the student the opportunity to experience the stimulation and satisfaction of successfully completing a series of reactions.

We have endeavored to maintain a good balance between the new and the old by retaining many classical procedures and synthetic experiments of proven value. For example, although we believe that an introduction to identification of organic compounds by spectroscopic methods must now be given in the first organic laboratory course, we have provided an adequate outline of the classical "wet" scheme of qualitative organic analysis, because of its pedagogical value and because students enjoy it—and thus learn a great deal of practical organic chemistry in the process.

The experimental procedures in this book are preceded by a rather complete discussion both of the theory upon which the experiment is based and of the important practical aspects—the how's and why's—involved in the experiment. In general, experimental directions are detailed and specific in the earlier chapters of the book but become less so as the student gains experience in the techniques of organic chemistry. In this way we hope to avoid the "cookbook" approach while at the same time providing sufficient information so that relatively little reference to other sources is required for a *complete* understanding of the experiments. In short, we have attempted to produce a laboratory *textbook* rather than a laboratory manual.

Acknowledgment is made to Sadtler Research Laboratories, Inc. for some of the spectra incorporated in this book. We are indebted to Professor R. Pettit and Dr. James C. Barborak for the procedure for tropylium iodide. We thank Professors Arthur Fry, Tom S. Ellis, James G. Traynham, and Frank Cartledge who read part or all of the text and tested some of the experiments. Special thanks are due to Professor Stanley H. Pine who provided us with a thorough critique and class testing of our book. Grateful acknowledgment is made to the Research and Development Center for College Instruction of Science and Mathematics of the University of Texas at Austin for financial support.

R.M.R. / J.C.G. / L.B.R. / A.S.W.

contents

experimental procedures

introduction

The laboratory part of an introductory course in organic chemistry is complementary to the lecture part; this is where you learn firsthand that the compounds and reactions described in lecture are not merely abstract notations. In addition, many of the theoretical concepts discussed in the lectures are amenable to experimentation at even an introductory level, and you will find that collecting and interpreting your own data will add reality to the theoretical framework of the subject.

Probably the most important purpose of the laboratory, however, is the introduction it provides to the experimental techniques of the practicing organic chemist. Learning to handle organic chemicals and manipulate apparatus is obviously a necessary part of a chemist's education. Less obvious, but equally important, is the learning of the scientific approach to laboratory work, which results in what is called good experimental technique. Some suggestions to aid in developing good technique are given in the following paragraphs.

Never begin any experiment unless you understand the overall purpose of the experiment and the reasons for each operation involved. This requires *studying* (not just reading) an experiment *before coming to the laboratory*. For this reason laboratory experiments are always assigned several days in advance, except under unusual circumstances. You will find that not only will your performance in laboratory be better if you are well-prepared, but also you will benefit much more from the experiments, in knowledge as well as in grades.

Neatness is an important part of good technique. Carelessness in handling chemicals may not only lead to poor results but is often unsafe. Similarly, careless assembly of apparatus is not only aesthetically displeasing but is also dangerous.

The procedures given in this book are meant to be followed closely. There is usually a reason why each operation is to be carried out exactly as described, although the reason may not, at first, be obvious to the beginning student. In the earlier chapters of the book, the operations required in the experimental procedures are described rather specifically, but in later chapters, after experimental techniques have become more familiar to you, the directions are less specific. In this way we hope to avoid your feeling that you are following "cookbook" directions. As you gain more practice in the techniques of the organic laboratory, you may even formulate alternative procedures for performing a reaction, purifying a product, etc. However, for the sake of safety it will be wise for you to check any original plans with your instructor. In more advanced courses you may expect to use your chemical initiative more freely.

Stopping points. In some of the experimental procedures, stars (*) have been inserted to indicate points at which the experiment may be interrupted without affecting the results unfavorably. This has been done to help instructors and students in planning to make the best use of laboratory time. For example, some experiments may be started toward the end of a laboratory period and carried through to a starred point (*) and then discontinued until the next laboratory period, the reaction mixture being safely stored in the desk in the interim. In cases where it is already obvious from the text that the experiment may or should be interrupted, stars are not inserted.

One of the most valuable characteristics that you can develop as an experimentalist is that of being *observant* during all stages of an experiment. Was there a fleeting color change when a drop of reagent was added to the solution? Was a precipitate formed? Is the reaction exothermic? Such observations may seem insignificant at the time, but may later prove to be vital to the correct interpretation of an experimental result. All such observations should be *recorded*, which brings us to the subject of notebooks.

A permanent record of all laboratory work should be kept in a *bound notebook*, approximately 8×10 inches, rather than in a loose-leaf or spiral notebook. Pages of a bound book are less likely to tear out accidentally and be lost. Always write in ink. It will be easier to keep the notes neat and legible if the pages are lined horizontally. (Although notebooks with vertical as well as horizontal lines are sometimes recommended for laboratory notes, particularly in physical chemistry, they are less desirable for organic chemistry laboratory notes. In the special cases where graphs are necessary, they may be pasted in.) Use the first page as a title page and reserve two additional pages to be filled in as a table of contents. The pages should be numbered throughout the notebook.

Your instructor will probably have specific directions regarding the format of your notebook which you should follow. These may include a preliminary "write-up" giving the title of the experiment, complete equations for the main and side reactions, calculations of the theoretical yield in synthetic experiments, etc.[1] You may be allowed some flexibility in style, but proper recording of experimental results does not allow great literary license.

The overriding requirements of good experimental description are accuracy and completeness of observation and recording. The results should usually be summarized and conclusions should be drawn from each experiment. If the results are obviously not those expected, an explanation should be given.

Safety in the Laboratory

The chemical laboratory is potentially a dangerous place; it contains flammable liquids, fragile glassware, and corrosive and poisonous chemicals. However, if proper precautions are taken and safe procedure is followed, it is no more dangerous than an

---

[1] See Appendix II for a suggested notebook format.

ordinary kitchen or bathroom. The accident rate in the chemical industry is actually lower than in many other types of industries. Unfortunately, however, there is usually less attention paid to safety measures in school laboratories than in industrial laboratories. Your own safety will largely depend on your knowledge of possible dangers and your adherence to proper procedures; some of these dangers and the ways to avoid them will be outlined here. Other specific precautions will be pointed out in the experimental directions at the appropriate places.

The most common types of danger in the organic laboratory may be divided into three categories: (1) fire and explosion, (2) chemicals, and (3) glassware. Your eyes are particularly susceptible to injury from all of these dangers. Therefore, *safety glasses must be worn at all times in the laboratory*.[2]

Fire and Explosion Precautions

1. Avoid using flames in the laboratory whenever possible.
2. If flames are used, observe the following precautions:

(a) Never heat a flammable liquid in an open container; *i.e.*, a flame should be used only when the container is protected by a condenser. Flammable solvents commonly used in the undergraduate laboratory include ethyl ether, carbon disulfide, hydrocarbons (pentane, hexane, benzene, toluene, etc.), alcohols (methanol, ethanol, 2-propanol), ketones (acetone, butanone), and esters (ethyl acetate). Use a steam bath or an electrical heating device if possible.

(b) When a flammable liquid is heated in a distillation or at reflux, make certain that all connections are tight and strain-free. When a very volatile liquid is being heated, it is good practice to connect a rubber tube to the atmospheric opening of the assembly and run the tube over the edge of the bench, away from your burner (and your neighbor's).

(c) Never pour flammable liquids from one container into another within several feet of a flame. Do not pour flammable liquids into the center trough of a laboratory bench, as the vapors may be carried near a flame farther down the bench.

(d) Never allow the flammable distillate from a condenser to drip free into a receiver several inches below it, particularly if it is near the flame of your neighbor; use an adapter to conduct the distillate into the receiver.

3. Never heat a closed, pressure-tight assembly of apparatus, even if it includes a condenser. The increase in pressure resulting from the heating may cause the assembly to blow apart or explode and, if a flame is being used, any flammable liquid present in the assembly may be ignited.

4. Whenever an exothermic reaction is being carried out, prepare a cold- or ice-water bath and have the reaction vessel positioned so that the bath may be placed around it if the reaction shows signs of getting out of control.

5. Familiarize yourself with the location of the nearest fire extinguisher and be sure you know how to use it. Your instructor will explain its operation. Also ascertain the location of any fire blankets, safety showers, or eye-wash fountains in the laboratory.

---

[2] Opinion is divided as to whether ordinary eye glasses afford satisfactory protection. Consult your laboratory instructor if you wish to wear only your own corrective glasses.

Chemical Precautions

1. Do not allow any chemicals to come in contact with your skin unnecessarily. Transfer solid materials by means of spatulas or other appropriate utensils. Should any chemical be spilled on the skin, wash the area immediately with copious amounts of soap and water. Inform your instructor if in doubt as to the toxicity of the chemical. Be particularly careful to avoid contact of open cuts or abrasions with poisons. Never use organic solvents such as acetone or alcohol to wash an organic chemical from the skin; such solvents may actually increase the rate of absorption of the chemical into the skin. Always wash your hands at the end of the laboratory period.

2. Never taste any chemical unless specifically instructed to do so.

3. Avoid inhalation of fumes and vapors of chemicals and solvents, as much as possible. The laboratory should be well ventilated. When working with volatile materials in open systems, use a fume hood if one is available. If a noxious gas or vapor (*e.g.*, hydrogen chloride) is produced in a reaction, attach an efficient gas trap.

Although the odor of an organic compound is often a useful criterion of its type or even of its identity, discretion should be used in smelling any substance. Some compounds are extremely irritating, if not toxic. Hold any unknown substance about six inches from the nose and, using your hand, gently waft the vapors toward you.

4. Clean up any spilled chemical immediately; if you wait, it may cause damage and the cleanup may be more difficult.

For further information on the toxicity and other physiological properties of organic compounds, consult the references at the end of the Introduction.

Glassware: Precautions and Cleaning

The cardinal rule in handling laboratory glassware is *never apply undue pressure or strain to any piece of glassware.*

This rule applies to insertion of thermometers or glass tubes into rubber stoppers, rubber tubes, or corks. "If you have to force it, don't do it!" Make the hole larger or use a smaller piece of glass. Lubricate the rubber or cork with a little water or glycerol. Always grasp the glass part very close to the rubber or cork part when pushing the glass into it.

Hopefully, these directions will become less necessary, as more and more laboratories are making the transition to all ground glass-jointed apparatus. However, the cardinal rule stated above applies to this glassware as well; take care that strain does not develop because of carelessly positioned components. Strained glassware will break when heated, or simply upon standing, at times.

If ground glass-jointed glassware is available, it is important that the joints be properly lubricated so that they do not "freeze" and become difficult, if not impossible, to separate. Proper lubrication can be accomplished by putting a thin layer of grease on opposite sides of the male joint, mating the two joints and then rotating them together in order to cover the surfaces of the joints with a thin coating of grease. The quantity of lubricant used is important since too much grease may cause ultimate contamination of the reaction whereas too little will permit the joints to freeze.

Glassware is most easily cleaned *immediately* after use. Most chemical residues can be removed by washing the glassware with detergent and water, or with common organic solvents such as toluene or acetone. (*Caution:* Do *not* use acetone to clean apparatus that contains residual amounts of bromine since a powerful lachrymator, bromoacetone, may thus be formed.)

More stubborn residues may require the use of powerful chemical cleaning solutions such as chromic acid (concentrated sulfuric acid and chromic anhydride) or potassium hydroxide in ethanol. Gentle warming of the organic solvent or cleaning solution on a steam bath will generally hasten the cleaning process. Before you resort to the use of any cleaning agents other than detergent and water, consult your instructor for permission and directions concerning their safe handling.

Brown stains of manganese dioxide on glassware may sometimes be encountered; these can generally be removed by rinsing the apparatus with a 30% aqueous solution of sodium bisulfite.

It is good practice to wipe off any lubricant from ground glass-jointed glassware with a towel or tissue wetted with a solvent such as acetone or hexane before washing with detergent and water. Otherwise the lubricant will stick to the brush used to scrub the glassware and be carried by the brush onto all surfaces it touches.

In Case of Accident

FIRE

Your first consideration is to remove yourself from any danger, *not* to extinguish the fire. Notify your instructor immediately. To help prevent spread of the fire, remove any containers of flammable solvents from the immediate area and turn off any burners, if this is possible. For the most effective use of a fire extinguisher, direct its nozzle toward the base of the flames.

If your clothing is on fire, **DO NOT RUN**; rapid movement will only fan the flames. Roll on the floor to smother the fire and to help keep the flames away from your head. Your neighbors can help to extinguish the flames by using fire blankets, laboratory coats or other items that are immediately available. Do not hesitate to aid your neighbor if he is involved in such an emergency, since a few seconds delay may result in serious injury.

A laboratory safety shower, *if close by*, can be used to extinguish burning clothing, as can a carbon dioxide extinguisher if the flames are not near the head. A carbon tetrachloride extinguisher, however, should **NOT** be employed.

If burns are minor, apply a burn ointment. In case of serious burns, do not apply any ointment but seek professional medical treatment at once.

CHEMICAL BURNS

Areas of the skin with which corrosive chemicals have come in contact should be immediately washed thoroughly with soap and water. If the burns are minor, apply burn ointment; for treatment of more serious burns, see a physician.

Bromine burns can be particularly serious. These burns should first be washed with soap and water and then thoroughly soaked with 10% sodium thiosulfate solution for three hours. Apply cod liver oil ointment and a dressing; see a physician.

If chemicals, particularly corrosive or hot reagents, come in contact with the eyes, immediately *flood* the eyes with water *from the nearest outlet*. A specially designed eye wash fountain is useful if it is available in the laboratory. *Do not touch the eye*. The eyelid as well as the eyeball should be washed with water for several minutes. If the injury appears to be serious, see a physician as soon as possible.

Cuts

Minor cuts may be treated by ordinary first-aid procedures; seek professional medical attention for serious cuts. If severe bleeding indicates that an artery has been severed, apply a tourniquet just above the injury.

A person who is injured severely enough to require a doctor's treatment *should be accompanied* to the doctor's office or infirmary, even if he protests that he is all right. Persons in shock, particularly after suffering burns, are often more seriously injured than they appear to be.

REFERENCES

1. N. I. Sax, *Dangerous Properties of Industrial Materials,* Reinhold Publishing Corp., New York, 1957.
2. *Merck Index of Chemicals and Drugs,* Merck and Co., Inc., Rahway, N. J., 8th ed., 1968.

chapter one
physical constants
of organic compounds

Physical properties such as melting point (mp), boiling point (bp), index of refraction (*n*), and density (*d*) have long been utilized in identification and characterization of organic compounds. More than one compound may exhibit the same constant for one or two of these properties; however, it would be highly fortuitous if more than one compound exhibited the same constant for each of the properties. It should be seen then that a list of physical constants is a highly useful characterization of a substance. In addition, the observed melting point or boiling point may give information about the purity of the substance under consideration. Other properties such as color, odor, and crystal form are also useful.

In Chapter 4 we shall see how more recently developed spectroscopic techniques, based on molecular-physical properties, are extremely useful in identification of organic compounds.

1.1 Melting Point of a Pure Substance

The melting point of a substance is defined as the temperature at which the liquid and solid phases exist in equilibrium with one another without change in temperature. If heat is added to a mixture of the solid and liquid phases of a pure substance at the melting point, ideally no rise in temperature will occur until all the solid has been converted to liquid (melted); if heat is removed from such a mixture, the temperature will not drop until all the liquid has been converted to solid (frozen). Thus the melting point and freezing point of a pure substance are identical. The relationship between phase composition, total supplied heat, and temperature for a pure compound is shown in Figure 1.1. It should be understood that *heat* is being supplied to the compound at a constant rate, and thus that the elapsed time of heating is a cumulative measure of the supplied heat. At the lower temperature (below the melting point) the compound exists in the solid phase and the addition of heat causes the temperature of the solid to rise. As the melting point is reached, the first small amount of liquid appears; equilibrium is

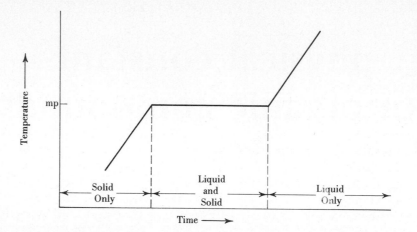

FIGURE 1.1   Phase changes with time and temperature.

established between the solid and liquid phases. As heat continues to be added, the temperature does not change, the additional heat causing conversion of solid to liquid with, however, both phases remaining in equilibrium. As the last of the solid melts, the subsequently supplied heat causes the temperature to rise linearly at a rate which depends on the rate of heat supply.

We may describe the interconversion of the liquid and solid phases in terms of their respective vapor pressures, which are directly related to the rates at which the molecules pass from one phase to the other. For simplicity we may consider only transfer of molecules between the solid and liquid phases. The term vapor pressure may then seem to be a misnomer, but the same "escaping tendency" which produces the measurable equilibrium vapor pressure is responsible for the direct transfer of molecules between the solid and liquid phases. Figure 1.2 represents these transfers between the three

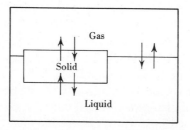

FIGURE 1.2   Phase equilibria.

phases by means of reversible arrows. At the melting point, the vapor pressures of the solid and liquid phases are equal, and no *net* transfer of molecules from one phase to the other occurs unless there is a change in the heat content of the system.

1.2   Effect of Impurities
on Melting Points; Mixtures

Consider a mixture of solid and liquid A at its melting temperature. If a small amount of a second pure material, B, is dissolved in the liquid A, solid A will begin to melt. This is because the addition of B lowers the vapor pressure of liquid A (Raoult's Law), and the vapor pressure of solid A remains the same; thus, the rate at which molecules of solid A pass into the liquid phase is greater than the rate of the reverse process. If the temperature is kept constant by supplying the heat required for this melting process, all of solid A will melt; if no heat is added, the temperature will drop because heat energy is consumed by the melting solid. Since the vapor pressure of a solid decreases more rapidly than the vapor pressure of its solution as the temperature drops, they will become equal at a lower temperature, and equilibrium will be established again at that temperature. This lower temperature then represents the freezing and melting point of the mixture of A and B, that is, "impure A." The more of the impurity B that is added, the lower will be the melting point of the mixture.

Binary mixtures of most organic substances exhibit this kind of behavior; that is, the addition of a second substance progressively lowers the melting point of the first, as may be represented by the melting point-composition diagram of Figure 1.3. In this diagram, a represents the melting point of pure substance A and b the melting point of

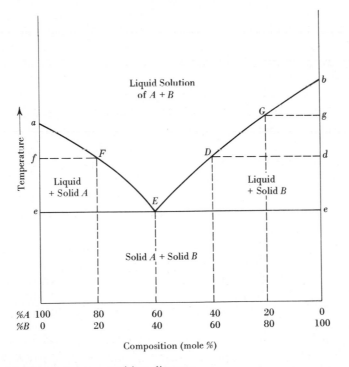

FIGURE 1.3   Melting point-composition diagram.

pure substance B. The point F represents the temperature (f) at which crystals of pure A are in equilibrium with a molten solution of B (20%) in A (80%). If a mixture of this composition is prepared and melted, and the temperature is then lowered to f, pure crystals of A will form in the presence of the melt. As additional heat is removed, A will continue to crystallize, reducing the percentage of A in the melt. The equilibrium temperature will correspondingly decrease because of the reduced vapor pressure of A in the molten solution. When the equilibrium temperature reaches e, at which time the melt has the composition indicated at E, then and *only then* does B, the impurity, also begin to crystallize. A and B will crystallize in the constant ratio 60% A to 40% B, and the melt composition will no longer change. During this stage, the melt of the specified composition is in equilibrium with both solid A and solid B. The temperature remains constant until the entire sample has crystallized and then again begins to fall as the solid cools. Point E is called the *eutectic point*, and it defines the composition at which A and B can cocrystallize (eutectic mixture) and the temperature (e) at which the eutectic mixture freezes (and melts).

Now consider the reverse process, the melting of a solid mixture of 80% A and 20% B, which is of more interest to the organic chemist. Heat is applied and the temperature rises. When it reaches e, A and B will melt together at a constant ratio (eutectic composition), and the temperature will remain constant (remember the eutectic melt is in equilibrium with both solid A and solid B). Eventually B will be entirely melted (as it is the minor component), leaving only solid A in equilibrium with a melt of eutectic composition. As more heat is applied, the remaining A will continue to melt, raising the percentage of A in the molten solution above the eutectic composition. Since the vapor pressure of A in the solution is thus increased, the temperature at which solid A is in equilibrium with the molten solution will rise, and the relationship between the equilibrium temperature and the composition of the molten solution will be represented by the curve EF. When the temperature reaches f the last of A melts. Thus this impure sample A exhibits a depressed melting "point" occurring over a relatively broad range (e–f). Notice that we have been considering the case of substance A with impurity B. If the composition of the solid had been to the right of point E in Figure 1.3 we would conversely speak of substance B with impurity A, and the rising temperature during the melting process would follow curve ED or EG, with melting ranges of e–d or e–g.

It may be noted that the actual melting *range* of a mixture containing 20% impurity may be broader than that of a mixture containing 40% impurity; for example, in Figure 1.3, compare the range e–g with e–d. However, in practice, particularly in the capillary tube method (see below), it is very difficult to observe the initial melting point. If the amount of impurity is small, the amount of liquid produced in the early stages (near the eutectic temperature) is very small. However, the temperatures at which the last crystals disappear (d, g) can be determined quite accurately, so that the mixture containing the smaller amount of impurity will have both a higher final melting point and a narrower (*observed*) range.

It should be noted that a sample whose composition is exactly that of the eutectic mixture will exhibit a *sharp* melting point at the eutectic temperature. Thus a eutectic mixture is sometimes mistaken for a pure compound. In such a case the sample may be identified by adding a small amount of either component (assuming one knows what the components are) and observing that the melting point rises.

The melting point depression produced by the introduction of an impurity into a pure compound may be used to advantage in identification of that compound. This important procedure is known as a *mixture-melting point* [1] determination. It is perhaps best presented in terms of an example. Assume that a compound melts at 133°, and you suspect it is either urea or *trans*-cinnamic acid (both melt at 133°). If the compound is mixed intimately with urea and the melting point of this mixture is found to be lower than either the pure compound or pure urea, then urea is acting as an impurity, and the compound cannot be urea. If the mixture-melting point is identical to that of the pure compound and of pure urea, then the compound is tentatively identified as urea. It should be noted that one can discard candidates by this method with far more certainty than one can elect them. Such a procedure is often used by the chemist when he encounters an identification problem and has on hand samples of likely candidates.

1.3 Other Kinds of Melting Behavior of Mixtures

Although the melting temperature behavior (eutectic formation) described above is typical of most binary mixtures, there are many exceptions to this general pattern. A second component does not always lower the melting temperature of an organic compound. Figure 1.4(a)–(c) illustrate some unusual melting point-composition diagrams.

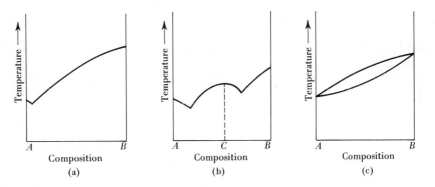

FIGURE 1.4 Unusual melting point-composition diagrams.

In Figure 1.4(a) the eutectic composition contains so little of component *B* that even a fairly small amount of *B* as impurity may actually raise the observed melting point of *A*. Figure 1.4(b) illustrates the formation of a compound (*C*) from *A* and *B*, and Figure 1.4(c) is representative of one type of solid solution formation. For further information consult Weissberger [2] or a physical chemistry text.

---

[1] Frequently referred to (ungrammatically) as a "mixed-melting point" determination.
[2] E. L. Skau and J. C. Arthur, Jr., in *Techniques of Chemistry*, A. Weissberger and B. W. Rossiter, editors, Wiley-Interscience, New York, 1971, Vol. 1, Part 5, Chapter 3, pp. 105 *f*.

1.4 Micro Melting Point Methods

The determination of an exact melting point of an organic compound requires enough material so that equilibrium can be established between the liquid and solid phases and the temperature at the equilibrium state measured, usually by means of repeated cooling and heating cycles which show a plateau in a temperature-time plot. The amount of material required for such procedures is often more than the chemist has available, so that micro methods have been developed which are not so exact but which are convenient and require negligible amounts of sample.[3] The most commonly used method is the capillary tube melting point procedure which is described in detail below. The *microscopic hot-stage* [4] and simpler variations of it are also popular; the equipment is somewhat more expensive, but even smaller amounts of crystals are required. In all micro methods the melting point is actually determined as a melting *range*, encompassed by the temperature at which the sample is first observed to melt and that at which the melting process is complete. When properly carried out with a *pure* crystalline compound, a range of no more than 0.5 to 1.0° should be observed. However, few commercially obtained compounds exhibit such a narrow melting range.

A wide variety of types of heating devices have been used in connection with capillary tube melting point determinations, ranging from a simple beaker containing a high-boiling liquid heated by a burner and stirred manually to elaborate electrically heated and mechanically stirred devices. Heated metal blocks are also used in connection with capillary tubes. Descriptions of these devices may be found in Weissberger. In our procedure we describe the use of a Thiele tube (Figure 1.5), but much of the description applies equally well to the use of other types of heating devices for capillary tubes. The Thiele tube is shaped so that heat applied by a burner to the sidearm is distributed to all parts of the vessel by convection currents in the heating fluid and no additional stirring is required.

EXPERIMENTAL PROCEDURE

A. CALIBRATION OF THERMOMETER

Determine the 0 and 100° [5] readings on your thermometer in the following ways.

Place about 50 ml of a mixture of finely chopped ice and water in a 100- or 150-ml beaker and stir it briskly with your thermometer (*Caution:* the bulb is made of thin glass; do not hit it against the beaker), taking care that the mercury bulb is entirely immersed in the ice-water mixture. Record the temperature reading when it becomes constant.

---

[3] It must be admitted that there is less emphasis on extremely precise melting point determinations since the development of spectroscopic methods for characterization of organic compounds (see Chapter 4).

[4] The Kofler hot-stage technique is described on p. 155 of the reference in footnote 2.

[5] In laboratories at altitudes of 1000 ft or more above sea level, a correction for the variation of boiling point with pressure (*cf.* p. 15) must be made. The boiling point of water is lowered approximately 1° by each 1000-ft increment in altitude above sea level, owing to the lowering of the atmospheric pressure.

Set up a simple distillation apparatus (see Figure 2.2) and place about 50 ml of distilled water in a 100-ml flask. Distil until the temperature becomes constant and record the reading. Be sure that the bulb of the thermometer is positioned correctly; the top of the mercury bulb should be about even with the sidearm of the flask or still head. The rate of distillation should be vigorous enough to insure equilibrium between liquid and vapor but not so fast as to produce superheating. Consult your instructor if in doubt.

Make a permanent record in your notebook of any corrections at the 0 and 100° readings and apply these in all future temperature measurements.

B. DETERMINATION OF CAPILLARY MELTING POINTS

Place the sample in a closed-end capillary tube of the commercial variety or one prepared from a piece of 10-mm soft glass tubing by heating and drawing to an out-side diameter of about 1–1.5 mm. (If you are to make your own capillaries, your instructor will demonstrate the procedure.) The easiest way to fill the capillary is to place a bit of the powdered sample on a small watch glass or the bottom of an inverted beaker and tap the open end of the capillary into the solid a few times. The solid may be made to "filter" down to the closed end of the tube by inverting the tube and scratching it gently with a file, or by tapping the closed end of the tube briskly on a solid surface. The solid should be tightly packed in the tube, and this can best be accomplished by finally dropping the capillary through a larger piece of glass tubing about two feet in length onto a hard surface. The size of the sample should be such as to fill the capillary to a depth of 2–3 mm after the compacting process. It should not be larger. Attach the capillary to a thermometer by means of a small rubber band (conveniently obtained as a slice of ordinary rubber tubing). The sample itself should be directly adjacent to the bulb of the thermometer. The rubber band should be positioned such that even at 200° it will remain above the level of the heating fluid (see Figure 1.5).[6] This accomplished, the thermometer is placed into the heating vessel and supported by means of a bored cork cut away on one side so as to make visible the thermometer markings in that vicinity. This cut also serves the purpose of making the apparatus an open system. *One should never heat a closed system.* By application of heat from a small Bunsen burner (microburner) raise the temperature of the heating fluid slowly (about 2° per minute). Note the temperature at which melting is first observed and the temperature at which the last of the solid melts and record these as the melting range of the solid. As one can obviously spend a great deal of time approaching the melting point of a high-melting solid this way, it is usually convenient to prepare two samples of the solid under consideration and determine the approximate melting point of the first by heating rapidly. Then let the heating fluid cool to 10–15° below this approximate point and insert the second tube and reheat slowly to the melting point.

The observed melting point is dependent on a number of factors: sample size, state of subdivision of the sample, and heating rate, as well as purity and identity of the

---

[6] It is dangerous to exceed 200° with a heating fluid such as mineral oil as this temperature is close to the flash point of this fluid. Other heating fluids such as some types of silicone oils can safely be taken to higher temperatures. Consult your instructor to learn the maximum temperature that can be safely reached with the particular heating fluid that you are using.

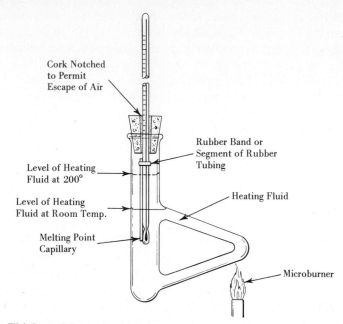

Cork Notched
to Permit
Escape of Air

Rubber Band or
Segment of Rubber
Tubing

Level of Heating
Fluid at 200°

Heating Fluid

Level of Heating
Fluid at Room Temp.

Melting Point
Capillary

Microburner

FIGURE 1.5   Thiele melting point apparatus.

sample. The first three cause the observed melting point to differ from the actual melting point due to the time lag in heat transfer from heating fluid to sample and conduction within the sample. Also, if heating is too fast, the thermometer reading will lag behind the actual temperature of the heating fluid.

Following the prescribed procedure, determine the melting point (or range) of the following:

1. Pure urea;
2. Pure *trans*-cinnamic acid;
3. Mixtures of the above two compounds in the proportions given below:

	urea	*trans*-cinnamic acid
mixture (a)	4	1
mixture (b)	2	1
mixture (c)	1	1
mixture (d)	1	2
mixture (e)	1	4

It is not necessary to weigh out each sample individually. Place a small amount (about 0.5 to 1.0 g) of each pure compound on watch glasses or smooth paper. Subdivide them with a spatula into small piles, estimating the proper ratios, mix them intimately with the spatula, and introduce them into capillary tubes.

Make a plot of melting temperatures (vertical) *versus* composition (horizontal); in the case of the mixtures, record the temperature of disappearance of the last bit of solid. Estimate the eutectic temperature and composition.

C. UNKNOWN

Determine the melting point of an unknown supplied by your instructor and identify it by consulting a list of possibilities and their melting points which will be provided.

1.5   Boiling Points

The molecules of a liquid are in constant motion. When at the surface, some of these molecules are able to escape into the space above the liquid. Consider a liquid contained in a closed *evacuated* vessel. The number of molecules in the gas phase can increase until the rate at which molecules reenter the liquid becomes equal to the rate of their escape; the rate of reentry is proportional to the number of molecules in the gas phase. At this point no further *net* change is observed in the system, and it is said to be in a state of dynamic equilibrium.

The molecules in the gas phase are in rapid motion and are continually colliding with the walls of the vessel, resulting in the exertion of pressure against the walls. The magnitude of this pressure at a given temperature is called the *equilibrium vapor pressure* of the particular liquid substance at that temperature. This vapor pressure is temperature dependent as shown in Figure 1.6. This dependence is easily understood in

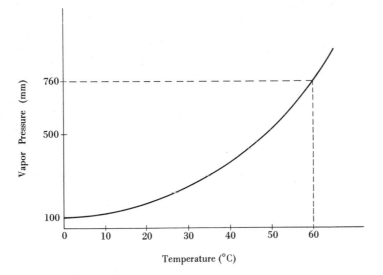

FIGURE 1.6 Dependence of vapor pressure on temperature.

terms of the escaping tendency of molecules from the liquid. As the temperature is increased, the average kinetic energy of the molecules increases, facilitating their escape into the gas phase. The rate of reentry also increases, and equilibrium is soon established at the higher temperature. However, the number of molecules in the gas phase is now larger than at the lower temperature, so the vapor pressure is greater.

Now consider a liquid sample at a given temperature, placed in an open container so that molecules of the vapor phase above the liquid may escape its confines. The vapor above such a sample is composed of molecules of air as well as of the sample. The *total* (external) pressure above the liquid is, according to Dalton's Law of Partial Pressures, equal to the sum of the partial pressures of the sample and of the air [see equation (1)]. The partial pressure of the sample is equal to its *equilibrium vapor pres-*

$$P_{\text{Total}} = P_{\text{Sample}} + P_{\text{Air}} \tag{1}$$

sure at the given temperature. If the temperature is raised (thus increasing the equilibrium vapor pressure of the sample), the number of molecules from the sample in the space immediately above the liquid will increase, in effect displacing some of the air. At the higher temperature the partial pressure of the sample will be a larger percentage of the total pressure. The same trend will continue as the temperature is further increased, until the equilibrium vapor pressure becomes equal to the external pressure, at which point all of the air will have been effectively displaced from the vessel. Further evaporation will have the effect of displacing gaseous molecules of the sample. Consideration of these facts leads to the conclusion that the equilibrium vapor pressure of the sample has an upper limit dictated by the external pressure. At this point the rate of evaporation increases dramatically (as evidenced by the formation of bubbles in the liquid), and the process commonly known as boiling occurs. *The temperature at which the vapor pressure of a liquid just equals the external pressure* is defined as the *boiling point* of the liquid. As the observed boiling point is obviously directly dependent on the external pressure, it is necessary in reporting boiling points to state the external pressure, for example, "bp 152° (752 mm)." The *standard boiling point* is typically measured at atmospheric pressure, 760 mm of Hg. The standard boiling point of the substance in Figure 1.6 is shown by the dotted lines to be 60°.

Boiling points are useful for identification of liquids and some low-melting solids. Higher melting solids usually boil at temperatures too high to be measured conveniently. Many compounds, notably the higher boiling ones, decompose at their standard boiling points.

Impurities, both volatile and nonvolatile, affect boiling points. These effects will be discussed in Chapter 2.

EXPERIMENTAL PROCEDURE

Micro boiling points. A simple micro boiling-point apparatus may be constructed from two 1-mm capillary melting-point tubes and 4-mm soft glass tubing in the following way. Seal the ends of two capillary tubes in a flame and join the tubes at the seal. Make a clean cut about 3–4 mm from the joint as shown in Figure 1.7(a). Seal a piece of 4-mm soft glass tubing at one end and cut it to a length about 1 cm shorter than the prepared capillary ebullition tube.

Attach the 4-mm tube to a thermometer with a rubber ring, with the rubber ring near the top of the tube and the bottom of the tube even with the mercury bulb of the thermometer. Place about two drops of the liquid whose boiling point is desired in the bottom of the tube by means of a capillary pipet. Introduce the capillary ebullition tube as shown in Figure 1.7(b). If the liquid level of the sample is below the joint-seal of the capillary, add enough more sample to bring it above the seal.

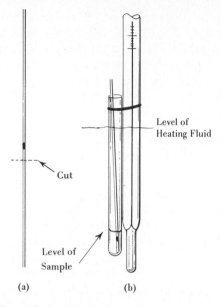

(a)　　　　　　　　(b)

FIGURE 1.7 Micro boiling-point apparatus.

Immerse the thermometer and attached tubes in a heating bath (Thiele tube or other melting-point apparatus), *taking care that the rubber ring is above the liquid level.* Raise the temperature rather quickly until a rapid and continuous stream of bubbles comes out of the capillary ebullition tube. (A decided change from the slow evolution of bubbles caused by thermal expansion of the trapped air will be seen when the boiling temperature of the liquid is reached.) Discontinue heating at this point. As the bath is allowed to cool down slowly, the rate of bubbling will decrease. At the moment the bubbling ceases entirely and the liquid begins to rise into the capillary, note the temperature of the thermometer. This is the boiling point of the liquid sample.

Remove the capillary ebullition tube and expel the liquid from the small end by gentle shaking. Replace it in the sample tube and repeat the heating and cooling procedure. With a little practice and care the observed boiling points may be reproduced to within 1 or 2°.

Your instructor will provide the liquids for which you should determine the boiling points.

1.6　Index of Refraction

The velocity of light propagation is not constant in all media. One direct and observable consequence of this variation is the bending of a beam of light as it passes through the interface of two different media. The light bends because its "speed" changes. (The effect is commonly observed; most people are aware that an object seen under water, a fish, for example, is not where it seems to be because of the refraction at the interface between the water and the air.) The angle of refraction (angle of bending) is dependent on the density of the media, the types of molecules present, the temperature of the media, and the wavelength of the light. If one medium such as air is chosen as a standard and all angles of refraction are measured at the same temperature with the same wavelength of light, then the measured angle is a property of the second medium at the interface and may be used as an identifying characteristic of the substance composing that medium.

The value that is actually reported as a physical constant of a liquid is not the angle of refraction, but the *index* of refraction (*n*), which is defined by the equation

$$n = \sin i / \sin p \qquad (2)$$

where *i* is the angle which incident light (in air) makes with a perpendicular to the interface and *p* is the angle which the refracted light (in the liquid) makes with the perpendicular (see Figure 1.8). Since monochromatic light gives more precise values than white light, most indices of refraction are reported for the *D* line of sodium ($\lambda = 589.3$ mμ.). This is indicated by a subscript, and the temperature by a superscript, so that a typical notation is $n_D^{20} = 1.3330$. In the instrument most frequently used by organic chemists, the Abbe refractometer, white light is used as a source, but compensation by a prism system gives indices for the sodium *D* line.

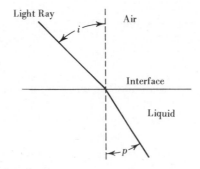

FIGURE 1.8 Refraction of light.

Impurities affect the index of refraction, and comparison of the observed index with the known index for the pure compound can be used as an indication of purity as well as for identification.

1.7 Density

The density of a substance is used less often than the physical constants discussed above for identification. Density is effectively a measure of the concentration of matter, and is generally measured and reported in units of grams per milliliter at 20°. The density of a liquid is rather easily measured, by making use of a small container, called a *pycnometer*, whose volume is accurately known. The container is weighed accurately and filled with the liquid whose density is to be determined. After bringing the pycnometer and its contents to 20° or some other desired temperature and readjusting the volume of the contents, if necessary, a second weighing will then yield the accurate weight of the substance whose volume is known, and the density may be calculated by dividing the weight in grams by the volume in milliliters.

Densities of solids are more difficult to determine. One way in which this measurement is often obtained is known as the *flotation method.* The density of the solid is determined as equal to the density of a liquid (measured as above) *in which* a single crystal of the solid will remain suspended, neither rising to the top nor settling to the bottom (Figure 1.9). Such a liquid may be obtained by first finding a suitable liquid *A*,

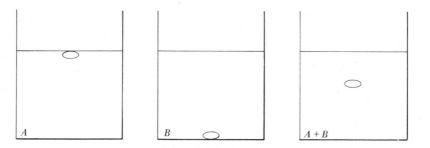

FIGURE 1.9   Density of solids by the flotation method.

in which the crystal rises to the top (liquid density is greater than the crystal density). A second liquid *B* is found in which the crystal sinks to the bottom. Of course only liquids in which the solid is insoluble are suitable. A mixture of *A* and *B* is then prepared in which the crystal remains suspended. The final measurement is made by determining the density of this liquid mixture, which is equal to the density of the solid.

EXERCISES

1. Describe errors in procedure which may cause an observed capillary melting point of a pure compound (a) to be *lower* than the correct mp; (b) *higher* than the correct mp; (c) *broad* in range (over several degrees).
2. Why is filter paper usually a poor material on which to powder a solid sample before introducing it into a capillary mp tube?
3. Criticize the following statements: (a) An impurity always lowers the mp of an organic compound. (b) A sharp mp for a crystalline organic substance always indicates a pure single compound. (c) If the addition of a sample of compound *A* to compound *X* does not lower the mp of *X*, *X* must be identical with *A*. (d) If the addition of a sample of compound *A* lowers the mp of compound *X*, *X* and *A* cannot be identical.
4. Give an explanation for the observed marked increase in rate of bubble evolution as the boiling point is reached in a micro bp determination.
5. Why is the micro bp taken as the temperature at the time the sample liquid begins to rise into the capillary tube? Justify this procedure in terms of the definition of boiling point.

REFERENCES

MELTING POINT

E. L. Skau and J. C. Arthur, Jr., in *Techniques of Chemistry,* A. Weissberger and B. W. Rossiter, editors, Wiley-Interscience, New York, 1971, Vol. 1, Part 5, Chapter 3, pp. 105 *f.*

INDEX OF REFRACTION

N. Bauer, K. Fajans, and S. Z. Lewin, in *Technique of Organic Chemistry,* A. Weissberger, editor, 3rd ed., Interscience Publishers, Inc., New York, 1960, Vol. 1, Part II, Chapter 18.

DENSITY

N. Bauer and S. Z. Lewin, in *Technique of Chemistry,* A. Weissberger and B. W. Rossiter, editors, Wiley-Interscience, New York, 1972, Vol. 1, Part 4, Chapter 2.

chapter two

separation and purification of organic compounds

1   Distillation, Recrystallization, and Sublimation

The practicing organic chemist seldom finds reactions that yield only the product (or products) he set out to obtain. This is because products of side reactions that proceed concurrently with the main reaction, varying amounts of unchanged starting materials, and solvents accompany the desired product in the reaction mixture. The chemist devotes a significant amount of effort to the separation of the desired products from such impurities. The purpose of this and the following chapter is to present an introduction to the theory and practice of the most important methods used by the modern chemist to separate and purify organic compounds.

The most commonly used method of purifying *liquids* is distillation. This process consists of vaporizing the liquid by heating and condensing the vapor in a separate vessel to yield a *distillate*. If the impurities present in the initial liquid are nonvolatile they will be left behind in the distillation residue, and a *simple distillation* will effect purification. If the impurities are volatile, a *fractional distillation* will be required.

2.1   Simple Distillation

When a nonvolatile impurity such as sugar, for example, is added to a pure liquid, it has the effect of reducing the vapor pressure of the liquid. This is because the presence of the nonvolatile component effectively lowers the concentration of the volatile constituent, *i.e.*, no longer are all of the molecules at the surface those of the volatile substance, and thus its ability to vaporize is lowered. The effect of a nonvolatile constituent on the vapor pressure of a mixture is shown in Figure 2.1. In this figure, curve 1 corresponds to the vapor pressure-temperature dependence for a pure liquid. The curve intersects the 760-mm line at 60°. Curve 2 is for the same liquid which now, however, contains a nonvolatile impurity. Note that the vapor pressure at any temperature is reduced by a constant amount by the presence of the impurity (Raoult's Law, see below). The temperature at which this curve intersects the 760-mm line is higher due to the lower vapor pressure, and the temperature of the boiling solution is higher, 65°.

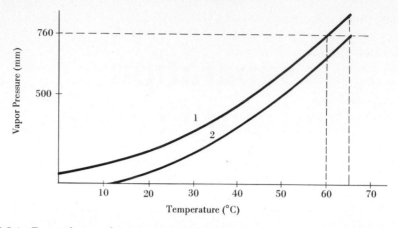
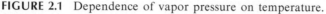

FIGURE 2.1 Dependence of vapor pressure on temperature.

A simple distillation may be carried out in an apparatus such as that in Figure 2.2. The thermometer is placed in the position shown for the purpose of determining the boiling temperature of the distillate (the material being collected in the receiver). In the case of a pure liquid, the temperature shown on the thermometer, the *head temperature,* will be identical to the temperature of the boiling liquid in the distilling vessel, the *pot temperature,* if the liquid is not superheated. The head temperature, which thus corresponds to the boiling point of the liquid, will remain constant throughout the process of distillation.

When a nonvolatile impurity is present in a liquid being distilled, the head temperature will be the same as for the pure liquid, since the material condensing on the

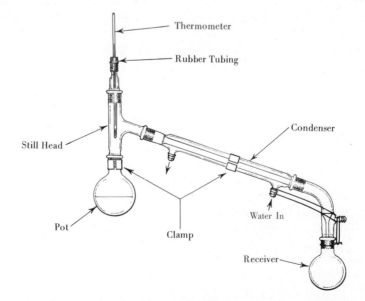

FIGURE 2.2 Typical apparatus for simple distillation, atmospheric pressure, or vacuum.

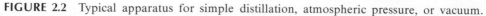

thermometer bulb is uncontaminated by the impurity. However, the pot temperature will be elevated owing to the decreased vapor pressure of the solution. The pot temperature will also rise during the course of the distillation since the concentration of impurity will increase steadily as the volatile component is distilled away, further lowering the vapor pressure of the liquid. However, the head temperature will remain constant, as in the case of a pure liquid.

The quantitative relation between vapor pressure and composition of homogeneous liquid mixtures (solutions) is known as Raoult's Law and may be expressed as in equation (1). P_R represents the partial pressure of component R and, at a given tem-

$$P_R = P_R^\circ N_R \tag{1}$$

perature, is equal to the vapor pressure of pure R at that temperature (P_R°) times the mole fraction of R in the mixture (N_R). The mole fraction of R is defined as the fraction of *all* molecules present that are molecules of R; it is obtained by dividing the number of moles of R in a mixture by the sum of the number of moles of all components [equation (2)].

$$N_R = \frac{n_R}{n_R + n_S + n_T + \ldots} \tag{2}$$

It should be noted that the partial vapor pressure of R above an ideal [1] solution containing R depends only on its mole fraction and is in no way dependent on the vapor pressures of the other components. If all components other than R are nonvolatile, the total vapor pressure of the mixture will be equal to the partial pressure of R, since the vapor pressure of nonvolatile compounds may be taken as zero. Thus the distillate from such a mixture will always be pure R. If, however, two or more of the components are volatile, then the total vapor pressure will equal the sum of the partial vapor pressures of each of these volatile components [Dalton's Law, equation (3), where R, S, and

$$P_{TOTAL} = P_R + P_S + P_T + \cdots \tag{3}$$

T refer to the volatile components only]. The process of distilling such a liquid mixture is significantly different from that of simple distillation discussed above, for here the distillate may contain each of the volatile components. Separation in this case will require fractional distillation.

2.2   Fractional Distillation

For simplicity we shall consider here only binary, ideal solutions, which contain two volatile components, R and S.[2] Ideal solutions are defined as those in which the interactions between *like* molecules are the same as those between *unlike* molecules.

---

[1] The definition of an ideal solution is given in the next paragraph.

[2] Solutions containing more than two volatile components are often encountered, but their behavior on distillation may be understood by extension of the principles given for binary systems.

Only ideal solutions obey Raoult's Law strictly, but many organic solutions approximate the behavior of ideal solutions. Some important examples exhibiting significant deviation from ideality are discussed in Section 2.4.

As vapor pressure is a measure of the ease with which molecules escape the surface of a liquid, the number of molecules of component R in a given volume of the vapor above a liquid mixture of R and S is proportional to the partial vapor pressure of component R. The same is true for component S, as may be seen from equation (4), where N'_R/N'_S is the ratio of the mole fractions of R and S in the *vapor* phase.

$$\frac{N'_R}{N'_S} = \frac{P_R}{P_S} = \frac{P^\circ_R\, N_R}{P^\circ_S\, N_S} \tag{4}$$

The mole fraction of each component may be calculated from the equations $N'_R = P_R/(P_R + P_S)$ and $N'_S = P_S/(P_R + P_S)$. The partial vapor pressures, P_R and P_S, are determined by the composition of the liquid solution (Raoult's Law). Since the solution will boil when the sum of the partial vapor pressures of R and S is equal to the external pressure, the boiling temperature of the solution is determined by its composition.

The relationship between temperature and the composition of the liquid and vapor phases of ideal binary solutions may be best illustrated by a diagram such as that of Figure 2.3 for mixtures of benzene (bp 80°) and toluene (bp 111°). The lower curve

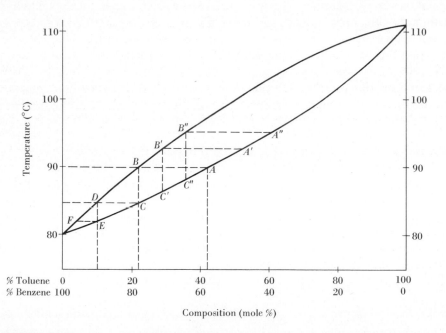

FIGURE 2.3 Boiling point-composition diagram for the system benzene-toluene.

gives the boiling points of all mixtures of these two compounds, and the upper curve, calculated using Raoult's Law, gives the composition of the vapor phase in equilibrium with the boiling liquid phase at the same temperature. For example, a liquid mixture with the composition 58 mole % benzene, 42 mole % toluene (*A*, Figure 2.3) boils at 90°, and the vapor phase in equilibrium with it has the composition 78 mole % benzene, 22 mole % toluene (*B*, Figure 2.3). Note that the *vapor phase is much richer in the more volatile component than the boiling liquid with which it is in equilibrium,* at any given temperature. For example, if component *R* is more volatile than component *S*, so that

$$\frac{P_R^\circ}{P_S^\circ} > 1$$

then from equation (4)

$$\frac{N_R'}{N_S'} > \frac{N_R}{N_S}$$

This is the basis of fractional distillation, as will be seen below.

If a mixture of composition *A* (Figure 2.3) is distilled, the first few drops of distillate (produced by *condensation* of some of the vapor phase) will have composition *B*, much richer in benzene than the original mixture from which it was distilled. Conversely, the liquid remaining in the distilling flask will be depleted in benzene and enriched in toluene, since more benzene than toluene was removed, and the boiling point will rise (from *A* to *A'*, for example). If distillation is continued, the boiling point of the mixture will continue to rise (from *A'* to *A''*, etc.) until eventually it approaches or reaches the boiling point of toluene. Meanwhile the composition of the distillate will change from *B* to *B'* to *B''*, etc., until at the end it will be nearly pure toluene.

Now let us return our attention to the first few drops of distillate which had the composition *B*. If these are collected separately and then redistilled, the boiling point will be the temperature at *C*, 85°. If only the first small fraction of distillate from *C* were collected, it would have the composition *D*, 90 mole % benzene, 10 mole % toluene. Repetition of this process could theoretically (but impractically) give a *very small amount* of pure benzene. Similarly, by collecting the *last* small fraction of each distillation and redistilling in the same stepwise manner, one could obtain a very small amount of pure toluene. By collecting larger amounts of initial and final distillation fractions, practical amounts of materials can be separated, but a proportionately larger number of individual simple distillations is required. Such a procedure is extremely tedious and time-consuming. Fortunately the repeated distillations can be accomplished virtually automatically in a single step by means of a *fractional distillation column,* the theory and use of which is described below.

2.3   Fractional Distillation Columns

There are a large number of types of fractional distillation columns, but all can be described in terms of a few fundamental characteristics. The column provides a vertical path, through which the vapor must pass from distilling flask to condenser; this path is significantly longer than in a simple distilling apparatus (Figure 2.4). As vapor from the distilling flask passes up the column, some of it condenses. *If the lower part of the column is maintained at a higher temperature than the upper part of the*

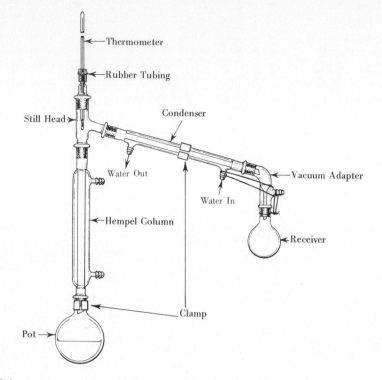

FIGURE 2.4 Apparatus for fractional distillation, atmospheric pressure or vacuum.

column, as the condensate drains down the column it will be partially revaporized. The uncondensed vapor, together with that produced by revaporization of condensate, rises higher in the column and goes through a series of condensations and vaporizations. These amount to repeated distillations [3] and, as described above with reference to Figure 2.3, the vapor phase produced in each step becomes richer in the more volatile component. The condensates, which drain down the column, are at each level richer in the less-volatile component than the vapor with which they are in contact. Under ideal conditions, equilibrium becomes established throughout the column between the liquid and vapor phases, with the vapor phase at the top consisting almost entirely of the more volatile component and the liquid phase at the bottom being very rich in the less volatile component. The most important requirements for producing this state are (1) intimate and extensive contact between the liquid and vapor phases in the column, (2) maintenance of the proper temperature gradient along the column, (3) sufficient length of the column, and (4) sufficient difference in the boiling points of the components of the liquid mixture.

[3] Each step along the route *A-B-C-D-E-F* of Figure 2.3 represents a single ideal distillation. Because one type of fractional distillation column, a "bubble-plate" column, was designed to bring about one such step for each "plate" it contains, the efficiency of any fractional distillation column is often described in terms of "theoretical plates."

If the first two requirements are well met, compounds having small differences in boiling points can be separated satisfactorily with a long column, since the length of the column required and the difference in boiling points of the components are reciprocally related. The most common way of providing the necessary contact between liquid and vapor phases is to fill the column with some inert material providing a large surface area, for example, glass, ceramic or metal pieces in a variety of shapes (helices, "saddles," woven mesh, etc.). A particularly effective way of achieving this liquid-vapor contact is to use a *spinning band* of metal or Teflon rotating at high speed within the column. This has the advantage of a low "hold-up" of material in the column, compared to "packed" columns with the same efficiency. (Hold-up refers to the amount of liquid and vapor required to maintain equilibrium conditions in the column.)

Maintenance of the proper temperature gradient in the column is a particularly important requirement for an effective fractional distillation. Ideally, the temperature at the bottom of the column is approximately equal to the boiling temperature of the solution in the pot, and it decreases continually in the column until it reaches the boiling point of the more volatile component at the head. The significance of the temperature gradient may be readily visualized by reference to Figure 2.3, where with each succeeding step the boiling temperature of the condensate decreases, *e.g.,* A (90°) to C (85°) to E (82°). The necessary temperature gradient from pot to still head will, in most distillations, be established automatically by the condensing vapors if the rate of distillation is properly adjusted. Frequently, this gradient can be maintained only by insulating the column with asbestos, glass wool, or, most effectively, with a silver-coated vacuum jacket. The insulation helps to reduce heat losses from the column to the atmosphere. For longer columns, particularly when poorer insulation is used, it is often necessary to supply additional heat by means of electrical resistance wire wrapped about the column. Even when insulated, if the rate of heating of the pot is too low, an insufficient amount of vapor may be produced to heat the column. In such a case, little or no condensate will reach the head and the rate of heating must be increased, but should be kept below that which will cause flooding of the column.

A factor intimately related to the temperature gradient in the column is the rate of heating of the pot and the rate at which vapor is removed at the still head. If the heating is vigorous and the vapor is removed too rapidly, the whole column will heat up almost uniformly and there will be no fractionation (separation of components). On the other hand, if the pot is heated too vigorously and if vapor is removed too slowly at the top, the column will "flood" with returning condensate. Proper operation of a fractional distillation column requires judicious control of heating and "reflux ratio"—the ratio of the amount of vapor condensed and returned down the column to the amount taken off as distillate at the still head in the same time period. In general, the higher the reflux ratio, the more efficient the fractionation.

The most convenient way to measure and control the reflux ratio is to use a still head equipped for total reflux and variable "take-off." For example, with a still head of this type (see Figure 2.5), the number of drops of condensate falling from the tip of the condenser (P) in a given length of time gives a measure of the total reflux rate and the number of drops allowed to pass through the stopcock (S) and fall into the receiver at

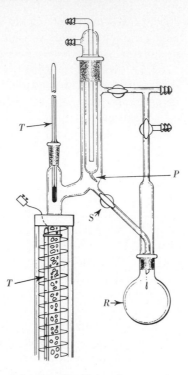

FIGURE 2.5   Total reflux-partial take-off still head.

R in the same time period is a measure of the take-off rate. If one drop of distillate is taken off for every ten drops of total reflux, the reflux ratio is 9 to 1. For very precise fractionation, reflux ratios of 100 to 1 may sometimes be used with an efficient column.

2.4   Fractional Distillation of Nonideal Solutions

Although most homogeneous liquid mixtures behave as ideal solutions, there are many examples known in which the behavior is nonideal; in these solutions the unlike molecules are not indifferent to one another's presence. The resultant deviations from Raoult's Law occur in either of two directions: Some solutions display greater vapor pressures than expected and are said to exhibit *positive deviation;* others display lower vapor pressures than expected and are said to exhibit *negative deviation.*

In the case of positive deviation, the forces of attraction between the different molecules of the two components are weaker than those between the identical molecules of each component, with the result that, in a certain composition range, the combined vapor pressure of the two components is greater than the vapor pressure of the pure, more volatile component. Hence, mixtures in this composition range (between X and Y in Figure 2.6) have boiling temperatures *lower* than the boiling temperature of either pure component. The *minimum-boiling* mixture in this range (the composition of Z, Figure 2.6) must be considered as if it were a third component. It has a constant

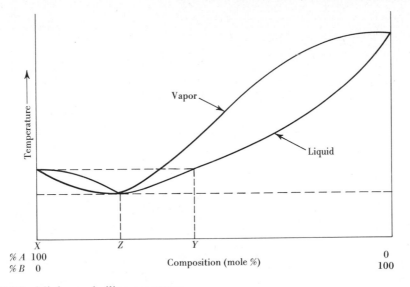

FIGURE 2.6 Minimum-boiling azeotrope.

boiling point because the vapor in equilibrium with the liquid has the same composition as the liquid itself. Such a mixture is called a *minimum-boiling azeotropic mixture*, or *azeotrope*.[4] Fractional distillation of such mixtures will not yield both of the components in pure form but only the azeotrope and the component present in excess of the azeotropic composition. It is for this reason that pure ethanol cannot be obtained by fractional distillation of aqueous solutions containing less than 95.57% ethanol (the azeotropic composition), even though the boiling point of this azeotrope is only 0.15° lower than the boiling point of pure ethanol.[5]

In the case of negative deviation from Raoult's Law, the forces of attraction between the different molecules of the two components are stronger than those between the identical molecules of each component, with the result that, in a certain composition range, the combined vapor pressure of the two components is less than the vapor pressure of the pure, less volatile component (see Figure 2.7). Hence, mixtures in this composition range (between X and Y in Figure 2.7) boil at even *higher* temperatures than the boiling temperature of the pure, higher-boiling component. There is one particular composition corresponding to a *maximum-boiling azeotrope* (Z, Figure 2.7). Fractional distillation of mixtures of any composition other than that of the azeotrope will result in elimination from the mixture of whichever of the two components is in excess of the azeotropic composition (Z), so that the composition of the pot residue will approach that of Z. As an illustration, the maximum-boiling azeotrope of formic

[4] This discussion is limited to binary mixtures. However, ternary mixtures are known that exhibit the same property, for example, water (7.4%), ethanol (18.5%) and benzene (74.1%).

[5] As optimum fractional distillations of aqueous solutions of ethanol containing less than 95.57% yield this azeotropic mixture, "95% ethyl alcohol" is the common form of commercial solvent ethanol. Pure or "absolute" ethanol can be obtained by chemical removal of the water or by a distillation procedure involving the use of a ternary mixture of ethanol, water and benzene.[4] This procedure is described in the references at the end of the chapter.

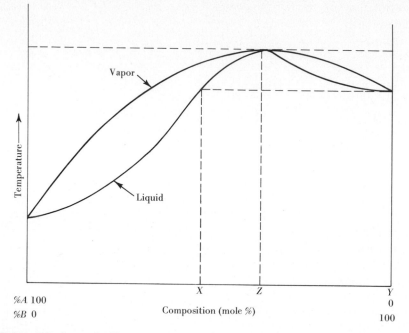

FIGURE 2.7 Maximum-boiling azeotrope.

acid (bp 100.7°) and water boils at 107.3°; any mixture having other than the azeotropic composition (77.5% formic acid, 22.5% water) will yield this mixture as residue
upon fractional distillation.

Data on other azeotropes, of both types, may be found in the references at the end
of the chapter; extensive tables are given in the handbooks listed there.

EXPERIMENTAL PROCEDURE

This experiment is designed to demonstrate in its several parts both the procedures
of carrying out simple and fractional distillations and the relative efficiencies of
different types of apparatus in separating mixtures of volatile components. Detailed
directions are given for the purification of benzene by simple distillation and for
the fractional separation of benzene-toluene mixtures. Take care to note the differences in procedure for the simple distillation of a liquid having only a single volatile
component and for the fractional distillation of a mixture of two or more volatile
components. In the former case, less complex apparatus may be used; compare, for
example, Figures 2.2 and 2.4. Furthermore, when fractional separation is not required, a much higher rate of distillation may be employed. The distillation of
benzene may be quite conveniently and satisfactorily carried out at the rapid rate
of 2 to 4 drops of distillate per second, whereas the fractional distillation of
benzene-toluene is effected at the more moderate rate of only 1 drop every 1 to
2 seconds. This slower rate is necessary to maintain the proper temperature gradient

in the distillation column. In simple distillation, the entire distillate is collected in a single receiving flask, whereas in fractional distillation receiving flasks are changed as the composition of the distillate changes, as indicated by the head temperature.

A. GENERAL INSTRUCTIONS:
ASSEMBLY OF GLASSWARE

The following general discussion pertains to the assembly of glassware for experiments throughout this text. Proper assembly of apparatus is an important part of good laboratory technique, particularly with regard to laboratory safety. For example, a loosely jointed distillation assembly may "leak" flammable vapors that are apt to be ignited by a nearby Bunsen burner; or a too rigidly clamped assembly may be stressed to the point of breaking during a laboratory operation, causing, in addition to the physical danger of broken glass, the spillage of flammable or caustic chemicals. It is possible here to provide only general guidelines to proper procedure; it is advisable in the first few experiments to ask the instructor to check your glassware for proper and safe assembly. The experience gained in these first experiments should provide you with the "feel" for correct usage of your equipment. Beyond the guidelines here, common sense is the watchword. Think about what you are to do before you do it.

In all experiments where the typical glassware assembly to be used is depicted in a figure, it should be noted that the "clamping points" have been marked. The general tendency is for beginning students to overclamp their apparatus; this frequently, although not necessarily, results in an unsafe situation because the more clamps that are used, the greater the likelihood of stressing and possibly breaking a glass joint between two clamps. Also, mispositioning of a clamp may not provide sufficient stability for the apparatus.

When applying a clamp, it is important that the jaws of the clamp be aligned parallel to the piece of glass being clamped; this will enable the clamp to be tightened without twisting the glass and either breaking it or pulling another joint loose. *Do not tighten the clamp until you are sure that the piece of glassware is correctly positioned and the clamp is properly aligned.*

It is generally most efficient to put together a complex assembly from the bottom up. For example, consider assembly of the fractional distillation apparatus shown in Figure 2.4. Clamp the distillation flask (pot) and then add the Hempel column and still head. Adjust the clamp so that the column is vertical. Now place the condenser in position and clamp it at the indicated position. Be sure that this clamp is properly aligned and vertically positioned *before* tightening it. Several minor adjustments may be necessary. It is not necessary that this clamp be *rigidly* tight, only that it prevent the joint between the condenser and still head from slipping under normal usage. The vacuum adapter may be added and held in place by means of a strong rubber band as shown in Figure 2.4. Likewise, by twisting a short piece of stiff wire about the neck of the receiving flask, it may be fastened to the vacuum adapter by a rubber band. Rubber has a tendency to deteriorate in the presence of laboratory vapors; be sure that the rubber bands are uncracked and retain sufficient elasticity and strength.

This use of rubber bands is acceptable in many instances; however, you should be careful not to depend on them to support relatively heavy weight. For example, additional support should be provided for flasks of 250-ml capacity or larger, as well as 100-ml flasks that may be expected to become more than half-full during the course of a distillation. In case of doubt, it is better, of course, to be overly cautious. Instead of using another clamp, use an iron ring holding a piece of wire gauze to support the flask from underneath.

B. EXPERIMENTS

1. Simple distillation.   In a 100-ml round-bottomed flask, place 40 ml of benzene and a few boiling chips to insure smooth boiling. (*Caution:* Do not pour flammable liquids within several feet of a flame! A list of flammable solvents is provided on page 3.) Arrange the apparatus for simple distillation according to Figure 2.2. Use a receiving flask of sufficient size to collect the anticipated volume of distillate. The position of the thermometer bulb is particularly important; note its position relative to the sidearm of the still head. Have your instructor check the assembly to be sure that all connections are tight. Place a wire gauze having an asbestos pad in the center *directly* beneath the pot (distilling flask). If this is not available, a square of asbestos board with a small (*ca.* 1-cm) hole in the center may be used in place of the wire-asbestos gauze.

Heat the flask[6] with a burner[7] placed so that the tip of the flame touches or is just below the wire gauze, and protect the flame from air currents so that you can adjust the heating as precisely as possible. As soon as the mixture starts to boil and the refluxing vapors have reached the thermometer, regulate the flame so that distillation continues steadily at a rate of *2 to 4 drops per sec.* Only a small flame will be required.

As soon as the distillation rate is adjusted and the thermometer has reached a steady temperature, note and record the temperature. Continue the distillation, periodically recording the head temperature, until only 2 or 3 ml of benzene remain in the distillation flask, and then turn off the flame. Note and record the distillation *range* of benzene that you have observed. Advise the instructor of your completion of the distillation and return the benzene, distilled and undistilled, to the bottle marked "Recovered Benzene." (It is not only economically advantageous to recover chemicals that may be reused, but also well worth the effort to avoid, insofar as is possible, the environmental abuse which results from discarding chemicals down the drain.)

[6]The apparatus being used is open to the atmosphere at the receiving end of the condenser. This allows for pressure equilibration. At no time in the laboratory should a *closed system* be heated. If pressure equilibration is not allowed for, material expansion within the system, resulting in elevated pressures, may cause the apparatus to *explode.*

[7]Better, the flask may be heated by immersion in an oil bath which is in turn heated either by an electric hot plate or by electric resistance coils. Alternatively, the flask may be heated with a "Glas-Col heating mantle," a heating device constructed of woven glass fiber and spun glass insulation in which electric resistance coils are embedded. Each mantle is form-fitted to accept a flask of given size. For a steam distillation, a steam bath may not be used for heating purposes as it will not supply enough heat.

2. Fractional distillation. Although this experiment provides detailed directions for the separation by fractional distillation of mixtures of benzene and toluene, similar results may be obtained with other mixtures such as carbon tetrachloride-toluene and methanol-water. The last mixture has the advantage of being less flammable than the other two. If you are assigned one of these latter mixtures, consult the information provided in footnote 8. Your instructor will inform you as to which procedure, I or II, below, you are to follow. Procedure II is advantageous if you are to repeat this experiment with different equipment, since it provides the data in graphic form. If you are to follow procedure II, prepare in your notebook, prior to coming to the laboratory, a table for recording approximately 30 successive measurements of temperature and total accumulated volume of distillate.

In a 100-ml round-bottomed flask, place 30 ml of benzene,[8,9] 30 ml of toluene,[8,9] and a few boiling chips to insure smooth boiling. (*Caution:* Do not pour flammable liquids within several feet of a flame! A list of flammable solvents may be found on page 3.) Arrange the apparatus for fractional distillation according to Figure 2.4 using a packed Hempel, or similar, column. The packing may consist of glass beads, broken pieces of glass tubing, or, more expensively, glass helices (listed in order of increasing efficiency). Alternatively, a very satisfactory substitute for glass helices may be made from a copper (or better, but more expensive, stainless steel) cleaning sponge. The sponge is pulled apart and stuffed into the glass column rather compactly with the aid of a pencil (*Note:* The rapid corrosion of copper by organic halides makes this type of packing material unsuitable for a mixture containing carbon tetrachloride.) Do not pack the column so tightly that vapors cannot pass through it.

In this assembly, the position of the thermometer bulb is particularly important; note its position relative to the side arm of the still head. Have your instructor check the assembly to be sure that all connections are tight. Place a wire gauze having an asbestos pad in the center directly beneath the pot (distilling flask). If this is not available, a square of asbestos board with a small (*ca.* 1-cm) hole in the center may be used in place of the wire-asbestos gauze.

Prepare three containers of 50-ml capacity or larger for use as receiving flasks and label them *A*, *B*, and *C*. If procedure I is to be followed, these may be bottles or Erlenmeyer flasks. If procedure II has been assigned, use three 50-ml graduated cylinders. In use, the tip of the vacuum adapter should extend inside the neck of these containers; do not leave vertical space between the adapter and the receiver since this will facilitate the escape of flammable vapors.

Heat the distillation flask[6] with a burner[7] placed so that the tip of the flame touches or is just below the wire gauze, and protect the flame from air currents so that you can adjust the heating as precisely as possible. As soon as the mixture starts to boil

---

[8]If you are to distil mixtures other than benzene-toluene, use 30 ml of each of the components. For carbon tetrachloride-toluene, change from *A* to receiver *B* at 80° and to receiver *C* at 109°. For methanol-water, change from receiver *A* to receiver *B* at 68° and to receiver *C* at 98°.

[9]The benzene and toluene should be dry. If you are not sure that they are, each may be dried by placing a 50-ml portion in a simple distillation apparatus and distilling until the distillate shows no cloudiness. Discontinue distillation and use the *residue* (not the distillate) in the fractional distillation experiment. This is a method of drying moist solvents that form minimum-boiling azeotropes with water.

and the vapors have reached the thermometer, regulate the flame so that distillation continues steadily at a rate of *only* about *1 drop of distillate every 1 to 2 sec*. Continue with either procedure I or II.

Procedure I.   Collect the first distillate in receiver *A*. When the head temperature reaches 83°,[8] change to receiver *B* and at 109°,[8] change to receiver *C*. Continue distillation until about 1 or 2 ml of liquid remain in the pot; then turn off the flame. Measure the volumes of the distillation fractions in receivers *A*, *B*, and *C* with a graduated cylinder and record them in your notebook. Allow the liquid in the column to drain into the pot; measure and record the volume of this residue. Submit 1-ml samples of each fraction *A, B,* and *C* for gas chromatographic analysis if you have been asked to do so.[10]

Procedure II.   Collect the first distillate in graduated cylinder *A*. As soon as the distillation rate is adjusted to a rate of approximately 1 drop every 1 to 2 sec, note and record the head temperature and the total volume of accumulated distillate in the receiver. Continue the distillation, recording the temperature and total volume for each additional increment of *ca.* 2 ml of distillate. When the temperature reaches 83°,[8] change to cylinder *B*, and when it reaches 109°,[8] change to cylinder *C*. Although receivers have been changed, continue to record the temperature and the *total* accumulated volume of distillate for 2-ml increments until about 1 or 2 ml of liquid remains in the pot, then turn off the flame. Record the volume of each of the three distillation cuts taken. Allow the liquid in the column to drain into the pot and measure and record the volume of this residue. Submit 1-ml samples of each fraction *A, B,* and *C* for gas chromatographic analysis if you have been asked to do so.[10] Plot, on graph paper, the head temperature versus total accumulated volume of distillate.

3. Fractional distillation using alternate equipment.   If you have been asked to compare the results of fractional distillation using different equipment, combine the three fractions obtained above with the pot residue, after setting aside 1-ml samples for analysis. Use this mixture for the second distillation. Repeat, exactly, the procedure provided above for fractional distillation using instead either the simple distillation apparatus, Figure 2.2, or the fractional distillation apparatus, Figure 2.4, *without the packing.* If you are following procedure II, it is most convenient to plot the data from two or more distillations on the same piece of graph paper, so that the curves obtained for each distillation involving different equipment will overlay one another. Compare the results for each type of distillation equipment, drawing any conclusions suggested by the data concerning the relative efficiencies of the various columns. When finished, return all of the benzene-toluene to the bottle labeled "Recovered Benzene-Toluene."

---

[10]Columns containing as the fixed phase either silicone gum rubber or a polyglycol such as Carbowax give good separation of benzene and toluene. (See Chapter 3.4 for discussion of gas chromatography.)

EXERCISES

1. Explain why a packed fractional distillation column is more efficient at separating two closely boiling liquids than an unpacked column.
2. If heat is supplied to the distillation flask too rapidly, the ability to separate two liquids by fractional distillation may be drastically reduced. In terms of the general theory of distillation presented in the discussion, explain why this is so.
3. Frequently in the distillation of benzene, the first few milliliters of distillate are "cloudy." To what might you attribute such an observation?
4. Explain why the column of a fractional distillation apparatus should be aligned as close to a vertical configuration as possible.
5. Explain the role of the boiling stones which are normally added to a liquid which is to be heated to boiling.
6. The bulb of the thermometer placed at the head of a distillation apparatus should be adjacent to the exit to the condenser. Explain the effect on the temperature reading of placement of the thermometer bulb (1) below the exit to the condenser and (2) above the exit.
7. (a) A mixture of 80 mole percent *n*-propylcyclohexane and 20 mole percent *n*-propylbenzene is distilled through a *simple distillation apparatus* (assume that no fractionation occurs during the distillation). The head temperature is found to be 157.3° as the first *small* amount of distillate is collected. The standard vapor pressures of *n*-propylcyclohexane and *n*-propylbenzene are known to be 769 mm and 725 mm, respectively, at 157.3°. Calculate the percentage of each of the two components in the first few drops of distillate.
   (b) A mixture of 80 mole percent benzene and 20 mole percent toluene is distilled under exactly the same conditions as in part (a). Using Figure 2.3, determine the distillation temperature and the percentage composition of the first few drops of distillate.
   (c) The standard boiling points of *n*-propylcyclohexane and *n*-propylbenzene are 156.9° and 159°, respectively. Compare the distillation results in parts (a) and (b). Which of the two mixtures would require the more efficient fractional distillation column for separation of the components? Why?
8. The still head of the fractional distillation assembly shown in Figure 2.5 is designed to allow receiving flasks to be changed during a vacuum distillation without disrupting the distillation by the readmission of air into the column. Determine the sequential manipulation of stopcocks necessary to accomplish this task. (*Hint:* See last paragraph of Section 2.5).

2.5 Distillation under Reduced Pressure

As was pointed out earlier, the boiling point of a liquid is that temperature at which the total vapor pressure is equal to the external pressure. It is most convenient to distil liquids under conditions such that the external pressure is the atmospheric pressure. However, in many instances, boiling temperatures at this pressure are higher than desirable, because the compound being distilled may decompose, oxidize, or undergo molecular rearrangement at temperatures below its normal boiling point. Sometimes impurities present may catalyze such reactions at higher temperatures. These problems may often be alleviated by carrying out the distillation at pressures less than atmospheric, for under these conditions the boiling temperature is, of course, lower.

Reduced pressures may be obtained by connecting an aspirator ("water pump") or a mechanical oil pump to the distillation apparatus. An aspirator will commonly reduce the pressure to about 25 mm[11] and an oil pump to below 1 mm; the exact pressure obtained in the lower range is highly dependent on the condition of the pump and its oil and on the tightness of the connections of the distillation apparatus.

Two useful *approximations* of the effect of lowered pressure on boiling points are the following:

1. Reduction from atmospheric pressure to 25 mm lowers the boiling point of a high-boiling compound (250–300°) by *about* 100–125°.

2. Below 25 mm, each time the pressure is halved, the boiling point is lowered about 10°. More accurate estimates of boiling points at various pressures may be made by the use of charts and nomographs found in references at the end of this chapter. Calculations may also be made using integrated forms of the Clausius-Clapeyron equation, as described in physical chemistry textbooks.

The typical apparatus for vacuum distillation is pictured in Figure 2.8. The distilling flask, which should not be filled more than half-full, is heated by means of an oil bath or some other suitable means. The flask should be immersed to a

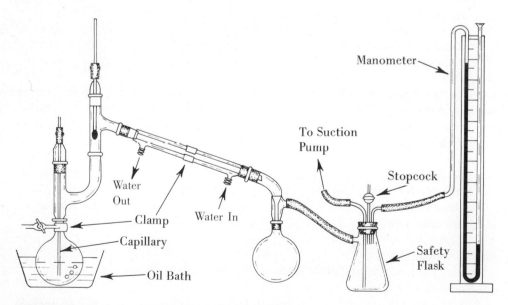

FIGURE 2.8   Typical apparatus for vacuum distillation.

[11] The vacuum produced by a water aspirator is limited by the vapor pressure, and hence by the temperature of the water used; in cold climates, pressures as low as 8–10 mm may be obtained.

depth above the level of the liquid in the flask as an aid in preventing bumping problems. The flask should *never* be heated directly with a flame as this causes localized hot spots and intensifies the problem of bumping. The oil bath may be heated with a burner or by means of electrical resistance coils immersed in the oil and connected to a variable transformer for temperature control. The vacuum adapter is connected to a safety flask which serves not only as a trap to prevent backup of water into the apparatus from the water pump in case of loss of water pressure, but also as a vacuum manifold, providing vacuum connections to the apparatus and manometer and bearing a vacuum release valve (stopcock). The manometer provides for measurement of the pressure at which the distillation is being effected. The reliability of such measurements of pressure is dependent on the rate of distillation, as discussed below. These measurements are an important and integral part of a reported boiling point at reduced pressure, *e.g.,* benzaldehyde: bp 180°(760 mm), 87°(35 mm), and should be taken with care.

There are two significant problems encountered in vacuum distillation which only rarely complicate atmospheric distillation. These are both created by the fact that the volume of vapor formed by the volatilization of a given amount of liquid is pressure dependent. Thus, for example, the volume of vapor formed from vaporization of a drop of liquid will be approximately 20 times as great at 38 mm as at 760 mm. Serious bumping problems may occur in the distilling flask during distillation under reduced pressure as *large* bubbles explosively escape from the liquid, giving rise to vigorous, and sometimes violent, splashing and splattering. The insertion of the Claisen connecting tube between the distillation flask and still head (Figure 2.8) is a partial solution to this problem. There is no direct "line-of-flight" path for liquid to splash from the flask into the side arm leading to the condenser. Boiling stones are generally ineffective in preventing bumping under reduced pressure. The usual method for promoting regular and even ebullition is to provide a thin, flexible capillary tube extending into the boiling flask. This tube allows a fine stream of air bubbles to be introduced at the bottom of the flask, where they serve as nuclei for the regular production of vapor bubbles. As the volume of air introduced is small compared to the evacuating capacity of a water or oil pump, this "leak" has no significant effect on the pressure of the system. The capillary is drawn from a piece of 6-mm soft glass tubing and should be fine enough to allow only a slow stream of fine bubbles when air is blown through it into a test tube containing acetone. Alternatively, if available, a magnetic stirring bar (a cylindrical bar magnet covered with Teflon or glass) in the flask may be employed with a magnetic stirrer. The stirrer consists of a larger, motor-driven bar magnet; magnetic coupling of the magnets allows the motor indirectly to spin the stirring bar in the flask. Other stratagems against bumping, although frequently not successful, include filling the volume of the distilling flask above the liquid with glass wool (finely spun glass), and insertion into the flask of relatively long (so that they will stand vertically) dry pine applicator sticks, such as those used for medicinal cotton swabs.

The second problem connected with lower vapor densities at reduced pressure is that the measured pressure is considerably affected by the rate of distillation. Since a drop of condensate forms from a much larger volume of vapor at low pressure than at atmospheric pressure, the *velocity* of vapor molecules entering the condenser, even at moderate distillation rates, is tremendously increased by a reduction of pressure. The back pressure occasioned by high velocities of vapor provides higher pressure in the distillation column than that read on the manometer, which is beyond the condenser and receiver and not affected by the precondensed vapor. The difference in actual *versus* apparent pressure is directly dependent on the rates of distillation and application of heat to the boiling flask. Therefore, important requirements for proper conduct of a vacuum distillation are to maintain a *slow,* but steady distillation rate, and to avoid superheating of the vapor by maintaining the oil bath at a temperature no more than 15–25° higher than the head temperature. This problem is accentuated at still lower pressures.

The following paragraphs provide the general procedure for carrying out a vacuum distillation using equipment such as that shown in Figure 2.8:

1. *Caution:* Never use glassware with cracks or thin-walled vessels, especially those with flat bottoms, such as Erlenmeyer flasks, in a system to be evacuated. Even with systems of only moderate size under water pump evacuation, pressures of many hundreds of pounds may be exerted on the exterior surfaces of the assembly. Weak points may yield to implosion; the in-rushing air will shatter the glassware in a manner little different from an explosion. A very real additional danger is that of burns from the hot oil of the bath. *Examine the glassware carefully and always wear safety glasses.*

2. Lubricate and seal all glass joints during assembly as an aid against air leaks. Check the rubber fittings holding the thermometer and capillary in place to insure that they are tight. The neoprene fittings normally used with the thermometer adapters may be replaced, if necessary, with short pieces of heavy-walled vacuum tubing. Rubber stoppers are not advisable since rubber in direct contact with the hot vapors during distillation may cause contamination. A three-holed rubber stopper in the safety flask should fit snugly and provide tight connections to the pieces of glass tubing. Heavy-walled tubing must be used for all vacuum connections. Test the completely assembled apparatus to make certain that the system is tight before placing liquid in the pot.

3. Place the liquid to be distilled in the flask. Make sure the capillary tip extends nearly to the bottom of the flask and turn on the water pump. The release valve on the safety flask should be *open.* Do not heat the flask until the system is fully evacuated. *Slowly* close the release valve, being ready to reopen it if necessary. If the liquid contains small quantitites of low-boiling solvents, and it very likely will if the liquid has been recently obtained from a solution by evaporation of solvent, foaming and bumping will almost certainly occur in the flask. If this happens, reopen the valve until the foaming abates. This may have to be done several times until the solvent has all been evaporated. As soon as the surface of the liquid is relatively quiet and the system is fully evacuated (check the manometer for constancy of pressure), begin the heating of the oil bath. Maintain a bath temperature as low as possible to provide a *slow* rate of distillation.

4. If, as in fractional distillation, it is necessary to use multiple receivers to collect fractions of different boiling ranges, the distillation will need to be interrupted to change flasks. Lower the oil bath with caution and allow the distillation flask to

cool a little. *Slowly* open the vacuum release valve to readmit air to the system. Change receivers, close the release valve, and when evacuation is complete raise the oil bath and continue. This operation frequently results in a change of pressure in the fully evacuated system. It is good practice periodically to monitor and record in the notebook the head temperature and the pressure, particularly just before and after changing flasks.

5. At the end of the distillation, lower the oil bath, allow the pot to cool somewhat, slowly release the vacuum, and turn off the water pump.

The most inconvenient aspect of this procedure is the disruption of the distillation attendant to changing receiving flasks. Improved apparatus, although seldom available in the introductory organic laboratory, is designed to eliminate this problem. For example, the "cow" receiver shown in Figure 2.9 bears four receiving flasks (one

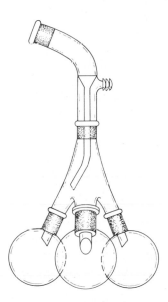

FIGURE 2.9   Multiple flasked receiver for vacuum distillation.

not in view). These flasks are successively employed by rotating them into the receiving position. A second example is shown as part of the fractional distillation apparatus in Figure 2.5. The "fraction cutter" is included as an integral part of the still head and cold-finger condenser. Although this type of receiver is a little more complex to use, it is more advantageous because it does not limit the number of receiving flasks that may be used without disrupting the distillation. The proper sequential manipulation of stopcocks allows: (1) isolation of the receiving end of the system while maintaining evacuation of the column; (2) release of vacuum in the receiver to change flasks; (3) reevacuation of the receiver while the column is isolated from the pump; and (4) reconnection of the receiver and column for continuation of the distillation (see Exercise 8, p. 35).

2.6   Steam Distillation

The separation and purification of *volatile* organic compounds which are immiscible with water or nearly so is often accomplished by steam distillation, a technique which involves the codistillation of a mixture of water and organic substances. The virtues and limitations of this technique can best be illustrated by consideration of the principles which underlie steam distillation.

The partial pressure, P_i, at a given temperature of each component, i, of a mixture of immiscible, volatile substances is equal to the vapor pressure, P_i°, of the pure compound at the same temperature [equation (5)] and does not depend on the mole frac-

$$P_i = P_i^\circ \tag{5}$$

tion of the compound in the mixture; that is, each component of the mixture vaporizes independently of the others. This behavior is in sharp contrast to that exhibited by solutions of miscible liquids, for which the partial pressure of each constituent depends on its mole fraction in the solution [Raoult's Law, equation (1)]. Now, the total pressure, P_T, of a solution (mixture) of gases, according to Dalton's Law [equation (3)] is equal to the sum of the partial pressures of the constituent gases so that the total vapor pressure of a mixture of immiscible, volatile compounds is given by equation (6).

$$P_\mathrm{T} = P_a^\circ + P_b^\circ + \cdots P_i^\circ \tag{6}$$

Note from this expression that the total vapor pressure of the mixture at any temperature is always greater than the vapor pressure of even the most volatile component at that temperature, due to the contribution of the vapor pressures of the other constituents of the mixture. The boiling temperature of a mixture of immiscible compounds must then be *lower* than that of the lowest boiling component.

Demonstration of the principles just outlined is available from discussion of the steam distillation of a mixture of water (bp 100°) and bromobenzene (bp 156°), substances which are insoluble in one another. The vapor pressure versus temperature plot for a mixture of these substances, Figure 2.10, along with the corresponding plots for the pure liquids, shows that the mixture should boil at about 95°, the temperature at which the total vapor pressure equals atmospheric pressure. As would be predicted from theory, this temperature is below the boiling point of water, the lowest boiling component in this example. The ability to distil a compound at the relatively low temperature of 100° or less by means of a steam distillation is often of great use, particularly in the purification of substances which are heat-sensitive and which would decompose at higher temperatures. It is useful also in the separation of compounds from reaction mixtures which contain large amounts of nonvolatile residues such as the notorious "tars" which so often are formed during the course of an organic reaction.

The composition of the condensate from a steam distillation depends upon the molecular weights of the compounds being distilled and upon their respective vapor pressures at the temperature at which the mixture distils. Consider a mixture of the

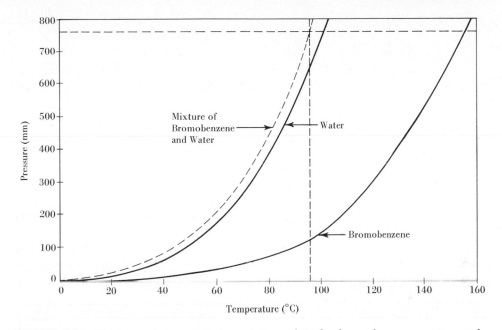

FIGURE 2.10 Vapor pressure versus temperature plots for bromobenzene, water, and a mixture of bromobenzene and water.

two immiscible components, A and B. If the vapors of A and B behave as ideal gases, the ideal gas law can be applied, and the following two expressions obtained:

$$P_A^\circ V_A = (g_A/M_A)(RT) \quad \text{and} \quad P_B^\circ V_B = (g_B/M_B)(RT) \tag{7}$$

where P° is the vapor pressure of the pure liquid, V is the volume in which the gas is contained, g is the weight in grams of the component in the gas phase, M is its molecular weight, R is the universal gas constant and T is the absolute temperature (°K). Dividing the first equation by the second, one obtains

$$\frac{P_A^\circ V_A}{P_B^\circ V_B} = \frac{g_A M_B (RT)}{g_B M_A (RT)} \tag{8}$$

As the (RT) factors in the numerator and the denominator are identical and because the volume in which the gases are contained is the same for both $(V_A = V_B)$, the above expression becomes

$$\frac{\text{grams of } A}{\text{grams of } B} = \frac{(P_A^\circ)(\text{Molecular weight of } A)}{(P_B^\circ)(\text{Molecular weight of } B)} \tag{9}$$

For a mixture of bromobenzene and water, which have vapor pressures of 120 and 640 mm, respectively, at 95° (see Figure 2.10), the composition of the distillate would be calculated from equation (9) as follows:

$$\frac{g_{\text{bromobenzene}}}{g_{\text{water}}} = \frac{(120)(157)}{(640)(18)} = \frac{1.64}{1}$$

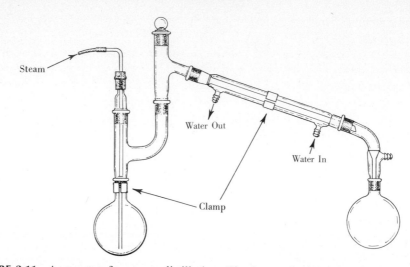

FIGURE 2.11 Apparatus for steam distillation. The long tube is replaced by a stopper if steam is generated *in situ*.

Consequently, on the basis of weight, more bromobenzene than water is contained in the steam distillate, even though the vapor pressure of the bromobenzene is much lower at the temperature of the distillation. Because organic compounds generally have molecular weights much higher than that of water, it is possible to steam-distil compounds having vapor pressures of only about 5 mm at 100° with a fair efficiency on a weight-to-weight basis. Even solids can often be purified by steam distillation.

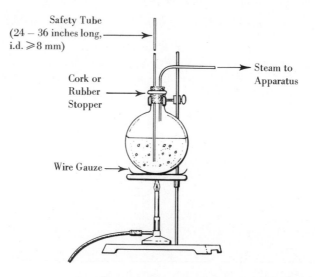

FIGURE 2.12 A steam generator. The round bottom flask is initially half-filled with water and boiling chips are added before heating. The safety tube serves to relieve internal pressure if steam is generated at too rapid a rate.

A steam distillation is generally accomplished in one of two ways. The first, and usually most efficient method, involves placing the organic compounds to be distilled in a round-bottomed flask fitted with a Claisen head which is equipped with a still head connected to a water-cooled condenser (Figure 2.11). The Claisen head helps to prevent splattering of the mixture into the condenser during distillation. Steam can be produced externally in a generator such as that shown in Figure 2.12 and then introduced into the bottom of the distillation vessel via a tube (Figure 2.11), or it can be obtained from a laboratory steam line. If the latter source is used, a trap

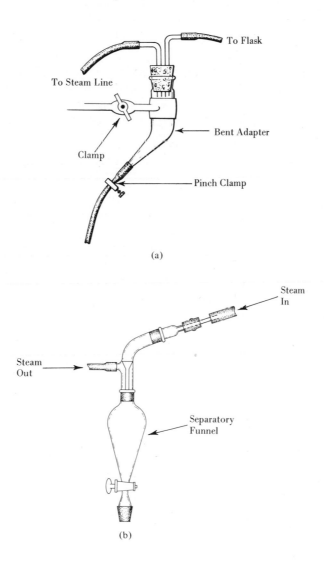

(a)

(b)

FIGURE 2.13   Water trap.

is usually placed between the line and the distillation flask to allow removal of any water present in the steam (Figure 2.13). When an external source of steam is used, water may condense in the distillation flask, filling it to undesirable levels. This problem can generally be circumvented by gently heating the flask with a Bunsen burner.[12]

If only a small quantity of steam is required to distil the mixture completely, a second method of steam distillation may be employed. This method involves combining water and the organic compounds to be distilled in the distillation flask. The flask is then heated directly with a Bunsen burner[12] to generate steam *in situ.* This method is, in general, not applicable for distillations in which large amounts of steam are needed; one would have to replenish the supply of water in the flask or use an inconveniently large flask.

In summary, steam distillation provides a method for separation of volatile liquid and solid organic compounds which are insoluble in water, or nearly so, from nonvolatile compounds under comparatively mild conditions. The technique obviously is not applicable to substances which decompose upon prolonged contact with steam or hot water, react with water, or have a vapor pressure of 5 mm or less at 100°.

2.7   Recrystallization

One of the most necessary and useful techniques to be mastered by an organic chemist is that of *recrystallization.* Many organic compounds are solids; the procedure of recrystallization is the technique of choice in most instances in effecting their purification. Even in those instances when an organic solid has been "purified" by sublimation (Section 2.8) or chromatography (Chapter 3), a careful worker will frequently, for added confidence, perform a final recrystallization of his material.

Essentially, the process is one in which the crystal structure is completely disrupted, either by fusion (melting) or by dissolution of the solid, and then crystals are allowed to regrow so that the impurities are left in either the melt or the solution. The fusion method is seldom used because the crystals are usually formed in the presence of a rather viscous oil (containing the impurities) from which they are difficult to separate. In solution recrystallization, a solvent is generally chosen in which the impurities are more soluble than the substance being purified, particularly in a cold solution. Crystals grown from such a solution will be of much greater purity than the original solid if the technique is performed properly and if the impurities are present only to the extent of a few percent. If impurities comprise more than a few percent of the original solid, two or more successive recrystallizations may be necessary.

Solution recrystallization as a technique involves several steps. These are (1) selection of the appropriate solvent, (2) dissolution of the solid to be purified in the solvent at or near its boiling point, (3) filtration of the hot solution to remove insoluble impurities, (4) crystallization from the solution as it cools, (5) separation

[12] See footnote 7 on page 32.

of the crystals from the supernatant solution, (6) washing the crystals to remove the adhering solution, and (7) drying the crystals.

Selecting a solvent. A solvent must satisfy certain criteria in order to be used as a recrystallization solvent. (1) Its temperature coefficients for the solute and impurities should be favorable; that is, the compound being purified should ideally be quite soluble in the hot solvent but somewhat insoluble in the cold (this will minimize losses), and the impurities should remain at least moderately soluble in the cold solvent. Another possibility here is that the impurities be insoluble in the hot solution, from which they may be filtered. (2) The boiling point of the solvent should be low enough so that it can be easily removed from the crystals in the final drying step. (3) It is generally preferable that the boiling point of the solvent be lower than the melting point of the solute. (4) The solvent should not react chemically with the compound being purified.

If the compound has been previously studied, consultation of the chemical literature will generally give information concerning a suitable solvent. If not, it will be necessary to resort to trial and error methods using *small* amounts of material. Some general solubility principles should be kept in mind if this needs to be done. Normally, polar compounds are insoluble in nonpolar solvents and soluble in polar solvents. Conversely, nonpolar compounds are more soluble in nonpolar solvents. These solubility relationships are frequently summarized with the phrase "like dissolves like." Therefore, a highly polar compound is unlikely to be soluble in a hot nonpolar solvent but may well be too soluble in a cold very polar solvent, so that a solvent of intermediate polarity will be optimum. Occasionally, mixed solvents are found to work well. In this case the solubility of a compound in one solvent is reduced by the addition of a second solvent in which it is much less soluble. Some frequently used mixed solvent pairs are alcohol-water, benzene-petroleum ether,[13] acetic acid-water, ether-alcohol, and ether-petroleum ether. Such solvent mixtures as benzene-alcohol are infrequently used, and then only with absolute alcohol, since the presence of water causes solvent separation, particularly on cooling; benzene and aqueous alcohol are not fully miscible.

Solution. The following operations are best carried out within a ventilation hood to avoid inhalation of solvent vapors. All organic solvents, excluding water, are either flammable or toxic to some extent, or both. The solid to be purified is placed in an appropriately sized Erlenmeyer flask, along with a few milliliters of the desired solvent. It is good laboratory technique to save a few crystals of the impure solid; they may be needed later to induce crystallization if problems are encountered at that step. With *constant* stirring, to prevent bumping of the boiling mixture (boiling chips are not very effective in the presence of quantities of undissolved solids), the mixture is then heated to boiling using either a hot plate or a steam bath. A Bunsen burner may be used, with care against too rapid heating, *but only if the solvent(s)*

---

[13] Petroleum ether is a mixture of aliphatic hydrocarbons obtained from petroleum refining. Such a mixture may vary in composition and boiling range, depending on the distillation "cut" taken. In order to define the liquid being used, the boiling range is usually given, *e.g.*, petroleum ether (bp 60–80°). The name *ligroin* is occasionally encountered in place of petroleum ether.

used are not flammable. More solvent is added in *small* portions to the boiling mixture until just enough boiling solvent is present to dissolve the solid. It is generally desirable at this point to add from 2 to 5% additional solvent to prevent premature crystallization during the hot filtration, if this step appears necessary. A large excess of solvent must be avoided in order to maximize the recovery of purified crystals. Solids remain soluble to some extent even in cool solution, and the recovery will be reduced by an amount which depends on both this solubility and the quantity of solvent present. If, near the end of the dissolution, it is apparent that additional solvent is not dissolving any more of the solid, *particularly when only a relatively small quantity of solid remains,* enough solvent has probably been added. The remaining solid is likely to consist of insoluble impurities and may be removed in the hot filtration step. Many beginning laboratory students obtain poor results in their recrystallizations because, when trying to dissolve the last traces of solid, quantities of solvent are added which are far in excess of that needed. During the dissolution of the impure solid, time should be allowed between each small addition of fresh solvent, for some solids dissolve only slowly.

When mixed solvents are used, they of course must be miscible. The crystals being purified should be soluble in one of the solvents, but insoluble or only slightly soluble in the second. The crystals are first dissolved in that pure, boiling solvent in which they are soluble. The second solvent is then added to the boiling solution until the solution turns *cloudy.* The second solvent decreases the dissolving ability of the solvent medium; when the solubility limit is reached, the solute begins to come out of solution, resulting in a cloudy appearance. If the second solvent has a lower boiling point than the first, cool the solution to below the boiling point of the second solvent before it is added. Finally, more of the first solvent is added *dropwise* until the solution just becomes clear again. Occasionally it is advantageous to carry out the hot filtration step before adding the second solvent to prevent crystallization during filtration.

If colored impurities are present, these may often be removed by adding a small amount of decolorizing carbon to the hot (*not boiling*) solution. Seldom is more decolorizing carbon needed than that which can be held on the tip of a small spatula. The impurities, especially colored ones, adsorb on the surface of the carbon particles and are removed during filtration. If too much carbon is used, some of the substance being purified will be adsorbed and subsequently lost in this step. After adding the carbon, the solution should be heated to boiling for a few minutes, while being continuously stirred or swirled to prevent bumping of the boiling mixture.

Hot filtration.   To remove insoluble impurities (including dust and decolorizing carbon, if used), the hot solution is filtered by gravity filtration. If no insoluble impurities are present, and the solution is clear, this step may usually be omitted. Suction filtration is not desirable, for evaporation of the hot solvent under reduced pressure will both cool and concentrate the solution, resulting in premature crystallization. A short-stemmed or stemless glass funnel and a fluted filter paper should be used for filtration into a second Erlenmeyer flask (Figure 2.14). The use of *fluted* filter paper allows for more rapid filtration. The top of the paper should not extend above the top of the funnel.

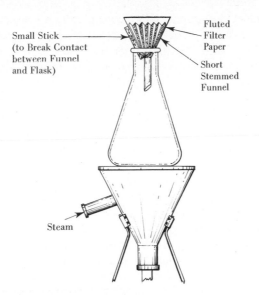

FIGURE 2.14   Apparatus for effecting hot filtration.

Be sure to pour the hot liquid on the upper portion of the filter paper in order to maximize the efficiency of the filtration. In this way, the solution will come in contact with a larger area of the filter paper, and thus will be filtered more rapidly. Be careful, however, not to allow any solution to pass between the edge of the paper and the funnel.

One of several possible ways of folding a fluted filter is depicted in Figure 2.15. Fold the paper in half, and then into quarters. Fold edge 2 into 3 to form edge 4, and then 1 into 3 to form 5 (a). Now fold 2 into 5 to form 6, and 1 into 4 to

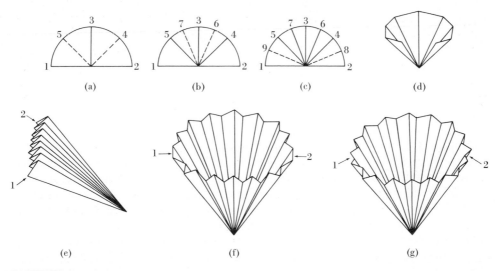

FIGURE 2.15   Folding of a fluted filter.

form 7 (b). Continue by folding 2 into 4 to form 8, and 1 into 5 to form 9 (c); the paper now appears as in (d). Note that all folds have been in the same direction. Do not crease the folds tightly at the center because this might weaken the paper causing it to tear during filtration. Now make new folds *in the opposite direction* between 1 and 9, 9 and 5, 5 and 7, etc., giving the paper a fan-like appearance (e). Open the paper (f) and fold each of the sections 1 and 2 in half with reverse folds to form (g). The paper is now ready to use.

Crystallization from the solution occasionally occurs in the filter paper or on the surface of the funnel. This is most conveniently avoided by adding 2 or 3 ml of the recrystallization solvent to the receiving flask and heating to boiling (Figure 2.14). The condensing vapors will heat the funnel, preventing crystallization. For low-boiling solvents, a steam bath may be used; for solvents boiling higher than about 80°, an electrically heated oil bath is preferable. A Bunsen burner can be used *only* if the solvent is nonflammable.

Decolorizing carbon, being very finely divided, is not always completely removed by this filtration. If not, a small amount of "filter-aid" (a finely divided silica that is easily filterable) may be added to the filtrate. The solution is heated again just to boiling, and refiltered using a fresh fluted filter paper. The carbon is trapped in the filter-aid and thus removed.

Crystallization. The hot filtrate is allowed to cool slowly by standing at room temperature. Crystallization should then ensue. Rapid cooling by immersion in water, etc., is undesirable because the crystals formed will tend to be quite small. Their large surface area may then facilitate *adsorption* of impurities from the solution. Generally the solution should not be agitated while cooling, since this will also lead to formation of small crystals. However, formation of very large crystals (larger than approximately 2 mm) may cause *occlusion* (trapping) of solution within the crystals. Such crystals are difficult to dry and will, when dried, have deposits of impurities in them. If large crystals seem to be forming, agitation may be used to lower the average crystal size. Judgment of proper crystal size will become easier with experience.

If crystallization does not occur after cooling, the super-saturated solution can usually be made to yield crystals by *seeding*. A tiny crystal of the original solid is added to the cooled solution. Crystals will usually form quite rapidly. Another method that may be used to induce crystal formation is to scratch the inside surface of the flask at or just above the surface of the solution with a glass rod.

Occasionally, the solute will separate as an "oil" rather than as crystals from the solution. As this type of precipitation is not as selective as crystallization, these oils generally contain significant amounts of impurities, and their formation is undesirable. There are two somewhat different types of oiling problems. (1) Oils may persist upon full cooling with no evidence of crystallization. The remedy in these cases which most frequently works is to scratch the oil against the side of the flask in the presence of the mother liquor with a glass stirring rod; this frequently will induce crystallization. Failing this, a few *small* seed crystals of the original impure solid may be added to the oil and the mixture allowed to stand for a time, perhaps until the next laboratory period. If this does not work it may be necessary to separate the oil from the mother liquor and crystallize it from a different solvent. (2) With certain compounds (acetanilide from water, for example), cooling of the hot concentrated solution results in the separation of an oil; at a lower temperature, this oil freezes into a compact mass, while, at this same temperature or a little lower, relatively pure crystals separate from the residual solution. This oil is not pure liquid solute, but a liquid solution of solute, solvent, and perhaps other impurities, whose freezing point lies below the temperature at which it separated from solution, even though the crystalline solute may have a melting point lying above the boiling temperature of the solution. When this occurs, the problem may usually be remedied by reheating the mixture to boiling, adding a few milliliters of additional solvent, and again allowing the solution to cool. This will either solve the problem or, at least, reduce the amount of oil formed. In the latter case, the procedure may be repeated.

After filtering the crystals (see below), a second "crop" of crystals can usually be obtained by further cooling with ice water, or by concentrating the solution by boiling away some of the solvent and cooling. The crystals obtained as a second (or even third) crop may not be as pure as the first. Melting points of the crystals obtained in various crops can be used to determine their purity.

Filtration.   The cool mixture of crystals and solution is now filtered by suction filtration using a Büchner funnel and a vacuum filter flask attached to an aspirator or house vacuum line through a trap as shown in Figure 2.16. The trap prevents

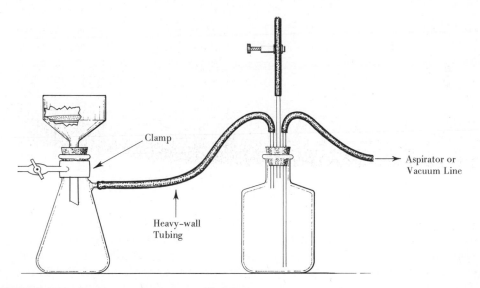

FIGURE 2.16   Apparatus for vacuum filtration.

water from the aspirator from backing up into the filter flask in case of loss of water pressure. The flask used for the trap should be a *heavy-walled* Erlenmeyer flask or bottle, or a second vacuum filter flask. If a filter flask is used, employ a two-holed stopper and attach the tube leading to the aspirator to the side arm. Before filtration, the filter paper, which should be of a size to lay *flat* on the funnel plate, should be wetted with the solvent in order to "seal" it to the funnel.

A stirring rod or spatula may be used as an aid in transferring the crystals to the funnel. The last quantity of crystals may be transferred by washing them out of the flask with some of the filtrate ("mother liquor"). As soon as the mother liquor has passed through the filter, release the suction by opening the screw clamp or stopcock on the trap. Wash the crystals to remove adhering mother liquor, containing impurities, by adding to the funnel *cold* fresh solvent just sufficient to cover the crystals. Close the trap to reapply suction and to remove the wash solvent from the crystals. Press the crystals as dry as possible on the filter plate under suction with a cork or spatula.

Drying the crystals.   Most of the solvent may be evaporated by allowing the aspirator to pull air through the mass of crystals on the funnel for a few minutes. By means of a spatula, the crystals are then transferred to a clean watch glass. Complete drying is accomplished by allowing them to air dry for a few hours. If necessary, the drying process may be accelerated by placing the watch glass in an oven (*Caution:* The temperature of the oven should be at least 20° below the melting point of the crystals), or by placing the crystals in a vacuum desiccator. A workable facsimile of a vacuum desiccator may be assembled by replacing the Büchner funnel (Figure 2.16) with a solid rubber stopper. The crystals may be laid on the bottom of the flask, or, better, if the quantity of crystals is not too great, they may be placed in a test tube around the mouth of which there is wrapped a piece of filter paper held in place with a rubber band. The test tube is placed in the desiccator. If the crystals were left on the bottom of the flask, take caution when releasing the pressure at the trap that the in-rushing air does not "blow" the crystals about the flask

Heat drying may not be used for crystals that sublime readily at atmospheric pressure (see Section 2.8).

EXPERIMENTAL PROCEDURE

A. Determination of solubilities

Place about 20 mg (a small spatula-tip full) of finely crushed resorcinol in a 10×75-mm test tube and add about 0.5 ml of water to the tube. Stir with a glass rod and determine whether resorcinol is soluble in water at room temperature. If it is not completely soluble, gently heat the test tube with a microburner to determine whether resorcinol is soluble in hot water. Record your observations of the solubility of resorcinol in water on the basis of the following definitions: soluble—20 mg of solute will dissolve in 0.5 ml of solvent; slightly soluble—some but not all of the 20 mg of solute will dissolve in 0.5 ml of solvent; insoluble—none of the solute appears to dissolve.

Repeat the solubility test using benzene and then petroleum ether (bp 60–80°).[13] *Do not heat these solvents with an open flame* as they are highly flammable! Rather, use a steam cone. Record the observed solubility of resorcinol in each of the solvents.

In like fashion determine the solubilities of napthalene, benzoic acid and anthranilic acid in water, in benzene and in petroleum ether. If any of these solutes is soluble in the hot solvent, but only slightly soluble or insoluble in the cold solvent, allow the hot solution to cool slowly to room temperature and compare the size, color, and crystal form of the resulting crystals with the original solid material. Note which solvents you would consider best suited for recrystallization of each of the solutes.

B. RECRYSTALLIZATION OF IMPURE SOLIDS

Read carefully the introductory paragraphs of this section to become familiar with the techniques for carrying out each of the steps in the recrystallization procedure. This information should allow you to avoid or surmount most of the problems which may be encountered in effecting the purification of impure solids by recrystallization. Only abbreviated directions are provided for the experiments below.

Obtain, as assigned, one or more samples of solids for recrystallization. Among those impure solids which might be assigned are: (1) benzoic acid; (2) acetanilide; and (3) naphthalene.

Benzoic acid.   Place 1 g of impure benzoic acid in a clean 50-ml Erlenmeyer flask. Measure 25 ml of water in a graduated cylinder and add a 10-ml portion of it to the benzoic acid. Heat to gentle boiling with a microburner. Add, as necessary, water in 1-ml portions until no more solid appears to dissolve in the boiling solution. Record the total volume of water used. No more than 20 ml should be required.

Since pure benzoic acid is colorless, a colored solution should be treated with decolorizing carbon. (*Caution:* Do not add decolorizing carbon to a *boiling* solution!) Cool the solution slightly, add approximately 0.1 g of carbon, and reheat to boiling for a few minutes. To aid in the filtration of the finely divided carbon, allow the solution to cool slightly, add a small amount of filter-aid, and reheat.

Perform a hot filtration according to the directions provided in the paragraph on that subject with reference to Figure 2.14. Rinse the empty flask with 1 or 2 ml of *hot* water and filter this wash solution into the main solution. If the filtered solution remains colored, repeat the treatment with decolorizing carbon. Cover the flask with a watch glass or inverted beaker,[★] and allow the filtrate to stand undisturbed until it has cooled to room temperature and no more crystals form. Then place the flask in ice water for at least 15 min to complete the crystallization.

Collect the white crystals by suction filtration and wash the filter cake with two small portions of *cold* water. Press the crystals as dry as possible on the funnel with a clean spatula or cork. Spread the crystals onto a piece of filter paper or, better, a watch glass and allow them to air dry completely. Determine the weight and melting point of the purified product. Calculate the percent recovery.

Acetanilide.   Place 5 g of impure acetanilide in a 250-ml Erlenmeyer flask. Measure 100 ml of water into a graduated cylinder and add a 50-ml portion to the crude acetanilide. Boil the mixture gently with the aid of a burner.

Note the formation of an oil layer consisting of a solution of water in acetanilide which forms a separate phase. This second liquid phase forms at temperatures only above 83° in mixtures whose compositions lie between 5.2 and 87% acetanilide.

Acetanilide, however, is sufficiently soluble in water at temperatures slightly above 100°,[14] the boiling point of the solution, to form solutions of greater than 5.2% acetanilide. Thus, a boiling solution prepared with the *minimum* quantity of water to effect solution will yield an oil upon cooling to 83°, and crystals below 83° (see the paragraph on Crystallization: discussion of oiling).

Continue adding water in small portions (3–5 ml) to the boiling solution until the oil has completely dissolved. Note that any solid present at this point must consist of insoluble impurities. (The interested student may wish to allow this solution to cool to about 50° to verify the phase properties of acetanilide-water discussed above. If this is done, reheat the solution to boiling following these observations and continue.) Once the acetanilide has just dissolved, add an additional 5 ml of water to prevent formation of oil during the crystallization step. If oil does form at that time, reheat the solution and add a little more water. Record the total volume of water used.

Allow the solution to cool below boiling and add *ca.* 0.1 g of decolorizing carbon; gently boil the solution for a few minutes, with stirring. Cool the solution, add a small amount of filter-aid, stirring thoroughly. Reheat to boiling, and perform a hot filtration according to the directions given in the paragraph on this subject with reference to Figure 2.14. Cover the flask with a watch glass or, better, an inverted beaker,★ and allow the filtrate to stand undisturbed while cooling to room temperature; when crystallization apparently is complete, cool the flask in ice water for at least 15 min to complete the crystallization.

Collect the crystals by suction filtration and wash the filter cake with two small portions of *cold* water. Press the crystals as dry as possible on the funnel with a spatula or cork. Spread the crystals onto a piece of filter paper or, better, a watch glass and allow them to air dry completely. Determine the weight and melting point of the purified acetanilide. Calculate the percent recovery.

Naphthalene.   Naphthalene may be conveniently recrystallized from either methanol, ethanol, or isopropyl alcohol. Because these solvents are all either toxic or flammable, or both, proper precautions should be taken. Operations through the hot filtration step should be carried out in a ventilation hood. If this is not available, an inverted funnel connected by tubing to an aspirator or vacuum line and positioned *no more than an inch or two* above the mouth of the flask in which the solvent is being heated will afford reasonable protection.

Place 5 g of impure naphthalene in a 250-ml Erlenmeyer flask and dissolve it in the minimum required amount of boiling alcohol. (*Caution:* Use a steam bath for heating; do not use a burner.) Add 2 or 3 ml of additional solvent. A colored solution should be treated with decolorizing carbon. Perform a hot filtration, and cover the flask containing the hot filtrate with either a watch glass or inverted beaker,★ and allow the filtrate to stand undisturbed while cooling to room temperature. Collect the crystals by suction filtration, wash with two small portions of *cold* solvent, and press dry. Transfer the crystals to a piece of filter paper or a watch glass and allow them to air dry. Determine the weight and melting point of the purified naphthalene. Calculate the percent recovery.

---

[14]The solubility of acetanilide in water is 5.5 g per 100 ml at 100° and 0.53 g per 100 ml at 0°.

EXERCISES

1. List each of the steps in the systematic procedure for recrystallization. Indicate briefly the functional purpose of each of these steps in accomplishing the purification of the originally impure solid.

2. The goal of the recrystallization procedure is to obtain *purified* material with a *maximized recovery*. For each of the items below, explain why this goal would be adversely affected.
   (a) In the solution step, an unnecessarily large volume of solvent is used.
   (b) The crystals obtained by suction filtration are not washed with fresh cold solvent before drying. This step is omitted.
   (c) The crystals referred to in (b) are washed with fresh *hot* solvent.
   (d) A large excess of decolorizing carbon is used.
   (e) Crystals are obtained by breaking up the solidified mass of an oil which originally separated from the hot solution.
   (f) Crystallization is accelerated by immediately placing the flask of hot solution in ice water.

3. A second crop of crystals may be obtained by concentrating the suction filtrate and cooling. Why is this crop of crystals probably less pure than the first crop?

4. Explain why the rate of dissolution of a crystalline substance may depend on the *size* of its crystals.

5. The solubility of benzoic acid at $0°$ is 0.02 g per 100 ml of water, and that of acetanilide is 0.53 g per 100 ml of water. If you performed either of these recrystallizations, calculate, with reference to the total volume of water used in preparing the hot solution, the amount of material in your experiment which was unrecoverable by virtue of its solubility at $0°$.

6. Assuming that either solvent is otherwise acceptable in a given instance, what advantages does ethyl alcohol have over *n*-octyl alcohol as a crystallization solvent? hexane over pentane? water over methyl alcohol?

7. Look up the solubility of benzoic acid in hot water. According to the published solubility, what is the minimum amount of water in which 1 g of benzoic acid could be dissolved?

8. Why is it important to:
   (a) break the suction before turning off the water pump when employing the equipment shown in Figure 2.16 (suction filtration)?
   (b) avoid the inhalation of vapors of organic solvents?
   (c) know the position and procedure of operation of the nearest fire extinguisher when employing benzene as a crystallization solvent?
   (d) use a *fluted* filter paper for hot filtration?

2.8 Sublimation

An alternative to crystallization for the purification of some solids is the process of *sublimation*. This method uses to advantage the differing vapor pressures of solids in a way analogous to simple distillation. The impure sample is vaporized directly from the solid state by heating it at a temperature below the melting point and the vapor is then condensed (crystallized) directly to the solid state on a cold surface; both processes occur *without* the intermediacy of the liquid state. Figure 2.17 shows a typical phase diagram, relating the solid, liquid, and vapor states of a substance with pressure and

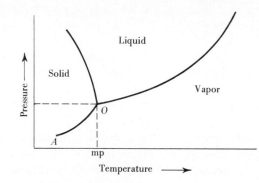

FIGURE 2.17   Single-component phase diagram.

temperature. Note that with conditions of temperature and pressure below those depicted at point O, the liquid state cannot exist (*i.e.*, is thermodynamically unstable). The vapor pressure of the solid at any temperature below the melting point is given by the curve OA. It is the equilibrium between solid and vapor, represented by this curve which is of importance for sublimation.

Two types of sublimation apparatus commonly used are pictured in Figure 2.18. They have two characteristics in common: (1) a chamber that may be evacuated by attachment to a vacuum pump (the impure sample is placed at the bottom of this

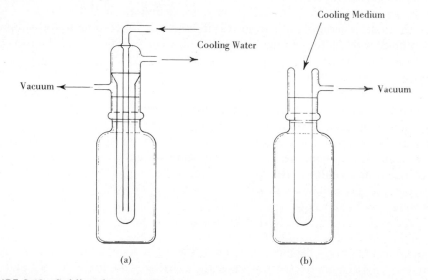

(a)   (b)

FIGURE 2.18   Sublimation apparatus.

chamber) and (2) a finger-like projection in the center of this chamber that may be cooled to provide a surface upon which the sublimed crystals may form. The "cold-finger" in (a) is cooled by circulating water, and that in (b) by a medium such as Dry-Ice in acetone, or ice water. Figure 2.19 shows a simple type of sublimation apparatus that may be assembled inexpensively from two test tubes (one with a side arm), rubber stoppers, and glass tubing.

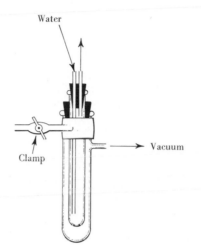

FIGURE 2.19   Test tube sublimator.

For effective purification using this technique, two criteria must be met: (1) the solid must have a relatively high vapor pressure, and (2) the impurities must exhibit vapor pressures substantially different from the primary material (preferably lower). Should the first criterion *not* be met, the time necessary to pass any significant quantity of material through the vaporization-crystallization sequence would usually be prohibitively long. In this case, either recrystallization (Section 2.7) or chromatographic techniques (Chapter 3) must be used.

The "mechanics" of purification by sublimation may be simply described as follows. The impure solid is heated to a temperature higher than that of the surface (cold-finger) upon which the pure material is to be condensed, but lower than the melting point. Since vapor pressure and thermodynamic instability of the solid vary directly with the temperature (see curve *OA*, Figure 2.17), material will be transferred via the vapor phase to the colder surface. Crystals developing on the cold surface will tend to be very pure, for the molecules of the impurities will usually not be incorporated into the growing crystal structure and therefore will not condense on the cold surface. Should the vapor pressure of an impurity be similar to that of the material being purified, sublimation will not be an effective method of purification, for crystals of both substances will tend to form on the cold surface (criterion 2 above).

Very few organic compounds exhibit vapor pressures adequate for sublimation at atmospheric pressure. Generally, reduced pressure (vacuum) is needed to increase the rate of evaporation of the solid. This process is analogous to the use of vacuum distillation for high-boiling liquids.

Sublimation is generally restricted to relatively nonpolar substances having fairly symmetrical structures. In these cases, crystal forces are lower and vapor pressures tend to be higher. The ease with which a molecule may escape from the solid to the vapor phase is determined by the strength of the intermolecular attractive forces between the molecules of the solid. The most important attractive force is of an electrostatic nature. Symmetrical structures will have relatively symmetrical distributions of electron density, and as such will have smaller dipole moments than less symmetrical structures whose electron distribution will be more polarized. A smaller dipole moment infers a higher vapor pressure. Van der Waals forces are also important, but are generally less important than electrostatic attractions. In general, van der Waals forces increase in magnitude with increasing molecular weight, and thus large molecules, even if symmetrical, are less adaptable to purification by sublimation.

REFERENCES

A. DISTILLATION
1. *Technique of Organic Chemistry,* A. Weissberger, editor, Interscience Publishers, New York, 1965, Vol. IV.
2. *Azeotropic Data,* American Chemical Society, Washington, D. C., 1952.
3. T. Earl Jordan, *Vapor Pressure of Organic Compounds,* Interscience Publishers, New York, 1954.

B. RECRYSTALLIZATION
R. Stuart Tipson, in *Technique of Organic Chemistry,* A. Weissberger, editor, 2nd ed., Interscience Publishers, New York, 1956, Vol. III, Part I, Chapter 3.

C. SUBLIMATION
R. Stuart Tipson, in *Technique of Organic Chemistry,* A. Weissberger, editor, Interscience Publishers, New York, 1965, Vol. IV, Chapter 8.

chapter three

separation and purification of organic compounds

2 Phase Distribution: Extraction and Chromatography

Two widely applicable and relatively simple methods of separating a desired compound from its impurities, or of isolating each of the individual components of a mixture, are extraction and chromatography, both of which are based upon the principle of *phase distribution*. A substance may establish an equilibrium distribution between two *insoluble* phases with which it is in contact, in a ratio dependent upon its relative stability in each of those phases. The techniques being discussed involve the *selective* removal of one or more of the components of a gaseous, liquid or solid mixture by contact of the mixture with a second phase.

The process upon which such a distribution depends may be one of two varieties: (1) *partitioning*, based upon differing relative solubilities of the components in immiscible solvents (selective dissolution); and (2) *adsorption*, based upon the selective adherence of the components of a liquid or gaseous mixture to the surface of a solid phase. The various techniques of chromatography involve both of these processes, whereas the techniques of extraction involve only the former.

3.1   Extraction

A most widely employed method of separating organic compounds from mixtures in which they are found or produced is that of liquid-liquid extraction. In fact, virtually every organic reaction requires extraction at some stage in the purification of its products.

In its simplest form, extraction involves the distribution of a solute between two immiscible solvents. The distribution is expressed quantitatively in terms of the *distribution* (or *partition*) *coefficient*, K [equation (1)]. This expression indicates that a solute, A, in contact with a mixture of two immiscible liquids, S and S', will be distributed (partitioned) between the liquids so that at equilibrium the ratio of concentrations of A in each phase will be constant, at constant temperature.

$$K = \frac{\text{concentration of } A \text{ in } S}{\text{concentration of } A \text{ in } S'} \tag{1}$$

Ideally, the distribution coefficient of A is equal to the ratio of the individual solubilities of A in pure S and in pure S'. In practice, however, this correspondence is generally only approximate since no two liquids are completely immiscible. The extent to which they dissolve in one another alters their solvent characteristics and thus slightly affects the value of K.

It is evident that for A to dissolve completely in one or the other of two immiscible liquids, the value of K must be infinity or zero.[1] Neither of these limiting values is actually attained. However, so long as K is larger than 1.0, and the volume of solvent S is equal to or larger than the volume of solvent S' (see below), the solute will be found in *greater amounts* in solvent S. The amount of solute that will remain in the other solvent, S', will depend on the value of K.

Equation (1) may be rewritten as shown in equations (2) and (3).

$$K = \frac{\text{grams of } A \text{ in } S/\text{ml of } S}{\text{grams of } A \text{ in } S'/\text{ml of } S'} \tag{2}$$

$$K = \frac{\text{grams of } A \text{ in } S}{\text{grams of } A \text{ in } S'} \times \frac{\text{ml of } S'}{\text{ml of } S} \tag{3}$$

Note that when the volumes of S and S' are equal,[2] the ratio of the grams of A in S and in S' will equal the value of K. If the volume of S is doubled, and the volume of S' kept the same, the ratio of the grams of A in S to the grams of A in S' will be *increased* by a factor of two. This must necessarily follow because K is a *constant*. Therefore, if A is to be recovered from solvent S, the amount of A recovered will be increased by using larger quantities of solvent S.

A further consequence of the distribution law [equation (1)] is of practical importance in performing an extraction. If a given total volume of solvent S is to be used to separate a solute from its solution in S', it can be shown to be more efficient to effect several successive extractions with portions of that volume than one extraction with the full volume of solvent. Thus, *more* butyric acid will be removed from water solution by two successive extractions with 50-ml portions of ether than will be removed in a single extraction with 100 ml of ether. Three successive extractions with 33-ml portions would be still more efficient (see exercises). There is, however, a point beyond which the further effort of additional extractions no longer yields a commensurate return. The larger the distribution coefficient, the fewer the number of repetitive extractions that are necessary to separate the solute effectively. This is an important consideration because it is desirable to keep the total volume of extracting solvent to a minimum, not only for reasons of expense, but also because of the time involved in eventual removal of the solvent by distillation.

---

[1] For simplicity, we shall define S as that solvent in which the solute is more soluble. Therefore, in the remaining discussion the value of K will by definition always be greater than 1.0.

[2] To be strictly correct, the volumes of *solution* should be used in this expression. However, when the solutions are reasonably dilute, volumes of *solvent* may be used without appreciable error.

Consider now a solution of two compounds in solvent S'. It should be evident that for effective separation of these two compounds by extraction with solvent S, the distribution coefficient of one should be significantly greater than 1.0, whereas the distribution coefficient of the other should be significantly smaller than 1.0. If these conditions are met, one compound will be mainly distributed in solvent S and the other in solvent S' at equilibrium. Physical separation of the two liquid layers would then result in at least a partial separation of the two compounds.

When the coefficients are of similar magnitude, separation by the extraction technique may be quite ineffective since the relative concentrations of the compounds in each of the two liquid phases may be little changed from those of the original mixture. In this case, other methods of separation should be used, such as adsorption chromatography (Section 3.5), fractional distillation (Section 2.2), fractional crystallization, etc. The specialized techniques of *fractional extraction* and *countercurrent distribution* can also be used to separate compounds with similar but not identical distribution coefficients. Details of these techniques are supplied for the interested student in the references at the end of the chapter.

Another type of experimental problem often encountered is that involving separation of one component that is only slightly soluble in the extracting solvent from a mixture whose other components are essentially insoluble. Large quantities of solvent would have to be used in order to effect the separation in only one or two extractions, and the handling of such quantities may be extremely unwieldy. Alternatively, it would be tedious to do a very large number of extractions with smaller quantities of solvent. The method of *continuous extraction*, in which a relatively small volume of extracting solvent is used, is a possible solution to this problem. By means of specialized apparatus, the solution of extracting solvent and solute is continuously separated into a boiling flask from the mixture being extracted. The solution is subjected to continuous distillation and the condensed distillate returned as fresh extracting solvent to the extraction vessel and reused. In the process, the extracted material builds up in an increasingly concentrated solution in the boiling flask. This is because more dilute solution is continuously draining into the flask, while, at the same time, the solvent is being distilled away.

For continuous liquid-liquid extraction, the solvent (as condensate from the condenser) is made to pass either up or down, depending on relative densities, through the solution containing the desired compound. If the solvent is less dense than the solution being extracted, an apparatus such as that shown in Figure 3.1 may be used. If it is more dense, apparatus such as that shown in Figure 3.2 may be used.

For separation of the components of a solid mixture by continuous solid-liquid extraction, a Soxhlet extraction apparatus (Figure 3.3) is convenient. The solid is placed in a *porous* thimble in the chamber as shown, and the extracting solvent in the boiling flask below. The solvent is heated to reflux, and the distillate, as it drops from the condenser, collects in the chamber. By coming in contact with the solid in the thimble, the liquid effects the extraction. After the chamber fills to the level of the upper reach of the siphon arm, the solution empties from this chamber into the boiling flask by a siphoning action. This process may be continued automatically and without attendance for as long as is necessary for effective removal of the desired component, which will then be contained with the solvent in the boiling flask.

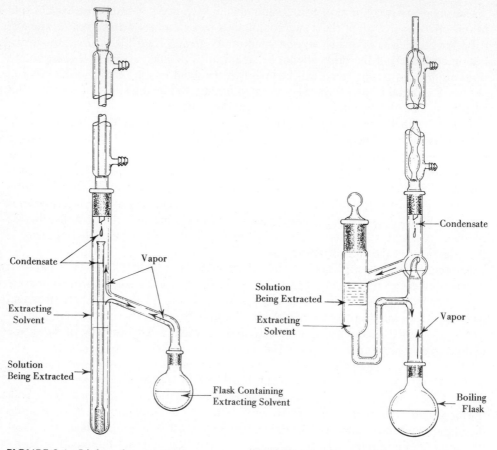

FIGURE 3.1 Light-solvent extractor.

FIGURE 3.2 Heavy-solvent extractor.

When choosing an extracting solvent for the isolation of a component from a solution there are some general principles to be kept in mind. (1) The solvent must, of course, be immiscible with the solvent of the solution. (2) It must be chosen to have the most favorable distribution coefficient for the component in question, and to have unfavorable coefficients for the impurities or other components. (Since distribution coefficients for most compounds in various solvents are not available, the solvents for the various experiments will be suggested to the student. As he gains experience in, and understanding about, the solubility principles of organic compounds, the student should be able to suggest appropriate solvents.) (3) A solvent must be chosen, just as in recrystallization, that does not react chemically in an undesirable manner with the components of the mixture. (4) After extraction, the solvent should be readily separable from the solute. Usually the solvent is removed by distillation.

The distribution coefficients of organic acids and bases may be quite dramatically affected by pH when one of the solvents is water. An organic acid that is insoluble in water at pH 7 may be quite soluble in dilute aqueous sodium hydroxide or sodium

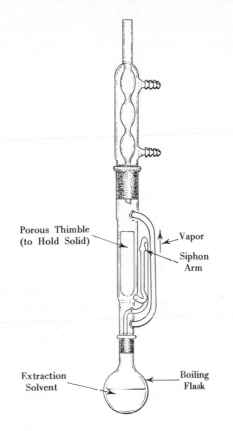

Porous Thimble
(to Hold Solid)

Vapor

Siphon
Arm

Extraction
Solvent

Boiling
Flask

FIGURE 3.3   Soxhlet extractor.

bicarbonate. In such a case, deprotonation of the acid affords its corresponding conjugate base

$$
\underset{\text{Organic Acid}}{R-\overset{\overset{\displaystyle O}{\|}}{C}-O-H} + HO^{\ominus}Na^{\oplus} \xrightarrow{H_2O} \underset{\text{Conjugate Base}}{R-\overset{\overset{\displaystyle O}{\|}}{C}-O^{\ominus}Na^{\oplus}} + H_2O \tag{4}
$$

which, because of its ionic character, is much more soluble in water, a polar solvent. Similarly, an organic base that is insoluble in water at pH 7 may be quite soluble in acidic media (pH less than 7) such as dilute hydrochloric acid. In this case, the increase in solubility relies on protonation by the aqueous acid of the organic base to produce the ionic, and therefore more water-soluble, conjugate acid.

$$
\underset{\text{Organic Base}}{R-NH_2} + H-Cl \xrightarrow{H_2O} \underset{\text{Conjugate Acid}}{R-\overset{\overset{\displaystyle H}{|}\,\oplus}{N}H_2\ Cl^{\ominus}} \tag{5}
$$

Thus, organic acids and bases may be selectively removed from nonpolar organic solvents, such as ether, dichloromethane, benzene, etc., by extraction with aqueous solutions having the appropriate pH.

Recovery of the original organic acid or base from the aqueous solution is accomplished by neutralization. Addition of aqueous base to an acidic solution will liberate the organic base whereas addition of aqueous acid to a basic solution will free the organic acid. If the organic acid or base thus formed is insoluble or only slightly soluble in water at pH 7, it will separate from the aqueous medium either in the form of a precipitate or as a liquid organic layer.

3.2   Technique of Simple Extraction

In the undergraduate organic laboratory, the student will often need to use extraction in the purification of the products of his reactions. Some of the common applications of extraction that will be encountered are (1) removal of organic reaction products from a water solution in which they are formed by extraction with an organic solvent; (2) removal of acid or base catalysts or inorganic salts from a reaction mixture by extraction with water; (3) separation of organic acids or bases from other organic compounds in an organic solvent by extraction with dilute aqueous base or mineral acid. Accordingly, the technique of simple extraction with a separatory funnel is described here in some detail.

Separatory funnels are available in several different shapes, from almost spherical to elongated pear shape (Figure 3.4). The more elongated the funnel, the longer the time

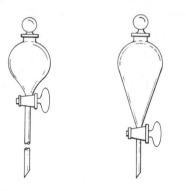

FIGURE 3.4   Separatory funnels.

required for the two liquid phases to separate once shaken together. When the liquids have similar densities, the flatter (more spherical) separatory funnels are preferable, for otherwise a needless amount of time may be wasted in waiting for the phases to separate. The funnels have a stopcock at the bottom through which the contents may be drained.

To perform an extraction, the solution is placed in a separatory funnel (with the stopcock closed!) and to this is added a quantity of the extracting solvent. The funnel should not be too full (three-fourths of the height of the funnel is a reasonable maximum). The upper opening of the funnel is stoppered either with a ground-glass stopper, for which most separatory funnels are fitted, or with a rubber stopper. The funnel is held during the shaking process in a rather specific manner, which will, when mastered, be found to be quite efficient. If right handed, set the stopper at the top of the funnel against the base of the index finger of the left hand and grasp the funnel with the first two fingers and the thumb. The funnel should be turned so that the first two fingers of the right hand can be curled around the handle of the stopcock. In such a manner, the stopper and stopcock can be held tightly in place during the shaking process (see Figure 3.5). Left-handed people would, of course, find it easier to use the opposite hand for each position.

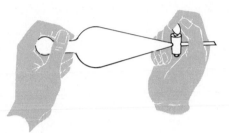

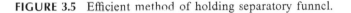

FIGURE 3.5 Efficient method of holding separatory funnel.

The funnel and its contents are shaken in a vigorous manner, so that the two immiscible liquids are mixed as intimately as possible. The purpose of the shaking process is to increase greatly the surface area of contact between the solvents so that the equilibrium distribution of solute between the solvents will be attained in a relatively short time. Every few seconds it is important to "vent" the funnel by inverting it (stopcock up) and carefully opening the stopcock to release any pressure that may have developed inside. This is particularly important when solvents of low boiling points are used, or if an acid solution is extracted with sodium bicarbonate solution (CO_2 is released). If this is not done, the stopper may well be blown free, and the contents of the funnel lost. If the funnel is held as described, the stopcock may be opened by twisting the fingers curled around it, without readjusting the grip on the funnel. At the end of the shaking period (if the shaking is vigorous, one or two minutes is usually sufficient), the funnel is vented a final time, supported on a ring [3] and the layers allowed to separate. The two liquid layers then are separated by carefully drawing the lower layer into a flask through the stopcock.

As a general rule, the layers will separate so that the solvent of greater density will be on the bottom. Thus knowledge of the densities of solvents to be employed is useful in identification of the layers. This, however, is not foolproof, because the nature and concentration of solute may be such as to invert the relative densities of two solvents.

[3] When using an iron ring to support a funnel, it is best to cover the ring with a length of rubber tubing to prevent breakage. This may be accomplished by slicing the tubing on one side and then slipping the tubing over the ring. Copper wire may be used to fix the tubing permanently in place.

A common mistake of students is to confuse the identity of the two layers in the funnel. Often one layer is discarded and it is later found that the wrong layer was saved. *It is suggested, as a point of caution, that both layers always be saved until there is no doubt about the identity of each.*

Occasionally, after shaking, the two immiscible liquids will not separate cleanly into two layers, but will instead form an emulsion. If one of the layers is water, often, but not always, the emulsion can be "broken" into separate layers by adding a small amount of acid or base or saturated aqueous sodium chloride to the mixture and reshaking gently. This method can be used only if the added material does not change the distribution coefficients in an unfavorable manner (which it usually does not unless pH is an important factor). In the case of difficult emulsions, it may be necessary to choose a different extracting solvent to avoid this problem.

Small quantities of insoluble solid material often collect at the phase interface, making it difficult to observe the layer boundary. Filtration of the heterogeneous mixture before separating the layers is the usual remedy for this problem.

EXPERIMENTAL PROCEDURE

This experiment will demonstrate the ability to separate various types of organic compounds by control of the pH of the extracting aqueous medium.

Prepare a *mixture* of 2 g each of the following compounds: benzoic acid, *p*-nitroaniline, and naphthalene. Dissolve this mixture in about 100 ml of dichloromethane. Extract this solution two times with 30-ml portions of 6 M hydrochloric acid. Combine the aqueous extracts and set them aside in a labeled flask. Reextract the organic layer from above two times with 30-ml portions of 3 M sodium hydroxide solution and combine these two aqueous extracts.

Dry the organic layer that is left at this point by adding 4–5 g of *anhydrous* sodium sulfate to the liquid contained in an Erlenmeyer flask;[4] let it stand for about 0.5 hr, swirling occasionally.

While the organic layer is drying, neutralize each of the aqueous extracts with either 6 M hydrochloric acid or 6 M sodium hydroxide, whichever is necessary. The 6 M hydrochloric acid extract should be cooled with an ice bath before neutralization. Upon neutralization, precipitates will be observed for each of the extracts. Cool each of the mixtures with an ice bath and then filter each by vacuum filtration. Each of the filtered solids should be washed *within the Büchner funnel* with *cold* distilled water. Pre-weigh (tare) three small labeled beakers, and place the solid materials thus far obtained in two of them. Either allow these to air dry until the next laboratory period[5] or place them in an oven held at 90° for 1.5 hr.

Filter the organic layer which has been drying over sodium sulfate by gravity filtration and remove the solvent by distillation, using a steam bath. Collect any solid residue, transfer it to the third tared beaker, and allow it to *air dry* (do not use oven).

---

[4] For a discussion of drying agents, see Appendix I.

[5] If these materials are allowed to air-dry in the laboratory desk until the next period, the beakers should be covered loosely with a piece of paper to prevent contamination of the compounds with dust.

Reweigh each of the beakers and obtain the weights of the solid materials. Determine the melting points of each of the materials obtained and identify each as one of the original compounds: benzoic acid, mp 121–2°; *p*-nitroaniline, mp 146–7°; naphthalene, mp 80–1°.

EXERCISES

1. From the results of the above experiment, what can you conclude about the properties of the compounds used? Write chemical equations showing any changes that occurred during the extraction procedure.
2. If, when extracting an aqueous solution with an organic solvent, you were uncertain of which layer in the separatory funnel was the organic layer, how could you quickly settle the issue?
3. Which layer (upper or lower) will each of the following solvents usually form if used to extract an aqueous solution: ethyl ether? chloroform? acetone? hexane? benzene?
4. Given 500 ml of an aqueous solution containing 8 g of compound *A*, from which it is desired to separate *A*, how many grams of *A* could be removed in a single extraction with 150 ml of diethyl ether? (Assume the distribution coefficient, diethyl ether: water, to equal 3.0.) How many *total* grams could be removed with three successive extractions of 50 ml each?

Note: In solving problems of this type, one should recognize that equation (3) applies to the situation pertaining *after* equilibrium is reached; *e.g.*, a practical form of the equation is $K = \dfrac{x}{a - x} \times \dfrac{\text{ml of } S'}{\text{ml of } S}$, where a = grams of *A* originally present in water (S') and x = grams of *A* present in ether (S) after extraction.

3.3 Chromatography

As mentioned in the introductory section, chromatography is based upon the general principles of phase distribution. Reduced to its fundamentals, the method involves the *selective removal* of the components of one phase from that phase as it is *flowing* past (or through) a second *stationary* phase. The removal of a component by the stationary phase is an equilibrium process, and the molecules of that component reenter the moving phase. Separation of two or more components in the moving phase will result when the equilibrium constants for the distribution of these components between the two phases differ. Expressed simply, the more tenaciously one component is held by the stationary phase, the higher the percentage of molecules of that component that are held *immobile*. A second component, less strongly held, will have a higher percentage of molecules in the *mobile* phase than will the first component. Therefore on the average, the molecules of the component that is held less strongly will move over the stationary phase (in the direction of flow) at a higher rate than the other, resulting in a migration of the components into separate regions (bands) of the stationary phase (see, for example, Figure 3.10).

The separation between the bands is linearly related to the distance traveled on the column. In general, the longer the distance, the greater the separation will be. For example, if one band moves 10 cm along the pathway and a second band moves 5 cm during the same time interval, the separation between the centers of the bands will be 5 cm. At a later time, if the first band has moved 100 cm, the second will have moved only 50 cm, and the separation will have increased to 50 cm. It should be remembered that the separation of a mixture by phase distribution requires that the components of the mixture have different distribution coefficients; if these coefficients are similar, only partial separation of the components into individual bands will occur, unless, of course, the path length is increased to give the components "time" to migrate apart.

There are four important types of chromatography based upon the principles discussed above. These are gas chromatography (more specifically, gas-liquid partition chromatography, glpc), column chromatography, thin-layer chromatography (tlc), and paper chromatography. These are discussed individually in the following sections.

3.4   Gas Chromatography

In gas chromatography, the mixture to be separated is *vaporized* and carried along a column by a flowing *inert* gas such as nitrogen or helium (the carrier gas). The gaseous mixture is the *mobile* phase. The column is packed with a solid, finely divided substance, on the surface of which is coated a liquid of relatively low volatility. This liquid serves as the *stationary* phase. Due to selective phase distribution of the components of the mixture between the mobile and stationary phases, these components may move through the column at different rates, and thus be separated. The physical process involved in the separation of the components of a mixture in the glpc column is the *partitioning* of the components between the gas and liquid phases.

A large variety of gas chromatographs of different types are commercially available. However, the basic features of these instruments are quite similar and are represented by the schematic shown in Figure 3.6. Parts 1–5 are needed to supply dry carrier gas at a controlled flow rate. The column (7) is connected to the gas supply and is contained within an oven (8); the temperature within the oven is controlled by a thermostat and heating elements. The sample to be separated is introduced into the flow system at the

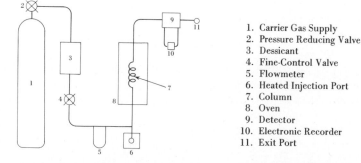

1. Carrier Gas Supply
2. Pressure Reducing Valve
3. Dessicant
4. Fine-Control Valve
5. Flowmeter
6. Heated Injection Port
7. Column
8. Oven
9. Detector
10. Electronic Recorder
11. Exit Port

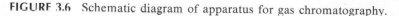

FIGURE 3.6   Schematic diagram of apparatus for gas chromatography.

injection port (6), which is individually heated to facilitate vaporization of the sample. The vaporized sample is then swept into the column by the carrier gas. As the sample passes through the column, its components separate into individual "bands" in the carrier gas, which then pass through the detector (9). The detector produces an electronic signal whose voltage is proportional to the amount of material different from the carrier gas itself present in the gas stream. The recorder (10) plots this voltage as a function of time to give the gas chromatogram (see Figure 3.8, for example). The vapors then pass from the detector into either the atmosphere or a collecting system at the exit port (11).

The elapsed time necessary for a given component to pass from the injection port to the detector is known as its *retention time*. Four important experimental factors affect the retention time of a compound: (1) the *nature* of the stationary liquid phase; (2) the *length* of the column; (3) the *temperature* at which the column is maintained; and (4) the *rate of flow* of the inert gas. *The retention time is independent of the presence (or absence) of other components in the mixture.* For any given column and set of conditions (temperature and flow rate), the retention time is a property of the compound in question, and may be used to identify it (see below).[6]

The choice of a stationary liquid phase best suited for a glpc experiment depends for the most part on the types of compounds to be separated. Although there are a very large number of liquid phases available, only a few (Table 3.1) are widely used. The best choice will be that liquid giving rise to the largest differences in the partition coefficients of the components to be separated. Obviously, then, the solubility principles of organic compounds need to be considered.

The solubility of a gas in a liquid depends to some extent on its vapor pressure. In general, lower boiling, more highly volatile components will move through the column faster and exit sooner than those of lower volatility because, to a rough approximation, the higher the vapor pressure of a gas the lower its solubility in a liquid. Factors other than volatility (and thus indirectly, molecular weight) are important, however, in determining the rate at which a compound will move through the column. The solubility of a substance in a liquid is also influenced by polar interactions between the molecules of solute and solvent, *e.g.*, hydrogen bonding and other electrostatic interactions.

An age old rule of chemistry is "like dissolves like." This would suggest that if it were desired to separate nonpolar types of compounds, a nonpolar liquid substrate should be used. On the other hand, if the components to be separated were of the polar variety, a nonpolar substrate would not be a very good choice, for none of the components would be very soluble and all would tend to pass unhindered through the column with little, if any, separation. Thus, for polar samples it is best to use polar liquid phases.

The most common means of supporting the liquid phase within the column is as a thin film coating on an inert solid support. The support should be of small, evenly meshed granules, so that a large surface area is available. This provides a correspond-

---

[6] For the accurate determinations necessary in many research laboratories, *retention volumes* are used rather than retention times. These measurements are more difficult to perform, but are more reproducible than retention times. They are a measure not of the time necessary for a component to pass through the apparatus, but instead, a measure of the *volume* of inert gas needed to carry the component through.

TABLE 3.1 GLPC Stationary Phases

Liquid Phase	Type	Property	Maximum Temp. Limit, °C	Used for Separating
Squalane	Hydrocarbon grease	Nonpolar	100	Hydrocarbons, general application
Apiezon-L	Hydrocarbon grease	Nonpolar	300	Hydrocarbons, general application
Carbowax 20M	Hydrocarbon wax	Polar	250	Alcohols, C_5-C_{18} aldehydes, sulfur compounds
DC-550	Silicone oil	Intermediate polarity	275	C_1-C_5 aldehydes, C_6-up hydrocarbons, halogen compounds
QF-1	Silicone (fluoro)	Intermediate polarity	250	Polyalcohols, alkaloids, halogen compounds, pesticides, steroids
SE-30	Silicone gum rubber	Nonpolar	375	C_5-C_{10} hydrocarbons, pesticides, steroids
Diethyleneglycol succinate (DEGS)	Polyester	Polar	190	Esters, fatty acids
Butanediol succinate	Polyester	Intermediate polarity	225	Esters, fatty acids

TABLE 3.2 Solid Supports

Chromosorb "P"	Pink diatomaceous earth (surface area: 4–6 m²/g)
Chromosorb "W"	White diatomaceous earth (surface area: 1–3.5 m²/g)
Crushed Firebrick	
Chromosorb "T"	40/60 mesh Teflon 6

ingly large film area in contact with the vapor phase, which is necessary for efficient separation. Some common types of solid supports are given in Table 3.2. The liquid is coated on the solid support by dissolving the liquid in a suitable low-boiling solvent and mixing this solution with the solid. The low-boiling solvent is then evaporated, leaving the solid granules evenly coated, and the column is then filled with these coated granules.

An alternative method of supporting the liquid phase is used in capillary columns of the Golay type. These are very long (300 ft is not unusual) and very small diameter (0.1–0.2 mm) columns. In these columns, the liquid is coated directly on the inner walls of the tubing. These types of columns are highly efficient (and relatively expensive).

In general, column efficiency increases with increasing path length and decreases with increasing diameter. As suggested in the introductory discussion above, the separation between bands increases with increasing path length. Therefore, with a longer column, the likelihood of separating two components will be increased (their retention time difference will be larger). The diameter of the column will affect the band *width*. A smaller diameter column will give rise to narrower bands and greater efficiency. It is evident that with a small band separation (measured from the band centers), wide bands are more likely to *overlap* [Figure 3.7(a)] than narrow bands [Figure 3.7(b)]. Therefore a smaller diameter column will effect a more efficient separation, i.e., the *resolution* will be greater.

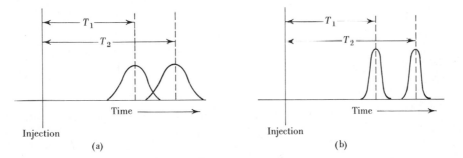

FIGURE 3.7   Effect of band width on resolution.

The temperature at which the column is maintained is another important factor in a glpc experiment. Retention times may be shortened by using higher temperatures. This is because the solubility of gases in liquids decreases with increasing temperature. The partition coefficients are thus affected, and the bands move through the column at a faster rate. This is sometimes desirable because of the convenience of a faster experiment, but it will also usually mean that the band separation will decrease, resulting in lower resolution.

For each liquid phase there is a maximum temperature limit. This point is determined by the stability and volatility of the liquid being used. At higher temperatures, the liquid phase may volatilize to some extent and be carried off the column by the carrier gas, which is of course undesirable. With some liquids there is a minimum temperature, governed by the melting point of the substance or its viscosity. Should the liquid be partially solidified or quite viscous, it will be very inefficient at dissolving the components of the gaseous mixture.

Just as with higher temperatures, higher flow rates of the carrier gas will also cause retention times to decrease. In spite of the decreased resolution obtained at higher temperatures and flow rates, these conditions are sometimes necessary for substances otherwise having very long retention times.

There are two basic types of experiments to which gas chromatography may be applied: (1) qualitative and/or quantitative analysis of the sample; (2) preparative experiments for the purpose of separating and purifying the components of the sample (preparative glpc).

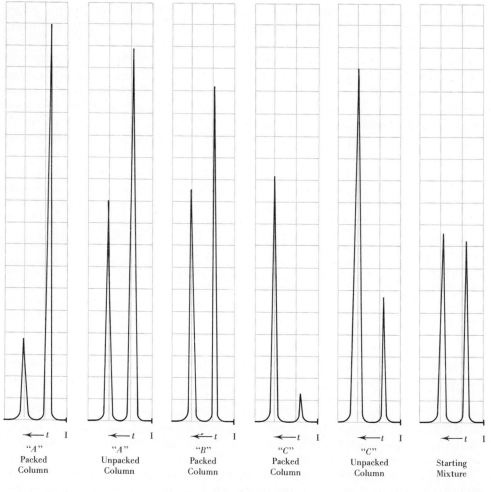

$\longleftarrow t$ I $\longleftarrow t$ I $\longleftarrow t$ I $\longleftarrow t$ I $\longleftarrow t$ I $\longleftarrow t$ I

"A"   "A"   "B"   "C"   "C"
Packed   Unpacked   Packed   Packed   Unpacked   Starting
Column   Column   Column   Column   Column   Mixture

(I = Injection Point; t = time)

FIGURE 3.8 Gas chromatographic analysis of the distillation fractions from the distillation experiment of Chapter 2.

Under a given set of conditions (column, temperature, flow rate, etc.), the retention time of a given compound is a property of that compound and may be used to identify it. Often, identification may be made by injecting and obtaining the retention time, under the same conditions, of material *known* to be the compound in question. If the retention time of the known compound matches that of a peak of the chromatogram, this is necessary evidence that the two compounds may be identical; it is not sufficient evidence, however, for more than one compound may exhibit the same retention time. Since one usually has some indication of what compounds are present in the mixture, the identification is reasonably certain. Furthermore, if the voltage output of the detector is proportional to the mole fraction of the material being detected in the vapor, the relative *areas* under two peaks of the chromatogram will equal their percentage ratio in the mixture. Thus, from retention times and peak areas, the qualitative and quantitative composition of a mixture can be obtained.

An example of the analytical use of glpc is shown in Figure 3.8. These sets of peaks represent gas chromatographic analyses of the distillation fractions of a benzene-toluene mixture similar to those obtained in the distillation experiment of Chapter 2. The notations *A*, *B* and *C* refer to the three fractions taken in that experiment. Comparison of retention times with those of pure benzene and pure toluene indicates that the first peak (lower retention time) in each case is benzene and the other is toluene. Comparison of the relative areas for the peaks of fractions *A* and *C* with the packed column with the results for fractions *A* and *C* for the unpacked column demonstrates the greater efficiency of the packed fractional distillation column.

In preparative glpc, one obtains the purified components of the mixture by collecting (and condensing) the vapors as they come from the exit port. The recorder indicates when the material is passing through the detector; by making allowance for the time required for the material to pass from the detector to the exit port, one can estimate when to collect the condensed vapor. As different peaks are observed, different collection vessels may be used.

Many instruments are designed specifically for analytical determinations. Relatively small columns are used, and thus only very small samples may be injected (*ca.* 0.1–5 *micro*liters). With such sample sizes, it is evident that a very large number of repetitive injections would be needed before any sufficient quantity of the components could be collected in a preparative experiment. Instruments designed specifically for preparative experiments are also available; these utilize much larger columns (some of the larger ones are 12″ in diameter), in order to handle larger samples. Larger quantities of purified materials can thus be obtained with fewer repetitions of the injection-separation-collection cycle.

3.5   Column Chromatography

Column chromatography involves distribution of substances between liquid and solid phases and therefore would be classified as a type of solid-liquid adsorption chromatography. The stationary phase is a solid, which separates the components of a liquid passing through it by selective adsorption on its surface. The types of interactions causing adsorption are the same as those that cause attractions between

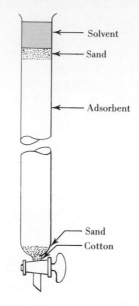

FIGURE 3.9   Chromatography column.

any molecules, *i.e.,* electrostatic attraction, complexation, hydrogen bonding, van der Waals forces, etc.

A column such as that shown in Figure 3.9 is used to separate a mixture by column chromatography. The column is packed with an "active" solid (the stationary phase) such as alumina or silica gel and a small liquid sample [7] is applied at the top. The sample will become adsorbed initially at the top of the column. An eluting solvent is then allowed to flow through the column; this mobile liquid phase will carry with it the components of the mixture. However, due to the selective adsorption power of the solid phase, the components may move down the column at different rates. A more weakly adsorbed compound will be eluted more rapidly than a more strongly adsorbed compound because the former will have a higher percentage of molecules in the mobile phase. The process is seen to be quite analogous to that operating in glpc. The progressive separation of the components is depicted in Figure 3.10.

The separated components may be recovered in two ways: (1) the solid packing may be extruded and the portion of the solid containing the desired band cut out and extracted with an appropriate solvent; (2) solvent can be passed through the column until the bands are eluted from the bottom of the column and collected in different containers, since they will be eluted at different times. The second method is the more commonly used; the first is rather difficult because the solid must be extruded from the column in one piece.

With colored materials, the bands may be directly observed as they pass down the column. The word "chromatography" was originally coined to describe this technique from such observations (Gr. *chromatos,* a color). With colorless materials, however,

---

[7] If the mixture is a solid, it must be dissolved in a minimum quantity of solvent and then applied.

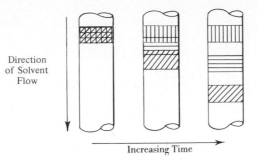

Direction
of Solvent
Flow

Increasing Time

FIGURE 3.10   Development of the chromatogram.

the changes cannot be observed directly. Many materials fluoresce when irradiated with ultraviolet light, however, and this provides a method of observing the bands in these cases. Usually, the progress of a column chromatographic experiment is followed by collecting a series of fractions of eluent of constant volume, for example, 25 ml. The solvent is then evaporated from each to see if any solute is present. If the volume of each fraction is kept relatively small—for example, less than 10% of the volume of the column—the different bands will usually be obtained in different flasks, although each component may be distributed among several flasks. Another convenient method of following the separation is to analyze the eluent at intervals by thin-layer chromatography (Section 3.7).

The composition of the material eluted from the column may be ascertained by determination of its physical constants (Chapter 1) or its spectra (Chapter 4) and comparison of these with those of known materials.

A few of the solid adsorbents commonly used include: alumina, silica gel, Florisil, charcoal, magnesia, calcium carbonate, starch, and sugar. The organic chemist usually finds alumina, silica gel and Florisil (fluorinated silicone polymer) of the greatest utility.

Alumina (Al_2O_3) is a highly active, strongly adsorbing, polar compound coming in three forms: neutral, and base or acid washed. Basic and acidic alumina offer good separating power for acids and bases, respectively. For compounds that are sensitive to chemical reaction under acidic or basic conditions, neutral alumina should be used. Being highly polar itself, alumina adsorbs polar compounds quite tenaciously, so that they may be difficult to elute from the column. The activity (adsorptivity) of alumina may be reduced by additions of small amounts of water; the weight percentage of water present determines the activity grade of the alumina. Silica gel and Florisil are also polar, but less so than alumina.

For the greatest effectiveness, the solid adsorbent should be of uniform particle size and of high *specific area*, a property which promotes more rapid equilibrium of the solute between the two phases. This is important for producing narrow bands. Good grades of alumina and silica gel have very high specific areas, of the order of several hundred m^2/g.

The strength of adsorption depends on the adsorbate as well as on the adsorbent. It has been found that the strength of adsorption for compounds having the following types of polar functional groups increases in the order noted on any given adsorbent:

$$Cl-, Br-, I- < \overset{\diagdown}{\diagup}C=C\overset{\diagup}{\diagdown} < -OCH_3 < -CO_2R < \overset{\diagdown}{\diagup}C=O < -CHO$$

$$< -SH < -NH_2 < -OH < -CO_2H$$

The nature of the liquid phases (solvents) to be used is an important consideration when designing a chromatographic experiment. The solvent may also be adsorbed on the solid, thereby competing with the solute for the adsorptive sites on the surface. If the solvent is more polar and more strongly adsorbed than the components of the mixture, these components will remain almost entirely in the mobile liquid phase and little separation will occur during the experiment. For effective separation, then, the eluting solvent must be significantly less polar than the components of the mixture. Furthermore, the components must be soluble in the solvent, for if they are not they will remain permanently adsorbed on the stationary phase of the column. The eluting powers of various solvents, *i.e.*, their ability to move a given substance down a column, are generally found to occur in the order shown.

hexane
carbon tetrachloride
toluene   increasing
benzene   eluting
dichloromethane   power
chloroform
ethyl ether
ethyl acetate
acetone
propanol
ethanol
methanol
water

In a *simple elution* experiment, the sample[8] is placed on the column and a *single* solvent is used throughout the separation. The optimum solvent choice will be that which produces the greatest band separation. Since the best solvent will most likely be found only by trial and error, it is sometimes convenient to use the techniques of thin-layer chromatography (Section 3.7) in the selection of a solvent for column chromatography. A series of thin-layer chromatographic experiments using various solvents may be performed in a relatively short time. The best solvent, or mixture of solvents, found in this way will usually be appropriate for the column chromatography.

A procedure known as *stepwise* (or *fractional*) *elution* is most commonly used. In this method, a *series* of increasingly more polar solvents is used to develop the chromatogram. Starting with a nonpolar solvent (usually hexane), one band may move down and off the column while the others remain very near the top. Ideally, then, the solvent would be changed to one of slightly greater polarity in the hope that one more band will be eluted while the others remain behind. If too large a jump in

[8] A sample size of approximately 1 g of sample per 25 g of adsorbent is satisfactory.

polarity is attempted, all of the remaining bands may come off at once. Therefore, small systematic increases in solvent polarity should be effected at each step. This is best accomplished not by changing solvents entirely, but by using *mixed* solvents. For example, an appropriate quantity [9] of hexane could be passed through the column, followed by a quantity of solvent having the composition 95% hexane—5% benzene. Solvent mixtures containing 10, 15, 20, 40, and 80% benzene could then be used in succession, followed by pure benzene. A similar series utilizing benzene with diethyl ether or chloroform could then be employed. This technique has proven to be quite efficient with regard both to column resolution and to time invested, and is very commonly employed in research laboratories.

The method of packing the column is exceedingly important, because a poorly packed column will have minimal resolution. The packing should be homogeneous and should not have entrapped air or vapor bubbles. The best way to achieve an evenly packed column is described in the experiment below. Column sizes are variable, but the same considerations hold as for glpc: the larger the diameter, the longer the column should be.

EXPERIMENTAL PROCEDURE

A. PREPARATION OF COLUMN

Clamp in a vertical position a 50-ml buret with its stopcock closed but ungreased. Fill the buret to approximately the 40-ml mark with 30–60° petroleum ether, and insert a small plug of glass wool into the bottom of the buret by means of a long piece of glass tubing. Introduce into the buret enough clean sand to form a 1-cm layer on top of the glass wool plug. As the sand settles through the liquid, any entrapped air bubbles will be allowed to escape. *Slowly* add 15 g of alumina to the buret while continually tapping the column with a "tapper" made from a pencil and a one-holed rubber stopper. The agitation of the column while the alumina is being sifted in assures an evenly packed column. With a little additional petroleum ether, wash down the inner walls of the buret to loosen any adhered alumina. In order to protect the packed alumina, introduce a 1-cm layer of sand at the top of the column. Open the stopcock and allow the solvent to drain just to the top of the top layer of sand.[10] The column, as shown in Figure 3.9, is now ready to receive the sample mixture to be separated.

B. SEPARATION OF SYN- AND ANTI-AZOBENZENES

Obtain from your instructor 1–2 ml of a half-saturated petroleum ether solution of *syn-* and *anti*-azobenzene. Prepare a pipet by drawing out a piece of soft glass tubing.

---

[9] A "rule-of-thumb" to use in determining the amount of solvent to pass through before changing to one of higher polarity is to use a volume equal to approximately three times the packed volume of the column. There are exceptions, but this is a reasonable suggestion in the absence of more specific knowledge.

[10] At no time during the experiment should the solvent be allowed to drain below this level. Air bubbles and channels would be introduced into the alumina which would give rise to ragged bands during the development of the column.

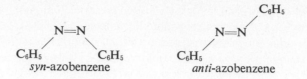

syn-azobenzene   anti-azobenzene

Apply 1.0 ml of the sample to the column with the pipet, allowing the solution to run evenly down the inside surface of the buret. Open the stopcock and allow the liquid to drain to the top of the sand. Introduce, in a like manner, 1–2 ml of fresh petroleum ether to wash down the inner surface of the buret. Again allow the liquid to drain to the top of the sand. Fill the buret with fresh petroleum ether. Open the stopcock; as the solvent runs through the column, a broad orange band of the *anti*-isomer should be observed to move slowly down the column. A narrow yellow band of the *syn*-isomer should remain very near the top of the column. Allow petroleum ether to run through the column until the orange band is eluted and collected at the bottom of the column in a small Erlenmeyer flask.[11] Remove the solvent from the solution of the *anti*-isomer on a steam bath in the hood. Air-dry the material obtained and determine its melting point.

After eluting the *anti*-isomer, allow the liquid in the buret to drain to the top of the sand and fill the buret with 5% methanol (by volume) in petroleum ether. Elute and collect the *syn*-isomer, remove the solvent, and, after air-drying, determine its melting point.

If desired by your instructor, obtain infrared spectra (Chapter 4) of the purified isomers and carefully note the differences.

EXERCISES

1. Why is it preferable to use an ungreased rather than a greased stopcock on the buret?
2. Why should care be used to ensure that there are no air bubbles in the packed column?
3. Why does *syn*-azobenzene move faster when a mixture of petroleum ether and methanol is used as eluent than when petroleum ether alone is used?

3.6 High-Pressure Column Chromatography

In recent years the utility of column chromatography has been tremendously increased by a number of technical developments. In the classical procedure, the liquid phase passes through the stationary phase under the influence of gravity, and the development of the chromatogram and subsequent separation of the components is usually a rather slow process.

It has been found that the speed and versatility of column chromatography can be greatly improved by pumping the liquid phase through the column at pressures

---

[11] If it is desired to elute the *anti*-isomer more rapidly, 20% benzene (by volume) in petroleum ether may be used for the second filling of the buret.

up to and sometimes exceeding 1000 psi inlet pressure. The gains in versatility have come about through the use of stationary phase packing material of uniform, smaller-diameter, high surface-area particles. Detectors have been developed which monitor and record the eluate continuously by means of differential refractometry and/or ultraviolet spectrometry. Automatic fraction collectors can be used conveniently since the eluate emerges at atmospheric pressure.

If the stationary phase consists of a solid surface which reversibly adsorbs solutes from the moving liquid phase, the process is referred to as *liquid/solid chromatography*. The stationary phase consists most commonly of polar materials such as silica gel and alumina, as described in Section 3.5. If the stationary phase is a solid, coated with a liquid which is immiscible with the moving liquid phase, the process is referred to as *liquid/liquid chromatography*. Stationary coating liquids are generally polar while the mobile phases are usually nonpolar.

High-pressure column chromatography may be used both analytically and preparatively. For example, a complex mixture obtained by synthesis or extracted from a natural substance may give a detector recording displaying a series of peaks similar to those produced by a gas chromatograph. A quantitative analysis of the mixture is indicated by the areas under the peaks. Since the detectors are located at the exit from the column, the various components in the eluate may be identified with the peaks on the detector recording and collected separately. By using larger columns or multiple injections on smaller columns, mixtures of from milligram to kilogram size may be separated into their components.

The improvement in operating time achieved by the new high-pressure systems is significant. Separations that would require many hours in gravity columns may be made in a matter of minutes in the high-pressure columns. There are advantages over other forms of chromatography as well. Efficiency is much higher than in thin-layer chromatography (Section 3.7). Whereas gas chromatography is not practical for high-molecular weight, relatively nonvolatile compounds (mw greater than 200), high-pressure column chromatography can be applied to high-molecular weight compounds with the same order of efficiency as that of gas chromatography for low-molecular weight compounds. Another valuable aspect is the fact that high-pressure liquid chromatography is normally carried out at room temperature, so that there is no danger of decompositions or molecular rearrangements which may be induced in thermally unstable compounds by gas chromatography.

High-pressure liquid chromatography is a new development in the analysis and separation of organic compounds which is predicted to be extremely valuable in the future, particularly in the field of natural products. It has been credited with making significant contributions to the synthesis of Vitamin B_{12}, a complex, high-molecular weight compound, and to the isolation of the sex attractant of the American cockroach.

3.7   Thin-Layer Chromatography

Thin-layer chromatography involves the same principles as column chromatography; it is a form of solid-liquid adsorption chromatography. However, in this case the solid adsorbent is spread as a thin layer (*ca.* 250μ) on a piece of glass or rigid plastic. A drop of the solution to be separated is placed near one edge of the plate and the plate is placed in a container (developing chamber) with enough of the eluting solvent to come to a level just below the "spot." The solvent migrates *up* the plate carrying with it the components of the mixture at different rates. The result may then be a series of spots on the plate, falling on a line perpendicular to the solvent level in the container [see, for example, Figure 3.11(b)].

This chromatographic technique is very easy and rapid to perform. It lends itself well to the routine analysis of mixture composition, and may also be used to advantage in determining the best eluting solvent for subsequent column chromatography.

The same solid adsorbents used for column chromatography may be employed for tlc, with silica and alumina being the most widely used. The adsorbent is usually mixed with a small amount of "binder," for example, plaster of paris, calcium sulfate or starch, to insure proper adherence of the adsorbent to the plate. The plates may be prepared before use, or commercially available prelayered plastic sheets may be used.

The relative eluting abilities of the various solvents that may be used are the same as those given earlier in Section 3.5. It should be remembered that the eluting power required of the solvent is directly related to the strength of adsorption of the components of the mixture on the adsorbent.

A distinct advantage of tlc is the very small quantity of sample required. A lower limit of detection of 10^{-9} grams is possible in some cases. However, sample sizes as large as 500 micrograms may be used. With the larger samples, preparative experiments may be conducted by scraping the individual spots from the plate and eluting (extracting) them with an appropriate solvent. Of course, it would be necessary to repeat this

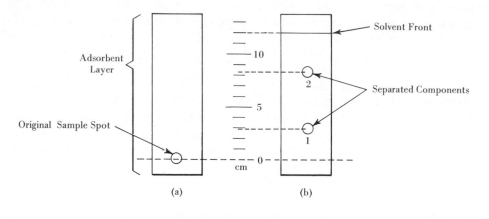

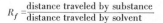

$$R_f = \frac{\text{distance traveled by substance}}{\text{distance traveled by solvent}}$$

$$R_f \text{ (compound 1)} = \frac{3.0 \text{ cm}}{12 \text{ cm}} = 0.25$$

$$R_f \text{ (compound 2)} = \frac{8.4 \text{ cm}}{12 \text{ cm}} = 0.70$$

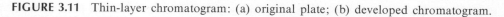

FIGURE 3.11   Thin-layer chromatogram: (a) original plate; (b) developed chromatogram.

process a number of times in order to obtain even several *milli*grams of material. Such a procedure would offer a method of identification of the various spots, however, for enough could be collected to obtain spectra of the individual components (see Chapter 4).

Detection of spots on the chromatogram is easy for colored materials, and a number of procedures are available for locating spots of colorless materials. For example, irradiation of the plate with ultraviolet light will permit location of the spots of compounds that fluoresce. Alternatively, the solid adsorbent may be impregnated with an otherwise inert, fluorescent substance. Spots of materials that absorb ultraviolet light, but do not fluoresce, will show up as black spots against the fluorescing background when the plate is irradiated with ultraviolet light. Other detecting agents are more often used. These agents may be sprayed onto the chromatograms, causing the spots to become readily apparent. Examples of detecting agents used in this way are sulfuric acid, which causes many organic compounds to char, and potassium permanganate solution. Iodine is another popular detecting agent. In this case, the plate is placed in a vessel whose atmosphere is saturated with iodine vapor. Iodine is adsorbed by many organic compounds, and their spots on the chromatogram become colored (usually brown).

Under a given set of conditions (adsorbent, solvent, layer thickness, and homogeneity) the rate of movement of a compound with respect to the rate of movement of the solvent front, R_f, is a property of that compound. The value is determined by measuring the distance traveled by a substance from a starting line to the middle of the spot and dividing this distance by the distance the solvent has traveled, as measured from the same starting line (Figure 3.11). The property has the same significance as retention volume[6] in a glpc experiment.

EXPERIMENTAL PROCEDURE

A. SEPARATION OF SYN- AND ANTI-AZOBENZENES

Obtain from your instructor a 10-cm strip of silica gel chromatogram sheet (without fluorescent indicator) and about 0.5 ml of a 10% benzene solution of commercial azobenzene. Place a spot of this solution on the tlc plate about 10 mm from one edge and about 20 mm from the bottom, using a capillary tube to apply the spot. The spot should be 1 or 2 mm in diameter. Allow the spot to dry and then expose the plate to sunlight for one or two hours. Alternatively, the plate may be placed beneath a sunlamp for about 20 min. After this time, apply another spot of the *original* solution on the plate at the same distance from the bottom as the first and leave about 10 mm between the spots. Again, allow the plate to dry.

A wide-mouth bottle with a tightly fitting screw top cap may be used as a developing chamber. Alternatively, a beaker covered with a watch glass may be used. Prepare an 8:1 (by volume) mixture of cyclohexane: benzene and place a 1-cm layer of this mixture in the bottom of the developing chamber. Fold a piece of filter paper as shown in Figure 3.12(a), and place it into the developing chamber as shown in Figure 3.12(b). Saturate the chamber with the vapors of the solvent by shaking. This inhibits the evaporation of solvent from the plate during the development of the chromatogram. The piece of filter paper aids in the maintenance of this saturated state.

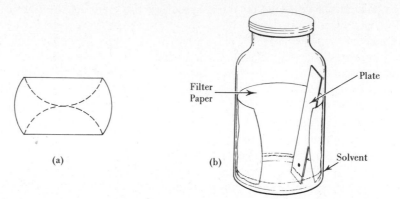

(a)   (b)

FIGURE 3.12   TLC chamber.

Place the chromatogram plate in the chamber, being careful not to splash the solvent onto the plate. Allow solvent to climb to within 15 mm of the top of the plate and then remove the plate and allow it to air-dry. Note the number of spots arising from each of the two original spots and compare the intensities of the two spots corresponding to *syn*-azobenzene (the spots nearest the starting point).

B. SEPARATION OF GREEN LEAF PIGMENTS

Place a few milliliters of a 2:1 mixture of petroleum ether and ethanol, along with a few green leaves[12] in a mortar and crush the leaves with a pestle. By means of a pipet, transfer the liquid extract to a small separatory funnel and *swirl* with an equal volume of water; shaking may cause formation of an emulsion. Separate and discard the lower aqueous phase. Repeat the water washing twice, discarding the aqueous phase each time. Transfer the petroleum ether layer to a small Erlenmeyer flask and add 2 g of *anhydrous* sodium sulfate. The solution may be concentrated, if necessary, by evaporation with a stream of dry nitrogen.

Obtain from your instructor a 10-cm strip of silica gel chromatogram sheet (without fluorescent indicator). Place a 1–2 mm spot of the pigment solution on the sheet about 1.5 cm from one end, using a capillary tube to apply the spot. Allow the spot to dry. Using 7:3 (v/v) benzene: acetone as a developing solvent, develop the chromatogram according to the general directions given in part A.

It is possible to observe as many as eight colored spots. In order of decreasing R_f values, these spots may be identified as the carotenes (two spots, orange), chlorophyll *a* (blue-green), chlorophyll *b* (green), and the xanthophylls (four spots, yellow).

[12] Green leaves from a variety of sources might be used, such as grass and various types of trees and shrubberies. The interested student may wish to complete this experiment with more than one type of leaf and compare the results.

EXERCISES

1. Which of the two isomers of azobenzene would you expect to be the more thermo-dynamically stable and why?
2. From the results of the tlc experiment with the azobenzenes, describe the role of sunlight.
3. In a tlc experiment, why must the *spot* not be immersed in the solvent in the developing chamber?
4. Explain why the solvent must not be allowed to evaporate from the plate during the development.

3.8 Dry-Column Chromatography

Experience with the solid-liquid chromatographic methods described in the previous two sections permits the following generalizations: (1) thin-layer chromatography has the advantage of being a convenient and rapid analytical technique but has the disadvantage of not being readily adaptable to preparative-scale separations; (2) column chromatography permits large-scale separations but suffers from the disadvantage that it is commonly quite time-consuming, 4 hours or more sometimes being required to elute the column. Dry-column chromatography is a technique that has recently been developed to take advantage of the speed afforded by thin-layer chromatography and of the scale offered by column chromatography.

In this method, the solid adsorbent is poured into a dry glass or nylon column and tamped to achieve uniform packing.[13] The mixture to be separated is adsorbed on a separate portion of adsorbent, and this is added to the top of the column. Solvent is then allowed to descend the column until it has just reached the bottom of the column. At this point, development of the column is terminated, the adsorbent is removed from the column, and the bands of separated components are isolated by extraction of the adsorbent with an appropriate solvent.

In general, it is not convenient to use a series of solvents with this type of chromatography; rather, use of a pure solvent or a single mixture of solvents is preferable. The choice of solvent is greatly facilitated by the observation that the solvent that provides the best separation of the components of mixture on a tlc plate will also give the best separation with a column. Thus, a number of pure solvents and mixtures of them can be tested rapidly on the plates to determine the solvent of choice for the dry-column separation.

EXPERIMENTAL PROCEDURE

Dissolve 0.5 g of a mixture of *cis*- and *trans*-azobenzene in a minimum amount of petroleum ether (bp 30–60°)[14] and add to this solution 3 g (about 3 ml volume)

---

[13] As a general rule, about 70 g of adsorbent is required per gram of mixture to be separated.

[14] See footnote 13 of Chapter 2.7 for a description of petroleum ether.

of activity III alumina.[15] Evaporate the solvent by stirring the resulting slurry under a gentle stream of air. When completely dry, the remaining solid should be granular and free-flowing.

Prepare the column necessary for the separation in the following way. Place a loose plug of glass wool in one end of a 40-cm $\times$ 17-mm section of clean and dry Pyrex glass tubing. With the aid of a funnel, add 35 g (ca. 35 ml) of activity III alumina through the other end of the tubing. Hold the filled column in a vertical position and tap the plugged end gently against a cushioned surface, e.g., a towel on top of the lab bench, to pack the alumina firmly in the column. Tap the sides of the tubing to aid in compacting the alumina in the column. Clamp the column in a *vertical* position and ascertain that the top of the alumina packing is level. If it is not, tap the column gently to achieve a level surface. (*Caution:* To avoid uneven development of solvent through the column, do *not* lay the column on its side after it has been packed.)

Add the mixture of alumina and azobenzene to the top of the column and tap the column gently to provide a level top. Complete the preparation of the column by adding enough fresh alumina to provide a level layer 6 mm in depth on top of the sample.

Develop the column in the following manner. Carefully pour a 1:1 mixture of toluene and petroleum ether (30–60°) onto the top of the column until a liquid layer of about 5 cm is reached. As the solvent percolates down the column, add fresh solvent so as to maintain the 5-cm head on the column. When the solvent is within about 2 cm of the bottom of the column, remove the excess solvent from the top with a pipet. When the solvent has advanced to within about 3 mm of the bottom, unclamp the column and lay it on its side to terminate development of the column. Immediately determine the R_f values for the two isomeric azobenzenes.

With the aid of a spatula, carefully scrape out the alumina containing the individual bands of products from the column, placing each band in an appropriately labeled beaker. Isolate the individual isomers by washing the alumina with two 20-ml portions of technical ether; after each wash, carefully decant the supernatant liquid into an appropriately sized Erlenmeyer flask. Evaporate the solvent to obtain the final product. (*Caution:* No open flames should be used during the evaporation step.)

Determine the melting point of each isomer and the total weight of the recovered azobenzenes. Calculate the percentage recovery for the process. If possible, obtain infrared spectra (Chapter 4) of the purified isomers and carefully note any differences in the spectra.

EXERCISES

1. Why is it important to have the packed column as nearly vertical as possible?
2. Why is it important to have the alumina in the column packed as compactly as possible?

[15] Activity III alumina suitable for dry-column chromatography can be prepared from commercially available activity I alumina by adding 6 wt % of water; or Woelm activity III alumina for dry-column chromatography can be obtained from Waters Associates, Inc., 61 Fountain Street, Framingham, Massachusetts 01701.

3. Which of the two isomeric azobenzenes has the greater R_f value in this experiment? Is this consistent with the relative R_f values found by tlc?
4. What would be the predicted effect on R_f values of the azobenzenes if the developing solvent in this experiment were pure benzene?

3.9 Paper Chromatography

Paper chromatography bears a resemblance to thin-layer chromatography (Section 3.7) but is slightly different in principle. Small spots of the mixture to be separated are placed near the bottom of a strip of paper; the end of the paper strip (but not the spot) is placed in solvent, and, as the solvent ascends the paper, the components of the mixture are separated into individual spots. Paper chromatography involves distribution of the sample between a polar liquid phase (normally water), which is strongly adsorbed on the cellulose fibers of the paper, and an eluting solvent. It is thus an example of liquid-liquid partition chromatography.

Compounds may be identified by comparison of their R_f values (defined in Section 3.7). R_f values are far more reproducible with paper chromatography than with tlc and thus are more valuable in this application.

The great utility of paper chromatography (and tlc as well) is in the rapid analysis of reaction mixtures rather than in separation of compounds on a useful preparative scale. The minute quantities of sample required to accomplish an analysis and the ease and rapidity with which qualitative analysis of reaction mixtures can be performed make this chromatographic technique a valuable tool to the organic chemist, and especially to the biochemist.

The application of paper chromatography to the separation and identification of α-amino acids is described in Chapter 23.2.

EXPERIMENTAL PROCEDURE

A. ANALYSIS OF INKS

Cut a 3 × 12-cm strip of filter paper. Apply spots of two different ink solutions (washable black, washable blue, and permanent emerald green are provided) to the paper at a distance approximately 1 cm from the end and 5 mm from each edge. The amount of solution used should be such that spots of only 1–2 mm in diameter are formed. Allow the paper to dry. Place a pencil mark at the edge of the strip at the level of these spots. Thread the paper through a slot in a stiff piece of cardboard, and hang the strip in a vertical configuration within a wide-mouth bottle which contains a 1-cm layer of water. The stiff piece of paper should cover the mouth of the bottle making the bottle as air-tight as possible (Figure 3.13). Do not let the strip of paper touch the sides of the bottle; only the bottom of the strip (the end having the spots of ink) is to be in contact with solvent. The spots should be above the water surface and not in contact with it. Leave the strip in the solvent until the solvent has climbed about 10 cm up the paper. Then remove the strip, place a pencil mark on it at the solvent front and allow the paper to dry. The developed chromatogram should exhibit several colored spots,

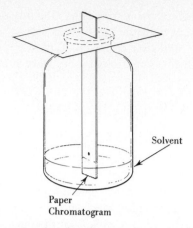

FIGURE 3.13   Paper chromatography chamber.

each of which corresponds to an organic dye which was a component of one or the other of the two inks. Calculate the R_f value for each dye.

Repeat this experiment with a *mixture* of the two inks. Calculate R_f values for each of the spots obtained. Using the R_f values obtained from the first part of the experiment, identify each of the spots of the chromatogram of the mixture of inks with one or the other of the original ink solutions.

B. ALTERNATIVE EXPERIMENT

As an alternative to the use of inks in part A, aqueous solutions of various food coloring agents may be used.

EXERCISES

1. Why would the spot on the paper not be below the level of solvent in the developing chamber?
2. What might be the consequence of putting a large spot of sample on the strip of paper?
3. Why is it important to have the developing chamber sealed as tightly as possible?

REFERENCES

A. EXTRACTION

L. C. Craig and D. Craig, in *Technique of Organic Chemistry,* 2nd ed., A. Weissberger, editor, Interscience Publishers, New York, 1956, Vol. III, Part I, Chapter 2.

B. CHROMATOGRAPHY

1. R. Stock and C. B. F. Rice, *Chromatographic Methods,* Reinhold Publishing Corp., New York, 1963.

2. *Chromatography,* 2nd ed., E. Heftmann, editor, Reinhold Publishing Corp., New York, 1967.
3. O. E. Schupp III, "Gas Chromatography," in *Technique of Organic Chemistry,* E. S. Perry and A. Weissberger, editors, Interscience Publishers, New York, 1968, Vol. XIII.
4. H. G. Cassidy, in *Technique of Organic Chemistry,* A. Weissberger, editor, Interscience Publishers, New York, 1957, Vol. X, Chapter 8.
5. J. G. Kirchner, "Thin-Layer Chromatography," in *Technique of Organic Chemistry,* E. S. Perry and A. Weissberger, editors, Interscience Publishers, New York, 1967, Vol. XII.
6. Louis F. Fieser, "Thin Layer Chromatography," *Chemistry,* **37,** 23 (1964).

chapter four

spectroscopic methods
of identification
and structure proof

An important part of the science of organic chemistry is determination of the exact structures of products formed in chemical reactions. This formerly was an exceedingly time consuming and often impossible task for the classical organic chemist because his primary method of determining the structures of compounds usually involved conversion, by chemical reaction, of his "unknown" substance to a compound having a "known" structure. The procedure is now known to have often led to erroneous conclusions concerning the identity of the "unknown" because the reactions employed to convert the "unknown" to a "known" compound often produced structural rearrangements which went unrecognized by the investigator. The modern organic chemist fortunately has at his disposal various spectroscopic techniques which supplement and even replace the more classical chemical methods of proof of structure and which permit much more rapid analysis of "unknowns." It is noteworthy that the development of spectroscopy as a tool for the elucidation of structures of organic compounds has, more than anything else, contributed to the extremely rapid growth of organic chemistry in the past twenty years, and, consequently, it is vitally important that the student learn how to analyze and to interpret the information available from infrared (ir), nuclear magnetic resonance (nmr) and ultraviolet (uv) spectroscopy and mass spectrometry (ms). Only the first two spectroscopic techniques will be discussed in this chapter because they are more readily available and of greater utility to the undergraduate. However, the interested student can profit by reading one or more of the references on ultraviolet spectroscopy and mass spectrometry cited at the end of the chapter.

4.1   Infrared Spectroscopy

The spectra most generally available for aid in elucidation of the structure of an organic compound are those obtained by irradiation of the compound with light from the infrared region, 5000 to 500 cm⁻¹ (reciprocal centimeters or wave numbers),[1] of the electromagnetic spectrum. The interaction of infrared light with an organic molecule can be explained qualitatively by imagining that molecular bonds between atoms are analogous to springs. These molecular springs are constantly undergoing stretching and bending motions (Figure 4.1) at frequencies which depend upon the masses of the

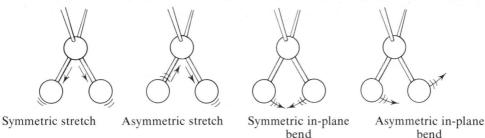

Symmetric stretch   Asymmetric stretch   Symmetric in-plane bend   Asymmetric in-plane bend

FIGURE 4.1   Some vibrational modes of a group of atoms.

atoms involved and upon the type of chemical bond joining the atoms. Since the frequencies of the various vibrations of the molecule correspond to those of infrared radiation, absorption of the radiation occurs, producing an increase in the *amplitude* of the molecular vibrational modes. No irreversible change in the molecule results, however, because the energy gained by the molecule in the form of light is soon lost in the form of heat. By plotting the percent transmittance, defined as the ratio of the intensity of the light passing through the sample, I, to the intensity of the light striking the sample, I_o, multiplied by 100 [equation (1)], *vs.* the frequency of the radiation, an

$$\%T - \frac{I}{I_o} \times 100 \qquad (1)$$

infrared spectrum such as that shown in Figure 4.2 is obtained.

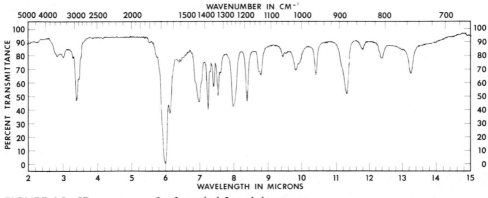

FIGURE 4.2   IR spectrum of a 3-methyl-2-cyclohexenone.

[1] The energy (in ergs) available from electromagnetic radiation is given by the familiar formula, $E = h\nu$, where h is Planck's constant (6.62×10^{-27} erg/second) and ν is the frequency of the radiation in cycles per second. The frequency of radiation expressed in cm⁻¹ is obtained by dividing ν by the speed of light, c, expressed in cm/sec. By dividing 10,000 by the frequency in cm⁻¹, the wavelength, in microns (μ), of the radiation is obtained. The position of an absorption band can thus be expressed either in microns or in wave numbers; the latter designation will be used in this text.

As a general rule, absorptions between 5000 and 1250 cm^{-1} involve vibrational excitations of particular functional groups, such as carbonyl, cyano, and nitro, present in the molecule. Thus, the stretching mode of the carbon-carbon double bond of an alkene usually appears in the region 1600–1680 cm^{-1} (see Figure 4.2). Analysis of the portion of the infrared spectrum from 5000–1250 cm^{-1} can therefore provide much information concerning the presence or absence of a number of functional groups. A tabulation of the normal frequency range of absorption for a number of groups is presented in Table 4.1. A more complete listing of infrared absorption bands is contained in Appendix IV.

TABLE 4.1   Infrared Absorption Ranges of Functional Groups

Bond	Type of Compound	Frequency Range, cm^{-1}	Intensity
C—H	Alkane	2850–2970	Strong
		1340–1470	Strong
C—H	Alkenes $\left(\diagup\diagdown{C}={C}\diagup^{H}\diagdown \right)$	3010–3095	Medium
		675–995	Strong
C—H	Alkynes (—C≡C—H)	3300	Strong
C—H	Aromatic rings	3010–3100	Medium
		690–900	Strong
O—H	Monomeric alcohols, phenols	3590–3650	Variable
	Hydrogen-bonded alcohols, phenols	3200–3600	Variable, sometimes broad
	Monomeric carboxylic acids	3500–3650	Medium
	Hydrogen-bonded carboxylic acids	2500–2700	Broad
N—H	Amines, amides	3300–3500	Medium
C=C	Alkenes	1610–1680	Variable
C=C	Aromatic rings	1500–1600	Variable
C≡C	Alkynes	2100–2260	Variable
C—N	Amines, amides	1180–1360	Strong
C≡N	Nitriles	2210–2280	Strong
C—O	Alcohols, ethers, carboxylic acids, esters	1050–1300	Strong
C=O	Aldehydes, ketones, carboxylic acids, esters	1690–1760	Strong
NO$_2$	Nitro compounds	1500–1570	Strong
		1300–1370	Strong

The absorption maxima observed between 1250 and 500 cm^{-1} in the infrared spectrum cannot usually be associated with vibrational excitation of a particular functional group, but rather are the result of a complex vibrational-rotational excitation of the *entire* molecule. Accordingly the spectrum from 1250–500 cm^{-1} is characteristic and unique for every compound, and the apt description, "fingerprint region," is often applied to this portion of the spectrum. It is considered highly unlikely that two different organic compounds would exhibit identical spectra in the fingerprint region even though their spectra might be quite similar from 5000–1250 cm^{-1} (see Figure 4.3).

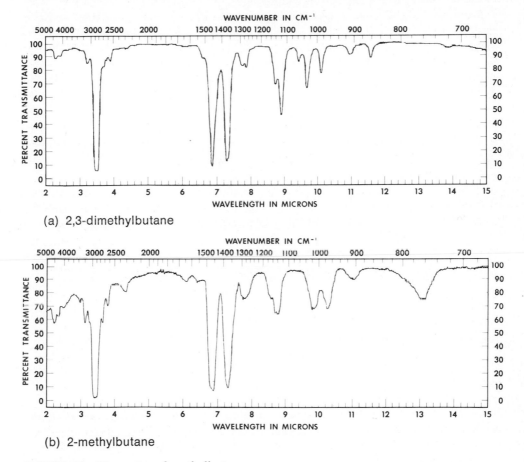

(a) 2,3-dimethylbutane

(b) 2-methylbutane

FIGURE 4.3 IR spectra of methylbutanes.

The apparently unique character of the infrared spectrum of each different organic compound has led to the acceptance of the hypothesis that if two spectra are completely identical as to positions and intensities of absorption maxima, that is, the spectra are completely superimposable, the spectra must be of the same compound. Comparison of spectra of "unknown" compounds with those of known compounds can therefore be one very useful technique for structure elucidation.

When the student is confronted with the infrared spectrum of a compound which he is attempting to identify, he should first attempt to determine what functional groups are present in the compound by careful analysis of the region, 5000–1250 cm^{-1}. Subsequently, he may be able to assign a few of the bands in the fingerprint region of the spectrum in order to define further the structure of the unknown substance. At this point, he should be in a position to formulate several possible structures for the unknown and may be able to differentiate between the several possibilities by comparison of their published spectra with that of the unknown. Several catalogs of infrared spectra are currently available, and some of the more useful and comprehensive are listed in the references at the end of this chapter.

A majority of the possibilities may also be eliminated by a comparison of the physical properties of the known, for example, mp, bp, index of refraction, with those of the unknown, or by their nuclear magnetic resonance spectra (see below). If, however, reference spectra of some of the substances considered as possibilities are not available, the information derived from the analysis of the infrared spectrum of the unknown may be utilized to suggest a specific chemical reaction which would provide a solid derivative, the melting point of which is published in the chemical literature. The identity of the derivative of the unknown with that of the known would then depend on a comparison of their melting points, or, ideally, a melting point of a mixture of the derivative of the unknown and an authentic sample of the same derivative obtained from the known compound.

Experimentally, infrared spectra can be obtained of samples contained in the gas, liquid (either of a pure, "neat," substance or of a solution) or solid phases. Only the general techniques necessary for the preparation for infrared spectra of samples in the liquid phase will be considered here since the apparatus required for obtaining spectra of gaseous or solid samples is somewhat too specialized for adaptation to the basic organic chemistry course. The student should see reference 2 for a thorough discussion of various ways of preparing samples for infrared analysis.

A brief discussion of the factors which determine the intensity of an infrared spectrum is appropriate here. The amount of absorption, termed absorbance, A, at a given frequency is defined by equation (2) in which I and I_o are, respectively, the

$$A = \log_{10} \frac{I_o}{I} = kcl \tag{2}$$

amounts of light of the given frequency transmitted by and impinging on the sample, k is the absorptivity of the sample at that frequency, l is the path length, in cm, of the cell in which the sample is contained, and c is the concentration, in g/1, of the solute in the solution. Although the absorptivity is a constant, its magnitude is frequency dependent which results in the observation of maxima and minima in the spectrum and which is somewhat characteristic of the functional group absorbing the radiation. Thus, the absorptivity of the stretching mode of a carbonyl group is generally greater than that of a carbon-carbon double bond, as can be seen by comparison of the intensities of the corresponding bands in the spectrum reproduced in Figure 4.2.

As the absorptivity of a given molecule is not subject to experimental modification, there are only two variables which can be changed in order to control the intensity of a spectrum: the path length, l, of the cell containing the sample and the concentration, c, of the solute in a solution. For spectra of neat samples, of course, only the former alternative is available, and it is found that path lengths of 0.025 to 0.030 mm are appropriate. As a general rule, solutions having 5 to 10 weight percent of solute and contained in a 0.1 mm cell yield suitable spectra.

The windows of most types of infrared cells are constructed of a clear fused salt such as sodium chloride or potassium bromide. As a result, contact of the cells with moisture must be scrupulously avoided or the windows will soon become cloudy and will transmit very little light. Thus, one should avoid breathing directly onto the windows, touching them with fingers, or filling them with samples which contain moisture. Furthermore, the cells should be cleaned with *dry* solvent (it is unwise to use any hydroxylic solvents, even alcohols, to wash the cells) and flushed with a slow stream of dry nitrogen or some other dry gas. The dried infrared cells should then be stored in a desiccator.

Infrared spectroscopy provides valuable information concerning the functional groups present in molecules and serves as a means of identifying an unknown compound if a reference spectrum of the compound is available.[2] In those cases, and there are many, in which the infrared spectrum alone does not provide sufficient information to permit determination of structure, other chemical or spectroscopic techniques must be employed. One such spectroscopic technique is discussed below.

EXERCISE

Assign as many bands as possible in the spectra presented in Figures 4.2 and 4.3 to the functional groups responsible for the absorptions.

4.2 Nuclear Magnetic Resonance Spectroscopy

Although infrared spectroscopy has been used profitably by organic chemists since the early 1950s, nuclear magnetic resonance (nmr) spectroscopy became generally available only in the early 1960s. This extremely important experimental technique is based upon the interesting property of *nuclear spin* exhibited by hydrogen nuclei.[3,4]

The hydrogen nucleus behaves as a spherical body having a uniform distribution of charge and a mechanical spin about an axis. The spin of the nucleus produces circulation of charge which, in turn, generates a nuclear magnetic dipole, the direction or moment of which is *colinear* with the axis of nuclear spin [Figure 4.4(a)].

When the spinning nucleus is placed in an external magnetic field, H_o, the magnetic dipole, and hence the axis of spin, becomes oriented either *with* the applied field, state E_1, or *against* it, state E_2 [Figure 4.4(b)]. The energy state E_1 represents the more stable alignment. It is very important to realize that the magnetic dipole does not become aligned either precisely parallel or antiparallel to H_o but rather traces out a circular path about the axis defined by H_o, in a process called *precession*. The phenomenon of

---

[2] In addition to its use as a tool for *qualitative analysis* of organic compounds, infrared spectroscopy is also utilized for *quantitative analysis* of mixtures of known substances. A discussion of this latter technique can be found in reference 2 at the end of this chapter.

[3] It will be convenient, during the presentation of this discussion, occasionally to refer to the hydrogen nucleus simply as a hydrogen or as a proton. These latter designations are not strictly correct since we are referring to the nucleus of a covalently bound hydrogen and not to the free atom or to the charged ion, H^+.

[4] The phenomenon is not limited to hydrogen nuclei but is observed for other nuclei, such as ^{19}F, ^{13}C, ^{31}P, as well. Our discussion, however, will treat only the case of hydrogen.

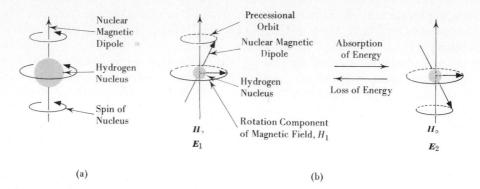

FIGURE 4.4 Spin properties of the hydrogen nucleus.

nuclear precession can be likened to the motion observed for a spinning gyroscope or toy top that has not been oriented exactly parallel to the earth's gravitational field.

The frequency, ω_o, of the precession is linearly related to the strength of the applied field, H_o, as is expressed in equation (3), in which the constant, γ_H, is the magnetogyric

$$\omega_o = \gamma_H H_o \tag{3}$$

ratio for the hydrogen nucleus. Thus, ω_o increases as H_o increases. Now, a transition between the energy states E_1 and E_2 can be accomplished if the nucleus, precessing at a frequency, ω_o, in a given field, H_o, is subjected to a second, oscillating magnetic field, H_1, which has a rotation component perpendicular to H_o Figure 4.4(b)]. The frequency of rotation of the magnetic component can be varied by changing the frequency, ν, of the oscillator used to produce the oscillating magnetic field, H_1.

When the frequency of rotation becomes equal to the precessional frequency, ω_o, *resonance* between the precessing nucleus and H_1 will be achieved. At this point, the nucleus in state E_1 can absorb the electromagnetic energy provided by the oscillator and undergo a "flip" of its spin into the higher energy state, E_2. The absorption of energy can be detected electronically and recorded as a peak on a chart.

The oscillator frequency required to produce the desired nuclear magnetic transition depends upon the strength of the applied field, H_o, as is expressed in equation (4). It

$$\nu = \frac{\gamma_H H_o}{2\pi} \tag{4}$$

turns out that if H_o is of an experimentally convenient magnitude, about 14,000 gauss, ν must be of the order of 60 megacycles per second (Mcps),[5] in order to produce resonance. This value is in the radiofrequency (rf) region of the electromagnetic spectrum so that an rf oscillator is required.

Inspection of equation (4) should make it apparent that the condition of resonance for a proton could be achieved either by holding H_o constant and varying ν, as is de-

[5] One megacycle is 10^6 cycles. Another designation, Hertz (Hz), is synonymous with cycles per second (cps) and is often used. The term, megaHertz (MHz), which is equivalent to Mcps, is also encountered.

scribed above, or by maintaining v at a constant value and changing H_o. The latter approach is more practical and is employed in all commercially available spectrometers.

A typical nmr spectrum, that of 1-nitropropane, is presented in Figure 4.5. In the spectrum, the strength of the applied magnetic field, H_o, *increases* as one goes from left to right, the *upfield* direction. Since the frequency of the radio oscillator is held constant, the *energy* required to produce the transition between the spin states, E_1 and E_2, also *increases* from left to right.[6] The three groups of peaks, *A*, *B*, and *C*, all result from absorption of energy by hydrogens, but the amount of energy required to realign the nuclei in the magnetic field, *i.e.*, the energy necessary to produce *resonance*, is clearly not identical for all the protons in the molecule. This is surprising as equation (4) indicates that all hydrogen nuclei should undergo spin flip at the same H_o (at a given value of v) so that only a *single* peak would be expected.

The qualitative explanation for the observation of more than one peak is as follows: The three groups, *A*, *B*, and *C*, represent three types of hydrogens which are rendered nonequivalent by virtue of the slightly different magnetic environments in which they exist. The magnetic environment of each proton in the molecule is dependent upon two major factors: (1) The externally applied field, H_o, which has a uniform strength over the entire molecule and therefore cannot produce nonequivalence of the different hydrogens; (2) a magnetic field which is induced by circulation of the electrons of the molecule and which, because the electron density varies over the molecule, is *not* uniform. This latter field results in the magnetic nonequivalence of the protons. Those hydrogen nuclei which experience higher fluxes of the induced magnetic field of the molecule will, because this field will act to oppose the external, applied field, be more highly shielded from the external field. In other words, more highly shielded nuclei will

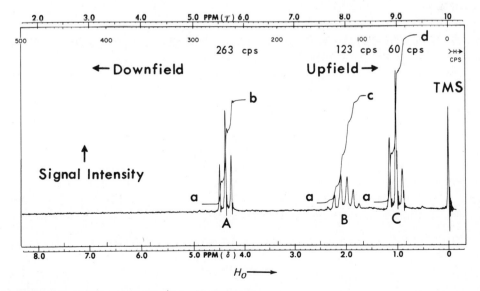

FIGURE 4.5   NMR spectrum of 1-nitropropane.

[6] Both v and H_o are linearly related to energy through the expression, $E = hv = \gamma_H H_o\, 2\pi h$.

experience an *effective* field, H_e, which is *less* than the applied field so that *higher* applied fields will be necessary to produce the desired nuclear spin transitions. The appearance of more than one group of peaks in the spectrum therefore reflects the difference in the flux of the magnetic field induced within the molecule by circulation of electrons.

The positions of peaks in the nmr spectrum of a compound are routinely measured relative to the position of the peak produced by the protons of a standard compound, tetramethylsilane (TMS), $(CH_3)_4Si$. In reference to the spectrum of Figure 4.5, the groups of peaks are found to be centered at 60, 123 and 263 cycles per second (cps) *downfield* from the TMS peak when an rf oscillator set at 60 Mcps is employed. These values represent the *chemical shifts* relative to TMS of the three types of hydrogen nuclei in 1-nitropropane.

The chemical shift of a hydrogen determined in units of cps is dependent on the frequency at which the oscillator is operating. It has been found useful to convert such shifts into units independent of the oscillator frequency. This is accomplished by dividing the chemical shift, expressed in cps, by the frequency, in cps, of the oscillator and then, to arrive at more convenient numbers, multiplying the result by 10^6 [equation (5)]:

$$\delta = \frac{\text{chemical shift in cps}}{\text{oscillator frequency in cps}} \times 10^6 \qquad (5)$$

Chemical shifts obtained in the above manner are values on a standard scale, termed the delta (δ) scale, which has the peak for TMS at 0δ, peaks downfield from TMS at values of $+\delta$, and peaks upfield from TMS at values of $-\delta$.[7] The oscillator frequency-independent chemical shifts, δ, of the groups of peaks observed for 1-nitropropane are as follows:

Group A: $(60 \text{ cps})/(60 \times 10^6 \text{ cps}) = 1.00 \times 10^{-6}$ or 1 part per million (ppm)
$\qquad (1.00 \times 10^{-6})(10^6) = 1.00 \ \delta_A$
Group B: $[(123)/(60 \times 10^6)](10^6) = 2.05 \ \delta_B$
Group C: $[(263)/(60 \times 10^6)](10^6) = 4.38 \ \delta_C$

A second standard scale, values of which are also independent of oscillator frequency, has been derived in which the TMS peak is defined as appearing at 10 tau (τ).[8] To convert a chemical shift given on the delta scale to a value on the tau scale, one simply subtracts the shift as measured on the delta scale from 10. The groups of peaks in the spectrum of 1-nitropropane, for example, appear at $10 - 1.0 = 9.0 \ \tau$, $10 - 2.05 = 7.95 \ \tau$ and $10 - 4.38 = 5.62 \ \tau$. As both these standard scales are currently in use, it is important to know how they are derived and how they can be interconverted.

---

[7] Actually, the sign convention should be exactly opposite to that given, so that peaks downfield from TMS would be at values of $-\delta$. However, common practice corresponds to the convention described above.

[8] The standard value of $10 \ \tau$ is chosen because the protons of most organic compounds absorb within 600 cps downfield of TMS when a 60-Mcps oscillator is used.

The magnitude of the chemical shift of a proton attached to a given type of functional group is relatively constant so that determination of the shifts of the various protons of an unknown compound will often provide valuable information concerning the functional groups present in the compound. Tables are available which correlate chemical shifts with structural features, and a partial compilation is presented in Table 4.2. Appendix III is a more extensive listing of the observed chemical shifts for a large variety of specific types of hydrogens.

TABLE 4.2   Typical Chemical Shifts of Hydrogens Attached to Various Types of Functional Groups

Type of Hydrogen	Chemical Shift, ppm		Type of Hydrogen	Chemical Shift, ppm	
	τ	δ		τ	δ
Cyclopropane	9.6–10.0	0.0–0.4	ICH	6–8	2–4
RCH_3	9.1	0.9	OCH (alcohols,		
R_2CH_2	8.7	1.3	ethers)	6–6.7	3.3–4
			O—CH	4.7	5.3
R_3CH	8.5	1.5	$\overset{|}{O}$		
C=CH	4.1–5.4	4.6–5.9	OCH (esters)	5.9–6.3	3.7–4.1
CCH	7–8	2–3	RO_2CCH	7.4–8	2–2.6
ArH	1.5–4	6–8.5	O		
ArCH	7–7.8	2.2–3	$\overset{\|}{RC}$CH	7.3–8	2–2.7
C=CCH$_3$	8.3	1.7			
C≡CCH$_3$	8.2	1.8	O		
FCH	5–5.6	4–4.5	$\overset{\|}{RC}$H	0–1	9–10
ClCH	6–7	3–4	ROH	4.5–9	1–5.5
Cl_2CH	4.2	5.8	ArOH	−2 to 6	4–12
BrCH	6–7.5	2.5–4	RCOOH	−2 to −0.5	10.5–12
O_2NCH	5.4–5.8	4.2–4.6	RNH_2	5–9	1–5

The peaks within each group of absorptions in the spectrum of 1-nitropropane are the result of *spin-spin splitting*, a phenomenon which results because there can be magnetic interaction, termed *coupling*, between magnetically nonequivalent hydrogen nuclei in the molecule. As a rule, only nonequivalent nuclei on the same carbon or on adjacent carbons will interact with one another to produce spin-spin splitting. Thus, if we label the three types of hydrogens as shown in Figure 4.6, we would expect the nuclei designated H_a to couple with those labeled H_b and these latter protons to interact with both the nuclei, H_a *and* H_c. However, the H_c protons should couple only with the H_b protons.

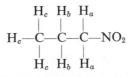

FIGURE 4.6 Types of hydrogens in 1-nitropropane.

The magnitude of the spin-spin splitting between hydrogen nuclei can be measured in cycles per second, and the values obtained are called *coupling constants, J.* In contrast to chemical shifts, the magnitude of coupling constants is *independent* of oscillator frequency, so units of cps or Hz[5] are used with such values. Some typical hydrogen-hydrogen coupling constants, J_{HH}, are given in Figure 4.7.

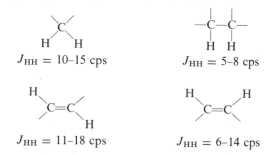

$J_{HH} = 10–15$ cps $J_{HH} = 5–8$ cps

$J_{HH} = 11–18$ cps $J_{HH} = 6–14$ cps

FIGURE 4.7   Some typical coupling constants.

The number of peaks into which the nmr absorption of a given hydrogen nucleus, H_i, will be divided due to spin-spin splitting depends in a linear manner on the number of protons coupling with H_i. Thus, if N protons, all of which have the same coupling constant, J, with a hydrogen, H_i, are coupled with this nucleus, the absorption of H_i will appear as a group of $N + 1$ peaks.[9] If the coupling constants are not identical, the spectrum may be considerably more complicated, but a detailed discussion of such cases is not appropriate here.

Applying the $N + 1$ rule to the different hydrogens of our reference compound, 1-nitropropane, the H_a protons should appear as a triplet, the H_b protons as a sextet, if $J_{ab} = J_{bc}$, and the H_c protons as a triplet. These expectations are confirmed in the observed spectrum. It should be apparent that the multiplicity of the nmr absorption of a given hydrogen, H_i, can be utilized to determine the number of protons adjacent to it, a fact which is quite useful for purposes of analyzing spectra.

The relative numbers of different types of protons can be determined by integration, electronic or manual, of the absorption bands of the spectrum since the areas under the bands are a linear function of the number of hydrogen nuclei producing the band. For example, in the spectrum of Figure 4.5, the relative areas of the three bands are obtained by measuring the distances, *a–b, a–c* and *a–d*. The ratio of these heights is (if measured in millimeters) 28:28:44, so that the relative abundance of the corresponding protons is determined to be 2.0:2.0:3.1. This is within experimental error of the absolute ratio of 2.0:2.0:3.0 anticipated for 1-nitropropane.

It should now be clear that analysis of an nmr spectrum can yield a wealth of information. Determination of the *chemical shift* of each peak or group of peaks allows one to speculate about the *type of functional group* to which the hydrogen nucleus producing the absorption is attached; the *spin-spin splitting* pattern provides knowledge con-

[9] This statement is an oversimplification and is strictly true only in those cases in which the ratio of the chemical shift, expressed in cps, between coupled protons and their coupling constant is greater than 5 to 10. For further discussion of this point, see, for example, references 3, 5, and 7.

cerning the *number of nearest neighbors* of the hydrogen; and *integration* permits an evaluation of the *relative numbers of each type* of hydrogen present in the molecule.

When attempting to analyze the nmr spectrum of an unknown compound, it is profitable to begin by determining the relative numbers of different types of hydrogen present by measurement of the integrated peak areas. If the molecular formula of the unknown is available, the *relative* ratios can be converted to *absolute* numbers of the hydrogens present.

The second step in the analysis is an attempt to define the functional groups present in the molecule by determination of the chemical shifts of the different hydrogens. Further knowledge concerning the molecular environment of a given type of hydrogen can be derived by analysis of the spin-spin splitting patterns. At this point, it should be possible to propose for the unknown one or more complete, or at least partial, structures which are consistent with the spectrum.

If an infrared spectrum of the unknown is also available, it can be analyzed as described in Section 4.1. This analysis should provide additional information concerning the nature of functional groups present, and this information may well be used to supplement or to confirm that derived from analysis of the nmr spectrum.

The structures of many "unknowns" whose molecular formulae are available can often be derived immediately by combination of the information available from the ir and nmr spectra.[10] However, proper interpretation of the spectra requires considerable practice on the part of the beginning student of spectroscopy. In order to provide an opportunity for such practice in interpretation, the spectra of many of the organic compounds encountered as reagents or products in reactions described in this text have been reproduced at the end of each experiment. As these spectra are labeled, you will not be attempting to assign a structure to the substance producing them, but will be trying to interpret the spectra in terms of the known structure. As complete analysis as is possible should be included as part of the write-up of each experiment.

As an example of the kind of analysis which can be done, consider the spectra of isobutyl alcohol, $(CH_3)_2CHCH_2OH$, given in Figure 4.8. Looking first at the ir spectrum [Figure 4.8(a)], three prominent bands are present in the region of 5000–1250 cm^{-1}. By referring to Table 4.1, it is possible to assign the absorptions at about 2900 and 1465 cm^{-1} to the C—H bonds of the molecule. The broad and intense band at 3250 cm^{-1} undoubtedly is due to the O—H function in the alcohol. The position and breadth of the band indicates that the alcohol is hydrogen bonded. Finally, the very strong absorption at 1050 cm^{-1} is characteristic of a C—O single bond. As would be expected on the basis of the structure of isobutyl alcohol, no absorptions are seen for carbonyl functions, aromatic rings or carbon-carbon multiple bonds.

Turning now to the nmr spectrum [Figure 4.8(b)], the first point of interest is the integration. The areas of the peaks, going *upfield*, are in the ratio (measured in millimeters), 2.8:5.6:3.0:15, which is equal to the ratio, 1.0:2.0:1.1:5.4. Since the total number of hydrogens in isobutyl alcohol is 10, the latter ratio is within experimental error of the absolute ratio.

---

[10] Structural assignments based solely on interpretation of spectra must be confirmed either by comparison of the spectra of the unknown with those of the known compound, which has been synthesized in an unequivocal fashion, or by more classical means such as determination of mixture-melting points of solid derivatives (see Section 1.2).

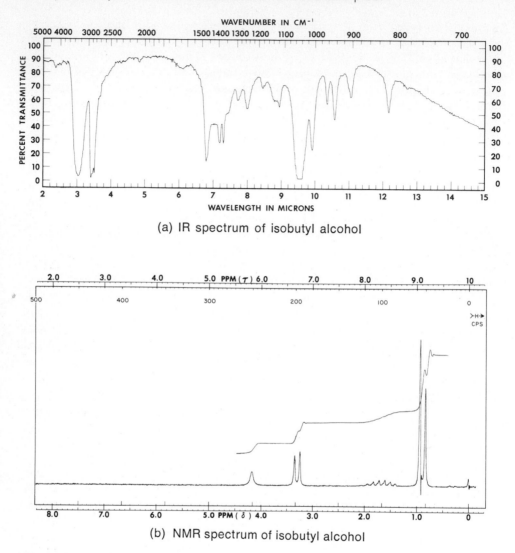

(a) IR spectrum of isobutyl alcohol

(b) NMR spectrum of isobutyl alcohol

FIGURE 4.8   Spectra of isobutyl alcohol.

It is now possible to complete the analysis of the spectrum by assigning absorptions to specific types of hydrogens, basing the assignment on chemical shift and integrated area. The doublet at about 0.85 δ has the appropriate relative area, 6, and chemical shift to be due to the six hydrogens of the two methyl groups of the alcohol. The multiplicity of the methyl absorption is due to coupling with the lone methine (tertiary) hydrogen, $(CH_3)_2CH$. The multiplet from 1.4–2.0 δ can be assigned to the methine hydrogen by consideration of the area, 1, the chemical shift, which is consistent with the nucleus being tertiary, and the multiplicity. The methine hydrogen is coupled with no less than eight other hydrogens. The doublet at about 3.3 δ has the area, 2, the chemical shift and multiplicity anticipated for the methylene group of isobutyl alcohol.

This leaves the somewhat broad singlet at 4.2 δ to be assigned to the hydroxylic hydrogen.[11]

For purposes of analysis by nmr, approximately 0.3 ml of a 10–20 percent solution (by weight) is normally required. Carbon tetrachloride is an excellent choice for the solvent since it has no protons which would produce absorption peaks in addition to those of the solute. For samples which are insoluble in carbon tetrachloride, benzene, 7.67 δ (chemical shift), chloroform, 7.27 δ, nitromethane, 4.33 δ, acetonitrile, 2.00 δ, or acetone, 2.17 δ, may be suitable solvents if their proton resonances do not occur in a region of absorption of the solute.[12]

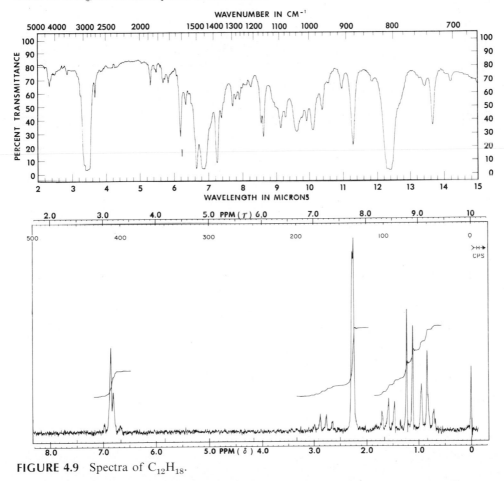

FIGURE 4.9   Spectra of $C_{12}H_{18}$.

[11] It may seem strange that the hydroxylic proton appears as a singlet, albeit a broad singlet, since one would expect it to be coupled to the methylene hydrogens. Such coupling can, in fact, be observed in highly purified alcohols, but for our purposes, one can assume that no spin-spin splitting will be detected between such a proton and protons on the carbon atom bonded to the oxygen.

[12] Many deuteriated solvents, such as chloroform-d (CDCl$_3$), acetone-d_6 (CD$_3$COCD$_3$), and benzene-d_6 (C$_6$D$_6$), are available. Use of these solvents circumvents the problem of resonance absorptions of solvent protons in the nmr spectrum because the chemical shift of deuterium (^{2}H) is vastly different from that of hydrogen.

EXERCISES

1. The nmr spectra of the three fractions obtained in the distillation of a mixture of benzene and toluene consisted of a sharp singlet and a multiplet in the region of 7.2 δ, and a singlet at 2.3 δ. The integrated intensities of the downfield and upfield absorptions are given below for each fraction.

	Downfield	Upfield
Fraction 1	165	9
Fraction 2	192	30
Fraction 3	167	81

Calculate the percentage of toluene present in each fraction.

2. The ir and nmr spectra of a compound, $C_{12}H_{18}$, are given in Figure 4.9(a), (b). Interpret both spectra as completely as possible and attempt to assign a structure or structures to the unknown on the basis of your interpretation.

REFERENCES

1. L. J. Bellamy, *The Infra-red Spectra of Complex Molecules,* 2nd ed., John Wiley & Sons, Inc., New York, 1958.
2. R. T. Conley, *Infrared Spectroscopy,* Allyn and Bacon, Inc., Boston, 1966.
3. J. R. Dyer, *Applications of Absorption Spectroscopy of Organic Compounds,* Prentice-Hall, Inc., Englewood Cliffs, N. J., 1965 (paperback).
4. L. M. Jackman and S. Sternhell, *Applications of Nuclear Magnetic Resonance Spectroscopy in Organic Chemistry,* 2nd ed., Pergamon Press, New York, 1969.
5. R. M. Silverstein and G. C. Bassler, *Spectrometric Identification of Organic Compounds,* 2nd ed., John Wiley & Sons, Inc., New York, 1967. Infrared, nuclear magnetic resonance and ultraviolet spectroscopy and mass spectrometry.
6. I. Fleming and D. H. Williams, *Spectroscopic Methods in Organic Chemistry,* McGraw-Hill Book Company, Inc., New York, 1966.

chapter five
alkanes

Free Radical Halogenation

5.1   Chlorination by Means of Sulfuryl Chloride

In general, saturated hydrocarbons are chemically inert even in the presence of very strong acids or bases, and the organic chemistry of these substances is rather limited as a consequence. Some chemical reactions have been developed, however, by which hydrocarbons can be converted to organic substances suitable for use in a variety of reactions. For example, a hydrocarbon, RH, can be transformed to a nitroalkane, RNO_2, or to a hydroperoxide, ROOH, upon reaction with nitrogen tetroxide or molecular oxygen, respectively.

This experiment involves the production of alkyl chlorides, RCl, from hydrocarbons. No reaction occurs when chlorine gas and a hydrocarbon are simply mixed at room temperature, but ultraviolet irradiation or heating (200–400°) of the mixture affords an alkyl chloride and hydrogen chloride [equation (1)]. The photochemical or thermal

$$RH + Cl_2 \xrightarrow[h\nu]{\text{heat or}} RCl + HCl \tag{1}$$

activation of the mixture is required in order to convert some molecular chlorine into chlorine atoms. The generation of chlorine atoms, which are free radicals, is essential to the initiation of the reaction between the hydrocarbon and chlorine to give the alkyl chloride.

It is somewhat easier experimentally to generate chlorine atoms by use of an initiator, a substance which will decompose to free radicals under relatively mild conditions [equation (2)]. The free radicals, In·, arising from decomposition of the initiator will

$$In—In \xrightarrow[h\nu]{\text{heat or}} In· + In· \tag{2}$$

$$In· + Cl—Cl \longrightarrow InCl + Cl· \tag{3}$$

react with molecular chlorine to produce InCl and chlorine atoms [equation (3)], and these chlorine atoms will then react with the hydrocarbon to produce an alkyl chloride.

In our experiment the usual procedure for chlorinating hydrocarbons is modified by the use of sulfuryl chloride, SO_2Cl_2, in place of chlorine gas, in the interest of safety and convenience in the undergraduate laboratory.

The mechanism of the free radical chlorination of hydrocarbons is fundamentally the same whether sulfuryl chloride or molecular chlorine is employed and can be divided into three different phases: *initiation, propagation,* and *termination.*

INITIATION

$$CH_3-\underset{\underset{CH_3}{|}}{\overset{\overset{CN}{|}}{C}}-N=N-\underset{\underset{CH_3}{|}}{\overset{\overset{CN}{|}}{C}}-CH_3 \xrightarrow{80-100°} N_2 + 2CH_3-\underset{\underset{CH_3}{|}}{\overset{\overset{CN}{|}}{C}}\cdot \qquad (4)$$

$$(CH_3-\underset{\underset{CH_3}{|}}{\overset{\overset{CN}{|}}{C}}\cdot = In\cdot)$$

$$In\cdot + Cl-\underset{\underset{O}{\|}}{\overset{\overset{O}{\|}}{S}}-Cl \longrightarrow InCl + \cdot\underset{\underset{O}{\|}}{\overset{\overset{O}{\|}}{S}}-Cl \qquad (5)$$

$$\cdot\underset{\underset{O}{\|}}{\overset{\overset{O}{\|}}{S}}-Cl \longrightarrow \underset{\underset{O}{\|}}{\overset{\overset{O}{\|}}{S}} + Cl\cdot \qquad (6)$$

PROPAGATION

$$Cl\cdot + RH \longrightarrow R\cdot + HCl \qquad (7)$$

$$R\cdot + ClSO_2Cl \longrightarrow RCl + \cdot SO_2Cl \qquad (8)$$

$$\cdot SO_2Cl \longrightarrow SO_2 + Cl\cdot \qquad (9)$$

TERMINATION

$$Cl\cdot + Cl\cdot \longrightarrow Cl_2 \qquad (10)$$

$$R\cdot + Cl\cdot \longrightarrow RCl \qquad (11)$$

$$R\cdot + R\cdot \longrightarrow R-R \qquad (12)$$

In our experiment, the first step of the initiation phase is the homolytic cleavage[1] of azobisisobutyronitrile[2] into nitrogen and the initiator free radicals $(CH_3)_2C(CN)\cdot$ [equation (4)], which is accomplished by heating at 80–100°. The initiator radicals then attack sulfuryl chloride molecules to produce SO_2 and chlorine atoms [equations (5) and (6)].

The propagation steps involve abstraction of a hydrogen atom from the hydrocarbon by a chlorine atom to produce a new free radical, $R\cdot$, which can attack the sulfuryl chloride to produce the alkyl chloride and to regenerate the radical, $\cdot SO_2Cl$, the precursor of chlorine atoms [equations (7)–(9)]. A chain reaction is thus produced.

---

[1] Homolytic cleavage may be defined as rupture of a bond in a fashion such that the bonding electrons become distributed equally between the atoms originally linked by the bond. The alternate manner of distribution in which the bonding electrons are not equally divided between the atoms is termed heterolytic cleavage.

[2] This compound is also known as 2,2′-azobis[2-methylpropionitrile].

Once initiated, the chain reaction could in principle continue until either of the principal reagents, sulfuryl chloride or the hydrocarbon, depending on which is the *limiting reagent*,[3] is exhausted. In practice, however, the termination reactions [equations (10)–(12), for example] intervene to interrupt the free radical-chain process so that initiation must be continued throughout the reaction.

It is worthwhile to note some characteristics of the various steps in a free radical-chain mechanism. In general, an *initiation step* involves the formation of a low concentration of free radicals from molecules and results in an increase in the concentration of free radicals present in the system. A *propagation step* produces no *net* change in the concentration of radicals present, whereas a *termination step* gives a decrease in their concentration.

Free radical halogenation of hydrocarbons generally produces mixtures of several isomers and polyhalogenated products which are difficult to separate into pure components. However, chlorination of hydrocarbons is a useful industrial process in cases where pure individual alkyl chlorides are not required. For example, *n*-dodecane may be chlorinated to give a mixture of monochlorododecanes

$$R\!-\!CH_2\!-\!CH_2\!-\!CH_2\!-\!CH_3 \xrightarrow{Cl_2} R\!-\!CH_2\!-\!CH_2\!-\!CH_2\!-\!CH_2Cl$$

$$+\ R\!-\!CH_2\!-\!CH_2\!-\!\underset{\underset{Cl}{|}}{CH}\!-\!CH_3 +\ R\!-\!CH_2\!-\!\underset{\underset{Cl}{|}}{CH}\!-\!CH_2\!-\!CH_3$$

$$(R = n\!-\!C_8H_{17})$$

$$+ R\!-\!\underset{\underset{Cl}{|}}{CH}\!-\!CH_2\!-\!CH_2\!-\!CH_3 + \cdots \qquad (13)$$

These alkyl chlorides may be converted by the Friedel-Crafts reaction (Chapter 8, Section 2) into a mixture of phenyldodecanes, which are useful as intermediates in the manufacture of biodegradable detergents.

For the preparation of pure individual alkyl halides, methods other than halogenation of saturated hydrocarbons are commonly used. The reaction of a hydrogen halide with an alkene

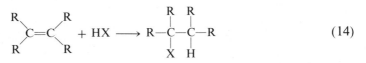

$$(14)$$

or an alcohol

$$R\!-\!OH + HX \longrightarrow R\!-\!X + H_2O \qquad (15)$$

provides the corresponding alkyl halide. Alcohols can also be transformed to alkyl halides by reaction with a thionyl halide, SOX_2 (X = Cl or Br), followed by thermal decomposition of the intermediate halosulfite ester

$$RO\!-\!H + X\!-\!\underset{\underset{O}{\|}}{S}\!-\!X \xrightarrow{-HX} RO\!-\!\underset{\underset{O}{\|}}{S}\!-\!X \xrightarrow{heat} R\!-\!X + SO_2 \qquad (16)$$

These reactions of alcohols are discussed further in Chapter 12.2.

---

[3] See Appendix II for an explanation of this term.

EXPERIMENTAL PROCEDURE

This experiment provides for the chlorination of three compounds: (A) cyclohexane, (B) heptane, and (C) 1-chlorobutane. Optimally, each student may do all

$$
\begin{array}{ccc}
& CH_2 & \\
CH_2 & & CH_2 \\
CH_2 & & CH_2 \\
& CH_2 & \\
\end{array}
\qquad CH_3(CH_2)_5CH_3 \qquad CH_3{-}CH_2{-}CH_2{-}CH_2Cl
$$

Cyclohexane   Heptane   1-Chlorobutane

three parts, since the number, type, and proportion of products from each of the three compounds illustrate important principles. Alternatively, the three compounds may be distributed among the class, and the students may compare their results. (*Caution:* Sulfuryl chloride is a lachrymator and should be weighed in a hood.)

A. CYCLOHEXANE

Fit a 100-ml round-bottomed flask with a water-cooled reflux condenser which is fitted at the top with a vacuum adapter connected as shown in Figure 5.1 to serve as a trap for the SO_2 and HCl produced in the reaction. Place 33.6 g (43 ml, 0.4 mole) of cyclohexane,[4] 27.0 g (16.2 ml, 0.2 mole) of sulfuryl chloride, and 0.1 g of azobisiso-butyronitrile in the flask, and weigh the flask and its contents. Connect the condenser and trap, turn on the aspirator so as to produce a gentle flow of air through the trap, and heat the mixture to gentle reflux with a micro burner for 20 min. Cool the reaction mixture, disconnect the flask from the condenser, and weigh it

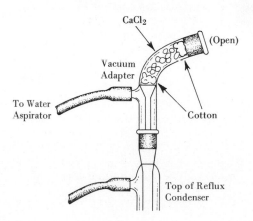

FIGURE 5.1   Details of gas trap arrangement.

---

[4] If the starting material has been stored in a metal container, it may be necessary to perform a simple distillation to remove traces of metal ions, which can inhibit the reaction.

and its contents.★ If the loss of weight is less than theoretical,[5] add another small portion of the azobisisobutyronitrile and heat the mixture to reflux for 10 min.

After the theoretical amount of weight has been lost, cool the reaction mixture and pour it into 50 ml of ice water. Transfer the resulting two-phase solution to a separatory funnel and separate the layers.★ If separation into two layers does not occur readily, add some sodium chloride to the separatory funnel and shake. Wash the organic layer in the funnel with 5% sodium carbonate solution until the washes are basic to litmus paper. Wash it once again with water and then dry it over about 3 g of *anhydrous* calcium chloride.[6]★ Filter the dried solution into a 100-ml flask fitted with a glass-packed column for fractional distillation and carefully distil in order to separate the chlorinated products from unchanged starting material. Suggested boiling ranges are as follows:

Fraction 1: Ambient-85°;
Fraction 2: 85–145°.

Weigh each of the distillation cuts and submit them for analysis by gas chromatography. Estimate the approximate yields of the chlorinated products, taking into account the amount of unchanged starting material recovered.

B. HEPTANE

Follow the directions given in part A, using 40.0 g (59 ml, 0.4 mole) of heptane[4] in place of cyclohexane. Suggested boiling ranges are as follows:

Fraction 1: Ambient-105°;
Fraction 2: 105–160°.

C. 1-CHLOROBUTANE

Follow the directions given in part A, using 37.0 g (42 ml, 0.4 mole) of 1-chlorobutane[4] in place of cyclohexane. Suggested boiling ranges are as follows:

Fraction 1: Ambient-82°;
Fraction 2: 82–165°.

EXERCISES

1. Suggest at least one termination process for halogenation with sulfuryl chloride reagent which has no counterpart in halogenation with molecular chlorine.
2. What is the reason for using less than the amount of sulfuryl chloride theoretically required to convert all of the starting materials to monochlorinated products?
3. From which of the three starting materials should it be easiest to prepare a single pure monochlorinated product? Why?
4. What factors determine the proportion of monochlorinated isomers of heptane? of 1-chlorobutane?

---

[5] The theoretical loss of weight is calculated on the basis that one mole each of HCl and of SO_2 is evolved for each mole of sulfuryl chloride consumed. This calculation should be made before coming to laboratory.

[6] See Appendix I for a discussion of drying agents.

5. Why is only a catalytic amount of initiator used?
6. Calculate the heat of reaction for the reaction between cyclohexane and chlorine

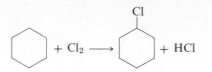

to yield chlorocyclohexane and hydrogen chloride, using the following bond energies (in kcal/mole):

$$C{-}H, 98.7; \quad C{-}Cl, 81; \quad Cl{-}Cl, 58.0; \quad H{-}Cl, 103.2.$$

7. Calculate the percent of each monochlorination product expected from *n*-heptane based on a relative reactivity of primary (1°): secondary (2°): tertiary (3°) hydrogens of 1.0:3.3:4.4. By referring to Figure 5.2, calculate the observed ratio of 1°:2° chloroheptanes and compare the result with the anticipated theoretical ratio.
8. Chlorination of propene leads to high yields of 3-chloro-1-propene to the exclusion of products of substitution of the vinyl hydrogens. Explain why this is so.
9. Write out the expected products of free radical reaction of SO_2Cl_2 with 1-methylcyclohexane. Predict which chlorinated isomer would be formed in highest yield.
10. By reference to Figure 5.3, calculate the percentage of each dichlorobutane

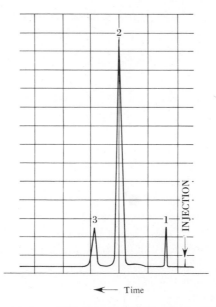

FIGURE 5.2 Gas chromatogram of fraction 2 resulting from chlorination of heptane. Peak 1: heptane; peak 2: 2-, 3-, 4-chloroheptanes; bp, 46° (19.5 mm), 48.3 (21 mm), and 48.9° (21 mm), respectively; peak 3: 1-chloroheptane bp 61.4 (27 mm); 158.5–159.5 (769 mm); column and conditions: 5 ft, 5% silicone elastomer on Chromosorb W; 80°, 40 ml/min.

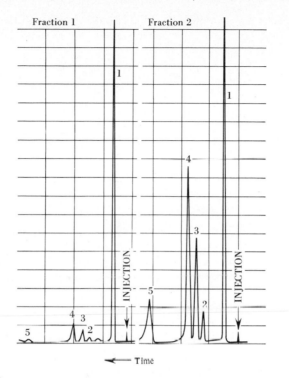

FIGURE 5.3 Gas chromatograms of fractions 1 and 2 resulting from chlorination of 1-chloro-butane. Peak 1: 1-chlorobutane; peak 2: 1,1-dichlorobutane (bp 114–115°); peak 3: 1,2-dichlorobutane (bp 121–123°); peak 4: 1,3-dichlorobutane (bp 131–133°); peak 5: 1,4-dichlorobutane (bp 161–163°); column and conditions: 5 ft, 5% silicone elastomer on Chromosorb W; 60°, 40 ml/min.

present in fraction 2 (do not include the area of the peak due to 1-chlorobutane when making the calculation). Explain why the observed ratio of products does not agree with that predicted by use of the relative reactivities given in exercise 7.

11. In Figure 5.4 is given the nmr spectrum of a mixture of dichlorobutanes obtained by free radical chlorination of 1-chlorobutane.

(a) Assign the multiplet at 5.80 δ, the doublet at 1.55 δ, and the triplet at 1.07 δ to the hydrogen nuclei of the three different isomers from which they arise. (Hint: Write the structures of the isomeric 1, x-dichlorobutanes and predict the multiplicity and approximate chemical shift of each group of hydrogens.)

(b) Calculate the approximate percentage of each isomer present in the mixture.

REFERENCES

1. M. S. Kharasch and H. C. Brown, *Journal of the American Chemical Society,* **61,** 2142 (1939).
2. P. C. Reeves, *Journal of Chemical Education,* **48,** 636 (1971).

SPECTRA OF STARTING MATERIALS AND PRODUCTS

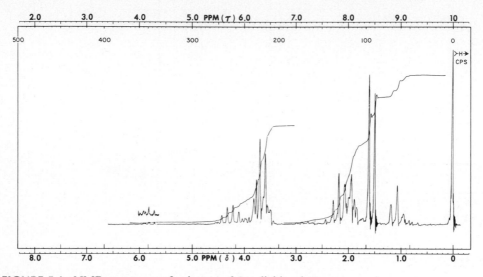

FIGURE 5.4   NMR spectrum of mixture of 1,x-dichlorobutanes.

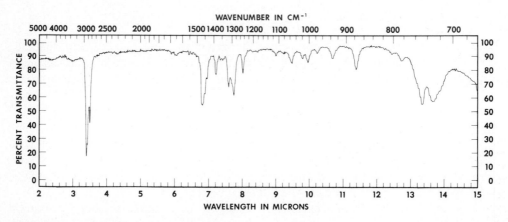

FIGURE 5.5   IR spectrum of 1-chlorobutane.

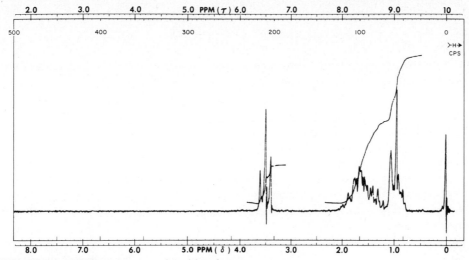

FIGURE 5.6 NMR spectrum of 1-chlorobutane.

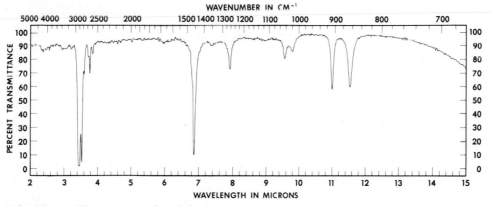

FIGURE 5.7 IR spectrum of cyclohexane.

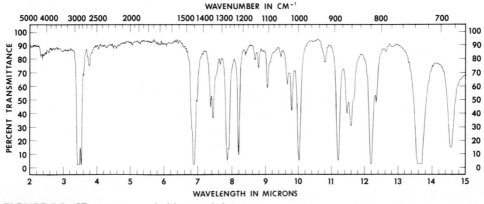

FIGURE 5.8 IR spectrum of chlorocyclohexane.

5.2   Bromination: Relative Ease of Substitution of Hydrogen in Different Environments

The preceding experiment allows a comparison of the reactivity of hydrogen atoms bonded to primary (1°) and secondary (2°) carbon atoms [7] toward substitution by chlorine atoms on the basis of the relative amounts of primary and secondary chloroheptanes produced from heptane. In the present experiment, the relative rates of reaction of bromine with several different hydrocarbons containing primary, secondary, and tertiary hydrogens will be determined. The hydrogens may also be described as being aliphatic, aromatic, or benzylic.[8] We may combine these terms and speak of primary, secondary and tertiary aliphatic or benzylic hydrogens. Only one type of aromatic hydrogen is possible. By careful consideration of your results from this experiment, you should be able to deduce an order of reactivity for seven different types of hydrogens.

Experimentally, the relative rates of bromination of the various hydrocarbons are measured by the lengths of time required for the bromine color to be discharged in the reactions, which are conducted under similar experimental conditions. Bromine is a less reactive and hence a more selective reagent than chlorine, so that the rates of bromination of the various hydrocarbons are sufficiently different to allow a qualitative order of reactivity to be determined quite easily.

The reactions are to be carried out in carbon tetrachloride solution under three different conditions: (1) at room temperature without special illumination, (2) at room temperature with strong illumination, and (3) at about 50° with strong illumination. Under all of these conditions, substitution of hydrogen by bromine occurs by a free radical mechanism analogous to that of the chlorination reactions of Section 5.1. In this case the initiation step is the thermal and/or photochemical dissociation of bromine molecules into bromine atoms (radicals) as shown in equation (17). The propagation steps are those of equations (18) and (19). The termination steps are exactly analogous to those of equations (10)–(12) of Section 5.1.

$$Br_2 \xrightarrow[\text{or } h\nu]{\text{heat}} 2 \ Br\cdot \qquad (17)$$

$$Br\cdot + RH \rightarrow HBr + R\cdot \qquad (18)$$

$$R\cdot + Br_2 \rightarrow RBr + Br\cdot \qquad (19)$$

EXPERIMENTAL PROCEDURE

1. Construct a table in your notebook with the following headings: *Hydrocarbon, Types of Hydrogen* (1°, 2°, etc.), *Conditions* (with the subheadings, 25°; 25°, hν; 50°, hν).

2. In each of seven 18 × 100-mm test tubes place a solution of 1 ml of hydrocarbon

[7] In the following discussion we shall follow the slightly inaccurate but convenient practice of referring to a hydrogen atom bonded to a primary carbon atom as a "primary hydrogen," etc.

[8] A benzylic hydrogen is one attached to a carbon atom bonded to an aromatic ring.

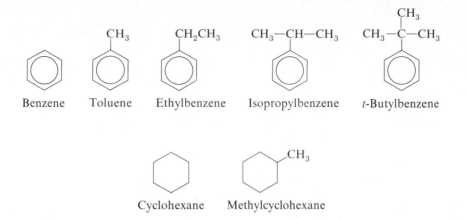

Benzene Toluene Ethylbenzene Isopropylbenzene *t*-Butylbenzene

Cyclohexane Methylcyclohexane

in 5 ml of carbon tetrachloride, using each of the following hydrocarbons: benzene, toluene, ethylbenzene, isopropylbenzene (cumene), *t*-butylbenzene, cyclohexane, and methylcyclohexane. In each of seven other test tubes place 1 ml of $1M$ bromine in carbon tetrachloride.[9] Add the bromine solutions to the hydrocarbon solutions in rapid succession, and shake each mixture after the addition.[10] Note and record the time required for the red bromine color to be discharged in each reaction mixture.

3. Carry out the experiment as in paragraph 2 above, but place the reaction tubes, immediately after mixing the reactants, about 4–5 in from an unfrosted 100- or 150-watt light bulb. Record the times of disappearance of color.

4. Carry out the experiment as in paragraph 2 above, but before mixing the hydrocarbon and bromine solutions, heat them all to about 50° in a large beaker of water placed on a steam cone or hot plate. Also, position the light bulb directly above the beaker containing the test tubes so that the reaction mixtures are all about 4–5 inches from the bulb. After mixing, keep the tubes containing the reaction mixtures in the beaker of warm water and maintain the temperature of the water as constant as possible. Record the times of disappearance of color as before.

EXERCISES

1. Arrange the seven hydrocarbons in increasing order of reactivity toward bromination.
2. On the basis of the order of reactivity of the hydrocarbons, deduce the order of reactivity of the seven different types of hydrogens found in these compounds; *i.e.*, (1) primary aliphatic, (2) secondary aliphatic, (3) tertiary aliphatic, (4) primary benzylic, (5) secondary benzylic, (6) tertiary benzylic, and (7) aromatic.
3. Clearly explain how you arrived at your sequence in question 2.

---

[9] Bromine, even in solution, is a potentially dangerous chemical. Do not breathe vapors of the carbon tetrachloride solution or spill the solution on your skin; work in the hood, if possible. In case of accident, see page 5.

[10] Because the rates of some of the brominations are quite high, it may be advisable to start only three or four of the reactions at one time and the remaining three or four at another time.

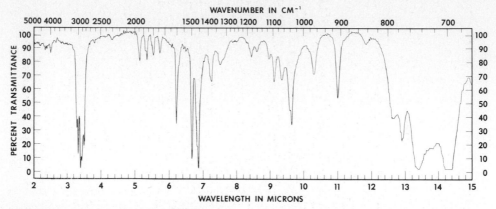

FIGURE 5.9 IR spectrum of ethylbenzene.

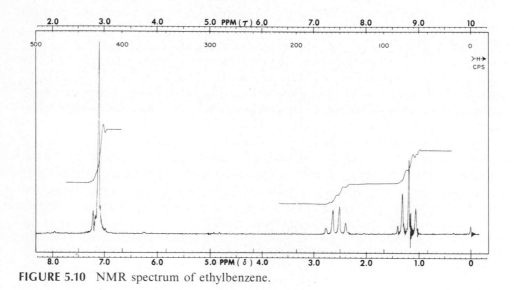

FIGURE 5.10 NMR spectrum of ethylbenzene.

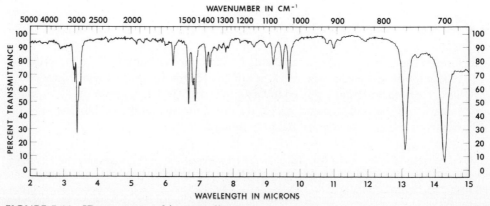

FIGURE 5.11 IR spectrum of isopropylbenzene.

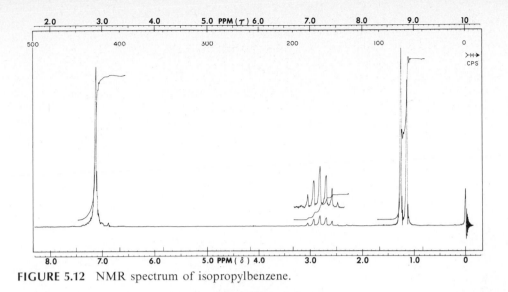

FIGURE 5.12   NMR spectrum of isopropylbenzene.

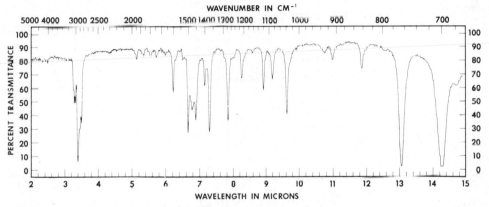

FIGURE 5.13   IR spectrum of *t*-butylbenzene.

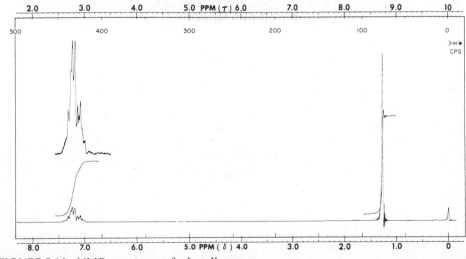

FIGURE 5.14   NMR spectrum of *t*-butylbenzene.

5.3 Triphenylmethyl Bromide, Triphenylmethyl Radical, and Triphenylmethyl Peroxide

This experiment demonstrates the use of direct halogenation for the production of an organic halide in practical yield. The starting material, triphenylmethane, is a compound containing one tertiary benzylic hydrogen [8] and fifteen aromatic hydrogens, thus assuring good selectivity (see Section 5.2) and freedom from significant amounts of isomers of the bromination product, triphenylmethyl bromide (**1**).

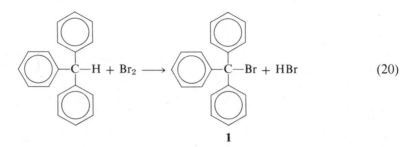

$$\text{C} - \text{H} + \text{Br}_2 \longrightarrow \text{C} - \text{Br} + \text{HBr} \qquad (20)$$

1

The first authentic free radical was identified by Moses Gomberg in 1900 as a product from triphenylmethyl chloride. Similar results can be obtained from the corresponding bromide. When triphenylmethyl bromide is treated with powdered zinc, the carbon-bromine bond is broken homolytically, and triphenylmethyl radicals (**2**) are produced [equation (21)].

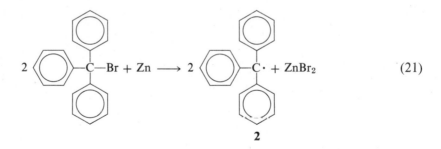

$$2 \; \text{C} - \text{Br} + \text{Zn} \longrightarrow 2 \; \text{C} \cdot + \text{ZnBr}_2 \qquad (21)$$

2

Some of these radicals combine (dimerize), but the reaction does not go to completion; the continuing presence of the radicals is indicated by a yellow color in the solution. From the time of Gomberg's observations until recently it was generally believed that the dimer in equilibrium with the triphenylmethyl radicals was hexaphenylethane. However, it has now been conclusively demonstrated that the dimer is not hexaphenylethane, but has the structure of **3** as shown in equation (22).[11]

---

[11] The evidence for the structure of **3** is based mainly on nmr and uv spectra. See H. Lankamp, W. Th. Nauta, and C. MacLean, *Tetrahedron Letters,* 249 (1968), and W. B. Smith, *Journal of Chemical Education,* **47,** 535 (1970).

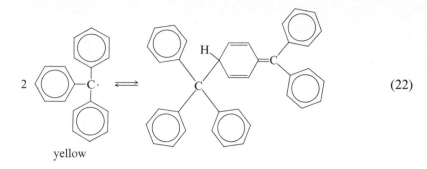

yellow

The dimerization of the triphenylmethyl radicals does not go to completion for two reasons: first, there is considerable steric interference by the bulky phenyl groups to bond formation by the radicals and, second, the radicals are stabilized by delocalization of the unpaired electron by the three benzene rings.

In the presence of oxygen, triphenylmethyl radicals react irreversibly to form a stable peroxide [**4**, equation (23)]. It should be noted that in the peroxide there is less steric strain than in the dimer because the bulky triphenylmethyl groups are spaced apart by the intervening oxygen atoms.

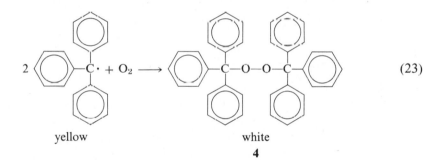

yellow   white
                                    4

EXPERIMENTAL PROCEDURE

A. Preparation
of triphenylmethyl bromide

(*Caution:* Bromine is a hazardous chemical. Do not breathe its vapors or allow it to come in contact with the skin. All operations involving transfer of the pure liquid into solvents or reaction flasks should be carried out in an efficient hood. In case of accident, see page 5. Refer to the caution on page 5 about allowing acetone to come in contact with bromine.)

The preparation of triphenylmethyl bromide is carried out in apparatus such as that pictured in Figure 5.15. Prepare a trap for hydrogen bromide by inserting a small pad of glass wool into a vacuum adapter, filling the adapter with pellets of sodium hydroxide, and holding the pellets in place with a second pad of glass wool.

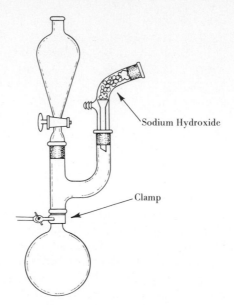

Sodium Hydroxide

Clamp

FIGURE 5.15   Apparatus of bromination of triphenylmethane.

Close the side tube of the vacuum adapter with a rubber dropper bulb. In a 100-ml round-bottomed flask, dissolve 2 g (0.009 mole) of triphenylmethane in 10 ml of dry carbon tetrachloride. Fit the flask with a Claisen connecting tube and attach the vacuum adapter to the bent arm and a separatory funnel to the straight arm of the connecting tube. Measure 0.5 ml (1.38 g, 0.009 mole) of bromine, dissolve it in 10 ml of carbon tetrachloride, and transfer the solution to the funnel. Place a 150-watt unfrosted light bulb near (4–6 in.) the round-bottomed flask and add the bromine solution dropwise to the solution of triphenylmethane as rapidly as the bromine color is discharged. Gentle swirling of the apparatus may be helpful in mixing the contents as the addition is made. After the addition of the bromine is complete, continue the irradiation until the reaction mixture is pale yellow (10–45 min, depending on the intensity of the light and its distance from the flask). Add a drop or two of cyclohexene to discharge the last trace of bromine.

Place a stopper in the open end of the vacuum adapter, close the stopcock of the dropping funnel, and attach the side arm of the vacuum adapter to an aspirator. Evaporate the solvent under aspirator vacuum, warming the flask by placing it in a pan or beaker of warm water.[12] Swirl the flask when applying the vacuum to keep the liquid from bumping and be careful not to warm the flask too much at first. Recrystallize the crystalline residue from petroleum ether to give pale yellow crystals of triphenylmethyl bromide (reported mp 153–155°). The yield should be 40–60%.

---

[12]Care should be taken not to overheat the product during the work-up, or decomposition to a black tar will occur. The product must also be protected from moisture until it is used in part B, as hydrolysis to triphenylmethanol occurs readily.

B. PREPARATION AND REACTIONS
OF TRIPHENYLMETHYL RADICAL

In a 25-ml Erlenmeyer flask, dissolve 1 g of triphenylmethyl bromide in 15 ml of benzene. Add 3 g of 30-mesh zinc, stopper the flask, and shake it vigorously for about 5 min. Filter the mixture through a fluted filter, collecting the filtrate in a 25-ml Erlenmeyer flask. Stopper the flask and shake the solution vigorously. Allow the flask to stand quietly for a few minutes. Note any color changes observed.

Remove the stopper and allow the flask to stand open for several minutes. Replace the stopper and shake the solution. Again, note any color changes. Repeat these procedures until no color change occurs when the closed flask is allowed to stand. This may require about 30 min. A white precipitate should have formed during this period.

Collect the white solid by vacuum filtration on a Hirsch funnel and wash it with a little ether. After allowing it to dry in air, take the melting point. The melting point of the white product obtained in this way by Gomberg was 185°.

EXERCISES

1. What is the purpose of the sodium hydroxide placed in the apparatus in part A?
2. Identify the white precipitate formed when the mixture of triphenylmethyl bromide and zinc is exposed to the atmosphere.
3. What is the purpose of adding cyclohexene to the colored reaction mixture from the preparation of triphenylmethyl bromide?
4. Explain the chemistry involved in the color changes of the reaction mixture from triphenylmethyl bromide and zinc which you observed when the flask was alternately opened and closed.
5. Write resonance formulas illustrating delocalization of the unpaired electron of the triphenylmethyl radical.
6. On the basis of electron delocalization energy, predict the relative stability of the following radicals:

$(C_6H_5)_3 C \cdot$ $\left(CH_3O - \bigcirc \right)_3 C \cdot$ $\left(\bigcirc - \bigcirc \right)_3 C \cdot$

7. How might bulky groups in the *o*-positions of phenyl groups affect the stability of triphenylmethyl radicals?

REFERENCE

J. Workentin, *Journal of Chemical Education*, **43**, 331 (1966).

SPECTRA OF STARTING MATERIAL

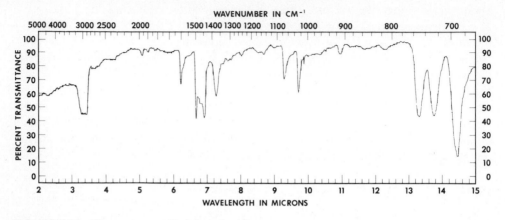

FIGURE 5.16 IR spectrum of triphenylmethane.

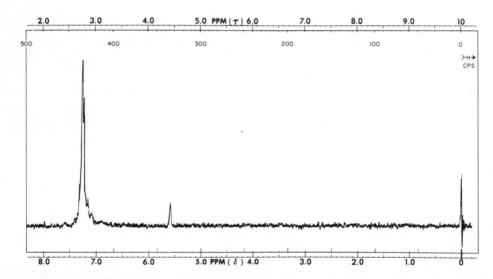

FIGURE 5.17 NMR spectrum of triphenylmethane.

alkenes

Preparation and Reactions

6.1 Generation of Alkenes by Elimination Reactions

Elimination reactions are often used by organic chemists for synthetic purposes and to aid in determination of molecular structure. Fundamentally, an elimination requires a *leaving group*, X, that may depart *with* its bonding electrons, and a group, usually hydrogen, on an (usually) adjacent carbon that may be abstracted *without* its bonding electrons. The product of the elimination may be either an alkene [equation (1)] or alkyne [equation (2)] depending on the reactant.

$$\begin{array}{c}\text{C}-\text{C} \\ | \\ \text{H}\end{array}\overset{X}{} \longrightarrow \quad \text{C}=\text{C} \; + \; \text{HX} \qquad (1)$$

$$\begin{array}{c}\text{C}=\text{C} \\ | \\ \text{H}\end{array}\overset{X}{} \longrightarrow \quad \text{C}{\equiv}\text{C} \; + \; \text{HX} \qquad (2)$$

An elimination may generally be classified according to one of three different mechanisms: (1) a two-step process in which X departs with its electrons leaving behind a carbonium ion which then in the second step loses a proton to form the double bond [E1 type, equation (3)]; (2) a one-step concerted process in which the proton

$$\begin{array}{c}\text{C}-\text{C} \\ | \\ \text{H}\end{array}\overset{X}{} \xrightarrow{-X^{\ominus}} \begin{array}{c}\text{C}-\overset{\oplus}{\text{C}} \\ | \\ \text{H}\end{array} \xrightarrow{-H^{\oplus}} \text{C}=\text{C} \qquad (3)$$

and X are lost simultaneously as the double bond forms [E2 type, equation (4)];[1] and

$$\begin{array}{c}\text{C}-\text{C} \\ | \\ \text{H}\end{array}\overset{X}{} + \; \text{B} \colon (\text{Base}) \longrightarrow \left[\begin{array}{c}\text{C}\cdots\text{C} \\ | \\ \text{H} \\ | \\ \text{B}\end{array}\overset{X}{}\right] \longrightarrow \text{C}=\text{C} \; + \; \text{BH} \; + \; \text{X} \colon \quad (4)$$

---

[1]Experiments involving eliminations following the E2 mechanism are found in Chapter 12.

(3) a two-step process in which base abstracts the proton to create a carbanion which then in the second step forms a double bond by ejection of the leaving group [equation (5)].

$$\begin{array}{c}\text{X}\\\searrow\text{C}-\text{C}\diagdown\\\text{H}\end{array}\longrightarrow\begin{array}{c}\widehat{\ \ }\text{X}\\\ominus\searrow\text{C}-\text{C}\diagdown\end{array}\longrightarrow\ \searrow\text{C}=\text{C}\diagup + \text{X}:^{\ominus}\qquad(5)$$

The acid-catalyzed dehydration of an alcohol can usually be classified mechanistically as an elimination of the E1 type. However, due to the relative instability of primary carbonium ions, many primary alcohols undergo dehydration via the E2 pathway, with a molecule of water or alcohol functioning as base. The E1 dehydration [equation (6)] involves an initial protonation of the oxygen atom to form an oxonium

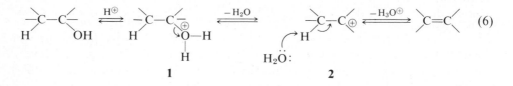

$$\qquad(6)$$

ion (1). The next step, kinetically a first-order process, involves the decomposition of the oxonium ion to a carbonium ion (2) and water. The carbonium ion then loses a proton from a carbon adjacent to the charged atom, yielding the alkene. This final loss of the proton is probably aided by a molecule of water or alcohol in the medium acting as a base.

It should be noted that each of the steps along the reaction pathway is reversible: Under the conditions of the experiment, the alkene may be rehydrated to alcohol. In order to carry the elimination to completion (100% conversion), the alkene may be distilled from the reaction mixture as it is formed. This has the effect of shifting the equilibrium to the right and maximizing the yield of alkene. *The constant removal from a reaction mixture of products as they are formed is a technique often utilized to afford high yields of products from equilibrium reactions.*

An E1 elimination is usually accompanied to some extent by substitution reactions. These competing reactions involve combination of the intermediate carbonium ion with the anion of the acid used for catalysis or some other anion. For example, with hydrochloric acid, the chloride ion may react to give some chloroalkane, a product in which chlorine has replaced the hydroxy group. The extent to which substitution products will be formed depends on the amount and nature of the acid used. However, since the formation of substitution products is frequently reversible under the reaction conditions, these side reactions will not always constitute a practical problem in affecting the yield of an elimination, assuming the alkene is distilled as it is formed.

In an E1 reaction, it is not at all uncommon for two or more isomeric alkenes to be formed. If the charge of the carbonium ion is unsymmetrically located on the carbon skeleton, and hydrogens are attached to two of the adjacent carbon atoms, different products may be formed, depending on which of the hydrogens is lost. This possibility is illustrated in equation (7).

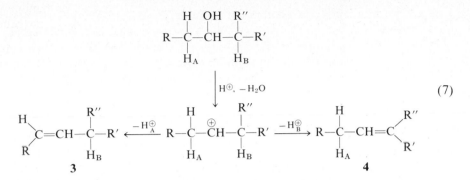

(7)

The proportion of the two alkenes (**3** and **4**) formed is dependent upon the relative activation energies of the two competing elimination processes. These activation energies are directly related to the free energies of the two transition states, which may be represented by the general formula

In these transition states the C—H bond is partially broken and the C—C double bond is partially formed. As each of the two transition states have partial double bond character, it is reasonable that their relative energies will parallel those of the corresponding alkenes, which of course have full double bonds. Therefore the more energetically favorable transition state will lead to the more stable alkene. Alkene stability usually increases as the number of alkyl substituents on the double bond is increased. Thus the more highly branched alkene (more highly substituted ethylene) is normally formed in higher yield when competing El eliminations are possible.

In many cases the number and kinds of alkenes formed as products in E1 reactions cannot be explained entirely by the type of reaction "choice" described above. In general, carbonium ions are highly susceptible to rearrangement if, by rearrangement, a more stable carbonium ion is produced. Indeed, some rearrangement may occur even when the new carbonium ion is less stable than the first, although the extent to which it occurs may be slight. Such rearrangements are accomplished by the migration of either an alkyl anion (R:⁻) or hydride ion (H:⁻) from an adjacent carbon atom to the carbon having the formal positive charge. Loss of a proton from the new carbonium ion intermediate may then lead to other isomeric alkenes.

Since carbonium ion stability increases with an increasing number of alkyl or aryl substituents, tertiary carbonium ions are more stable than secondary, which in turn are more stable than primary ions. Therefore, a migration that leads from a secondary or

$$CH_3-\underset{\oplus}{\overset{CH_3}{C}}-CH_3 > CH_3-\underset{\oplus}{\overset{H}{C}}-CH_3 > CH_3-\underset{\oplus}{\overset{H}{C}}-H$$

decreasing stability
(increasing free energy)

primary ion to a tertiary carbonium ion will usually be quite favorable. For example, the neopentyl cation (5), a primary carbonium ion, rearranges by methyl migration to give the *t*-pentyl cation (6) as the predominant species. The same "driving force" (tendency to minimize free energy) will also cause the isobutyl cation (7), a primary carbonium ion, to rearrange to the *t*-butyl cation (8) by hydride migration. In this case

$$
\underset{\mathbf{5}}{\overset{\mathrm{CH_3}}{\underset{\mathrm{CH_3}}{\mathrm{CH_3\!-\!\underset{}{\overset{.\,\rightarrow\,\oplus}{C}}\!-\!CH_2}}}} \xrightarrow{\ \sim\mathrm{CH_3:}^{\ominus}\ } \underset{\mathbf{6}}{\overset{\mathrm{CH_3}}{\underset{\mathrm{CH_3}}{\mathrm{CH_3\!-\!\overset{\oplus}{C}\!-\!CH_2}}}} \tag{8}
$$

one should expect very little methyl migration, for this would lead to a secondary ion, whereas hydride migration leads to the even more stable tertiary ion.

$$
\underset{\mathbf{7}}{\overset{\mathrm{CH_3}}{\underset{\mathrm{H}}{\mathrm{CH_3\!-\!\underset{.\,\rightarrow\,\oplus}{C}\!-\!CH_2}}}} \xrightarrow{\ \sim\mathrm{H:}^{\ominus}\ } \underset{\mathbf{8}}{\overset{\mathrm{CH_3}}{\underset{\mathrm{H}}{\mathrm{CH_3\!-\!\overset{\oplus}{C}\!-\!CH_2}}}} \tag{9}
$$

Experimental procedures for the dehydration of 4-methyl-2-pentanol and cyclo-hexanol are presented below. 4-Methyl-2-pentanol yields a mixture of isomeric alkenes including 4-methyl-1-pentene (9), *trans*-4-methyl-2-pentene (10), *cis*-4-methyl-2-pentene (11), 2-methyl-2-pentene (12), and 2-methyl-1-pentene (13). Products 12

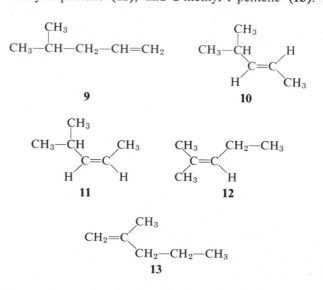

and 13 are typically found to comprise approximately 45% of the product mixture showing that rearrangement is an important aspect of carbonium ion chemistry. These products arise by loss of a proton from the carbonium ion formed by two successive hydride migrations [equation (10)].

$$CH_3-\overset{\overset{\displaystyle CH_3}{|}}{\underset{\underset{\displaystyle H}{|}}{C}}-\overset{\overset{}{}}{\underset{\underset{\displaystyle H}{|}}{CH}}-\underset{\oplus}{CH}-CH_3 \longrightarrow CH_3-\overset{\overset{\displaystyle CH_3}{|}}{\underset{\underset{\displaystyle H}{|}}{C}}-\underset{\oplus}{CH}-\overset{}{\underset{\underset{\displaystyle H}{|}}{CH}}-CH_3 \longrightarrow$$

$$CH_3-\overset{\overset{\displaystyle CH_3}{|}}{\underset{\underset{\oplus}{}}{C}}-\overset{}{\underset{\underset{\displaystyle H}{|}}{CH}}-\overset{}{\underset{\underset{\displaystyle H}{|}}{CH}}-CH_3 \quad (10)$$

The rather complex reaction mixture may be analyzed by gas chromatography (see Chapter 3). By comparing retention times of authentic samples of the methylpentenes with those of the components of the mixture, the peaks in the gas chromatogram of the mixture can be assigned to the different methylpentenes. By measuring the relative areas under the peaks, the percentage of each component of the mixture may be obtained.

Cyclohexanol undergoes acid-catalyzed dehydration without rearrangement to yield a single product, cyclohexene. The reaction product may be identified, after purification, by comparing its infrared spectrum with that of pure cyclohexene.

$$\quad (11)$$

It is frequently desirable to determine whether or not a compound contains carbon-carbon multiple bonds. Two simple qualitative tests for such unsaturation are described below.[2]

Bromine will add to the carbon-carbon double bond of alkenes to produce dibromoalkanes [equation (12)].[3] When this reaction occurs, molecular bromine is de-

$$\quad (12)$$

stroyed and its characteristic color disappears (if bromine is not present in excess). The rapid disappearance of the color of bromine is a positive test for unsaturation. However, the test is not unequivocal, for some alkenes do not react with bromine and some react very slowly. In case of a negative test, moreover, it is usually good practice to carry out the permanganate test (see below).

A second qualitative test for unsaturation, the Baeyer test, depends on the ability of potassium permanganate to oxidize the carbon-carbon double bond to give alkanediols [equation (13)] or the carbon-carbon triple bond to give carboxylic acids [equation (14)]. The permanganate ion is destroyed in the reaction and a brown precipitate of MnO_2 is produced. The disappearance of the characteristic color of the permanganate

---

[2] More detailed descriptions of these qualitative tests may be found in Chapter 8 of reference 1 following Chapter 25.3.

[3] Addition reactions of alkenes are discussed in detail in Section 6.2.

ion is a positive test for unsaturation. However, care must be taken, as compounds containing certain other types of functional groups (for example, aldehydes, containing the $-CH{=}O$ group) also decolorize permanganate ion.

$$\underset{\diagup}{\diagup}C{=}C\underset{\diagdown}{\diagdown} + MnO_4^\ominus \xrightarrow{\ H_2O\ } -\underset{|}{\overset{\overset{\displaystyle HO}{|}}{C}}-\underset{|}{\overset{\overset{\displaystyle OH}{|}}{C}}- + MnO_2 \qquad (13)$$

$$R{-}C{\equiv}C{-}R' + MnO_4^\ominus \xrightarrow{\ H_2O\ } R{-}CO_2H + HO_2C{-}R' \qquad (14)$$

EXPERIMENTAL PROCEDURE

A. Dehydration of 4-methyl-2-pentanol

Assemble a fractional distillation apparatus as has been described earlier (Figure 2.4). To minimize losses during distillation, the receiver, a test tube, should be cooled by an ice-water bath. Place 20 g (24.8 ml, 0.196 mole) of 4-methyl-2-pentanol in a 100-ml round-bottomed flask and to this add 10 ml of 9 M sulfuric acid.[4] Thoroughly mix the contents of the flask by swirling or stirring with a stirring rod. Add two or three boiling stones and connect the flask to the distillation apparatus. Distil the mixture until about 10 ml of the original mixture remains in the distillation flask. The rate of distillation should be such that the head temperature never exceeds 90°. (*Caution:* The distillate and vapors are flammable.) Transfer the distillate to a small separatory funnel, add 5–10 ml of 3 N sodium hydroxide solution, and shake well. Allow the layers to separate and drain off the aqueous layer. Transfer the organic layer to a dry 50-ml Erlenmeyer flask and add 1–2 g of *anhydrous* calcium chloride.★ Occasionally swirl the mixture during a period of 10 min and then decant the dried organic mixture into a 100-ml distilling flask. Add boiling stones and distil the mixture through a simple distillation apparatus. Collect the fraction boiling between 53 and 69° in a preweighed, ice-cooled receiver. Weigh the product and calculate the yield obtained. If carried out properly, a yield of 75–85% may be expected.★ The expected products of the reaction and their boiling points are 4-methyl-1-pentene (53.9°), *trans*-4-methyl-2-pentene (58.6°), *cis*-4-methyl-2-pentene (56.4°), 2-methyl-1-pentene (61°), and 2-methyl-2-pentene (67.3°).

Test the distillate for unsaturation by each of the following methods: (1) To 1 or 2 ml of 2% bromine solution in carbon tetrachloride add 1 or 2 drops of the distillate. Rapid disappearance of the bromine color to give a colorless solution is a positive test for unsaturation. Perform this same experiment using cyclohexane instead of the distillate. (2) Perform the Baeyer test for unsaturation by dissolving 1 or 2 drops of the distillate in 2 ml of ethanol. Then add a 0.1 M potassium permanganate solution dropwise, observing the results. Count the number of drops added before the permanganate color persists. For a *blank determination*, count the number of drops that

---

[4] 85% phosphoric acid is less satisfactory. The reaction is slower and the yields are lower. Use 5 ml of this acid should your instructor desire you to do so.

may be added to 2 ml of ethanol before the color persists. A significant difference in the number of drops required in the two cases is a positive test for unsaturation. Carry out the same experiment using cyclohexane in place of the distillate.

Submit a small sample of your product for glpc analysis. Figure 6.1 shows a typical glpc tracing of the products of this reaction.[5] Calculate the percentage composition of your reaction mixture. Also submit another small sample for infrared spectroscopic analysis. By comparing the absorption peaks from the infrared spectrum obtained with those from spectra of each of the pure authentic methylpentenes, identify as many of the components as you can.

B. DEHYDRATION OF CYCLOHEXANOL

Place 20 g (21.2 ml, 0.20 mole) of cyclohexanol and 10 ml of 9 M sulfuric acid[4] in a 100-ml round-bottomed flask and mix well. Add boiling stones and distil the mixture through a fractional distillation apparatus as described above for 4-methyl-2-pentanol until about 10 ml is left in the distillation flask. The receiver, a test tube, should be cooled in an ice-water bath. Transfer the distillate to a separatory funnel and shake well with 10 ml of 3 N sodium hydroxide solution. Separate the aqueous layer, pour the organic layer into a 50-ml Erlenmeyer flask, and add 1-2 g of *anhydrous* sodium sulfate.★ Occasionally swirl the flask during about 5 min and then decant the liquid into another dry 50-ml Erlenmeyer flask. Add another 1-2 g of *anhydrous* sodium sulfate, swirl for 5 min, and filter the liquid into a 100-ml round-bottomed flask. (The product *must be dry* at this stage in order to obtain pure cyclohexene since water and cyclohexene form a minimum-boiling azeotrope.) Distil the crude cyclohexene through a simple distillation apparatus and collect the fraction boiling between 80 and 85° (12–14 g) in a preweighed ice-cooled receiver. Calculate the yield of the reaction.★ Perform each of the tests for unsaturation described under part A and submit a sample for infrared analysis. Compare your spectrum with that of authentic cyclohexene.

EXERCISES

1. Suggest the reason for washing the distillates in each of the above experiments with dilute sodium hydroxide.
2. Why, during the above dehydration experiments, is the distillation temperature kept below 90°?
3. Which of the following primary alcohols would be most likely to dehydrate via the E1 mechanism? the E2 mechanism? Explain.

$$CH_3CH_2OH \qquad (CH_3)_3CCH_2OH$$

---

[5] The peaks in Figure 6.1 have been assigned as follows: (1) 4-methyl-1-pentene (14.4%); (2) *cis-* and *trans-*4-methyl-2-pentene (combined 52.1%); (3) 2-methyl-1-pentene (6.5%); (4) 2-methyl-2-pentene (26.9%); and (5) unidentified.

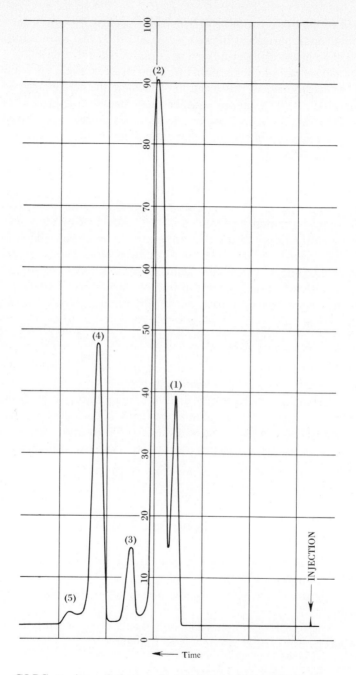

FIGURE 6.1   GLPC tracing of the product mixture from dehydration of 4-methyl-2-pentanol.[5,6]

[6]This analysis was made with an 8-ft × ⅛-in. column packed with 5% SF-96 on 60/80 Chromosorb W at 35°.

4. Give a detailed mechanism explaining each of the products obtained from the dehydration of 4-methyl-2-pentanol.
5. Near the end of the dehydration of 4-methyl-2-pentanol, a white solid may precipitate from the reaction mixture. What is this solid likely to be?
6. Give structures for the products of dehydration of each of the following alcohols. For each, order the products with respect to preference of formation.

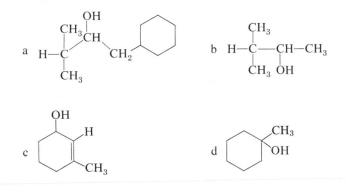

7. Suggest a reason for the necessity of a blank determination when performing the Baeyer test for unsaturation.

SPECTRA OF STARTING MATERIALS AND PRODUCTS

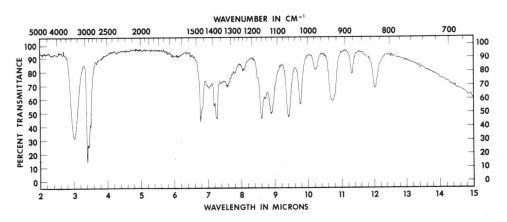

FIGURE 6.2   IR spectrum of 4-methyl-2-pentanol.

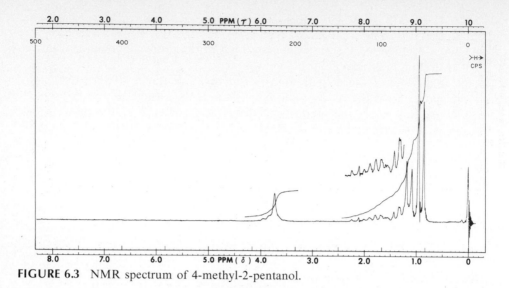

FIGURE 6.3 NMR spectrum of 4-methyl-2-pentanol.

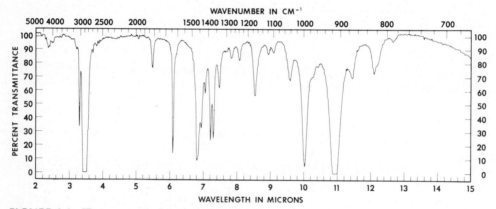

FIGURE 6.4 IR spectrum of 4-methyl-1-pentene.

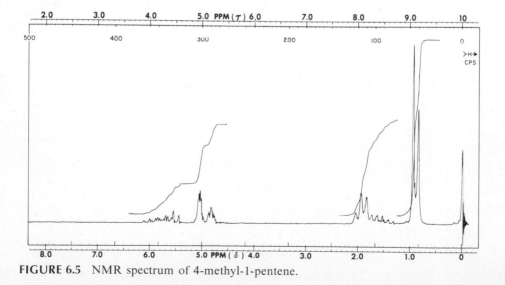

FIGURE 6.5 NMR spectrum of 4-methyl-1-pentene.

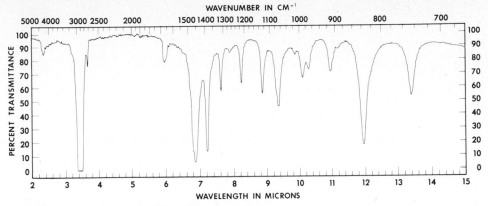

FIGURE 6.6 IR spectrum of 2-methyl-2-pentene.

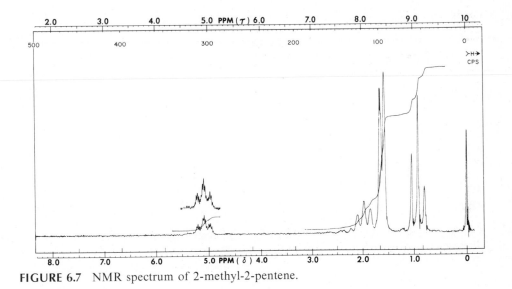

FIGURE 6.7 NMR spectrum of 2-methyl-2-pentene.

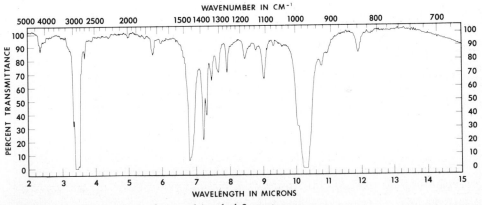

FIGURE 6.8 IR spectrum of *trans*-4-methyl-2-pentene.

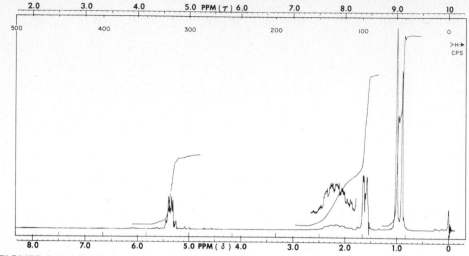

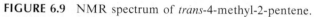

FIGURE 6.9 NMR spectrum of *trans*-4-methyl-2-pentene.

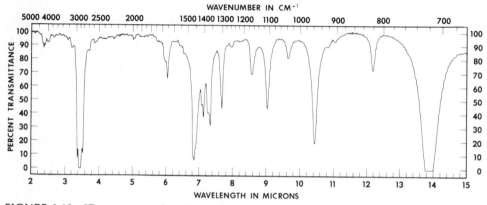

FIGURE 6.10 IR spectrum of *cis*-4-methyl-2-pentene.

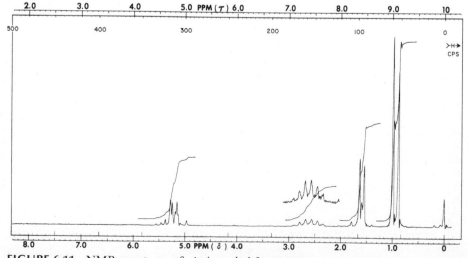

FIGURE 6.11 NMR spectrum of *cis*-4-methyl-2-pentene.

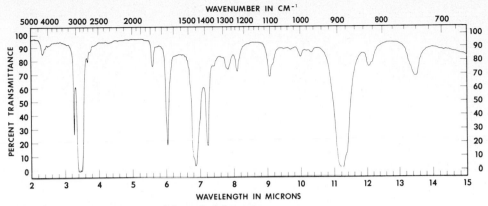

FIGURE 6.12 IR spectrum of 2-methyl-1-pentene.

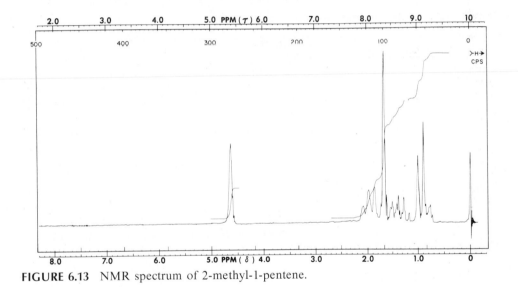

FIGURE 6.13 NMR spectrum of 2-methyl-1-pentene.

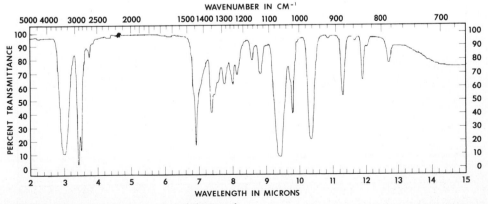

FIGURE 6.14 IR spectrum of cyclohexanol.

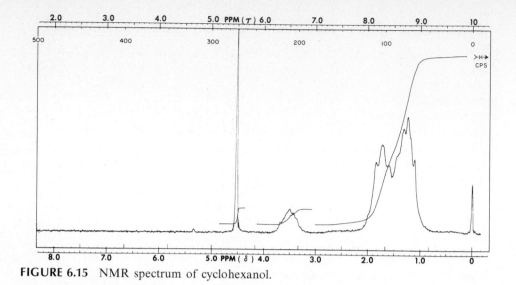

FIGURE 6.15   NMR spectrum of cyclohexanol.

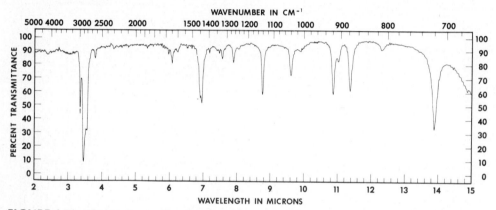

FIGURE 6.16   IR spectrum of cyclohexene.

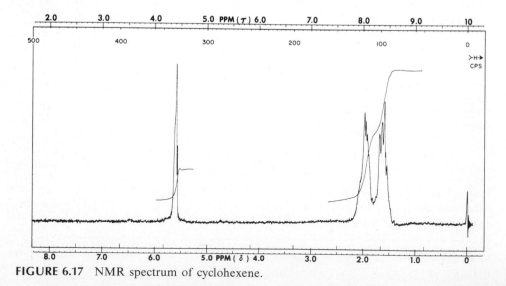

FIGURE 6.17   NMR spectrum of cyclohexene.

6.2 Addition Reactions of Alkenes

As a class, alkenes are exceedingly useful starting materials for organic syntheses since these compounds undergo a large variety of reactions involving the carbon-carbon double bond. In general, much of the chemistry of the double bond of alkenes can be viewed as the addition of a reagent, $X-Y$, across the pi bond so that a saturated molecule results [equation (15)]. The mechanism through which the addition proceeds depends upon both the nature of $X-Y$ and the conditions under which the reaction is accomplished.

$$R_2C{=}CR_2 + X{-}Y \longrightarrow R_2C{-}CR_2 \qquad (15)$$
$$\underset{X \quad Y}{|\quad|}$$

One of the more common mechanisms of addition of a reagent to an alkene involves the attack of an electrophile, E^+, an electron-deficient species, on the double bond to give an intermediate onium ion **(14)** which reacts with a nucleophilic reagent, Nu^-, to give the product **15** [equation (16)]. For unsymmetrical $E-Nu$ compounds, Y is the more electronegative of the two atoms. The bromination of alkenes, a reaction used both as a qualitative test for unsaturation as described in Section 6.1 and as a quantitative measure of the amount of unsaturation, generally proceeds by this

$$R_2C{=}CR_2 + E^{\oplus} + Nu^{\ominus} \rightleftharpoons \left[R_2C{-}CR_2 \atop \underset{\oplus}{E} \right] \xrightarrow{Y^{\ominus}} R_2C{-}CR_2 \atop E \quad Nu \qquad (16)$$
$$\underset{E \quad Nu}{\Updownarrow} \qquad\qquad \mathbf{14} \qquad\qquad \mathbf{15}$$

mechanism. The direction of addition of $E-Nu$ to an unsymmetrical alkene is such as to result in bond formation between the electrophile, E^+, and the carbon atom bearing the larger number of hydrogen atoms [Markownikoff's Rule, *e.g.*, equation (17)]. The stereochemistry of the addition is such that E and Nu are *trans* to one another, as shown by the observation that *trans*-1,2-dibromocyclohexane **(17)** is obtained from the addition of bromine to cyclohexene.

$$R_2C{=}CH_2 + HBr \xrightarrow{CCl_4} R_2C{-}CH_2 \qquad (17)$$
$$\underset{Br \quad H}{|\quad|}$$
$$\mathbf{16}$$

17

A related stepwise mechanism, which is observed less commonly, involves initial attack on the alkene by a nucleophile, Nu^-, to produce the carbanion **18**, which then reacts with E^+ to provide **19** [equation (18)]. In fact, addition by this mechanism occurs readily only if the double bond bears substituents such as cyano ($-C{\equiv}N$), nitro ($-NO_2$), or carbonyl ($-\overset{\|}{\underset{O}{C}}-$), which are capable of stabilizing a negative charge. The

$$R_2C{=}CR_2 + Nu^{\ominus} \rightleftharpoons R_2\overset{\ominus}{C}\text{—}\underset{\underset{\textbf{18}}{Nu}}{CR_2} \xrightarrow[\text{(or E—Nu)}]{E^{\oplus}} R_2\underset{\underset{\textbf{19}}{Nu\ \ E}}{C}\text{—}CR_2 \qquad (18)$$

1,4-addition of aniline to benzalacetophenone [equation (19)], an experiment described in Chapter 16, follows a mechanism of this type.

$$C_6H_5CH{=}CHCOC_6H_5 + C_6H_5NH_2 \longrightarrow C_6H_5\overset{\overset{NHC_6H_5}{|}}{C}H\text{—}CH_2COC_6H_5 \qquad (19)$$

A reagent, X—Y, can often be added to an alkene by a free radical, stepwise process if free radical initiators are present or if the reaction mixture is exposed to light. In such reactions, the radical, X·, adds to the double bond to yield an intermediate radical **(20)**. The subsequent reaction between **20** and X—Y provides the product **(15)** and regenerates X· [equation (20)]. With unsymmetrical alkenes, X becomes bonded to that

$$R_2C{=}CR_2 + X\cdot \longrightarrow R_2\underset{\underset{\textbf{20}}{X}}{C}\text{—}\dot{C}R_2 \xrightarrow{X\text{—}Y} R_2\underset{\underset{\textbf{15}}{X\ \ Y}}{C}\text{—}CR_2 + X\cdot \qquad (20)$$

carbon bearing the larger number of hydrogen atoms to yield the more stable of the two possible radicals as an intermediate [equation (21)].

$$R_2C{=}CH_2 + HBr \xrightarrow[\text{(initiator)}]{\text{peroxides}} R_2\underset{\underset{\textbf{21}}{H\ \ Br}}{C}\text{—}CH_2 \qquad (21)$$

Thus, the products **16** and **21**, respectively, of polar and of free radical addition of hydrobromic acid to an alkene are different, since bromine atom, Br·, is the chain carrier in the latter process. One must therefore carefully control reaction conditions in order to obtain pure products when free radical and polar addition can produce different products. Such control may still be important even in those instances in which the *direction* of addition is of no importance, as the *stereospecificity* of free radical addition is generally low, and products resulting from both *trans* and *cis* addition of X—Y to the double bond can be obtained.

Reaction conditions which favor *electrophilic* addition are low temperatures, use of polar solvents such as water and alcohols, the presence of ionic salts like sodium or potassium bromide in the bromination reaction, and the absence of light and of peroxides. In contrast, *free radical* addition is favored by performing the reaction in the gas phase or in nonpolar solvents in the presence of strong light or peroxides.

A final point concerning additions to alkenes is noteworthy. We have now formulated three general mechanisms of addition of a reagent, X—Y, to an alkene, mechanisms which are distinguished by the type of species, electrophile, nucleophile, or free radical, involved in the initial step of the reaction. This first step, at least in principle, could be *reversible*, and, in the case of appropriately substituted alkenes, could lead to geometric isomerization [equation (22)]. Thus, the intermediates **18** and **20**, which are generated by addition of a nucleophile and of a free radical, respectively, to an alkene,

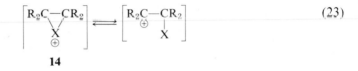

$(* = \odot, \oplus, \ominus)$

are free to undergo *rotation* about the single bond *between* the carbons which were originally joined by a double bond. Regeneration of the alkene could produce a mixture of geometric isomers. Geometric isomerization might also be a consequence of reversible addition of an electrophile to an alkene if the intermediate so formed, **14**, exists in equilibrium with a carbonium ion [equation (23)] which would be free to rotate about the necessary carbon-carbon bond to produce isomerization [equation (22), $* = \oplus$].

$$\left[\begin{array}{c} R_2C-CR_2 \\ \diagdown / \\ X \\ \oplus \end{array} \right] \rightleftharpoons \left[\begin{array}{c} R_2C-CR_2 \\ \oplus \quad | \\ X \end{array} \right] \qquad (23)$$

14

As a specific example, consider the possible consequences of the addition of bromide ion to dimethyl maleate **(22)**, a *cis* isomer [equation (24)]. This process would give the stabilized anion **23** which could undergo rotation about the carbon-carbon bond to afford **24**. Loss of bromide ion from **24** provides the geometric isomer of **22**, dimethyl fumarate **(25)** [equation (24)]:

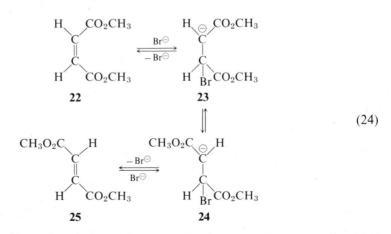

(24)

As described below, the conversion of dimethyl maleate to dimethyl fumarate is accomplished by treatment of the maleate with a dilute solution of bromine in carbon tetrachloride. A potentially important reaction, the attack of bromonium ion, Br^+, on the double bond of either the maleate or fumarate, is relatively unimportant because the presence of two electron-withdrawing carbomethoxy groups, —CO_2CH_3, on the double bond decreases the electron density of the bond sufficiently to preclude facile attack of an electrophile. Therefore the mechanism of the geometric isomerization may be either nucleophilic or free radical in character. Observations made in the experiment should provide a basis for a decision between the two.

A fourth type of mechanism for the reaction of the reagent X—Y with an alkene is concerted addition [equation (25)]. As might be anticipated, the stereochemistry of the reaction is such as to provide *cis* addition of the reagent. Some reactions of this general mechanistic type are hydroxylation, hydroboration, ozonolysis, hydrogenation, and the Diels-Alder reaction. All of these reactions are of considerable synthetic utility and examples of some of them are described elsewhere in this manual (see Experimental Procedure, below, and Chapter 10).

$$R_2C{=}CR_2 + X{-}Y \longrightarrow \left[\begin{array}{c} R_2C{=}CR_2 \\ \mid \quad\quad \mid \\ X{-}{-}Y \end{array} \right] \longrightarrow \begin{array}{c} R_2C{-}CR_2 \\ \mid \quad\quad \mid \\ X \quad\ Y \end{array} \tag{25}$$

In an experiment described below, an alkene will be converted to an alkane by hydrogenation (X = Y = H). This reaction can be accomplished in several ways, but generally it involves use of hydrogen gas either at atmospheric pressure or greater and a noble metal such as palladium, platinum or nickel as the catalyst. For our purposes, it will be convenient to utilize sodium borohydride as the *in situ* source of hydrogen [equation (26)] and to generate the catalyst by reduction of platinic chloride with hydrogen in the presence of decolorizing carbon, which serves as a solid support for the finely divided metallic platinum produced [equation (27)].

$$NaBH_4 + HCl \xrightarrow{\ 3\,H_2O\ } 4\,H_2 + B(OH)_3 + NaCl \tag{26}$$

$$H_2PtCl_6 + 2\,H_2 \longrightarrow Pt° + 6\,HCl \tag{27}$$

EXPERIMENTAL PROCEDURE

A. GEOMETRIC ISOMERIZATION
OF DIMETHYL MALEATE

Place 1.5 ml of dimethyl maleate in each of three 150-mm test tubes and with a dropper add enough of a 10% solution of bromine in carbon tetrachloride to *two* of the tubes so that an orange solution results. Add an equal volume of carbon tetrachloride to the third test tube. Stopper all the test tubes and place one of the tubes containing bromine in the dark and expose the other two tubes to strong light. If decoloration of a solution should occur, add an additional portion of the bromine solution. After 30 min cool all three solutions in ice water, observe in which test tube(s) crystals appear, and isolate the precipitate by vacuum filtration. Wash the crystals free of bromine with a little *cold* carbon tetrachloride and press them as dry as possible on the filter disk. Recrystallize the product from ethanol and determine its melting point and weight. The reported melting point for dimethyl fumarate is 101–102°.

B. HYDROGENATION

OF 4-CYCLOHEXENE-*cis*-1,2-DICARBOXYLIC ACID

Caution: NO FLAMES! Hydrogen is highly flammable.

Prepare a reaction vessel for hydrogenation by tying a heavy-walled balloon to the sidearm of a 125-ml filter flask. Also prepare a 1 *M* aqueous solution of sodium borohydride by dissolving 0.4 g (0.01 mole) of sodium borohydride in 10 ml of water and adding 0.1 g of sodium hydroxide as a stabilizer. Place 10 ml of water, 1 ml of a 5% solution of chloroplatinic acid ($H_2PtCl_6 \cdot 6H_2O$), and 0.5 g of decolorizing carbon in the reaction flask and add, with swirling, 3 ml of the 1 *M* sodium borohydride solution. Allow the resulting slurry to stand for 5 min to permit formation of the catalyst. During this time, dissolve 1 g (0.006 mole) of 4-cyclohexene-*cis*-1,2-dicarboxylic acid (see Chapter 10) by heating it with 10 ml of water.

Pour 4 ml of *concentrated* hydrochloric acid into the catalyst-containing reaction flask, and then add the hot aqueous solution of the diacid to the flask. Seal the flask with a serum cap and wire the cap securely in place. Draw 1.5 ml of the 1 *M* sodium borohydride solution into a plastic syringe, push the needle of the syringe through the serum cap, and inject the solution dropwise while swirling the flask. If the balloon on the sidearm of the flask becomes inflated, stop the addition of sodium borohydride solution until deflation occurs. When the syringe is empty, remove it from the flask and refill it with an additional 1.5 ml of solution. The dropwise addition of this further quantity of sodium borohydride should cause the balloon to inflate, and to remain inflated, indicating a positive pressure of hydrogen in the system. Remove the syringe from the serum cap and allow the flask to stand with occasional swirling for 5 min. Heat and swirl the flask on the steam cone until the balloon deflates to a constant size and then heat for an additional 5 min. A total of no more than 20 min of heating is required.

Release the pressure from the reaction flask by pushing the needle of a syringe from which the barrel has been removed through the serum cap, filter the hot reaction mixture by suction, and place the filter paper in a container reserved for "recovered catalyst."* Cool the filtrate and extract it with three 25-ml portions of technical ether. Combine the extracts, wash them with 10 ml of saturated sodium chloride solution and then filter the organic solution through a cone of sodium sulfate into a tared 125-ml Erlenmeyer flask. Evaporate the ether on a steam cone in the hood in order to isolate the saturated diacid. Determine the weight and melting point of the crude diacid.

Transfer the bulk of the crude diacid by scraping it into a 25-ml Erlenmeyer flask. Add about 2 ml of water to the larger flask, heat to dissolve any residual diacid, and pour the hot solution into the smaller flask. Bring all the diacid into solution at the boiling point by adding no more than two additional milliliters of water. After solution has been accomplished, add three drops of *concentrated* hydrochloric acid to decrease the solubility of the diacid in the solution, allow the mixture to cool to room temperature, and then place the flask in an ice-water bath to effect more complete crystallization of product. Isolate the product and determine its weight and melting point. The reported melting point is 192°. Ascertain that the product is saturated by performing the qualitative tests for unsaturation described in Section 6.1. Also determine whether hydrogenation has affected the acidic nature of the molecule by testing an aqueous solution of the diacid with litmus paper.

EXERCISES

A. GEOMETRIC ISOMERIZATION

1. Write a reasonable mechanism for the bromine-catalyzed geometrical isomerization of dimethyl maleate to dimethyl fumarate observed in this experiment.
2. What is the function of light in the isomerization reaction?
3. What is the purpose of exposing a sample of dimethyl maleate containing no bromine to light?
4. Suggest a reason why decoloration of the solution of bromine and dimethyl maleate is slow.
5. Offer an explanation for the observation that maleic anhydride (26) does not isomerize to 27 under the influence of bromine and light.

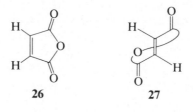

26   27

6. Figure 6.18 is an nmr spectrum of a mixture of dimethyl fumarate and dimethyl maleate. Calculate the percentage of each isomer in this mixture.

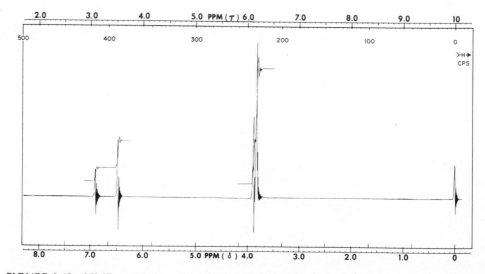

FIGURE 6.18 NMR spectrum of a mixture of dimethyl maleate and dimethyl fumarate.

B. HYDROGENATION
1. Draw the structure of the product of catalytic hydrogenation of
    (a) 1,2-dimethylcyclohexene
    (b) *cis*-2,3-dideuterio-2-butene
    (c) *trans*-2,3-dideuterio-2-butene.
2. Is the same product obtained from (b) and (c) of question 1?
3. Why does the addition of hydrochloric acid to an aqueous solution of a carboxylic acid decrease the solubility of the carboxylic acid in water?

SPECTRA OF STARTING MATERIALS AND PRODUCTS

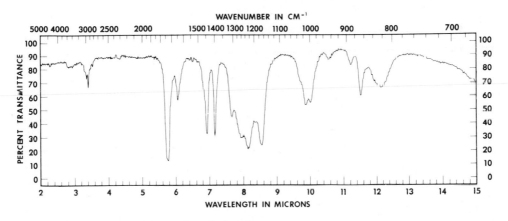

FIGURE 6.19   IR spectrum of dimethyl maleate.

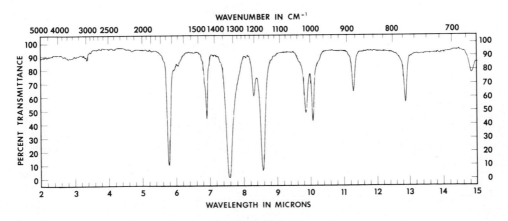

FIGURE 6.20   IR spectrum of dimethyl fumarate.

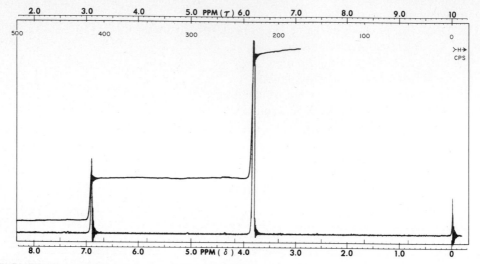

FIGURE 6.21   NMR spectrum of dimethyl fumarate.

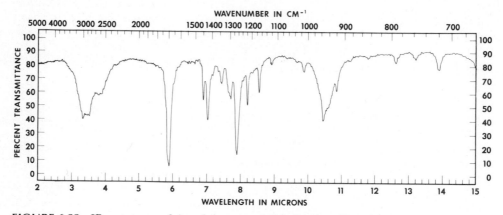

FIGURE 6.22   IR spectrum of 4-cyclohexene-*cis*-1,2-dicarboxylic acid.

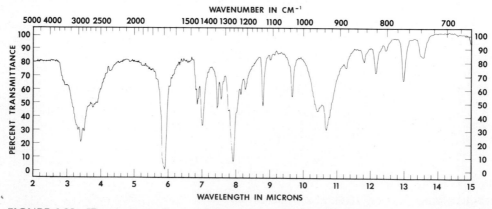

FIGURE 6.23   IR spectrum of *cis*-cyclohexane-1,2-dicarboxylic acid.

chapter seven

studies of
isomerism with
molecular models

Isomers are compounds which have the same composition, *i.e.*, the same molecular formula, but have different properties. There are two main classes of isomers: (1) *constitutional isomers,* and (2) *stereoisomers.* These two main classes have been subdivided in various ways by different chemists, which may lead to confusion, as we shall note.

We shall discuss isomers and isomerism in terms of the following outline:

A. Constitutional isomers
    1. Skeletal isomers
    2. Positional isomers
    3. Functional isomers
B. Stereoisomers
    1. Conformational isomers (conformers)
    2. Configurational isomers
        (a) Dissymmetric (chiral) molecules
        (b) Symmetric (achiral) molecules

The three types of *constitutional isomers* are closely related, and all of them are sometimes referred to as "structural isomers." This is ambiguous, however, for the way in which stereoisomers differ from one another and from constitutional isomers is also a matter of structure.

As an example of *skeletal isomers,* consider *n*-butane and isobutane:

$$CH_3—CH_2—CH_2—CH_3 \qquad CH_3—CH \Big\langle {{CH_3} \atop {CH_3}}$$

n-Butane   Isobutane

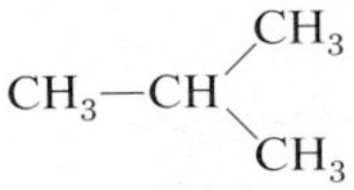

Both have the molecular formula C_4H_{10}, but they are *constituted* differently in that *n*-butane has a continuous chain of four carbon atoms in its carbon skeleton, whereas isobutane has a skeletal structure which is branched.

n-Propyl chloride and isopropyl chloride (both C_3H_7Cl) are examples of *positional*

$$CH_3—CH_2—CH_2—Cl \qquad CH_3—\underset{\underset{Cl}{|}}{CH}—CH_3$$

n-Propyl Chloride   Isopropyl Chloride

isomers. The position of the chlorine atom in the three-carbon chain is different in the two isomers. The similarity to skeletal isomerism is obvious; it is only the difference in the position of a chlorine atom, rather than a methyl group, which produces different structures. The distinction between skeletal and positional isomers becomes clear upon noting that the *carbon* skeletons of positional isomers are identical.

Functional isomers may be illustrated by ethyl alcohol and methyl ether (both C_2H_6O).

$$CH_3—CH_2—O—H \qquad CH_3—O—CH_3$$

Here it is the difference in the position of the oxygen atom which produces isomeric molecules with *different functional groups.* Thus, skeletal and positional isomers will display similar chemical properties, whereas functional isomers differ not only in molecular constitution but also may vary widely in their chemical behavior.

Stereoisomers are those which have the same composition and constitution but differ in the orientation of their atoms in space. One way in which the atoms of covalent molecules may assume different positions in space is by the rotation of parts of the molecule about single bonds, for example, the C—C bond in ethane. Two of the infinite number of *conformations* possible for ethane molecules are illustrated by perspective and Newman projection formulas.

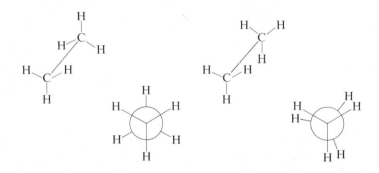

The differences in energy separating *conformational isomers* ("conformers") are usually too small to allow their separation. Compounds capable of assuming two or more distinct conformations are typically found as equilibrium mixtures of these conformational isomers.

Configurational isomers are those in which the different spatial arrangements of the isomeric molecules cannot be produced by rotation about single bonds but require that bonds be broken and remade. The tetrahedral covalency of carbon leads to two different arrangements in space (configurations) of four different atoms (or groups of atoms) attached to a single carbon atom. Such a carbon atom is called *asymmetric*, because the molecule lacks all elements of symmetry. The configurational isomers resulting from the two arrangements have the relationship of an object and its mirror image and are called *enantiomers*. Enantiomers have the unusual physical property of rotating the plane of polarized light in opposite directions. For this reason they have been called "optical isomers." The enantiomers of 2-chlorobutane illustrate this type of configurational isomers.

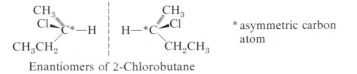

* asymmetric carbon atom

Enantiomers of 2-Chlorobutane

There are some organic molecules which do not contain asymmetric carbon atoms, and yet they may have two different configurations that are nonsuperimposable mirror images of one another. In all such cases the stereoisomers will be optically active. Any molecule whose mirror image is nonsuperimposable with itself is said to be *dissymmetric*; this is the necessary requirement for optical activity.[1] Dissymmetric molecules are also called *chiral*, from the Greek word for hand, because the hands are perhaps the best-known illustration of mirror image dissymmetric objects.

Symmetric molecules may also have different configurations. In contrast to chiral molecules, some of these may be represented adequately in two dimensions. Ethylenic *cis,trans* isomers belong to this category.

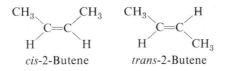

cis-2-Butene   trans-2-Butene

They are also often called *geometric isomers*, which is somewhat misleading, because all stereoisomers differ in their geometry. Certain symmetrical cyclic compounds exhibit isomerism which is very similar to that of ethylenic *cis,trans* isomers, *e.g.*, 1,3-dimethylcyclobutane.

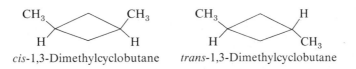

cis-1,3-Dimethylcyclobutane   trans-1,3-Dimethylcyclobutane

---

[1]A dissymmetric molecule may or may not be asymmetric. An example of a dissymmetric molecule that has no asymmetric atoms and that actually has an axis of symmetry is an appropriately substituted

allene such as $\overset{Cl}{\underset{H}{\diagdown}}C=C=C\overset{H}{\underset{Cl}{\diagup}}$.   See footnotes 5 and 6.

Of course, cyclic molecules may also be dissymmetric, resulting in an apparent overlap between *cis,trans* isomers and "optical isomers."[2] For example, *cis*-1,2-dimethylcyclobutane is a symmetric (achiral) molecule having a plane of symmetry as shown, but *trans*-1,2-dimethylcyclobutane is dissymmetric (chiral) and exists in two mirror-image, optically active forms (enantiomers).

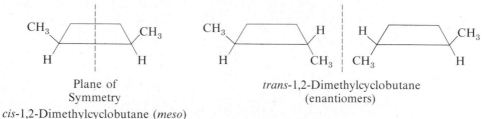

Plane of
Symmetry
cis-1,2-Dimethylcyclobutane (*meso*)

trans-1,2-Dimethylcyclobutane
(enantiomers)

Thus, it may be seen that it is possible for molecules with multiple asymmetric carbon atoms to be symmetric, and hence not optically active. Such compounds are called "*meso.*" *cis*-1,2-Dimethylcyclobutane is an example. Tartaric acid provides the classic example; there are three stereoisomers of tartaric acid, a pair of optically active enantiomers and the achiral *meso* form.

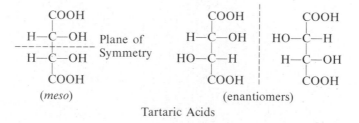

(*meso*)   (enantiomers)
Tartaric Acids

It should be noted that the relationship between *cis*-1,2-dimethylcyclobutane and the optically active *trans*-isomers is similar to that between *meso* and optically active tartaric acids. Stereoisomers which are not enantiomers are called *diastereomers*. This definition includes *cis,trans* isomers of the ethylenic as well as the cyclic types. Thus, *cis*- and *trans*-2-butene are diastereomers, as are *cis*-1,2-dimethylcyclobutane and either of the *trans*-1,2-dimethylcyclobutane enantiomers, and *meso*-tartaric acid and either of the two optically active tartaric acid enantiomers.

In the examples of diastereomers given in the preceding paragraph, one of the isomers is *meso*. There are many cases, however, in which none of the diastereomers is *meso*; for example, there are four stereoisomers of 3-chloro-2-butanol, two pairs

---

[2] It is for this reason that we have not used "optical isomers" as a class name. The term "optical isomers," if it is used at all, should be reserved for those stereoisomers which are optically active; for example, *meso* compounds are often described ambiguously under the heading of "optical isomers" (see the discussion of *meso* compounds below).

of enantiomers, and all are optically active. The *erythro*-isomers are diastereomers of the threo-isomers.[3]

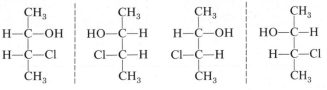

erythro-3-Chloro-2-butanol   threo-3-Chloro-2-butanol

EXPERIMENTAL PROCEDURE

A. Constitutional isomers

The laboratory instructor will assign one or more of the following problems to each student:

1. All hexane isomers (C_6H_{14}).
2. All dichlorobutanes ($C_4H_8Cl_2$) produced by introduction of another chlorine atom into 2-chlorobutane.
3. All dichloropentanes ($C_5H_{10}Cl_2$) produced by introduction of another chlorine atom into:
   (a) 2-chloropentane
   (b) 3-chloropentane
4. All isomers of C_5H_{10}.
5. All isomers of $C_4H_{10}O$.

In your notebook, draw in condensed form all of the possible constitutional isomers possible in each case. Then, for each isomer, construct a three-dimensional molecular model.[4] Draw in your notebook, for each isomer, a three-dimensional picture of the molecule. An example is shown at the top of p. 146.

[3]The terms *erythro* and *threo* are derived from the structurally related four-carbon sugars erythrose and threose.

CHO   CHO   CHO   CHO
H—C—OH   HO—C—H   HO—C—H   H—C—OH
H—C—OH   HO—C—H   H—C—OH   HO—C—H
CH₂OH   CH₂OH   CH₂OH   CH₂OH

D- and L-Erythrose   D- and L-Threose

[4]We have no particular preference for the type of model kit to be used. Several different types which are satisfactory and are priced inexpensively are available. Examples are: Framework Molecular Models, sold by Prentice-Hall, Inc., Englewood, New Jersey; Student Molecular Models from E. H. Sargent & Co.; and IIGS Molecular Models from W. A. Benjamin, Inc., New York.

$$CH_3CH_2CH_2CH_3 \qquad CH_3\underset{\underset{\displaystyle CH_3}{|}}{C}HCH_3$$

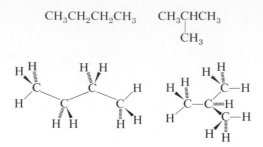

For each problem, indicate which pairs of isomers are skeletal, positional, and functional isomers.

B. STEREOISOMERS

1. **Conformational isomers.**   The laboratory instructor will assign one or more of the following molecules for examination:
   (a) $ClCH_2$—CH_2Cl
   (b) $CHBr_2$—$CHBr_2$
   (c) $(CH_3)_2CH$—CH_2CH_3
   (d) cyclohexane and *cis*-1,2-dimethylcyclohexane
   (e) cyclohexane and *trans*-1,3-dimethylcyclohexane
For (a), (b), or (c) construct a model and study the interactions between nonbonded atoms which result from rotation about the bond shown by a line in those formulas. In your notebook, do exercises (i) through (iii) below for the molecule.

If (d) or (e) is assigned, construct models for each molecule and do exercises (iv) and (v) in your notebook.

(i) Draw a potential energy versus angle of rotation curve, which shows your best estimate of the relative energies of the conformations assumed by the molecule upon rotation about the indicated bond. It will not be possible to compute these relative energies exactly; however, it should be possible to estimate qualitatively the relative heights of peaks and depths of troughs on your curve.

(ii) Using the models as guides, draw Newman-type projection formulas for each of the conformations of the molecule which correspond to a peak or trough on the rotation curve from (i).

(iii) From these results, explain the reasoning that led you to establish, from among many possibilities, the conformation you have shown to be the most stable (lowest-energy). Explain, in like fashion, your choice of the least stable (highest-energy) conformation.

(iv) Examine the model for cyclohexane in the chair, boat, and twist conformations. Using the model as a guide, draw Newman-type projection formulas for each of these conformations. A Newman projection for the chair form of cyclohexane is shown below. Note carefully the conformational changes which occur when the chair and boat forms are interconverted, and also when the boat and twist forms are interconverted.

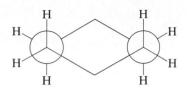

(v) Examine the model for *cis*-1,3-dimethylcyclohexane and/or *trans*-1,2-dimethylcyclohexane in each of two possible chair conformations. Draw Newman projection formulas for both chair conformations which clearly demonstrate any important steric interactions in the molecules. Indicate which chair form is the more stable and show clearly in your drawings what interactions create any difference in stability. Also study the molecule in each of the possible boat conformations. Draw Newman projection formulas for the most stable and the least stable conformations. Show clearly in your drawing the interactions which create this difference in stability.

2. **Configurational isomers.**   The instructor will assign for examination one or more of the problems listed below.
  (a) 2-chloropropane and 2-chlorobutane
  (b) 2-methylpentane and 3-methylpentane
  (c) 1,3-dichlorocyclopentane and 1,3-dichlorocyclobutane
  (d) 2,3,4-trichloropentane

Assemble models for the molecule(s) provided in the assigned problem. Examine the structure in a search for elements of symmetry[5] other than simple axes of symmetry.[6] If none is evident, cause the model to adopt other conformations. Can you locate elements of symmetry, other than simple axes, in *any* of the other attainable conformations?

(i) If you have been unable to locate in your model elements of symmetry which are by definition absent in dissymmetric objects,[6] construct also a model of the apparent mirror image of the original model. Can these models adopt any conformations in which they are superimposable? Draw, in your notebook, three-dimensional formulas for one or two (depending on the results of the above tests) configurational isomers of the compound under study.

(ii) Continue by breaking bonds in the model and reassembling them in a search for as many other *configurational isomers* of the compound as you are able to find. Be careful that all structures assembled remain constitutionally equivalent, *i.e.,* are not constitutional isomers of one another. For each apparently new configurational

[5] A *plane of symmetry* is determined to be present when a line drawn perpendicular to an internal plane of the molecule from any atom encounters an identical atom at an equal displacement on the other side of the plane; *i.e.,* one-half of the molecule is the mirror image of the other half. A *center of symmetry* is determined to be present if a line drawn through any atom and also through the point passes through an identical atom at an equal displacement on the other side of the point. An *n*-fold *simple axis of symmetry* is present if the molecule, after rotation about the axis by $360/n$ degrees, is identical to the original molecule.

[6] Dissymmetric molecules do not contain a plane or center of symmetry. They may, however, contain a simple axis of symmetry. Note also the symmetry axis which bisects the front and rear bonds of the ring in the perspective formula of *trans*-1,2-dimethylcyclobutane, p. 144. See footnote 1.

isomer, apply the tests described above in a search for symmetry or dissymmetry, and for equivalence of the "new" structure to any already recorded in the notebook. Draw three-dimensional formulas in the notebook for any additional configurational isomers located.

(iii) Label, as appropriate, those structures which are *meso;* note which pairs of structures are diastereomeric and/or enantiomeric. Label those which are optically active.

(iv) By reference to a textbook in which the rules for assignment of absolute configurations are given, assign the terms R and S to the configurations at each of the asymmetric centers in the structures that have been drawn.

chapter eight
arenes

Electrophilic Aromatic Substitution

8.1 Introduction to Electrophilic
Aromatic Substitution

Electrophilic aromatic substitution is a very important part of organic chemistry, for the introduction of many functional groups onto an aromatic ring is accomplished in this manner. The general form of this reaction may be represented by equation (1), in which Ar—H is an aromatic compound (an "arene") and E^+ is any one of a number

$$Ar{-}H + E^{\oplus} \rightarrow Ar{-}E + H^{\oplus} \tag{1}$$

of different electrophiles that may replace H. Although equation (1) represents the overall net reaction correctly, it is greatly oversimplified. For example, the electrophile must usually be generated from the starting materials during the course of the reaction.

The chemical kinetics of electrophilic aromatic substitution reactions have been studied extensively. For many, but not all,[1] cases the rate of substitution has been found to be first order in arene and first order in electrophile:

$$\text{Rate} = k_2[Ar{-}H]\,[E^{\oplus}] \tag{2}$$

Because of its bimolecular nature, it is often termed an S_E2 reaction (S = substitution, E = electrophilic, and 2 = bimolecular). A general mechanism for the S_E2 reaction is given below. This mechanism has been substantiated by considerable evidence, especially that coming from isotope effect studies.

Step 1: Formation of Electrophile (an equilibrium reaction):

$$E{-}Nu \xrightleftharpoons{\text{catalyst}} E^{\oplus} + Nu^{\ominus} \tag{3}$$

---

[1] No attempt will be made here to differentiate specifically between the systems which follow the second-order rate expression and those which do not; in general, however, activated arenes do not follow this rate expression.

Step 2:   Reaction of Electrophile with Arene (slow step):

$$\text{Ar}-\text{H} + \text{E}^{\oplus} \underset{\text{slow}}{\rightleftarrows} \left[\text{Ar}\overset{\oplus\diagup\text{H}}{\underset{\diagdown\text{E}}{}}\right] \tag{4}$$

intermediate

Step 3:   Loss of Proton to Give Product (fast step):

$$\left[\text{Ar}\overset{\oplus\diagup\text{H}}{\underset{\diagdown\text{E}}{}}\right] \underset{\text{fast}}{\rightleftarrows} \text{Ar}-\text{E} + \text{H}^{\oplus} \tag{5}$$

The electrophile is most often produced by the reaction between a catalyst and a compound which contains a potential electrophile [equation (3)]. The second-order nature of the reaction may be seen to be related to equation (4), in which *two* molecules react to give the intermediate. If, indeed, the kinetics are second-order, then this must be the slow step in the reaction; the loss of a proton [equation (5)] which follows must be fast relative to the reaction of equation (4).

A listing of various common electrophiles, the conditions under which they are produced, and the reactions which they undergo is given in Table 8.1, along with references to other parts of this book where experiments utilizing them are given. The specific mechanisms, along with detailed information and side reactions, will accompany each of the experiments in the text.

Table 8.1   Examples of Electrophilic Aromatic Substitution Reactions

Type of Reaction	Electrophile, E^+	Electrophile Precursor	Catalyst, if Any [a]	Structure of Product	Reference to this Reaction
Friedel-Crafts Alkylation	R^+	$R-X$	AlX_3	$Ar-R$	Chapter 8.2
Friedel-Crafts Acylation	$R-\overset{\oplus}{C}=O$	$R-\overset{\overset{O}{\|}}{C}-X$	AlX_3	$Ar-\overset{\overset{O}{\|}}{C}-R$	Chapter 19.1, 20
Chlorination	Cl^+	$Cl-Cl$	None	$ArCl$	Chapter 19.3
Bromination	Br^+	$Br-Br$	None	$ArBr$	Chapter 18.3, 19.3
Iodination	I^+	$I-Cl$	None	ArI	Chapter 19.3
Nitration	NO_2^+	$HO-NO_2$	H_2SO_4	$ArNO_2$	Chapter 19.2
Sulfonation	SO_3	$Cl-SO_3-H$	None	$ArSO_3H$	Chapter 19.2

[a]Catalysts listed are those which are used in the experiments referred to in the table; in those reactions where no catalyst is indicated, other examples than those specifically referred to frequently require catalysts.

8.2   Friedel-Crafts Alkylation of Benzene with Butyl Chlorides

The Friedel-Crafts alkylation reaction is one of the classic types of electrophilic aromatic substitution and as such has been subjected to extensive mechanistic study. It is also of great practical importance as the most versatile method for attaching alkyl side chains to aromatic rings. The two main limitations to this reaction as a synthetic tool are (1) the difficulty of preventing the introduction of more than one alkyl group onto an aromatic ring (owing to the activating effect of the first group introduced) and (2) the occurrence of rearrangements of the alkyl group. The first difficulty can often be resolved by using a large excess of the arene (note the proportion of benzene to butyl chloride used in the experiments below). There has been considerable confusion about the occurrence and extent of alkyl group rearrangement; it has sometimes been exaggerated and sometimes overlooked. The following experiments should give some indication of the types of alkylations which take place with and without accompanying rearrangements.

The generally accepted mechanism of alkylation [equations (6), (7), and (8)] involves a carbonium ion or partially ionized complex, shown for simplicity as R^+, as the electrophile.

$$R \quad Cl + AlCl_3 \longleftrightarrow R^{\oplus} + AlCl_4^{\ominus} \tag{6}$$

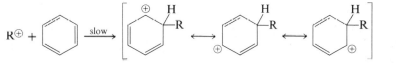

(7)

this resonance-stabilized intermediate may be represented by the single symbol:

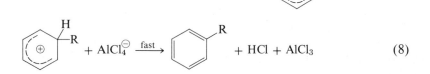

(8)

The reaction is catalyzed by a Lewis acid ($AlCl_3$) which assists in pulling the chlorine atom from the alkyl chloride to give the carbonium ion (R^+) and $AlCl_4^-$. Observe that the $AlCl_3$ does indeed serve as a catalyst, for it is regenerated in the final step.

Since a carbonium ion intermediate is involved, rearrangement is to be expected in some but not necessarily all cases (for carbonium ion rearrangements accompanying elimination reactions, see Chapter 6). For example, rearrangement of an unstable (high energy) carbonium ion to a more stable (lower energy) carbonium ion is to be expected, but a rearrangement which involves little decrease in energy or requires a higher-energy intermediate is not necessarily to be expected. The extent of rearrangement is also dependent on other factors such as the nature of the arene (when other than benzene), the temperature, the solvent (if any), and the nature and concentration of the catalyst.

This type of rearrangement can be illustrated by the reaction between *n*-propyl chloride, aluminum chloride, and benzene. In this reaction, the *n*-propyl cation (a primary carbonium ion) can rearrange to the isopropyl cation (a secondary carbonium ion), which is more stable. The rearrangement is not complete, as is seen by examination of the products; both *n*-propyl- and isopropylbenzene are formed, in the ratio of 35:65. These facts are explained by the following formulation:

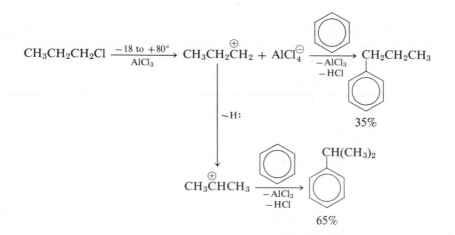

From the experiments which are given in this section, you will be able to demonstrate that this type of rearrangement is observed in alkylations with some isomeric butyl chlorides.

Rearrangements of the side chain may also occur after attachment to the aromatic ring. By this we do not mean the possible sequence of reversal of alkylation (dealkylation), rearrangement of carbonium ion, and realkylation, but rather a direct rearrangement of the intact alkyl benzene. For example, the rearrangement of *t*-pentylbenzene to 2-methyl-3-phenylbutane is explained by a carbonium ion chain mechanism, as follows:

$$
\underset{\underset{CH_3}{\overset{CH_3}{|}}}{C_6H_5-C-CH_2-CH_3} + R^{\oplus} \longrightarrow \underset{\underset{CH_3}{\overset{CH_3}{|}}}{C_6H_5-C-\overset{\oplus}{C}H-CH_3} \xrightarrow{\sim CH_3:} \underset{\underset{CH_3}{\overset{CH_3}{|}}}{C_6H_5-\overset{\oplus}{C}-CH-CH_3}
$$

$$+ \; RH$$

$$\downarrow RH$$

$$
R^{\oplus} + \underset{\underset{CH_3}{\overset{CH_3}{|}}}{C_6H_5-CH-CH-CH_3}
$$

The R$^+$ may be produced initially from the reaction of RX or a trace of alkene with AlCl$_3$.

The following experiments with the butyl chlorides provide illustrations typical of the kinds of behavior described above: alkylation without rearrangement, alkylation with rearrangement at the carbonium ion stage, and rearrangement of the alkylbenzene after side chain attachment. Analysis of the products of the various reactions by means of gas chromatography (glpc) and nmr spectroscopy should allow you to decide what has occurred in each experiment.

The infrared spectra of the four butylbenzene isomers are given in Figure 8.1. Gas chromatography with several types of column materials (silicone gum rubber, apiezon, or polyesters are satisfactory) will provide good separation of *n*-butylbenzene and *t*-butylbenzene from *sec*-butylbenzene and isobutylbenzene (Figure 8.2). The latter two isomers cannot be separated readily by glpc, but they can easily be distinguished from one another by their infrared or nmr spectra, and the approximate composition of mixtures of these isomers can be estimated.

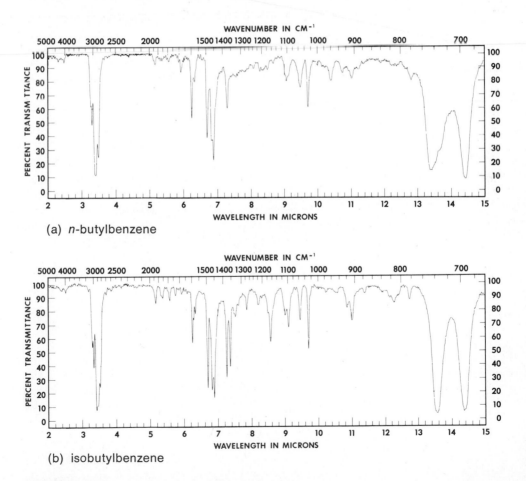

(a) *n*-butylbenzene

(b) isobutylbenzene

FIGURE 8.1 IR spectra of the butylbenzenes.

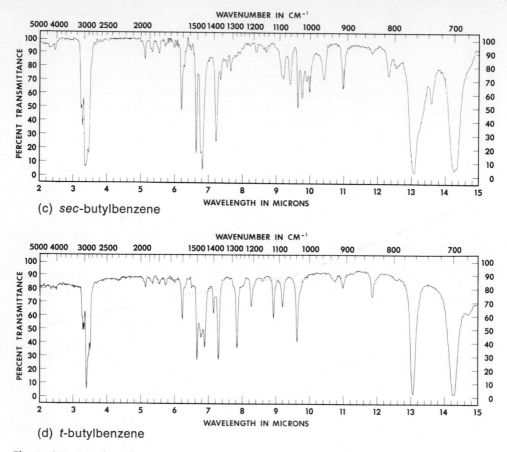

(c) *sec*-butylbenzene

(d) *t*-butylbenzene

Figure 8.1 (continued)

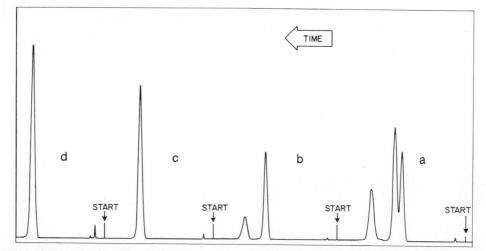

FIGURE 8.2 GLPC data on butylbenzenes. (a) Synthetic mixture of four isometric butyl-benzenes. (b) Fraction II from reaction of *n*-butyl chloride, benzene, and $AlCl_3$ at 27°. (c) Fraction II from reaction of *sec*-butyl chloride, benzene, and $AlCl_3$ at 27°. (d) Fraction II from reaction of *sec*-butyl chloride, benzene, and $AlCl_3$ at 80°.

EXPERIMENTAL PROCEDURE

(*Caution:* If flames are used for heating in this experiment, be careful to keep them away from open flasks containing benzene or the butylbenzenes, as they are highly flammable.)

A. *n*-BUTYL CHLORIDE AT ROOM TEMPERATURE

Provide a 250-ml round-bottomed flask with a reflux condenser equipped at the top with a one-hole rubber stopper carrying a tube leading to a simple gas trap such as shown in Figure 5.1. Clamp the flask high enough on the ring stand so that a cold water bath may be used to cool the reaction mixture as needed.

In the flask place 2.7 g (0.020 mole) of anhydrous powdered aluminum chloride and immediately cover it with 50 ml of dry benzene.[2] Measure 21 ml (18.5 g, 0.20 mole) of *n*-butyl chloride into a 100-ml graduated cylinder. Add about 3 ml of this through the top of the condenser, and reattach the gas trap to the condenser as rapidly as possible. Swirl the reaction flask to mix the reactants; evolution of hydrogen chloride should start almost at once. If the evolution becomes too vigorous, raise the cooling bath momentarily around the flask. As soon as the reaction subsides, add another 3-ml portion of *n*-butyl chloride through the condenser. Continue the addition intermittently with swirling and cooling, if necessary, so that all of the *n*-butyl chloride is added within about 15 min. Swirl the reaction mixture frequently during an additional 15 min, and then pour it into a 250-ml beaker containing 40 g of crushed ice. After stirring the mixture until all the ice has melted, pour it into a 125-ml separatory funnel and drain off the lower (aqueous) layer and discard it. Pour the organic layer into an Erlenmeyer flask containing about 3 g of *anhydrous* calcium chloride (4–6 mesh) and swirl it for about 5 min.★ Filter the dried solution into a 100-ml round-bottomed flask and distil it through a short Vigreux column using a water-cooled condenser. Place a wire gauze beneath the flask and, after making sure that all connections are tight, use a burner flame for heating. Distil excess benzene and forerun into one receiver (I) until the distillation temperature reaches 167°, then extinguish the flame and allow the liquid in the Vigreux column to drain into the flask.★ Replace the Vigreux column by a simple distilling head and resume distillation, collecting the fraction boiling between 167° and 185° in a second receiver (II). (If any forerun boiling below 167° is obtained in the second distillation, add it to the first receiver.) When the distillation temperature reaches 185°, stop the heating, allow the residue to cool, and pour it into running water in the sink.

Record the weights of fractions I and II. Submit samples of each for gas chromatographic analysis and infrared spectroscopic analysis. Using the weights of fractions I and II and the analytical results, calculate the yield of butylbenzene(s) and estimate the proportions of any isomers identified.

---

[2] If dry benzene is not provided, you may dry ordinary benzene by azeotropic distillation. Take an amount about 25% in excess of the total amount of benzene you need for these experiments and distil it, using a dry flask. After the excess benzene has been distilled out, or whenever all signs of turbidity (due to water) in the drops of distillate have disappeared, the main part of the benzene (in the *distilling flask*, not the receiver) may be used in the alkylation reactions. Be sure it is cool (room temperature) before using it.

B. *sec*-BUTYL CHLORIDE AT ROOM TEMPERATURE

Carry out this experiment exactly as in A, except use *sec*-butyl chloride in place of *n*-butyl chloride.

C. *sec*-BUTYL CHLORIDE AT 80°

Carry out this experiment exactly as in B except for the following modifications. Attach a Claisen connecting tube at the top of the reflux condenser. Insert a separatory funnel in the opening of this tube directly above the reflux condenser, and connect the gas trap of Figure 5.1 to the other opening. Heat the mixture of benzene and aluminum chloride to gentle reflux, and add the *sec*-butyl chloride dropwise from the separatory funnel during the course of 15 min. Continue heating an additional 30 min after the addition is complete. Allow the reaction mixture to cool to room temperature and then treat it as in A.

D. ISOBUTYL CHLORIDE AT ROOM TEMPERATURE

Carry out this experiment exactly as in A, except use isobutyl chloride in place of *n*-butyl chloride and use different distillation temperature ranges, as follows: fraction I, up to 162°, fraction II, 162–172°.

EXERCISES

1. Prepare in chart form a listing of the different products which are formed from each of the reactions above. Below each of these products, indicate the approximate amount which is produced. Account for the formation of these possible products using mechanisms. Pay particular attention to possible carbonium ion rearrangements of the types outlined in the discussion section and indicate where you think these are important.

2. What method would you use to prepare pure *n*-butylbenzene, *i.e.*, without the necessity of separating it from isomers by chromatography?

3. What products would you expect from alkylation of benzene with aluminum chloride and each of the following alkyl chlorides at 25°?
   (a) *t*-butyl chloride
   (b) 2-chloropentane
   (c) neopentyl chloride
   (d) 3-methyl-1-chlorobutane

4. Figure 8.2 shows gas chromatograms obtained in the authors' laboratories: (a) a synthetic mixture of the four isomeric butylbenzenes; (b) fraction II from reaction of *n*-butyl chloride, benzene, and $AlCl_3$ at 27°; (c) fraction II from reaction of *sec*-butyl chloride, benzene, and $AlCl_3$ at 27°; (d) fraction II from reaction of *sec*-butyl chloride, benzene, and $AlCl_3$ at 80°.

   Figure 8.3 gives the infrared spectra of fraction II from (a) reaction of *n*-butyl chloride, benzene, and $AlCl_3$ at 27°; (b) reaction of *sec*-butyl chloride, benzene, and $AlCl_3$ at 27°; (c) reaction of *sec*-butyl chloride, benzene, and $AlCl_3$ at 80°.

   Figure 8.4 gives the nmr spectra of fraction II from (a) reaction of *sec*-butyl chloride, benzene, and $AlCl_3$ at 27° and (b) reaction of *sec*-butyl chloride, benzene, and $AlCl_3$ at 80°.

   From the data presented in Figures 8.2, 8.3, and 8.4, determine the identity and approximate proportions of the products from these alkylations with *n*-butyl chloride and *sec*-butyl chloride.

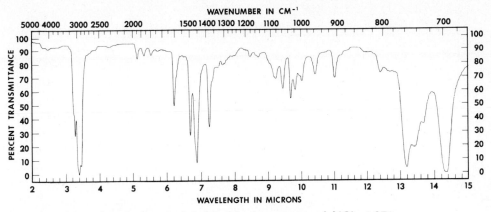

(a) reaction mixture from *n*-butyl chloride, benzene, and AlCl₃ at 27°

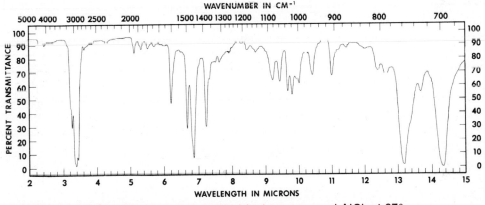

(b) reaction mixture from *sec*-butyl chloride, benzene, and AlCl₃ at 27°

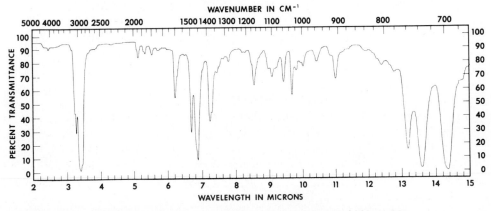

(c) reaction mixture from *sec*-butyl chloride, benzene, and AlCl₃ at 80°

FIGURE 8.3   IR spectra of reaction mixtures from alkylations with butyl chlorides.

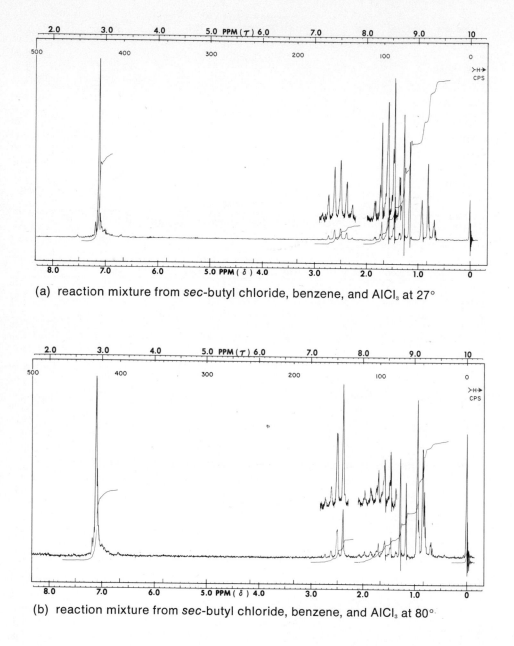

(a) reaction mixture from *sec*-butyl chloride, benzene, and AlCl₃ at 27°

(b) reaction mixture from *sec*-butyl chloride, benzene, and AlCl₃ at 80°

FIGURE 8.4 NMR spectra of reaction mixtures from alkylations with butyl chlorides.

8.3   Relative Rates of Electrophilic
Aromatic Substitution

Electrophilic aromatic substitution reactions are among the best understood of all organic reactions from the mechanistic standpoint. Most of the information which has been obtained about aromatic substitution is internally consistent. The qualitative principles of the reaction are usually discussed in organic chemistry textbooks. These include the effect of a substituent already present in an aromatic compound upon the reactivity of that compound towards a new, entering substituent and upon the orientation of the entering substituent (*ortho*, *meta*, or *para*). Normally, however, the subject of reactivity is discussed in terms of activation or deactivation of the molecule with respect to benzene, and little, if any, information is given about the *quantitative* differences in rates and reactivities of substituted aromatic compounds. The purpose of this discussion and accompanying experimental work is to give the student some feeling for the magnitude of the actual differences in reactivity.

Table 8.2[a]   Relative Rates of Reactivity of Toluene toward Electrophilic Substitution

Type of Reaction	*Relative Rate* [b]
Bromination	605
Chlorination	340
Acetylation	128
Nitration	24.5
Sulfonation	31.0
Methylation	2.9

[a] From L. M. Stock and H. C. Brown, *Advances in Physical Organic Chemistry*, V. Gold, editor, Academic Press, New York, 1963, Vol. I, pp. 50–52.
[b] The rates here are relative rates, where the rate of substitution on benzene has been taken as 1.0. Mathematically, the relative rate may be defined as:

$$\text{Relative Rate} = \frac{\text{Rate of Substitution on Toluene}}{\text{Rate of Substitution on Benzene}}$$

The values quoted were obtained at *ca.* 25° or have been corrected to 25° by extrapolation.

Table 8.2 lists some data which have been collected for some common types of electrophilic substitution reactions. Most of the reactions were carried out under identical conditions, so that a direct comparison of the numbers is reasonably valid. As mentioned in the footnote of Table 8.2 the rates are for toluene *relative* to benzene; for example, bromination is about 605 times more rapid for toluene than for benzene. Indeed, it can be seen that all electrophilic substitution reactions of toluene are faster than those for benzene, which supports the idea presented in most textbooks that the methyl group *activates* an aromatic ring and makes it more reactive toward an electrophile.

Table 8.3 gives some data for the bromination of different monosubstituted benzene derivatives. The rates are again relative to benzene (R = H in C_6H_5R), so that these results indicate the relative reactivities of the various substituted benzenes. Those exhibiting a rate greater than 1.0 have substituents already present which activate the aromatic ring toward further substitution, whereas those whose rate is less than 1.0 have substituents already present which deactivate the ring.

It can be seen from the data in Table 8.3 *how much* more activating one group is compared to another. In particular, it should be pointed out that phenol (R = OH) is *about 10^{16} times more reactive* toward bromine than is nitrobenzene (R = NO_2). Differences in reactivity can be seen only from data of this sort, and the student is advised to examine Table 8.3 carefully.

As has been stated before, the two most important steps in electrophilic aromatic substitution reactions are (1) the attack of the electrophile on the electron-rich aromatic ring and (2) loss of a proton [see equations (4) and (5) in Section 8.1]. The intimate details of the mechanism, particularly with regard to the rate expression, differ with various substrates and reaction conditions. Part of the problem lies in the role and concentration of the catalyst in these reactions, and some differences in mechanism are believed to be related to the speed with which the electrophile, E^+, is generated and consumed.

The measurement of reaction rates makes it necessary to review some of the factors

Table 8.3[a]   Rates of Bromination for Various Aromatic Compounds

Substituent R in C_6H_5R	Relative Rate [b]
—OH	6.1×10^{11}
—OCH_3	1.8×10^9
—$NHCOCH_3$	2.1×10^8
—OC_6H_5	1.7×10^7
—C_6H_5	1.0×10^3
—CH_3	6.05×10^2
—C_2H_5	4.6×10^2
—$CH(CH_3)_2$	2.6×10^2
—$C(CH_3)_3$	1.38×10^2
—H	1.0 (standard)
—X (*e.g.*, Cl, Br)	0.5 (estimated value only)
—CO_2H	7.5×10^{-3}
—$\overset{+}{N}(CH_3)_3$	1.6×10^{-5}
—NO_2	1.6×10^{-5}

[a] From L. M. Stock and H. C. Brown, *Advances in Physical Organic Chemistry*, V. Gold, editor, Academic Press, New York, 1963, Vol. I, pp. 62–80.

[b] Rates are relative rates, where the rate of substitution on benzene has been taken as 1.0. The values were obtained at *ca.* 25° with bromine in aqueous acetic acid or have been corrected to 25° by extrapolation.

which influence them (see Chapter 12.3 for a brief summary). Some of the variable factors are temperature, concentration of reactants, solvent composition, and nature and concentration of catalysts. Since it is possible to change any of these variables while keeping all other factors constant, the effect of all of the variables can be determined.

In this experiment, it will be possible for the student to measure rates and rate differences for an electrophilic aromatic substitution reaction. The reaction chosen for study here is the bromination by elemental bromine of a series of typical aromatic compounds. The rate expression for this reaction is given by equation (9), in which Br_2 has

$$\text{Rate} = k_2[\text{ArH}][\text{Br}_2] \tag{9}$$

been inserted in place of E^+ of equation (2) in Section 7.1, since bromine is clearly the precursor to the actual electrophile. The rate expression could be more complex under certain conditions. Bromination of less reactive aromatic compounds requires the use of a Lewis acid catalyst for reaction at a reasonable rate. Water, in *trace* amounts, has been found to accelerate the rate of these electrophilic reactions. However, by avoiding the use of a catalyst and by carrying out the brominations in a mixed solvent composed of 90% acetic acid and 10% water,[3] both of these complications are excluded from the rate studies. The rate expression of equation (9) states that the rate of the reaction is proportional to the concentration of the aromatic substrate and to the concentration of bromine, and as written is a second-order rate expression. While a student could, in principle, work with this rate expression, the calculations are much more difficult and time-consuming than are those for first-order reactions. Thus, the experimental procedure is designed to provide a "pseudo-first-order" reaction to simplify the calculations. This is accomplished by maintaining the concentration of the aromatic substrate at a high level so that it does not change appreciably throughout the course of the reaction. Under these conditions the rate expression becomes

$$\text{Rate} = k_1[\text{Br}_2] \tag{10}$$

where k_1 is a new rate constant which will contain all of the factors of constancy for the reagents and solvents.[4]

The integrated form of equation (10) is

$$[\text{Br}_2]_t = [\text{Br}_2]_o\, e^{-k_1 t} \tag{11}$$

---

[3] This much water in the solvent insures a concentration which remains essentially constant during the course of the reaction and hence does not enter into the rate expression.

[4] Rate constant k_1 is called a *pseudo*-first-order rate constant. Since all other factors are being held constant, the reaction rate will *appear* to depend only on the concentration of bromine, as the rate expression indicates. *Pseudo*-rate constants in general refer to reactions in which the concentration of one or more reactants which appear in the rate expression are chosen so that they do not appreciably change as the reaction proceeds. Thus they remain constant and are automatically incorporated in the pseudo-rate constant.

which can be rearranged to

$$\ln \frac{[Br_2]_o}{[Br_2]_t} = k_1 t \tag{12}$$

or

$$2.303 \, \log \frac{[Br_2]_o}{[Br_2]_t} = k_1 t \tag{13}$$

where $[Br_2]_o$ = initial concentration of bromine, and $[Br_2]_t$ = concentration of bromine at any time t.[5] From the experimental data, one can prepare a graph of $\log[Br_2]_o/[Br_2]_t$ vs. time, t. If the reaction is first-order, the graph should show a reasonably straight line with a slope of $k_1/2.303$ if common logarithms are used. It is important to note that $[Br_2]_t$ is the concentration of bromine *remaining* at time t, and not the amount of bromine which has reacted.

The reaction temperature and initial concentration of reactants must be controlled. The use of a large water bath in the experiment and careful preparation of the reagents will insure that these factors are kept constant.

One of the main reasons for choosing bromination reactions for carrying out rate studies is that it is easy to follow the rate of disappearance of the bromine color as the reagent reacts with the aromatic substrate. The virtue of choosing reaction conditions so that the rate of the reaction is proportional only to the concentration of the bromine should become quite apparent.

An important limitation must be placed on the types of aromatic compounds to be studied. As no Lewis acid catalyst is present, the study is by necessity limited to compounds which are more reactive than benzene. Benzene and compounds of similar or lower reactivity would react so slowly with bromine in the absence of a catalyst that it would not be possible to obtain good kinetic data in a single laboratory period. This limitation presents no serious problem, for it will still be possible to determine the relative reactivities of a series of compounds and to illustrate the principles which are involved in these sorts of measurements.

Experiments are described which allow one to obtain qualitative differences in reactivity. Depending on the facilities available, either semi-quantitative or quantitative measurements of absolute rates may also be determined.

The qualitative experiments involve preparing solutions of various substrates in 90% acetic acid containing bromine. From the length of time required for the bromine color to be discharged, one can deduce the relative reactivities.

The semi-quantitative experiments involve preparing a series of solutions of 90% acetic acid which contain known concentrations of bromine. If one then prepares a solution containing an excess of substrate and a known concentration of bromine, a rough estimate of the rate may be made by visual comparison of reaction mixture color with those of the known solutions.

---

[5] See Chapter 12.3 for a more detailed discussion of these equations for first-order rate constants. The equations are very similar to those which are presented for the hydrolysis of tertiary halides.

The quantitative experiments require the use of an inexpensive visible/ultraviolet spectrophotometer,[6] which allows accurate measurement of the *rate of disappearance of bromine*. This instrument is non-recording, but the data can be read from a meter. If a compound strongly absorbs light at a given wavelength, then the ability of light to pass through a sample of that compound is diminished. The spectrophotometer measures the amount of light which has been absorbed, this value being called absorbance. Absorbance is defined as

$$\text{Absorbance } (A) = \log \frac{I_o}{I} \tag{14}$$

and is often called *optical density*. In equation (14), I_o is the intensity of incident light and I is the intensity of emergent light at a given wavelength. If a sample absorbs strongly, the ratio of I_o/I will be large and A will be large. The absorbance also depends on variables other than the nature of the absorbing species, these being the *concentration* of the absorbing species and the *length* of the cell which contains the solution. The absorbance is *proportional* to the cell length and to the concentration, and this relationship, called the *Lambert-Beer Law*, may be written as

$$A = kcl \tag{15}$$

where A is the absorbance, k the proportionality constant, c the concentration of absorbing species, and l the path length of the solution. In these experiments you will use the same cell throughout, and the absorbing species (bromine) will remain the same, so that l and k will remain constant. Thus, if the absorptivities, A_1 and A_2, of two solutions of concentrations c_1 and c_2 are measured, equations (16a) and (16b) may be written and combined to give equation (17).

$$A_1 = kc_1l \tag{16a}$$

$$A_2 = kc_2l \tag{16b}$$

$$\frac{A_1}{A_2} = \frac{kc_1l}{kc_2l} = \frac{c_1}{c_2} \tag{17}$$

It may then be seen that if the concentration of one solution is known and the measurements of absorptivity made, the concentration of the other solution may be determined.

Additional instructions about treatment of data and use of equations will be given in the experimental procedure.

EXPERIMENTAL PROCEDURE

In the following experiments, it would be best to have two students working together.

A. QUALITATIVE RATE COMPARISONS

Prepare a water bath using a one-liter beaker, and adjust the temperature to about 35°. During the experiments which follow, keep the temperature at $35 \pm 2°$. Place

[6] Several relatively inexpensive instruments are available, such as the Bausch and Lomb "Spectronic 20" (about $250.00) which can be used in this experiment.

2.0 ml of 0.2 M solutions of each of the following substrates (contained in 90% acetic acid) in separate small test tubes: (a) phenol; (b) anisole; (c) phenyl ether; (d) acetanilide; (e) p-bromophenol; (f) benzene; and (g) α-naphthol. If stock solutions are provided, use special care to insure that one solution is not contaminated by another. Suspend the carefully labeled test tubes partially in the water bath by looping a piece of copper wire around the neck of the test tube and over the rim of the beaker.

Prepare some droppers with a capacity of 2.0 ml. Calibrate these by comparison with known volumes, and mark the droppers by scratching the glass with a file. These will be used to introduce the bromine solution into the test tubes containing the substrates. It is essential that the addition be done as rapidly as possible and that the volumes be fairly accurate. Transfer about 45 ml of a solution containing 0.05 M bromine in 90% acetic acid to an Erlenmeyer flask and allow the solution to equilibrate in the water bath for a few minutes. Add 2.0 ml of the bromine solution to one of the test tubes containing a substrate. Make the addition rapidly, mix the solution quickly, and note the exact time of addition. Observe the reaction mixture and note how long it takes for the bromine color to become faint yellow or to disappear. Repeat this procedure with each of the substrates, making sure that you use the same end-point color for each one.

When the reaction is slow; *i.e.*, no decolorization occurs within 5 min; go on to another compound while waiting for the end-point to be reached. Record the reaction times, and on the basis of these observations, arrange the compounds in order of increasing reactivity towards bromine.

If it is impossible to determine the relative rates accurately at 35°, repeat the experiment with the compounds in doubt at 0°, using an ice-water bath.

B. SEMI-QUANTITATIVE RATE MEASUREMENTS

Obtain stock solutions of each of the following prepared with 90% acetic acid: 0.05 M bromine, 0.5 M p-nitrophenol, 0.5 M acetanilide, and 0.5 M phenyl ether. Prepare standard color comparison solutions by diluting the 0.05 M stock bromine solution according to the directions in Table 8.4. Fill small test tubes with *ca.* 4 ml of each of these solutions. Adjust the levels of solution in each tube so that all are filled to a common depth. Arrange the tubes in order of decreasing concentration

TABLE 8.4   Preparation of Standard Bromine Solutions for Color Comparison

| Solution | Mix | | Concentration |
	Bromine Solution	Solvent[a]	
A	3.0 ml of stock[b]	3.0 ml	$2.50 \times 10^{-2}\ M$
B	3.0 ml of stock	9.0	1.25×10^{-2}
C	6.0 ml of B	6.0	6.25×10^{-3}
D	6.0 ml of C	6.0	3.13×10^{-3}
E	6.0 ml of D	6.0	1.56×10^{-3}
F	6.0 ml of E	6.0	7.81×10^{-4}
G	3.0 ml of F	3.0	3.91×10^{-4}

[a] 90% acetic acid in water.
[b] Stock bromine solution (0.05 M).

of bromine in a test tube rack; place a white background beneath them as an aid in color discernment.

Calibrate an additional test tube of the same size, measuring the volume to the nearest 0.1 ml necessary to fill it to the same depth as the comparison tubes. Add to this tube one-half the calibrated volume of one of the stock substrate solutions. In a second tube place this same volume of the 0.05 M bromine solution. Working with the aid of a partner, pour the measured bromine solution into the calibrated test tube containing the substrate solution and stir with a small stirring rod to insure homogeneity; note and record the time of mixing to the nearest second. Continue with one partner marking the time that the color of the reaction mixture most closely matches each standard bromine solution, and the second partner reading and recording the times to the nearest second. The colors of the solutions should be compared by viewing the tubes vertically through the length of the tubes. Repeat, using each of the other substrate solutions, and any others that the instructor may assign. As time allows, perform duplicate determinations on each substrate; to minimize temperature effects, carry out all measurements during the same laboratory period.

Treatment of data

1. From the color comparison data and the bromine concentrations in each standard solution given in Table 8.4, plot, for each substrate, the molar concentration of bromine remaining (vertically) versus time (seconds, horizontally). Include as an additional point the initial concentration of bromine, $2.5 \times 10^{-2} M$, at time, $t = 0$.
2. Draw the best curve that you can through the points which you have drawn. Draw a line tangent to the curve where it crosses the $t = 0$ point. Determine the slope of this line. These slopes, when compared for the various substrates, give an estimate of the *initial relative rates* and thus of the relative reactivities.
3. The rate expression for bromination [equation (10)] may be rewritten as follows:

$$\text{Rate of reaction} = \frac{\Delta[\text{Br}_2]}{\Delta t} = k_1[\text{Br}_2]$$

where $\Delta[\text{Br}_2]/\Delta t$ = the rate of disappearance of bromine.

When the reaction has just started, the percentage change in the bromine concentration is very small compared to the total amount present. Under these conditions, it is possible to substitute the *initial rate* determined in part 2 for $\Delta[\text{Br}_2]/\Delta t$. Since we know the initial rate and we know the initial concentration of bromine, it is possible to substitute these values into the above rate equation and solve for k_1, the rate constant. Do this for each substrate.

C. QUANTITATIVE MEASUREMENTS

Obtain three sample tubes for use with the spectrophotometer (these tubes resemble small test tubes, but the dimensions and glass thickness are carefully controlled; they are called "Cuvette" tubes). As all solutions absorb light to some extent, though they appear transparent to the naked eye, it is necessary to determine the absorbance of 90% acetic acid alone (without bromine or substrate) so that the absorbance of the solvent may be canceled out in all future measurements. Add 4 ml of 90% acetic acid to one Cuvette and stopper it. This is to be used as a blank for the experiment; your instructor will show you how to use the spectrophotometer and how to set it to zero absorbance for the blank. Determine the blank reading and all future readings at 520 mμ, the wavelength at which elemental bromine absorbs light.

Have at your disposal stock solutions (0.5 M) of acetylsalicylic acid (aspirin), phenyl ether, and acetanilide in 90% acetic acid, as well as a 0.02 M solution of bromine in 90% acetic acid. It is necessary, in preparing for and executing the following experiments, that you be well organized so that you can work quickly. It may take several tries before you are able to work fast enough to get reasonable results. Before starting, equilibrate several 10-ml samples of each of the stock solutions at 35° (using a water bath).

Allow 2.0 ml of the stock solution of aspirin to equilibrate at 35° in a stoppered Cuvette. Now, rapidly and in the order listed, carry out the following operations: (a) open Cuvette; (b) add 2.0 ml of the equilibrated bromine solution; (c) stir quickly, one time only; (d) replace cork tightly; (e) remove Cuvette from water bath and dry it with paper towel; (f) place Cuvette in spectrophotometer, alongside the blank. A second person should note the exact time of mixing, so that this can be used as zero time.

Take readings at 30, 60, 90, 120, 150, and 180 sec. If an error is suspected, repeat the experiment since it will take only a little time. Remove the Cuvette and place it in the water bath. Take absorbance readings on the sample every 15 min for the remainder of the laboratory period.

Repeat this experiment for phenyl ether and acetanilide. With these substrates, absorbance readings should be made every 10 sec for the first minute and every 30 sec for the next 4 min. Continue taking readings as long as there is a change in absorbance; however, this will normally not last more than 5 min. If possible, perform a duplicate run on the latter two substrates.

Treatment of data

1. Plot, for each substrate, the data which you have obtained, with absorbance (vertically) *vs.* time (seconds, horizontally). Extrapolate each curve to zero time; the point at which the curve crosses the zero time line is the A_o value (absorbance at time $t = 0$).

2. Divide the values of A for each datum point by A_o to obtain A/A_o ratios. Do this for each substrate, being careful not to intermix A and A_o values from the different runs.

3. Plot on a single sheet of graph paper A/A_o *vs.* time (seconds, from 0 to 500) for each substrate. Draw tangents to each curve at the point of zero time, and from these tangents estimate the *relative initial rates*. (See a discussion of relative initial rates in part B.)

4. Calculate, for each substrate, A_o/A values. (These correspond to $[Br_2]_o/[Br_2]_t$.) Determine ln A_o/A or log A_o/A, and plot these values *vs.* time. The slope is the rate constant in the case of natural logarithms, and 2.303 times the rate constant for common logarithms. The fact that a straight line is obtained here is further proof that the reaction is indeed first-order in bromine. Determine the rate constants for the bromination of each substrate studied and determine the relative reactivities from these.

chapter nine

alkynes

Hydration

Alkynes are quite similar to alkenes (see Chapter 6) insofar as their reactions are concerned. In particular, alkynes undergo electrophilic *addition* reactions with reagents such as E—Nu [equation (1)]. Using carefully controlled reaction conditions, it is possible to stop the addition reaction at the alkene stage in some cases. As with alkenes, the

$$R-C\equiv C \quad R' \xrightarrow[1\ \text{mole}]{E-Nu} R-\underset{\underset{E}{|}}{C}=\underset{\underset{Nu}{|}}{C}-R' \xrightarrow[1\ \text{mole}]{E-Nu} R-\underset{\underset{E}{|}}{\overset{\overset{E}{|}}{C}}-\underset{\underset{Nu}{|}}{\overset{\overset{Nu}{|}}{C}}-R' \qquad (1)$$

electrophilic addition of unsymmetrical reagents to terminal alkynes follows Markownikoff's Rule, and the less electronegative atom of the reagent becomes attached to the carbon bearing the hydrogen [equation (2)]. Note that in the addition of the second mole of E—Nu to the alkene, the same orientation is followed as in the initial addition of E—Nu to the alkyne [equations (1) and (2)].

$$R-C\equiv C-H \xrightarrow[1\ \text{mole}]{E-Nu} R-\underset{\underset{Nu}{|}}{C}=\underset{\underset{E}{|}}{C}-H \xrightarrow[1\ \text{mole}]{E-Nu} R-\underset{\underset{Nu}{|}}{\overset{\overset{Nu}{|}}{C}}-\underset{\underset{E}{|}}{\overset{\overset{E}{|}}{C}}-H \qquad (2)$$

Acetylene, the simplest alkyne, is a widely used commercial organic compound, but because it is a gas at room temperature, it is not experimentally suitable for investigating the properties and reactions characteristic of the carbon-carbon triple bond. We shall study an alkyne having a terminal triple bond and therefore having chemical properties quite similar to those of acetylene. This compound, 2-methyl-3-butyn-2-ol (**1**) has the advantages of being inexpensive and readily available commercially, as well as easily handled.[1]

---

[1] Although this molecule contains a hydroxyl group, its presence has little effect on the chemical properties of the carbon-carbon triple bond. The main effect of the hydroxyl group is on the physical properties of the molecule, such as melting and boiling point. Acetylenic *hydrocarbons* having the same molecular weight would melt and boil much lower.

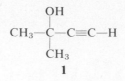

1

As with alkenes, alkynes readily decolorize a solution of bromine in carbon tetra-chloride. Either one or two moles of bromine may react per mole of alkyne, with two moles being required for complete reaction. Another common reaction which is used to detect the presence of unsaturation is the Baeyer test, in which a solution of potassium permanganate is decolorized.

The acetylenic hydrogen of a terminal alkyne ($-C{\equiv}C-H$) is relatively acidic compared to the hydrogens in alkanes and alkenes. The acidity of hydrocarbons has been studied extensively, and it has been found that the order of decreasing acidity is as follows:

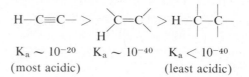

This acidity is an important property of terminal alkynes, and a simple test for this group is the ready formation of a precipitate on addition of a solution containing silver ammonia complex (silver nitrate dissolved in ammonia). Terminal alkynes give a solid silver salt, which results when the terminal hydrogen is removed by ammonia acting as a base and replaced by silver ion. This reaction is shown in equation (3) in general terms, and compound **1** is a representative example of a compound which would so react.

$$R-C{\equiv}C-H + Ag(NH_3)_2^{\oplus} \longrightarrow \underline{R-C{\equiv}C-Ag} + NH_3 + NH_4^{\oplus} \qquad (3)$$
a silver acetylide

This unique reaction of the terminal alkynes provides an easy method for separating them from nonterminal alkynes. The former form insoluble silver salts, whereas the latter do not and remain in solution. The silver acetylide salts can be reconverted to the terminal alkynes on treatment with hydrochloric acid, as shown in equation (4). *Dry* silver salts of this type are quite sensitive to shock and in this condition tend to decompose *explosively*. Care should be taken to insure that they never dry out before decomposition by hydrochloric acid.

$$\underline{R-C{\equiv}C-Ag} + HCl \longrightarrow R-C{\equiv}C-H + \underline{AgCl} \qquad (4)$$

A very useful reaction involving the carbon-carbon triple bond is its hydration to give an aldehyde or ketone. For example, the hydration of acetylene is a commercial method of preparation of acetaldehyde. The addition of water to acetylene is catalyzed by mercuric sulfate in the presence of sulfuric acid. The reaction proceeds via the initial formation of vinyl alcohol, which is unstable and tautomerizes to acetaldehyde, as shown by equation (5). With substituted acetylenes, hydration occurs in accordance

$$H—C\equiv C—H + H_2O \xrightarrow[\text{HgSO}_4]{\text{H}_2\text{SO}_4} \left[CH_2{=}C\begin{smallmatrix}OH\\\\H\end{smallmatrix} \right] \rightleftharpoons CH_3—C\begin{smallmatrix}O\\\\H\end{smallmatrix} \qquad (5)$$

vinyl alcohol
(unstable)

with Markownikoff's Rule (the proton going to the carbon atom already bearing the greater number of hydrogens), and the product is a ketone. 2-Methyl-3-butyn-2-ol **(1)** is a monosubstituted acetylene which, under typical hydration conditions, will give a ketone, 3-hydroxy-3-methyl-2-butanone **(2)**, as shown by equation (6).

$$\underset{\textbf{1}}{CH_3\overset{HO}{\underset{CH_3}{—C—}}C{\equiv}C—H} + H_2O \xrightarrow[\text{HgSO}_4]{\text{H}_2\text{SO}_4} \underset{\text{(unstable)}}{\left[CH_3\overset{HO}{\underset{CH_3}{—C—}}\overset{OH}{C}{=}CH_2 \right]} \rightleftharpoons \underset{\textbf{2}}{CH_3\overset{HO}{\underset{CH_3}{—C—}}\overset{O}{C}—CH_3} \quad (6)$$

The hydration reactions often require specific catalysts, some of which are cuprous, mercuric, and nickelic ions, but of these mercuric ion is used most often. The role these catalysts play in the reaction is not completely understood. It is felt that they may form a π-complex with the triple bond and thus render the alkynes more soluble in the polar aqueous solvent. It is also possible that these metal ions actually *add* directly to the triple bond and are then removed in a subsequent displacement reaction.

The hydration of **1** will be carried out using mercuric sulfate as the catalyst in the presence of sulfuric acid.[2] *Dilute* acid is used to eliminate the problem of having a highly exothermic reaction.

It has been asserted above that the addition of water (an unsymmetrical reagent) to **1** (an unsymmetrical alkyne) will occur according to Markownikoff's Rule to give **2** as the final product. This can readily be verified by identifying the product of the hydration experiment. If the water added in an anti-Markownikoff fashion, 3-methyl-3-hydroxybutanal **(3)** would be formed.

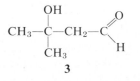

3

In order to distinguish between the two possible products, a solid derivative of the product can be prepared for identification purposes. (See Chapter 25 for additional information concerning identification of organic compounds through preparation of

[2] The experimental procedure is a modification of that reported by N. C. Rose in *Journal of Chemical Education*, **43**, 324 (1966).

derivatives.) One suitable solid derivative for carbonyl-containing compounds is the semicarbazone (Chapter 16.1). Compounds **2** and **3** have been prepared by independent methods, and their semicarbazone derivatives have been made; the semicarbazone (**4**) of ketone **2** melts at 162.5°, whereas aldehyde **3** first dehydrates under the conditions of semicarbazone formation, and the resulting unsaturated aldehyde, $(CH_3)_2C\text{=}CHCHO$, then reacts to form a semicarbazone (**5**), which melts at 222–

$$(CH_3)_2\underset{\underset{HO}{|}}{C}\text{—}\underset{\underset{CH_3}{|}}{C}\text{=}\underset{\underset{O}{\|}}{N}NHCNH_2 \qquad (CH_3)_2C\text{=}CH\text{—}\underset{\underset{H}{|}}{C}\text{=}\underset{\underset{O}{\|}}{N}NHCNH_2$$

4   **5**

223°. Thus, if the semicarbazone of the product is prepared and its melting point determined, the principal mode of hydration may be deduced.

EXPERIMENTAL PROCEDURE

A. QUALITATIVE TESTS FOR THE TRIPLE BOND

Reaction with bromine in carbon tetrachloride.   See the experimental procedure given in Chapter 6.1. Follow the same procedure for the qualitative test for the carbon-carbon triple bond in 2-methyl-3-butyn-2-ol.

Baeyer test for unsaturation.   See the experimental procedure given in Chapter 6.1, and follow it for the reaction of 2-methyl-3-butyn-2-ol with potassium permanganate.

Reaction with silver ammonia complex; formation of a silver acetylide and its decomposition.   Prepare a solution of silver ammonia complex from 5 ml of 0.1 M silver nitrate solution by adding ammonium hydroxide solution dropwise. Brown silver oxide forms first; add *just* enough ammonium hydroxide to dissolve the silver oxide. Dilute the solution by adding 3 ml of water. Add 3 ml of the diluted silver ammonia complex solution to about 0.1 ml of 2-methyl-3-butyn-2-ol.

Note the formation of the silver acetylide salt. (If you have at your disposal a disubstituted acetylene, treat it similarly and note any reactions which occur.) Filter the silver salt from the aqueous solution; be careful not to let it dry, for the dry salt is quite explosive. Treat the silver salt with a small amount of dilute hydrochloric acid, and observe what changes occur, especially in the color and form of the precipitate. Ultimately, destroy *all* solid salt by treatment with hydrochloric acid.

B. PREPARATION OF 3-METHYL-3-HYDROXY-2-BUTANONE;
THE HYDRATION OF 2-METHYL-3-BUTYN-2-OL

Add 18 ml of *concentrated* sulfuric acid *carefully* to 115 ml of water contained in a 500-ml round-bottomed flask. Dissolve 1 g of mercuric oxide in the resulting warm solution [3] and then cool the flask to about 50°. Attach a reflux condenser to the flask,

---

[3] It has been found that the quality (purity) of the mercuric oxide can greatly affect the yield of the hydration reaction. Use a good grade of mercuric oxide or prepare fresh mercuric oxide by adding sodium hydroxide to mercuric chloride solution; ask your instructor for further details.

and then add *in one portion* 12.6 g (0.15 mole) of 2-methyl-3-butyn-2-ol through the condenser. The precipitate, which forms immediately, is presumably the mercury complex of the alkyne. Shake the reaction flask to mix the contents. When this is done, an exothermic reaction ensues as the precipitate dissolves, and the solution turns light brown. Allow the reaction to proceed by itself for about two minutes and then heat the mixture to reflux.[4] As soon as the reflux begins, remove the heat and cool the reaction mixture to 50°. Add an additional 12.6 g of alkyne in one portion through the condenser. A precipitate again forms, and the contents of the flask should be mixed by shaking the flask. After several minutes, heat the mixture to reflux and continue heating for 15–20 min. Equip the flask for simple distillation, add 100 ml of water to the reaction mixture, and distil. Be especially careful during the collection of the first 75 ml of distillate, since foaming occurs. Continue distillation until a total of about 150 ml of distillate has been collected.* Transfer the distillate to a separatory funnel, add about 25 g of potassium carbonate sesquihydrate to the distillate, and then saturate the solution with sodium chloride. (A second layer may form; if so, do not separate it but continue with the extraction as described.) Extract the mixture with three 35-ml portions of benzene (technical grade) and combine the extracts. Dry the organic extracts over 8 g of *anhydrous* potassium carbonate, filter, and distil. After removal of the benzene, collect the product boiling between 138 and 141°. The average yield is between 10 and 15 g; the product should be a colorless liquid.

C. Identification of the Product: Spectroscopic Methods

The infrared and nmr spectra of both the starting material (1) and the product (2) are provided at the end of this chapter. While it is possible to identify the product by converting it to a known solid derivative (see below), it is also quite easy to prove the structure using spectral methods. An examination of the infrared spectra provided will indicate the absence of the C=O group in 1, whereas an intense absorption for this group is clearly present in 2. As an exercise and with the aid of Table 4.1, identify the absorption attributed to this functional group. The nmr spectra will also show a terminal acetylenic hydrogen for 1, whereas the hydrogens of the newly formed CH_3 group in CH_3C=O will be observed for 2. Identify the peaks in the nmr spectra, using Table 4.2.

D. Identification of the Product: Semicarbazone Formation

Add 1 ml of the product to a solution prepared from 1 g of semicarbazide hydrochloride and 1.5 g of sodium acetate dissolved in 5 ml of water. Shake or stir vigorously. The solid which forms is the crude semicarbazone. Recrystallize about one-third of the crude solid as follows. Add to it about 5 ml of benzene, heat the mixture on a steam bath, and add 95% ethanol dropwise until all of the solid has dissolved. Cool and collect the crystals by filtration. Determine the melting point and use this information to identify the product obtained by hydration of 2-methyl-3-butyn-2-ol.

---

[4] It may be that the exothermic reaction will not begin on its own; if it does not, heat the mixture until it starts to reflux. Then proceed as directed.

172   Alkynes

EXERCISES

1. It was stated that 2-methyl-3-butyn-2-ol is commercially available and is inexpensive. Suggest a method of preparation of this compound from two simple inexpensive organic compounds.
2. How might you separate, and ultimately obtain in pure form, 1-octyne and 2-octyne from a mixture containing both of them?
3. Give the structures for the products which you would expect to obtain on hydration of 1-octyne and of 2-octyne.
4. It is well known that alcohols can be oxidized to ketones by oxidizing agents such as potassium permanganate. An examination of alkyne 1 will show that it has a hydroxyl group as well as a triple bond. Therefore, a student might conclude that the positive permanganate test is due to the oxidation of the hydroxyl group and not to reaction of the triple bond. Could this be the case? If so, what additional "control" experiments might you suggest to eliminate this possibility from further consideration?
5. Consider the compounds hexane, 1-hexene, and 1-hexyne. What similarities and what differences would you expect in the reactions of these compounds with: (1) bromine in carbon tetrachloride; (2) an aqueous solution of potassium permanganate; (3) an aqueous solution of sulfuric acid; (4) a solution of silver ammonia complex? Give the structures of the products, if any, which would be obtained from these reactions.
6. Suggest a reason for adding the potassium carbonate and saturating the solution with sodium chloride in the work-up procedure of part B.

SPECTRA OF STARTING MATERIAL AND PRODUCT

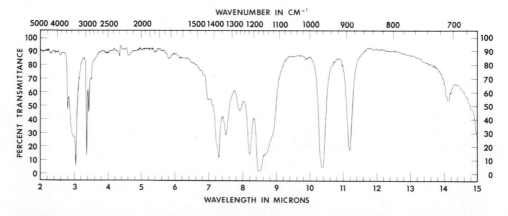

FIGURE 9.1   IR spectrum of 2-methyl-3-butyn-2-ol.

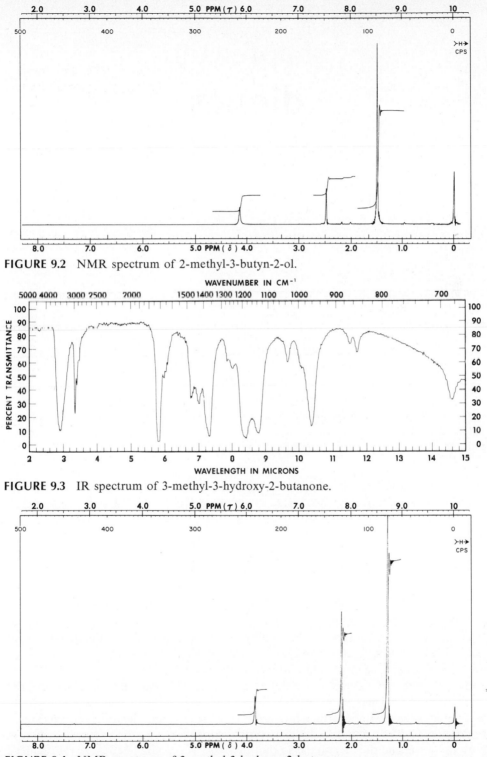

FIGURE 9.2 NMR spectrum of 2-methyl-3-butyn-2-ol.

FIGURE 9.3 IR spectrum of 3-methyl-3-hydroxy-2-butanone.

FIGURE 9.4 NMR spectrum of 3-methyl-3-hydroxy-2-butanone.

chapter ten

dienes

The Diels-Alder Reaction

One of the more useful synthetic methods available to the organic chemist is the reaction between a 1,3-diene and an alkene, sometimes referred to as the dienophile (Gr., *philos*, loving), to produce a derivative of cyclohexene [equation (1)]. If an alkyne is used as the dienophile, instead of an alkene, a derivative of 1,4-cyclohexadiene will be formed [equation (2)]. The reaction, which is termed the Diels-Alder reaction in honor of its primary developers, Otto Diels and Kurt Alder, results in the formation of new carbon-carbon bonds between the two π-bonded carbons of the dienophile and the 1- and 4-carbons of the diene. Thus, there is an overall 1,4-addition of the dienophile to the diene.

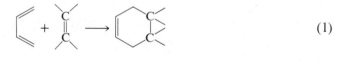

$$(1)$$

$$(2)$$

The Diels-Alder reaction is of quite general utility, and many different dienes and dienophiles have been employed to afford good yields of adducts. The presence of electron-releasing substituents such as alkyl and alkoxy (—OR) on the diene and/or electron-withdrawing groups like cyano (—CN) and carbonyl (—CO—) on the dienophile seems to enhance the yield obtained in the reaction.

Even though considerable research has been directed toward the elucidation of the mechanism of the Diels-Alder reaction, a detailed description of the reaction pathway is still not available. In deriving a mechanism for the reaction, the following facts must be considered. The reaction exhibits first-order dependency upon the concentration of both the diene and the dienophile so that the process is second-order overall. Kinetic measurements show that the rate of the reaction is normally not significantly changed by the addition of catalysts such as acids, bases, or free radicals, by irradiation with light of various wavelengths,[1] or by the phase (gas or liquid) in which the reaction is performed. The evidence suggests that the transition state of the reaction is constructed from a single molecule of each of the reactants, and that neither highly polar intermediates, *e.g.*, carbonium ions, nor free radicals are involved in the mechanistic pathway linking starting materials and products. A mechanism consistent with these observations is one in which bond-breaking and bond-making occur simultaneously in the transition state, so that little polarity or free-radical character is developed [equation (3)].[2]

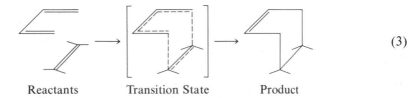

$$\tag{3}$$

| Reactants | Transition State | Product |

There is one important limitation to the Diels-Alder reaction which is a result of the concerted nature of the process, and it is the requirement that the diene must be capable of attaining an *s-cis* conformation,[3] which is the geometry needed to give a *cis* double bond in the cyclic product. Reaction of a dienophile with a diene in an *s-trans* conformation would lead to a *trans* double bond which is not possible in a six-membered ring for geometric reasons (models are helpful here).

s-cis   *s-trans*

The strict orientation in the transition state of the dienophile relative to the diene might lead one to expect high stereoselectivity in the Diels-Alder reaction, and this expectation is fulfilled. As an example, the adduct resulting from reaction of *trans, trans*-2,4-hexadiene and a dienophile is exclusively *cis*-3,6-dimethylcyclohexene [equation (4)], whereas the product obtained when *cis, trans*-2,4-hexadiene is utilized is *trans*-3,6-dimethylcyclohexene.

[1] There are important exceptions to this, particularly with compounds in which the diene grouping is part of an aromatic system. As an example, it is possible to accomplish a Diels-Alder reaction between maleic anhydride and benzene by *irradiation* of a mixture of the two; note that it is generally not possible to cause benzene to participate in thermally induced Diels-Alder reactions because its aromatic stabilization is lost if a dienophile adds 1,4- to the ring. Students interested in investigation of the above-mentioned photochemical Diels-Alder reaction should consult reference 6 at the end of this chapter.

[2] Theoretical analysis of the nature of the molecular orbitals in the diene and the dienophile and in the product resulting from their condensation has shown that a continuous transformation of the ground-state orbitals of the reagents into those of the Diels-Alder adduct is possible. Thus, the reaction is said to be allowed as a concerted thermal process, and its mechanism is said to be controlled by orbital symmetry. For a more complete discussion of the relationships between the symmetry of molecular orbitals and the mechanisms of organic reactions, see reference 7 at the end of this chapter.

[3] The *s-cis* conformation of a 1,3-diene has the two conjugated double bonds on the same side of the single bond linking them. They are on opposite sides in the *s-trans* conformation.

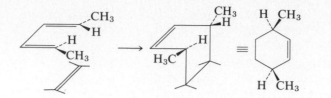

(4)

A second type of stereoselectivity is observed for the reaction. The addition of a dienophile such as maleic anhydride (**1**) to a *cyclic* diene like cyclopentadiene (**2**) could, in principle, provide two products, **3** and **4** [equation (5)]. However, only a single adduct, **3,** is observed experimentally. The explanation for this stereoselectivity is not entirely clear, but one attractive theory is that stabilization, *i.e.,* a lowering in energy, of the transition state **5** leading to formation of **3,** is provided by interaction between the *p*-orbitals of the diene and those of the dienophile. Analogous stabilization is not possible in the transition state **6** required for production of **4.**

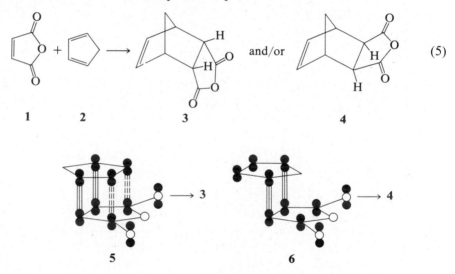

(5)

It should be noted that all Diels-Alder reactions are not as stereoselective as that between cyclopentadiene and maleic anhydride so that mixtures of products are obtained. The product resulting from interaction of the diene and dienophile in a transition state analogous to **5** generally predominates over that arising from a transition state similar to **6.**

The Diels-Alder synthesis is remarkably free of complicating side reactions, and the yields of the desired product are often nearly quantitative. Probably the single most important side reaction observed is dimerization of the diene used, since one molecule of the diene can serve as the dienophile and another as the diene [equation (6)]. Such a reaction is usually important only in those instances in which a poor dienophile like ethylene is employed.

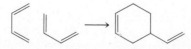

(6)

EXPERIMENTAL PROCEDURE

A. PREPARATION OF *endo*-NORBORNENE-*cis*-5,6-DICARBOXYLIC ANHYDRIDE

The cyclopentadiene required for this experiment is not commercially available as such because it readily dimerizes at room temperature to produce dicyclopentadiene [equation (7)]. Fortunately the equilibrium between the monomer and the dimer can be established at the boiling point of the dimer (170°) so that pure cyclopentadiene can be isolated by fractional distillation. The diene must be kept cold in order to prevent extensive dimerization before it can be used for a Diels-Alder reaction.

$$2 \quad \overset{\text{heat}}{\rightleftharpoons} \qquad\qquad\qquad\qquad (7)$$

Place 20 ml of dicyclopentadiene in a 100-ml flask and attach the flask to an apparatus set for fractional distillation into an ice-cooled receiver. Gently heat the dimer until brisk refluxing occurs, and the monomer begins to distil in the range 40–42°. Distil the monomer as rapidly as possible, but do not permit the temperature of the vapor to exceed 43–45°. Approximately 6 g of monomer should be obtained after distillation for *ca.* 30 min. The distillation should be terminated when approximately 5 g of residue remains in the flask. If the distilled cyclopentadiene is cloudy due to condensation of moisture in the cold receiver, add about 1 g of *anhydrous* calcium chloride to dry the diene.

While the distillation is in progress, place 6 g (0.061 mole) of maleic anhydride in a 125-ml Erlenmeyer flask and dissolve it in 20 ml of ethyl acetate by heating on a steam bath. Add 20 ml of petroleum ether[4] (bp 60–80°) and cool the solution thoroughly in an ice-water bath. To this cooled solution, add 4.8 g (6 ml, 0.073 mole) of dry cyclopentadiene and swirl the resulting solution until the exothermic reaction is complete and the adduct separates as a white solid.★ Heat the mixture on the steam bath until the solid has dissolved and then allow the solution to cool slowly to room temperature.★ Filter the solution and determine the yield and melting point of the solid anhydride obtained. The reported melting point is 164–165°. The yield should be about 80% of theoretical. Test the product for unsaturation (Chapter 6.1).

The anhydride can be converted to the corresponding diacid in the following manner. Place 4 g of the anhydride and 25 ml of distilled water in a 125-ml Erlenmeyer flask, and swirl the flask over a Bunsen burner until boiling occurs and all the oil which initially forms has dissolved. Allow the solution to cool to room temperature and then scratch the inner wall of the flask at the air-liquid interface to induce crystallization.★ After crystallization has begun, cool the flask in ice to complete the process, and filter

---

[4] See footnote 13, p. 45, Chapter 2.

the solution. Determine the yield and melting point of the product. The reported melting point of the expected dicarboxylic acid is 180–182°. Perform appropriate tests to show whether the hydrolysis has destroyed the carbon-carbon double bond. Also test a saturated aqueous solution of the diacid with litmus paper and record the result.

B. REACTION OF CYCLOPENTADIENE WITH *p*-BENZOQUINONE

p-Benzoquinone (7) has two carbon-carbon double bonds which can serve as dienophiles. In this experiment, the Diels-Alder adduct resulting from reaction of two moles of cyclopentadiene with one mole of *p*-benzoquinone will be prepared [equation (8)].

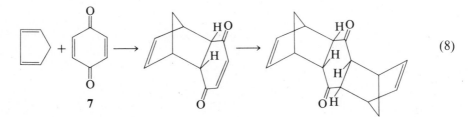

$$(8)$$

Prepare dry cyclopentadiene as described in the preceding experiment. Add a solution of 3.2 g (4 ml, 0.049 mole) of cyclopentadiene and 10 ml of toluene to 2.7 g (0.025 mole) of *p*-benzoquinone[5] dissolved in 20 ml of toluene and swirl the resulting solution until the mildly exothermic reaction has subsided.★ After one hour, cool the reaction mixture in an ice-water bath and filter to isolate the solid adduct.

In order to collect a second crop of product, concentrate the mother liquor to one-third of its original volume by heating on a steam bath, cool the solution, collect the precipitate, and wash it with a few milliliters of ice-cold toluene. Combine the two crops of the diadduct and recrystallize by dissolving the product in a minimum amount of boiling acetone (*Caution:* No flames!) and then adding water until turbidity begins to appear in the boiling mixture. Clarify the solution by addition of a *small* amount of acetone and allow the solution to cool slowly to room temperature. Complete the recrystallization by cooling in an ice-water bath.

The pure adduct forms beautiful irridescent white needles or platelets and has a reported melting point of 157–158°. The yield should be 60–70%.

C. REACTION OF 1,3-BUTADIENE AND MALEIC ANHYDRIDE

1,3-Butadiene is a gas at room temperature (bp −4.4°). Diels-Alder reactions in which this diene is utilized are normally performed in closed steel pressure vessels

---

[5]Commercial *p*-benzoquinone is generally rather impure and may require purification before being used in this experiment. However, a grade of this chemical having a melting point of 113–115° can be used successfully without further purification. *p*-Benzoquinone can be purified, if necessary, by sublimation in an apparatus such as that pictured in Fig. 2.19. Consult your instructor for details of this method of purification.

(autoclaves) into which the diene is introduced under pressure. However, it has been discovered that 1,3-butadiene may be generated conveniently by the thermal decomposition of 3-sulfolene [equation (9)] and that the diene will then react with a dienophile present in the reaction vessel. Such an *in situ* preparation of 1,3-butadiene will be used in this experiment to prepare 4-cyclohexene-*cis*-1,2-dicarboxylic anhydride [equation (10)].

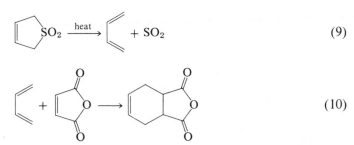

$$\begin{array}{c} \text{(9)} \\ \\ \text{(10)} \end{array}$$

Place 10 g (0.085 mole) of 3-sulfolene, 6 g (0.061 mole) of finely pulverized maleic anhydride and 4 ml of dry xylene in a 100-ml round-bottomed flask equipped with a water-cooled condenser, which is fitted with the gas trap described in Chapter 5 (Figure 5.1). Warm the flask gently while swirling to effect solution, then heat the solution to a gentle reflux with a Bunsen burner; continue the heating for about 30 min. Cool the solution, add 30 ml of benzene and about 1 g of decolorizing carbon, and heat on a steam cone with constant swirling for a few minutes. Filter the hot mixture (*Caution:* No flames!) through a short stemmed funnel, preferably using a fluted filter, and carefully add petroleum ether to the filtrate until cloudiness develops. Set the solution aside to cool to room temperature.★ Collect the crystals by suction filtration and dry them thoroughly. Record their weight and determine the melting point of the product. The reported melting point is 103–104°. Perform qualitative tests for unsaturation and hydrolyze 4 g of the anhydride to the diacid according to the procedure described in part A above. Determine the yield and melting point of the diacid and test it for unsaturation. The reported melting point of 4-cyclohexene-*cis*-1,2-dicarboxylic acid is 164–166°.

EXERCISES

1. Why should 3-sulfolene and maleic anhydride be completely dissolved in the xylene before decomposition of the sulfolene is attempted?
2. Suggest a reason for the observation that cyclopentadiene dimerizes much more readily than do acyclic dienes.
3. Explain why cyclopentadiene reacts more rapidly with *p*-benzoquinone than with another molecule of itself.
4. Write the structure of the product resulting from the reaction of *cis, trans*-2,4-hexadiene and tetramethylethylene.

5. Write the structures of the products expected in the following reactions. If no reaction is to be anticipated, write N.R.

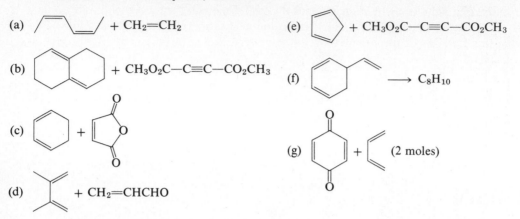

(a) $\diagup\!\!\!\diagdown\!\!\diagup$ + $CH_2{=}CH_2$

(e) ⬠ + $CH_3O_2C{-}C{\equiv}C{-}CO_2CH_3$

(b) (naphthalene-type bicyclic) + $CH_3O_2C{-}C{\equiv}C{-}CO_2CH_3$

(f) (cyclohexenyl vinyl) $\longrightarrow$ C_8H_{10}

(c) (cyclohexadiene) + (maleic anhydride)

(g) (benzoquinone) + (butadiene) (2 moles)

(d) (isoprene) + $CH_2{=}CHCHO$

6. The "cracking" of dicyclopentadiene to two moles of cyclopentadiene [equation (7)] is an example of a reverse Diels-Alder reaction. Predict the products to be anticipated from an analogous reaction with the compounds shown below.

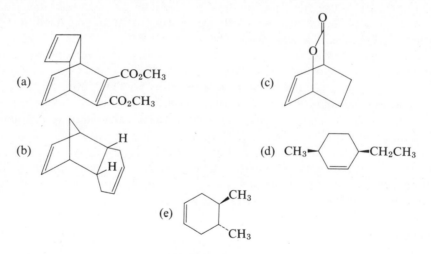

(a) (bicyclic diester with CO_2CH_3, CO_2CH_3)

(b) (bicyclic with H, H)

(c) (bicyclic lactone with O, O)

(d) $CH_3{-}$⬡${-}CH_2CH_3$

(e) (cyclohexene with CH_3, CH_3)

REFERENCES

1. J. Sauer, *Angewandte Chemie, International Edition*, **5**, 211 (1966). A recent general review.
2. M. Kloetzel, in *Organic Reactions*, John Wiley & Sons, Inc., New York, 1948, Vol. IV, Chapter 1.
3. H. L. Holmes, *ibid.*, Chapter 2.
4. L. W. Butz and A. W. Rytina, *ibid.*, Vol. V, 1949, Chapter 3.
5. T. E. Sample, Jr. and L. F. Hatch, *Journal of Chemical Education*, **45**, 55 (1968).
6. R. E. Bozak and V. E. Alvarez, *Journal of Chemical Education*, **47**, 589 (1970).
7. R. B. Woodward and R. Hoffmann, *The Conservation of Orbital Symmetry*, Academic Press, Inc., New York, 1970.

SPECTRA OF STARTING MATERIALS AND PRODUCTS

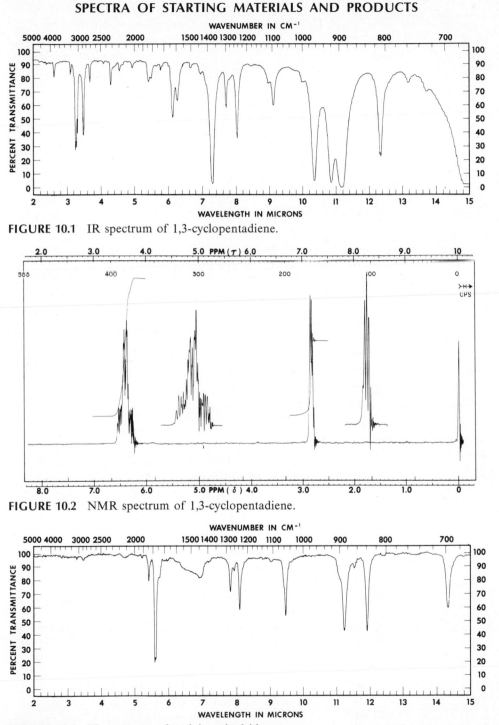

FIGURE 10.1   IR spectrum of 1,3-cyclopentadiene.

FIGURE 10.2   NMR spectrum of 1,3-cyclopentadiene.

FIGURE 10.3   IR spectrum of maleic anhydride.

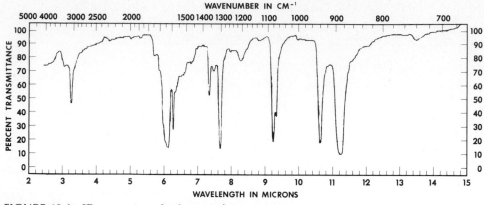

FIGURE 10.4   IR spectrum of *p*-benzoquinone.

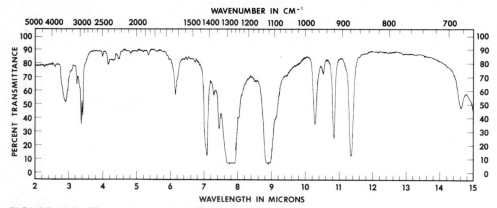

FIGURE 10.5   IR spectrum of 3-sulfolene.

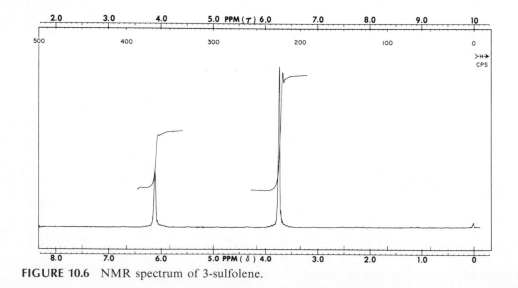

FIGURE 10.6   NMR spectrum of 3-sulfolene.

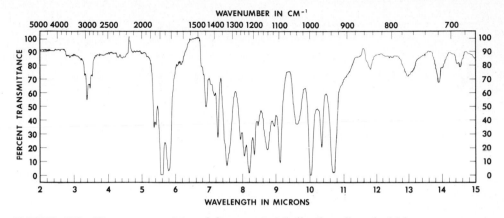

FIGURE 10.7   IR spectrum of 4-cyclohexene-*cis*-1,2-dicarboxylic anhydride.

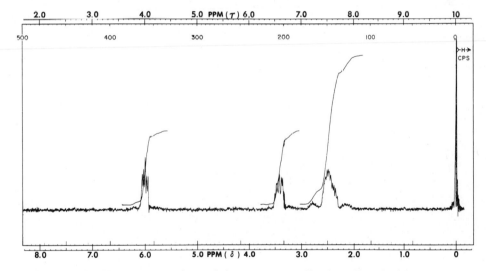

FIGURE 10.8   NMR spectrum of 4-cyclohexene-*cis*-1,2-dicarboxylic anhydride.

chapter eleven

kinetic and equilibrium control of a reaction

In predicting which of two competing reactions will predominate, a useful "rule of thumb" is to choose the reaction which is more exothermic. This rule is based on the fact that *most commonly* the more exothermic reaction will have the smaller heat of activation, and thus will have the greater rate of reaction. For example, Figure 11.1

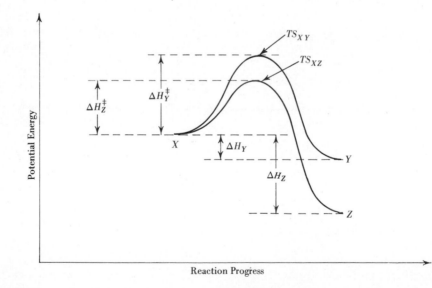

FIGURE 11.1 Typical reaction profile for competing reactions.

depicts the potential energy changes involved in the conversion of compound X into the products Y and Z by two competing reactions [equations (1) and (2)]. The more

$$X \rightleftarrows Y \qquad (1)$$

$$X \rightleftarrows Z \qquad (2)$$

184

exothermic reaction [equation (2)] produces the more stable product, Z (the one lying at the lower energy level). Note that the energy of the transition state leading to Z (TS_{XZ}) is lower than the energy of the transition state leading to Y (TS_{XY}); *i.e.*, the relative energies of the transition states are in the same order as the relative energies of the products to which they lead. Hence, the heat of activation required for production of Z ($\Delta H_Z^\ddagger$) is lower than the heat of activation required for production of Y ($\Delta H_Y^\ddagger$), so that the rate of the reaction producing Z is greater than the rate of the reaction producing Y.

Most organic reactions are either practically irreversible or they are carried out under conditions such that equilibrium between products and starting materials is not attained, so that the yields of products are determined by the relative rates of competing reactions as described above. The major product of such a reaction is said to be the product of *kinetic control*. However, when the experimental conditions are favorable for equilibrium to be reached between starting materials and products, the product that predominates initially because of kinetic control *may not* be the major product after equilibrium has been attained. In some reactions, the product of kinetic control is less stable than another product that is formed at a lower reaction rate. An example of such a system [equations (3) and (4)] is represented by Figure 11.2, in which product

$$X \rightleftharpoons A \qquad\qquad (3)$$
$$X \rightleftharpoons B \qquad\qquad (4)$$

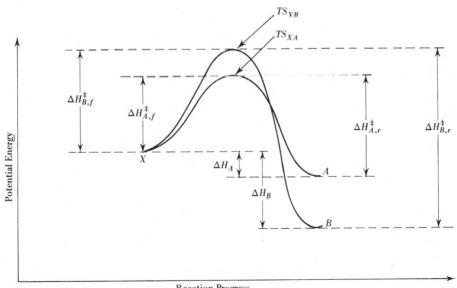

FIGURE 11.2   Atypical reaction profile for competing reactions.

B is shown to be more thermodynamically stable than product A. It should be noted that, in contrast to the more usual relationships depicted in Figure 11.1, the relative energies of the transition states are not in the same order as the relative energies of the products to which they lead. Thus, although product A will predominate initially in the reaction mixture, [1] if the reactions are allowed to come to equilibrium, the more stable product B will be found to be the major product. For this reason it is called the product of *equilibrium control* (or thermodynamic control).

There are many well known reactions of various mechanistic types in which kinetic and equilibrium control lead to different major products under different experimental conditions. For example, the addition of hydrogen bromide to 1,3-butadiene at low temperatures gives the 1,2-adduct, 3-bromo-1-butene, as the main product [equation (5a)], whereas at room temperature the main product is the 1,4-adduct, 1-bromo-2-

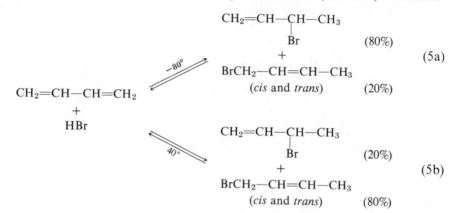

butene [equation (5b)]. The 1,2-adduct is the product of kinetic control, since equilibrium is not attained at low temperatures. The 1,4-adduct, although formed more slowly, is the more stable product, so it predominates after equilibrium has been established at higher temperatures.

Sulfonation of naphthalene at 120° gives mainly 1-naphthalenesulfonic acid [equation (6a)]. However, at temperatures of 160° or above, or with prolonged heating at

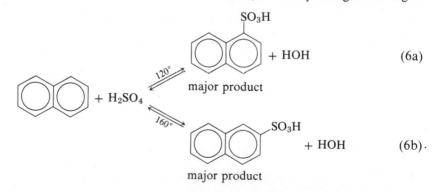

[1] It is interesting to note that although A is produced initially more rapidly than B because of the relative heats of activation for the competing forward reactions ($\Delta H^{\ddagger}_{A,f} < \Delta H^{\ddagger}_{B,f}$), A is also reconverted to starting material (X) more rapidly than B because the heats of activation for the reverse reactions have the same relationship ($\Delta H^{\ddagger}_{A,r} < \Delta H^{\ddagger}_{B,r}$), owing to the greater stability of B than A. This is not true for competing reactions that have the more usual energy relationships represented by Fig. 11.1, so that even when equilibrium conditions are attained in these systems, the same product (Z) will predominate; *i.e.*, the product of kinetic control is also the product of equilibrium control, although the proportion of products is not necessarily the same under all experimental conditions.

lower temperatures, the major product is 2-naphthalenesulfonic acid, the more stable isomer [equation (6b)]. A similar behavior is observed in the sulfonation of phenol, which gives mainly the *ortho*-isomer at lower temperatures [equation (7a)] and mainly the *para*-isomer at higher temperatures [equation (7b)].

Another aromatic substitution reaction which is reversible, and hence subject to either kinetic or equilibrium control, is Friedel-Crafts alkylation. Reaction of toluene with methyl chloride and aluminum chloride at low temperatures gives chiefly *o*- and *p*-xylene [equation (8a)]; at higher temperatures the chief product is *m*-xylene [equation (8b)].

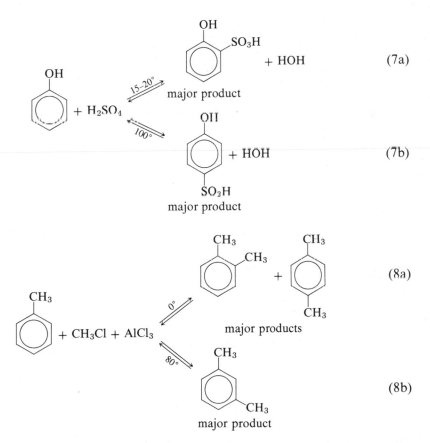

$$+ \text{HOH} \tag{7a}$$

$$+ \text{HOH} \tag{7b}$$

$$\tag{8a}$$

$$\tag{8b}$$

The elimination reaction that occurs when 2-phenyl-2-butyl acetate is heated in acetic acid gives a very different proportion of products when a small amount of a strong acid is added to the reaction mixture [equations (9a) and (9b)]. Apparently the strong acid catalyzes the reaction so that equilibrium is attained, and the equilibrium product mixture contains a much higher proportion of the *cis*- and *trans*-isomers of the internal alkene, which are more thermodynamically stable than the terminal alkene.

The particular example chosen to illustrate the principles of kinetic and equilibrium control of products involves the competing reactions of two carbonyl compounds, cyclohexanone [**2**, equation (10)] and 2-furaldehyde [**4**, equation (11)], with semi-carbazide (**1**). The organic products of these reactions (**3** and **5**), which are known as

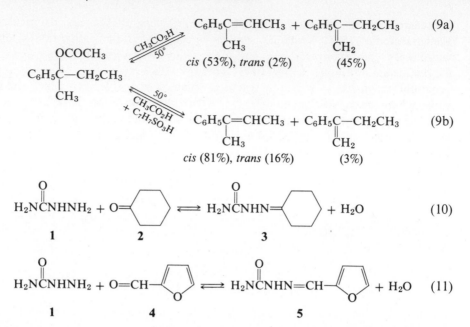

$$C_6H_5\overset{|}{\underset{CH_3}{C}}=CHCH_3 + C_6H_5\overset{|}{\underset{CH_2}{C}}-CH_2CH_3 \quad (9a)$$

$$cis\ (53\%),\ trans\ (2\%) \qquad (45\%)$$

$$C_6H_5\overset{|}{\underset{CH_3}{C}}=CHCH_3 + C_6H_5\overset{|}{\underset{CH_2}{C}}-CH_2CH_3 \quad (9b)$$

$$cis\ (81\%),\ trans\ (16\%) \qquad (3\%)$$

$$\underset{\textbf{1}}{H_2N\overset{O}{\overset{\|}{C}}NHNH_2} + \underset{\textbf{2}}{O=\text{⬡}} \rightleftharpoons \underset{\textbf{3}}{H_2N\overset{O}{\overset{\|}{C}}NHN=\text{⬡}} + H_2O \qquad (10)$$

$$\underset{\textbf{1}}{H_2N\overset{O}{\overset{\|}{C}}NHNH_2} + \underset{\textbf{4}}{O=CH-\text{⌬}} \rightleftharpoons \underset{\textbf{5}}{H_2N\overset{O}{\overset{\|}{C}}NHN=CH-\text{⌬}} + H_2O \quad (11)$$

the semicarbazones of the respective carbonyl compounds, are crystalline solids and have distinctive melting points by which they may be identified easily. For the purposes of the experiment, semicarbazide may be taken to be X of either Figure 11.1 or Figure 11.2, and the problem to be solved is identification of the semicarbazones of cyclohexanone and 2-furaldehyde with either Y and Z or A and B; that is, the experimental determination of which compound is the kinetically controlled product and which is the equilibrium controlled product (they might be the same as in Figure 11.1 or different as in Figure 11.2).

The reactions of carbonyl compounds with nucleophiles such as semicarbazide are subject to significant effects of pH on rates and equilibrium constants.[2] This is because of the different way in which the reactants are affected by reaction with hydrogen ions. The addition of a hydrogen ion to a carbonyl compound [equation (12)] makes the carbonyl carbon atom *more* electrophilic because of the partial positive charge, as shown in the resonance hybrid of the conjugate acid **6**. On the other hand, addition of a hydrogen ion to the nucleophilic semicarbazide [**1**, equation (13)] converts it to its

$$R_2C=O + H^{\oplus} \rightleftharpoons \left[R_2\overset{\oplus}{C}-OH \longleftrightarrow R_2C=\overset{\oplus}{O}H \right] \qquad (12)$$
$$\textbf{6}$$

$$\underset{\textbf{1}}{H_2N\overset{O}{\overset{\|}{C}}NHNH_2} + H^{\oplus} \rightleftharpoons \underset{\textbf{7}}{H_2N\overset{O}{\overset{\|}{C}}NH\overset{\oplus}{N}H_3} \qquad (13)$$

---

[2] Actually, these reactions are affected not only by the concentration of hydrogen ions (of which pH is a measure) but also by the concentration (and nature) of any weak acids that may be present. A discussion of these effects is beyond the scope of this book; further information may be found in textbooks on physical-organic chemistry, in connection with specific and general acid catalysis.

conjugate acid (7) which is *not* nucleophilic. At very low pH (high H$^+$ concentration) the concentration of the nucleophile (1), is reduced and at high pH (low H$^+$ concentration) the concentration of the activated electrophile (6) is reduced. Accordingly, for the reaction of each carbonyl compound with each nucleophilic reagent there is an optimum pH at which the product of the concentrations of the conjugate acid of the carbonyl compound and the nucleophilic reagent is maximized. In order to produce and maintain this optimum pH for reactions of aldehydes and ketones with reagents such as semicarbazide, phenylhydrazine, and hydroxylamine, these reactions are carried out in *buffered* solutions.

In parts A–C of the experimental procedure, the use of a phosphate buffer, which provides reaction rates that are most satisfactory for demonstrating the principles of kinetic and equilibrium control, is recommended. In part F a bicarbonate buffer is employed, providing a higher pH than the phosphate buffer. The results from the experiments of part F should be compared with those obtained in the experiments of part C.

EXPERIMENTAL PROCEDURE

A. In a 50-ml Erlenmeyer flask, dissolve 1.0 g of semicarbazide hydrochloride and 2.0 g of dibasic potassium phosphate (K$_2$HPO$_4$) in 25 ml of water. This should give a solution having a pH of about 6.1–6.2.[3] Using a 1-ml graduated pipet, deliver 1.0 ml of cyclohexanone into a test tube and dissolve it in 5 ml of 95% ethanol. Pour the ethanolic solution into the aqueous semicarbazide solution and swirl or stir the mixture immediately. Allow 5 or 10 min for crystallization of the semicarbazone to reach completion, then collect the crystals by suction filtration and wash them on the filter with a little cold water. Dry the crystals in air and determine their weight and melting point. The reported melting point of cyclohexanone semicarbazone is 166°.★

B. Prepare the semicarbazone of 2-furaldehyde by following the procedure of part A exactly, except use 0.8 ml of 2-furaldehyde [4] instead of 1.0 ml of cyclohexanone. The reported melting point of 2-furaldehyde semicarbazone is 202°.★

C. Dissolve 3.0 g of semicarbazide hydrochloride and 6.0 g of dibasic potassium phosphate in 75 ml of water.[3] This solution will be referred to below as *solution W*.

Prepare a solution of 3.0 ml of cyclohexanone and 2.5 ml of 2-furaldehyde in 15 ml of 95% ethanol. This solution will be referred to below as *solution E*.★

(1) Cool a 25-ml portion of solution W and a 5-ml portion of solution E *separately* in an ice bath to 0–2°, then add solution E to solution W and swirl them together; crystals should form almost immediately. Place the mixture in an ice bath for about 3–5 min, and then collect the crystals by suction filtration and wash them on the

---

[3] Sodium acetate (2.0 g) may be substituted for dibasic potassium phosphate in parts A and B of this experiment and may be used for the general preparation of semicarbazones of aldehydes and ketones as needed for qualitative organic analysis (Chapters 16 and 25). The solution prepared from semicarbazide hydrochloride and sodium acetate will have a pH of about 4.9–5.0. Sodium acetate *must not* be substituted for dibasic potassium phosphate in part C of this experiment.

[4] For best results the 2-furaldehyde should be redistilled just before use.

filter with a little cold water. Dry the crystals and determine their weight and melting point.

(2) Add a 5-ml portion of solution E to a 25-ml portion of solution W at room temperature; crystals should be observed in about 1–2 min. Allow the mixture to stand at room temperature for 5 min, cool it in an ice bath for about 5 min, and then collect the crystals by suction filtration and wash them on the filter with a little cold water. Dry the crystals and determine their weight and melting point.

(3) Warm a 25-ml portion of solution W and a 5-ml portion of solution E *separately* on a steam bath or in a water bath to 80–85°, then add solution E to solution W and swirl them together. Continue to heat the mixture for 10–15 min, cool it to room temperature, and then place it in an ice bath for about 5–10 min. Collect the crystals by suction filtration and wash them on the filter with a little cold water. Dry the crystals and determine their weight and melting point.★

D. In a 25-ml Erlenmeyer flask, place 0.3 g of cyclohexanone semicarbazone (prepared in part A), 0.3 ml of 2-furaldehyde, 2 ml of 95% ethanol, and 10 ml of water. Warm the mixture until a homogeneous solution is obtained (about 1 or 2 min should suffice) and then an additional 3 min. Cool the mixture to room temperature and then in an ice bath. Collect the crystals on a filter and wash them with a little cold water. Dry the crystals and determine their melting point.★

E. Repeat the experiment in part D, but use 0.3 g of 2-furaldehyde semicarbazone (prepared in part B) and 0.3 ml of cyclohexanone in place of the cyclohexanone semicarbazone and 2-furaldehyde.★

F. Dissolve 2.0 g of semicarbazide hydrochloride and 4.0 g of sodium bicarbonate in 50 ml of water. (The pH of this solution should be about 7.1–7.2.) Prepare a solution of 2.0 ml of cyclohexanone and 1.6 ml of 2-furaldehyde in 10 ml of 95% ethanol. Divide each of these solutions into two equal portions.

Mix one half of the aqueous solution and one half of the ethanolic solution at room temperature. Allow the mixture to stand at room temperature for 5 min, cool it in an ice bath for 5 min, and then collect the crystals by suction filtration and wash them on the filter with a little cold water. Dry the crystals and determine their weight and melting point.

Warm the other portions of the aqueous and ethanolic solutions *separately* on a steam bath or in a water bath to 80–85°, then combine them and continue heating the mixture for 10–15 min. Cool the solution to room temperature and then place it in an ice bath for 5 to 10 min. Collect the crystals by suction filtration and wash them on the filter with a little cold water. Dry the crystals and determine their weight and melting point.

EXERCISES

1. On the basis of the results from the experiments in parts C, D, and E state which semicarbazone is the product of kinetic control and which is the product of equilibrium control. Are they the same or different?
2. On the basis of the results from the experiments of parts C, D, and E draw a

diagram similar to either Figure 11.1 or 11.2, whichever is appropriate, and clearly label the products corresponding to Y and Z or A and B.

3. Explain the differences, if any, in the results of the experiments of parts D and E.
4. On the basis of the results from the experiments of part F, explain the effect of the higher pH on the reactions between semicarbazide and the two carbonyl compounds.
5. What results might be expected if sodium acetate buffer (which provides a pH of *ca.* 5) were used in experiments analogous to those of part C? Explain.

SPECTRA OF STARTING MATERIALS

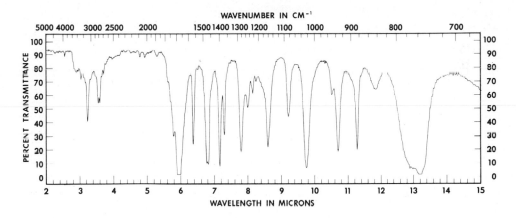

FIGURE 11.3   IR spectrum of 2-furaldehyde.

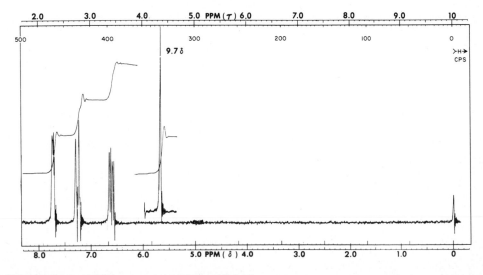

FIGURE 11.4   NMR spectrum of 2-furaldehyde.

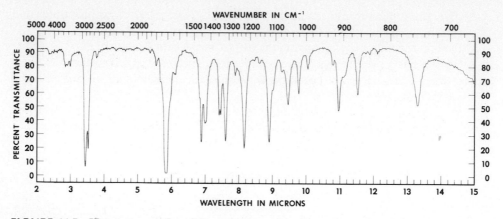

FIGURE 11.5 IR spectrum of cyclohexanone.

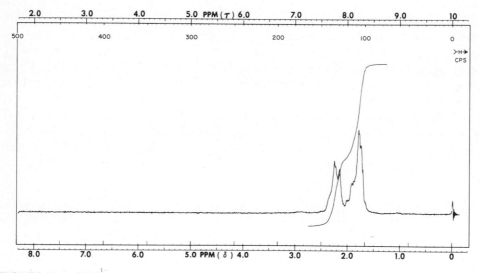

FIGURE 11.6 NMR spectrum of cyclohexanone.

chapter twelve
elimination and substitution reactions

12.1 Base-Induced Elimination Reactions

The elimination reaction was introduced in Chapter 6.1. It was noted that all elimination reactions do not proceed via the same mechanism. In that chapter, the E1 mechanism was discussed and experiments were performed which demonstrated both the techniques of performing a dehydration (an elimination which is often of the E1 variety) and the principles of carbonium ion rearrangements. The following discussion and experiments are concerned with E2 elimination reactions.

E2 eliminations are *base-induced* and proceed by a direct attack of the base, B:, on the substrate molecule, with concurrent formation of the double bond and loss of the leaving group, L, as shown in equation (1). The dotted lines in the transition state

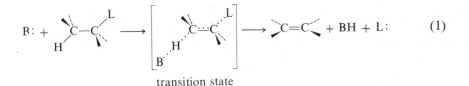

transition state

represent bonds partially formed or broken. It is important that the atoms HCCL are pictured in a *trans*-coplanar arrangement, for E2 eliminations occur most readily when this geometry can be attained. Generally, *strong* bases are employed, such as alkoxide ion (RO⁻) in alcohol (ROH), amide ion (NH₂⁻) in either ethyl ether or benzene, or potassium hydroxide in ethanol solution. Types of leaving groups frequently encountered are the halogens (—Cl, —Br, and —I), the sulfonate ester groups (1), the dimethylsulfonium group (2), and the trimethylammonium group (3). All of these, after

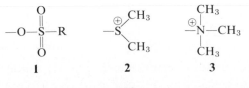

1   2   3

departing, becomes species that are relatively *weak* bases. The ability of a group to serve as a leaving group is dependent on its ability to accept the electron pair with which it is bonded to the molecule, and this ability varies inversely with the base strength of the species formed.

Substitution reactions may sometimes compete with elimination reactions [equation (2)], decreasing the yield of the desired alkene. The degree to which this side reac-

$$
\begin{array}{c}
\underset{H}{\overset{L}{\underset{B}{\mid}}} C{=}C \longrightarrow \quad C{-}C \quad + \; L{:} \qquad (2)
\end{array}
$$

tion is important depends on the nature of the substrate and the base employed. The competition is most important when the carbon bearing the leaving group is primary, becomes less so when it is secondary, and is usually unimportant when it is tertiary. With strong bases such as those listed above, substitution reactions are seldom competitive. The competition between substitution and elimination is discussed in more detail in Section 12.2.

In cases where the leaving group is unsymmetrically located on the alkyl carbon skeleton, elimination may occur in two different directions. Thus, for example, *t*-pentyl chloride yields both 2-methyl-2-butene **(4)** and 2-methyl-1-butene **(5)**.

$$
\underset{\substack{\mid \\ H}}{CH_2}{-}\underset{\substack{\mid \\ Cl}}{\overset{\overset{\textstyle CH_3}{\mid}}{C}}{-}\underset{\substack{\mid \\ H}}{CH}{-}CH_3 \xrightarrow{\text{base}} \underset{\substack{\mid \\ H}}{CH_2}{-}\overset{\overset{\textstyle CH_3}{\mid}}{C}{=}CH{-}CH_3 \; + \; CH_2{=}\overset{\overset{\textstyle CH_3}{\mid}}{C}{-}\underset{\substack{\mid \\ H}}{CH}{-}CH_3 \qquad (3)
$$

$$\qquad\qquad\qquad\qquad\qquad\qquad\qquad\quad \mathbf{4} \qquad\qquad\qquad\qquad \mathbf{5}$$

As these reactions are *irreversible* under the experimental conditions, the alkenes **4** and **5** may be considered to be the products of two competing eliminations, and their proportions thus subject to kinetic control (Chapter 11). Their proportions are therefore determined by the relative free energies of their respective transition states. The transition states for elimination in each direction for *t*-pentyl chloride are shown below (**4‡** and **5‡**).[1] When there are no complicating factors (primarily steric, see below),

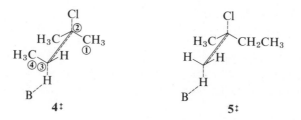

the predominant product in an E2-type elimination is found to be the more highly substituted ethylene. [Recall that increasing the number of alkyl substituents on the double

---

[1] The symbol ‡ is frequently used by organic chemists to refer to a transition state.

bond increases the stability of alkenes (decreases free energy).] This is interpreted to mean that the free energy of the partial double bond is the major contributor to the total free energy of the transition state. Thus, the free energy of activation for the formation of **4** is less than that for the formation of **5**, and **4** should.be the preferred product in the *t*-pentyl chloride example. Prediction of the major product is not always this simple, however, for there may be other complicating factors.

The relative free energies of the transition states of the competing reactions may also be influenced by steric crowding. This crowding will tend to increase the free energy of the transition state and thus the energy requirements of the reaction. If steric effects are more important along the pathway to one of the products than along the pathway to the other, the proportion of products formed may be quite dramatically affected. This is particularly so when the pathway to the more highly substituted ethylene suffers from the greater steric crowding, in which case it may well be the *minor* product.

Unfavorable steric interactions in the transition state may be caused by steric interference of the substituents on the HCCL grouping [*cf.* equation (1)] to the approach of the base (B:), particularly when the base and/or substituents are bulky. Substituents attached to the carbon from which the hydrogen is removed will cause greater hindrance to the base than will substituents on the carbon bearing the leaving group. Therefore, because of the methyl group attached to carbon 3, the transition state **4**‡ should be more adversely affected than **5**‡ if a more bulky base is used to effect the elimination. If, then, the base were large enough, the lesser substituted ethylene could become the major product.

Similar trends toward favoring the lesser substituted ethylene would also be observed by using a larger leaving group L,[2] or by changing the substrate so that the size of the group(s) crowding the base is increased.

The following series of experiments is designed to demonstrate the steric effect as a molecular parameter of the E2 reaction and the general techniques of performing base-induced elimination reactions.[3]

EXPERIMENTAL PROCEDURE[4]

A. ELIMINATION WITH ALCOHOLIC POTASSIUM HYDROXIDE
Place 0.075 mole of potassium hydroxide and 50 ml of *absolute* ethanol in a dry 100-ml round-bottomed flask. Equip the flask for fractional distillation as shown

---

[2] Steric effects caused by increasing the size of the leaving group make it more difficult to attain the *trans*-coplanar arrangement of HCCL leading to the more substituted ethylene. In examples in which the dimethylsulfonium or trimethylammonium group functions as the leaving group, the lesser substituted ethylene is usually the major product, regardless of the nature of the base.

[3] These experiments are based in part on the work of H. C. Brown and I. Moritani, *Journal of the American Chemical Society*, **75**, 4112 (1953).

[4] It is suggested that students do the experiment with different bases that are assigned at random. The results may then be collected, averaged for each base, and these results presented to the class in order to observe the trend of the effect with base size.

in Figure 2.4.[5] In order to increase the efficiency of the Hempel column as a reflux condenser, fill it with pieces of broken glass tubing or pieces of a copper cleaning sponge. Fit the vacuum adapter holding the receiving flask with a calcium chloride drying tube and immerse the receiving flask in an ice-water bath. Circulate water through the jacket of the Hempel column during the period of reflux for the reaction. Heat the mixture with a burner until the potassium hydroxide dissolves. Cool the solution with a water bath, remove the flask, and add to it 0.05 mole of 2-chloro-2-methylbutane (*t*-pentyl chloride) and a few boiling stones. Reattach the flask to the reflux condenser and heat at gentle reflux for 2 hr.[6] By the end of this time, some solid should have precipitated and may cause some bumping.★

At the end of the reflux period, cool the reaction mixture with an ice bath, turn off the cooling water, and remove the water hoses from the Hempel column allowing the water to drain from the jacket. Reconnect the cooling water leads to the condenser so that the apparatus is now set for fractional distillation. If any of the low-boiling products has condensed in the receiving flask during the reflux period, allow it to remain, while continuing to cool that flask in an ice bath. Distil the product mixture, collecting all distillate boiling below 45° (2-methyl-1-butene, bp 31°; 2-methyl-2-butene, bp 38°). Transfer the product to a preweighed sample bottle with a tight fitting stopper or cap and determine the yield. Perform qualitative tests which will demonstrate the presence of alkenes in the distillate (Chapter 6, Section 1). Submit a sample of your product for glpc analysis and, after receiving the results, calculate the relative percentage of the two isomeric alkenes formed. Typical glpc tracings of the products from this elimination and from one in which potassium *t*-butoxide was used as base are shown in Figure 12.1. Figure 12.8 shows the nmr spectrum of the product mixture from a typical experiment.

B. ELIMINATION WITH SODIUM METHOXIDE

Prepare a solution of sodium methoxide in *anhydrous* methanol by either of the following methods. (1) If solid sodium methoxide is available, prepare the base solution by dissolving 0.075 mole of the solid in 50 ml of *anhydrous* methanol contained in a dry 100-ml round-bottomed flask. Use a clean *dry* spatula for transferring the methoxide, taking care to perform the weighing operation as rapidly as possible to minimize the reaction between the methoxide and atmospheric water vapor [equation (4)]. The bottle from which the sodium methoxide is taken should

$$NaOCH_3 + H_2O \rightleftharpoons HOCH_3 + NaOH \qquad (4)$$

be kept *tightly* closed. (2) The sodium methoxide solution may also be prepared by dissolving 0.075 mole of sodium metal in 50 ml of *anhydrous* methanol. (*Caution: No flames; hydrogen is evolved!*) In order to weigh the sodium metal, place a 50-ml beaker containing 15 ml of *dry* toluene on a triple beam balance and determine

---

[5]If you are using glassware equipped with ground glass joints, be sure to lubricate the joint connecting the Hempel column to the flask with a hydrocarbon or silicone grease. Lubrication of ground glass joints is good laboratory practice at all times; however, it is particularly important in this experiment because the strong bases being used may cause the joints to "freeze."

[6]A 2-hr reaction time is sufficient to obtain about 95% yield of product.

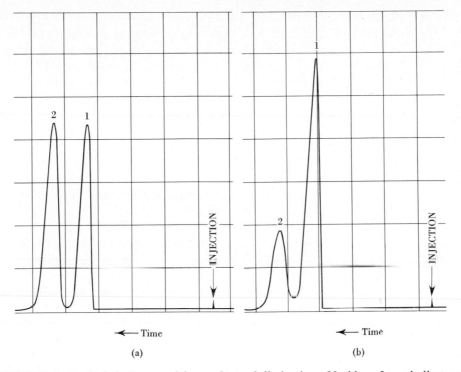

FIGURE 12.1   Typical glpc traces of the products of elimination of 2-chloro-2-methylbutane.[7] Assignments: peak 1: 2-methyl-1-butene; peak 2: 2-methyl-2-butene. (a) Elimination with KOH, showing approximately 44% 2-methyl-1-butene. (b) Elimination with $KOC(CH_3)_3$, showing approximately 76% 2-methyl-1-butene.

its weight. Add an amount equal to the weight of sodium desired to this weight on the beam. Sodium metal is normally stored under mineral oil for protection from the air (water and oxygen). Use a small knife or spatula (*dry*) to cut a small piece of sodium metal about the size of a pea. Impale this piece of sodium on the tip of the knife and "swish" it in a second beaker containing toluene to remove the mineral oil. Remove the piece of sodium, blot it with a *dry* paper towel, and transfer it to the beaker on the balance. Continue until the correct amount of sodium has been obtained. Place 50 ml of *anhydrous* methanol in a 100-ml round-bottomed flask and, with forceps, remove a piece of sodium from the beaker of toluene, again blotting it to remove toluene, and add it to the methanol. Wait until the reaction subsides, then continue adding the sodium one piece at a time until it is all added. Attach a calcium chloride drying tube to the flask and allow time for the sodium to go completely into solution.

If the base solution prepared by either of the above methods is warm, cool it to room temperature in a water bath, and add 0.05 mole of 2-chloro-2-methylbutane to the flask. Add a few boiling stones, connect the flask to the apparatus as described in part A, and continue from this point following the directions of part A.

C. ELIMINATION WITH SODIUM ETHOXIDE

Prepare a solution of sodium ethoxide in *absolute* ethanol by dissolving 0.075 mole of sodium metal in 50 ml of *absolute* ethanol contained in a 100-ml round-bottomed flask. The procedure for handling the sodium is described in part B. Cool the solution to room temperature and add 0.05 mole of 2-chloro-2-methylbutane to the flask. Add a few boiling stones, connect the flask to the apparatus described in part A, and continue from this point following the directions of part A.

D. ELIMINATION WITH POTASSIUM *t*-BUTOXIDE

Place 0.075 mole of solid potassium *t*-butoxide and 50 ml of *anhydrous* *t*-butyl alcohol in a *dry* 100-ml round-bottomed flask. Use the same precautions for handling potassium *t*-butoxide as used for solid sodium methoxide described in part B. Attach the flask to the reflux apparatus described in part A and heat the mixture at gentle reflux with a *water bath* for about 10 min in order to dissolve the solid. The base may not all dissolve. Cool the mixture to room temperature and add 0.05 mole of 2-chloro-2-methylbutane to the flask along with a few boiling stones. Reattach the flask to the reflux apparatus and continue from this point following the directions of part A with the following exception: Use a water bath instead of a burner for heating during the reflux period. A typical glpc tracing[7] and a typical nmr spectrum of the products of this elimination are shown in Figures 12.1 and 12.9, respectively.

EXERCISES

1. What would be the expected results of the alkoxide-induced eliminations if the alcohol solvents employed were wet?
2. What is the solid material that precipitates as the eliminations proceed?
3. Why does the excess of base used in these eliminations favor the E2 elimination as opposed to the E1 elimination? [*Hint:* Consider the rate law expressions for the E1 and E2 processes.]
4. If all of the elimination reactions in the experimental section had proceeded by the E1 mechanism, would the results have been different from those actually obtained? Why?
5. From the results of the above experiments and/or from the data in Figure 12.1, what conclusions can be drawn concerning the relative sizes of the bases employed?
6. What differences might you expect in product distributions for the eliminations of 2-chloro-2-methylbutane and 2-chloro-2,3-dimethylbutane with excess sodium methoxide in methanol solution?
7. If the leaving group in the 2-methyl-2-butyl system were *larger* than a methyl group, why would 2-methyl-1-butene be expected to be formed in greater amounts than if the leaving group were smaller than methyl, regardless of the base used? Use "sawhorse" structural formulas in your explanation.

[7] These analyses were performed at 45° on a 10-ft column of 30% Silicone Gum Rubber supported on Chromosorb P.

8. What is the reason for packing the Hempel column with pieces of broken glass tubing in these experiments?
9. Referring to Figures 12.5, 12.7, 12.8, and 12.9, calculate the percentage compositions of the mixtures of isomeric methylbutenes obtained from reaction of 2-chloro-2-methylbutane with potassium hydroxide and potassium *t*-butoxide.

SPECTRA OF STARTING MATERIALS AND PRODUCTS

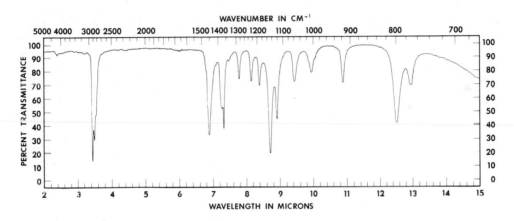

FIGURE 12.2   IR spectrum of 2-chloro-2-methylbutane.

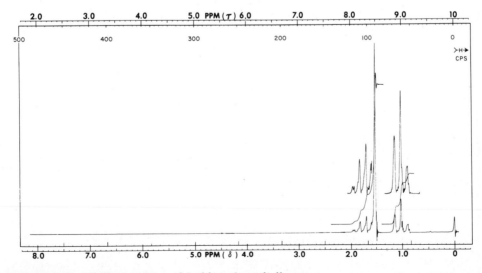

FIGURE 12.3   NMR spectrum of 2-chloro-2-methylbutane.

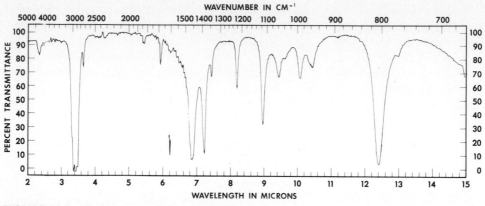

FIGURE 12.4   IR spectrum of 2-methyl-2-butene.

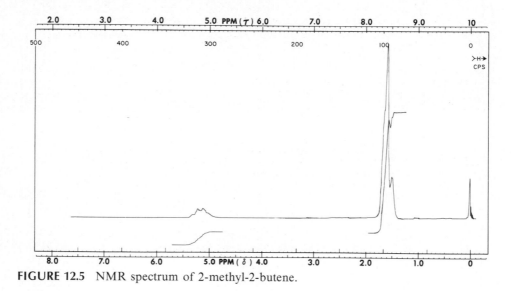

FIGURE 12.5   NMR spectrum of 2-methyl-2-butene.

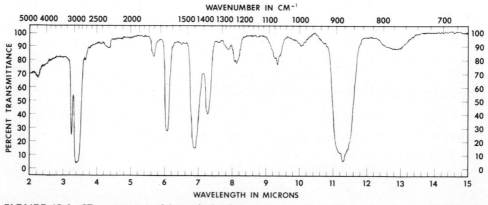

FIGURE 12.6   IR spectrum of 2-methyl-1-butene.

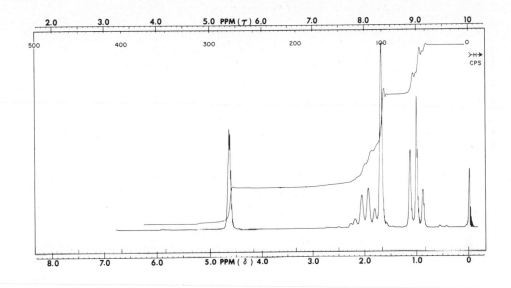

FIGURE 12.7 NMR spectrum of 2-methyl-1-butene.

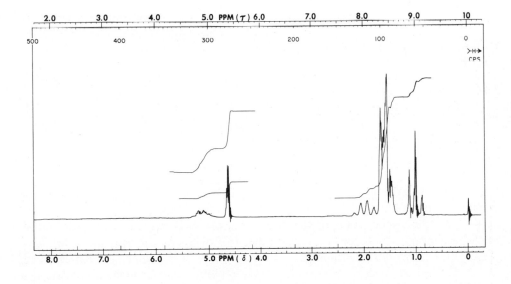

FIGURE 12.8 NMR spectrum of the product mixture from the elimination of 2-chloro-2-methylbutane with potassium hydroxide.

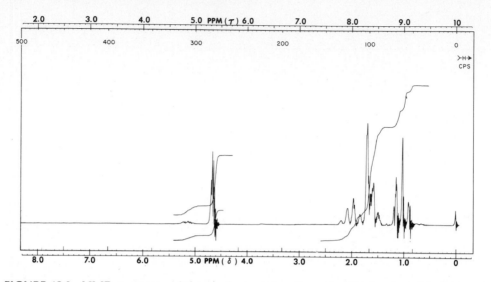

FIGURE 12.9   NMR spectrum of the product mixture from the elimination of 2-chloro-2-methylbutane with potassium *t*-butoxide.

12.2   Nucleophilic Substitution at Saturated Carbon

Many organic reactions involve the *substitution* of one group for another at a saturated carbon atom [equation (5)]. Substitutions are also known to occur at unsaturated carbon atoms (*e.g.*, benzene derivatives, alkenes, etc.); these reactions are discussed elsewhere (see Chapters 8 and 19). In equation (5), L represents the *leaving group* and

$$R\text{—}L + Nu\text{:} \rightarrow R\text{—}Nu + L\text{:} \tag{5}$$

Nu: represents a molecule or ion which has *nucleophilic* character (*i.e.*, Cl⁻, Br⁻, HO⁻, CN⁻, H_2O, NH_3, etc.). These substitution reactions are discussed in the following paragraphs with regard to application and mechanism.

Types of nucleophilic substitution.   Although nucleophilic substitution is a very general reaction for aliphatic compounds (R—L), the mechanism followed in the conversion is highly dependent upon the structure of the alkyl group and upon the nature of the leaving group. It has been found that two distinctly *different* mechanisms of substitution may operate, depending on the structure of the starting material. These are designated as the S_N1 (**S**ubstitution, **N**ucleophilic, and **1** for **uni**molecular) and S_N2 (**S**ubstitution, **N**ucleophilic, and **2** for **bi**molecular) mechanisms [equations (6) and (7), respectively].

S_N1 MECHANISM:

$$R_3C\text{—}L \xrightleftharpoons{\text{slow}} R_3C^{\oplus} + L\text{:}^{\ominus} \tag{6a}$$

$$R_3C^{\oplus} + Nu\text{:}^{\ominus} \xrightarrow{\text{fast}} R_3C\text{—}Nu \tag{6b}$$

S$_N$2 MECHANISM: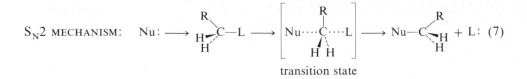

transition state

It is seen that the S$_N$1 mechanism occurs in two steps. The first step of the reaction is the departure of the leaving group taking the pair of bonding electrons with it; the carbon from which L departs becomes a carbonium ion. The second step involves a nucleophilic attack on the carbonium ion; this is an exceedingly fast process, as all ionic reactions are, in general. Normally the concentration of the nucleophile is large compared to that of L:$^-$, so that the reverse of reaction (6a) is relatively unimportant in most cases. It is important to note, for purposes of classifying the mechanism as S$_N$1, that the reaction rate is dependent only on the concentration of R$_3$C—L, and independent of the nature and concentration of the nucleophile

The S$_N$2 mechanism, on the other hand, is represented as a direct attack of the nucleophile on the backside (with respect to L) of the carbon atom. The C—L bond is broken concurrently with the formation of the C—Nu bond. As the substrate (RCH$_2$—L) and nucleophile are both involved in the transition state, this is a bimolecular reaction. Therefore, the rate of an S$_N$2 reaction is dependent on the natures and concentrations of both substrate and nucleophile.

In light of the two basic mechanisms, S$_N$1 and S$_N$2, outlined for nucleophilic substitution, it is not surprising to find that two factors play an important role in dictating the preferred mode of reaction for a particular substrate: (1) In going from CH$_3$—L to R$_3$C—L the number of alkyl groups attached to the central carbon atom is increased, making it more difficult for the backside attack by the nucleophile because of steric hindrance; this serves to decrease the ease of S$_N$2 reaction; (2) as alkyl groups are added to the central carbon, the incipient carbonium ion of the S$_N$1 reaction becomes more stable, facilitating the reaction along the S$_N$1 pathway. (The relative stability of substituted carbonium ions has been discussed earlier in Chapter 6.1.) These two effects reinforce one another in producing the general trends shown in the following diagram:

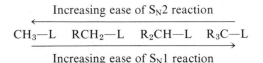

Increasing ease of S$_N$2 reaction

CH$_3$—L   RCH$_2$—L   R$_2$CH—L   R$_3$C—L

Increasing ease of S$_N$1 reaction

Competition between substitution and elimination. In all reactions in which nucleophilic substitution occurs, there is the possibility of competing elimination reactions which produce alkenes. For the most part, bimolecular elimination reactions (E2) compete with S$_N$2 substitution, and unimolecular elimination reactions (E1) compete with S$_N$1 substitution. E1 reactions have been discussed in detail in Chapter 6.1 and E2 reactions in Section 12.1.

The competition in the two cases may be represented by the following equations. Generally, elimination is favored when strongly basic and only slightly polarizable

S_N2 *vs.* E2:

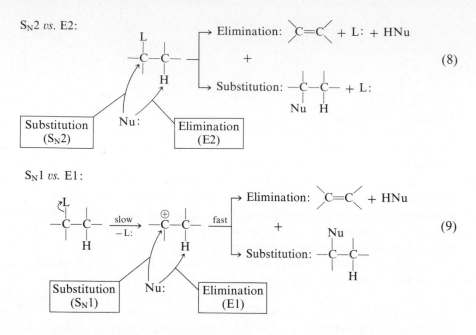

S_N1 *vs.* E1:

species are employed (*i.e.*, RO^-, NH_2^-, H^-, HO^-, etc.), and substitution is favored when weakly basic and highly polarizable species are used (*i.e.*, I^-, Br^-, Cl^-, H_2O, $CH_3CO_2^-$, etc.). Polarizability is a measure of the ease of distortion of the electron cloud of the species by a nearby charged center (positively charged in this case).

The preparation of *n*-butyl bromide; an S_N2 reaction. A common method of converting a primary alcohol to an alkyl halide is shown in equation (10). This method

$$\text{HX} + \text{R—OH} \underset{\text{heat}}{\overset{H_2SO_4}{\rightleftharpoons}} \text{R—X} + H_2O \qquad (10)$$

involves the establishment of equilibrium between an alcohol and a mineral acid (HX = HCl, HBr, HI). The best yield of alkyl halide is of course obtained when the position of equilibrium lies well to the right (see below).

The preparation of *n*-butyl bromide may be accomplished by heating *n*-butyl alcohol with concentrated hydrobromic acid in the presence of concentrated sulfuric acid [equation (11)]. The mechanism is best written in two steps, which are given in equa-

$$CH_3CH_2CH_2CH_2\text{—OH} + \text{HBr} \underset{\text{heat}}{\overset{H_2SO_4}{\rightleftharpoons}} CH_3CH_2CH_2CH_2\text{—Br} + H_2O \qquad (11)$$

tion (12). The first step involves the protonation of the alcohol to give oxonium ion **6**,

$$n\text{-}C_4H_9\text{—}\overset{..}{\underset{..}{O}}H + H^{\oplus} \rightleftharpoons n\text{-}C_4H_9\text{—}\overset{\oplus}{O}\underset{H}{\overset{H}{\diagup}} \qquad (12a)$$

$$Br^{\ominus} + n\text{-}C_3H_7\text{---}CH_2\text{---}\overset{\oplus}{O}\begin{smallmatrix}H\\ \\H\end{smallmatrix} \rightleftharpoons n\text{-}C_4H_9\text{---}Br + H_2O \qquad (12b)$$

6

an example of Lewis acid-Lewis base complex formation. Once formed the oxonium ion undergoes displacement by the bromide ion to form the alkyl bromide and water [equation (12b)]. In this S_N2 reaction, *water is the leaving group* and *bromide ion is the nucleophile*. The sulfuric acid serves two distinct and important purposes: (1) as a dehydrating agent to reduce the activity of water and shift the position of equilibrium to the right, and (2) as an added source of hydrogen ions to increase the concentration of oxonium ion **6**. The use of *concentrated* hydrobromic acid also helps to establish a favorable equilibrium.

The mixture of hydrobromic acid and sulfuric acid may be prepared in two ways. One method is to add concentrated sulfuric acid to concentrated hydrobromic acid. The second way is to generate the hydrobromic acid *in situ* by adding concentrated sulfuric acid to sodium bromide [equation (13)]. These methods work well and give

$$NaBr + H_2SO_4 \rightleftharpoons HBr + NaHSO_4 \qquad (13)$$

high yields of the bromide when low-molecular-weight alcohols are used. With higher-molecular-weight alcohols, however, it has been found that the sulfuric acid is best omitted; this is because of the decreased solubility of the alcohol when the reaction is carried out in concentrated salt solutions. With very high-molecular-weight alcohols, the addition of hydrogen bromide gas to the alcohol at elevated temperatures appears to give the best results.

Although the added sulfuric acid is desirable as discussed above, it may also give rise to two important side reactions. The alcohol may react with sulfuric acid to form a hydrogen sulfate ester [**7**, equation (14)]. This ester formation is reversible and the

$$ROH + H_2SO_4 \rightleftharpoons RO\text{---}SO_3H + H_2O \qquad (14)$$

7

alcohol is regenerated (position of equilibrium shifted to the left) as its concentration is reduced in the production of alkyl bromide. The formation of the hydrogen sulfate ester would not directly decrease the yield, were it not able to undergo reaction in two other ways to give undesired products. (1) On heating, it can undergo elimination to

$$RO\text{---}SO_3H \xrightarrow{\text{heat}} \text{alkenes} + H_2SO_4 \qquad (15a)$$

$$RO\text{---}SO_3H + ROH \xrightarrow{\text{heat}} R_2O + H_2SO_4 \qquad (15b)$$

give alkenes [equation (15a)], and (2) it can react with another molecule of alcohol to give a dialkyl ether by an S_N2 reaction in which the nucleophile is ROH [equation (15b)]. Both of these side reactions consume alcohol and result in a decreased yield of alkyl bromide. For primary alcohols, the temperatures which are used for the substitution reaction are generally not high enough to cause these side reactions to be of great importance.

In the preparation of secondary bromides from secondary alcohols, the method involving the use of concentrated sulfuric acid cannot be employed. This is because secondary alcohols are far more easily dehydrated by sulfuric acid to give the corresponding alkenes [see Chapter 6 and equation (15a)] than are primary alcohols. One means of circumventing this problem is to use 48% hydrobromic acid without the sulfuric acid. However, the secondary halides are generally best prepared by use of phosphorus trihalides [PX_3, X = Cl or Br, equation (16)] or thionyl halides [SOX_2, X = Cl or Br, equation (17)]. The latter reagents are required when the alcohol is susceptible to *carbonium ion rearrangements* (Chapter 6.1) so that the presence of acid is to be avoided. In these cases, pyridine is used as the solvent, since this base reacts with the acid (HX) as it is formed in the reaction and serves to neutralize it [equation (17)].

$$3 \text{ ROH} + PX_3 \longrightarrow 3 \text{ RX} + H_3PO_3 \tag{16}$$

$$\text{ROH} + SOX_2 \xrightarrow{C_5H_5N} \text{RX} + SO_2 + \text{(pyridinium)} \; X^{\ominus} \tag{17}$$

The preparation of tertiary alkyl chlorides; an S_N1 reaction. Although the same general approaches as described above can be used for conversion of tertiary alcohols to their corresponding halides, only one method is found to give acceptable yields. This method involves the use of a mineral acid with the alcohol in a reaction which is typically S_N1. Phosphorus trihalides or thionyl halides are not applicable for the preparation of *t*-alkyl halides from the alcohols because elimination reactions predominate.

The reaction of *t*-butyl alcohol with hydrochloric acid [equation (18)] is illustrative

$$\underset{\underset{CH_3}{|}}{\overset{\overset{CH_3}{|}}{CH_3-C-OH}} + HCl \rightleftharpoons \underset{\underset{CH_3}{|}}{\overset{\overset{CH_3}{|}}{CH_3-C-Cl}} + H_2O \tag{18}$$

of the S_N1 reaction. The mechanism, shown in equation (19), involves three distinct steps. The second step [equation (19b)] is the slowest step in this sequence although it is faster than the corresponding S_N2 attack of Cl$^-$ on oxonium ion **8**. The reverse is

$$\underset{\underset{CH_3}{|}}{\overset{\overset{CH_3}{|}}{CH_3-C-\ddot{O}H}} + H^{\oplus} \rightleftharpoons \underset{\underset{CH_3}{|}}{\overset{\overset{CH_3}{|}}{CH_3-C-\overset{\oplus}{O}}} \overset{H}{\underset{H}{\diagdown}} \tag{19a}$$

8

$$CH_3-\overset{\overset{\displaystyle CH_3}{|}}{\underset{\underset{\displaystyle CH_3}{|}}{C}}\overset{\oplus}{\underset{\underset{\displaystyle H}{}}{O}}\overset{H}{\diagdown} \quad \overset{slow}{\rightleftharpoons} \quad CH_3-\overset{\oplus}{C}\overset{\diagup CH_3}{\diagdown CH_3} + H_2O \qquad (19b)$$

relatively stable
tertiary carbonium ion

$$CH_3-\overset{\oplus}{C}\overset{\diagup CH_3}{\underset{\diagdown CH_3}{}} + Cl^{\ominus} \quad \overset{fast}{\rightleftharpoons} \quad CH_3-\overset{\overset{\displaystyle CH_3}{|}}{\underset{\underset{\displaystyle CH_3}{|}}{C}}-Cl \qquad (19c)$$

true for the oxonium ions of primary and secondary alcohols. This reactivity difference between tertiary oxonium ions and primary and secondary oxonium ions is a
reflection of the relative stabilities of the three types of carbonium ions which result
from the loss of a water molecule. Although *relatively* slow compared to the other
steps in the sequence, reaction (19b) is quite a rapid one from an *absolute* viewpoint.
Thus, with concentrated hydrochloric acid a favorable equilibrium giving high yields
of *t*-butyl chloride is established in a few minutes at room temperature. Concentrated
hydrochloric acid is necessary to establish a favorable equilibrium [equation (18)].

The principal side reaction is the E1 elimination, resulting from the loss of a proton
from the incipient carbonium ion to give 2-methylpropene [isobutylene, equation (20)].
Under the reaction conditions, however, the reaction is readily reversible through the
addition of HCl to the alkene to give *t*-butyl chloride [according to Markownikoff's
Rule, equation (21)]. Therefore, this is not a serious complication.

$$CH_3-\overset{\oplus}{C}\overset{\diagup CH_2-H}{\underset{\diagdown CH_3}{}} \quad \longrightarrow \quad CH_3-C\overset{\diagup CH_2}{\underset{\diagdown CH_3}{}} + H^{\oplus} \qquad (20)$$

$$CH_3-C\overset{\diagup CH_2}{\underset{\diagdown CH_3}{}} + H^{\oplus} \longrightarrow CH_3-\overset{\oplus}{C}\overset{\diagup CH_3}{\underset{\diagdown CH_3}{}} \overset{Cl^{\ominus}}{\longrightarrow} CH_3-\overset{\overset{\displaystyle CH_3}{|}}{\underset{\underset{\displaystyle CH_3}{|}}{C}}-Cl \qquad (21)$$

EXPERIMENTAL PROCEDURE

***n*-Butyl bromide.** Place 0.30 mole of sodium bromide in a 250-ml round-bottomed
flask and add about 35 ml of water and 0.30 mole of *n*-butyl alcohol. Mix thoroughly
and cool the flask in an ice bath. *Slowly* add 35 ml of *concentrated* sulfuric acid to the
cold mixture with swirling and continuous cooling. Remove the flask from the ice
bath. Fit it with a reflux condenser and warm the flask gently until most of the salts
have dissolved. Add boiling stones and heat the mixture to gentle reflux. A noticeable
reaction occurs, and two layers form, the upper layer being the alkyl bromide; the
inorganic layer has a high concentration of salts and thus is much more dense than
n-butyl bromide. Continue heating at reflux for 45 min.*

Equip the flask for simple distillation. Distil the mixture rapidly and collect the distillate in an ice-cooled receiver. Codistillation of n-butyl bromide and water occurs. Continue the distillation until the distillate is clear. The head temperature should be around 115° at this point. (The increased boiling point is caused by codistillation of sulfuric acid and hydrobromic acid with water.)★

Transfer the distillate to a separatory funnel. Add about 30 ml of water and shake gently, with venting. Note that two layers are formed; decide which of these is the organic layer by adding a couple of drops of water to the heterogeneous mixture and seeing in which layer it dissolves.[9] Remove the organic layer and transfer it to a separatory funnel. Add an equal volume of *cold, concentrated* sulfuric acid and shake the mixture gently. Remove the organic layer. If it is difficult to determine where the interface of the two layers is, view the mixture from several different angles. Wash the organic layer, first with about 20 ml of water, then with 10 ml of 2 N sodium hydroxide solution, and finally with 20 ml of water.

Transfer the cloudy n-butyl bromide layer to a small Erlenmeyer flask, and dry it over granular *anhydrous* calcium chloride.★ The drying process may be hastened by warming slightly on a steam bath and by swirling gently. Decant the dry liquid into a round-bottomed flask, add boiling stones, and equip for simple distillation. Distil and collect the material boiling at 99–103°. Weigh the product and compute the yield. The yield should be 70–82%.

t-**Butyl chloride.**   Place 0.5 mole of *t*-butyl alcohol and 1.5 moles of hydrogen chloride provided by reagent grade *concentrated* (12 N) hydrochloric acid in a 250-ml separatory funnel. After mixing, swirl the contents of the separatory funnel gently, without the stopper on the funnel. After about a minute of swirling, stopper and invert the funnel; after inverting, vent to release excess pressure by carefully opening the stopcock (be careful to avoid shaking until this is done). Shake the funnel for several minutes, with intermittent venting. Allow the contents to stand until the mixture has separated into two distinct layers; these should be completely clear.

Separate the layers and wash the organic layer with 50 ml of *saturated* sodium chloride solution and then with 50 ml of *saturated* sodium bicarbonate solution. On initial addition of the bicarbonate solution, vigorous gas evolution will ensue; gently swirl the *unstoppered* separatory funnel until the vigorous effervescence ceases. Stopper the funnel and invert gently; vent immediately to release gas pressure. Shake the separatory funnel gently, with *frequent* venting; then shake vigorously, again with frequent venting. Separate the organic layer (the density of a saturated sodium bicarbonate solution is about 1.1), wash it with about 40 ml of water, and then with 40 ml of *saturated* sodium chloride solution. Carefully remove the aqueous layer.

Transfer the *t*-butyl chloride to a small Erlenmeyer flask and dry with *anhydrous* calcium chloride; shaking expedites the drying process.★ When the organic layer is

[9] The drop of water will be readily soluble in the inorganic layer, but will be insoluble in the organic layer. If the organic layer is on top, the water will fall through it and dissolve in the inorganic layer; if the inorganic layer is on top, the water will dissolve directly in it. A common misfortune that students encounter is in finding that the organic layer has been thrown away and the inorganic layer has been kept. A good, general practice in the laboratory is to keep *both* layers, until you are sure which is the one you are to keep. It is recommended that the student do this throughout the experiments.

clear, transfer the product by decantation to a small round-bottomed flask and distil, collecting the distillate in an ice-cooled receiver. Make sure that the distilling flask and condenser used are *dry*. Collect the fraction boiling from 49–52°. If any lower boiling material is obtained, it should be redistilled and any material boiling from 49–52° retained. The yield of *t*-butyl chloride should be 78–88%.

_t_-Pentyl chloride. *t*-Pentyl chloride may be prepared from *t*-pentyl alcohol using the same general procedure as that described for *t*-butyl chloride, using 0.5 mole of *t*-pentyl alcohol. The *t*-pentyl chloride will be obtained as the product on final distillation; material boiling between 83–85° should be collected. The yield will be about 75%.

EXERCISES

1. Observe that some water is added to the initial reaction mixture. How might the yield of *n*-butyl bromide be affected by the failure on the part of the student to add the water? What products would be obtained? How might the yield of *n*-butyl bromide be affected by adding twice as much water as is called for?
2. In the purification process, the impure *n*-butyl bromide is "washed" with concentrated sulfuric acid. What impurities would a wash of this sort remove? Why?
3. Suppose that a student added 0.3 mole of trimethylene glycol, $HOCH_2CH_2CH_2OH$, instead of 0.3 mole of *n*-butyl alcohol. In principle, what product would you expect to be produced from the glycol? Write an equation for its formation. Do you think that it would be produced in good yield under the conditions of this experiment (and using the same quantities of reagents which are called for herein)? Why or why not?
4. Draw the structures of the other alcohols which are isomeric with *t*-butyl alcohol. Arrange these alcohols in order of *increasing* reactivity toward concentrated hydrochloric acid. Which, if any, of these alcohols would you expect to give a reasonable yield of the corresponding alkyl chloride under such conditions?
5. The work-up procedure in the *t*-butyl chloride preparation calls for using sodium bicarbonate. From your experience in the laboratory, you know that this work-up is accompanied by vigorous gas evolution which increases the difficulty of handling and requires considerable caution. Alternatively, you might consider using a dilute solution of sodium hydroxide instead of the sodium bicarbonate. Comment on the relative advantages and disadvantages of using these two basic solutions in the work-up. Based on these considerations, why were you instructed to use sodium bicarbonate, even though it is more difficult to handle?

SPECTRA OF STARTING MATERIALS AND PRODUCTS

The ir and nmr spectra of *n*-butyl chloride (Figures 5.5 and 5.6, respectively) are quite similar to those of *n*-butyl bromide, so that the spectra of the latter compound are not included. For ir and nmr of *t*-pentyl chloride (2-chloro-2-methylbutane), see Figures 12.2 and 12.3, respectively.

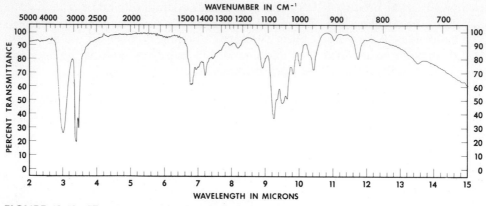

FIGURE 12.10 IR spectrum of *n*-butyl alcohol.

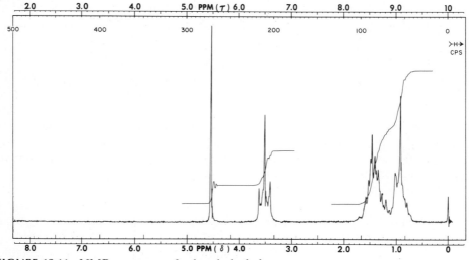

FIGURE 12.11 NMR spectrum of *n*-butyl alcohol.

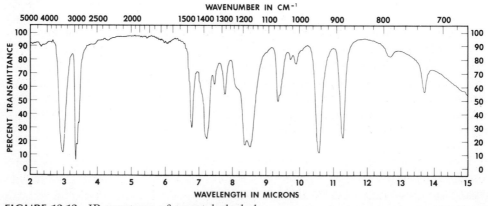

FIGURE 12.12 IR spectrum of *t*-pentyl alcohol.

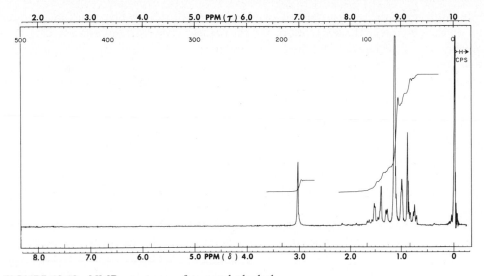

FIGURE 12.13 NMR spectrum of *t*-pentyl alcohol.

12.3   Chemical Kinetics: Evidence for Nucleophilic Substitution Mechanisms

Some of the most important criteria in support of the S_N1 and S_N2 mechanisms have been obtained from kinetic studies on various types of substrates. For primary and most secondary compounds, it is found that the *rate* of substitution is proportional to the concentrations of the substrate *and* the nucleophile. This situation is represented by the following rate law:

$$\text{Rate} = k_2[\text{R–L}][\text{Nu:}] \tag{22}$$

The proportionality constant, k_2, relating rate and concentrations is known as the *rate constant*. This equation indicates that the transition state involves both the substrate and the nucleophile, and that the reaction can be represented mechanistically as shown in equation (7), *i.e.*, as an S_N2 reaction.

On the other hand, the rates of substitution of tertiary and some secondary compounds are found to be proportional to the concentration of the substrate but *independent* of the concentration of the nucleophile. Therefore the transition state of the slow step in these cases involves only the substrate and not the nucleophile.[10] This is interpreted to indicate an initial slow ionization of the substrate to yield a carbonium ion as represented in the S_N1 mechanism [equation (6)]. For this reaction, then, the rate law is as given in equation (23).

$$\text{Rate} = k_1[\text{R–L}] \tag{23}$$

[10] Molecules of solvent are involved as well, since the reaction is performed in solution. For simplicity, the involvement of the solvent (often called "solvation") will not be considered since any *changes* due to solvation are insignificant when a large excess of solvent is present.

In addition to the concentration dependence, rates of substitution are also dependent on such factors as the temperature and the nature of the reactants and solvent. The dependency of the rate on these factors may be considered, for the sake of simplicity, as being reflected in the value of the rate constant; *e.g.*, if a temperature change results in a doubling of the rate (at the same concentration of reactants) then the rate constant is larger by a factor of two.

The S_N1 reaction and its rate law. S_N1 reactions are first-order reactions in that the rate of reaction is proportional to the first power of the concentration of the substrate [equation (23)].[11] It should be evident that a graphical plot of *rate* vs. *concentration* in such a case would yield a straight line of slope k_1, and that doubling the concentration would double the rate. The rate constant is independent of the concentration.

When a substance is being consumed in a first-order reaction, its concentration decreases exponentially with time. If C_o is the *initial* concentration of the substrate (at time $t = 0$, or t_o), and C_t is its concentration at any elapsed time t (where t is measured in any unit of time, *e.g.*, seconds, minutes, hours, etc.) after the reaction is started, the relationship between these variables is

$$C_t = C_o \, e^{-k_1 t} \tag{24}$$

The rate constant, k_1, has the units of (time)$^{-1}$, *e.g.*, (sec)$^{-1}$. Equation (24) may be rewritten as

$$\ln (C_o/C_t) = k_1 t \tag{25}$$

or

$$2.303 \log (C_o/C_t) = k_1 t \tag{26}$$

If the initial concentration of the reactant (C_o) is known, and if one measures the concentration of the reactant (C_t) at a series of measured time intervals while the reaction is proceeding, then the rate constant may be determined in either of two ways: (1) The values of C_t and t measured at each point during the reaction may be substituted into equation (26), which is then solved for k_1. Thus several values of k_1 are determined and these values are then averaged. The rate constant (the *correct* one) is not easily obtained by this method, for the average value will be affected without bias by any points (determinations) that may be "bad" (incorrect due to experimental error). (2) An alternative and better method of calculation of k_1 from the experimental data consists of preparing a plot of $\log (C_o/C_t)$ *vs.* the time t. Assuming the reaction to be first-order and the data to be fairly accurate, a reasonably straight line should be obtained. This line is drawn on the graph in such a way as to lie closest to the largest

---

[11] In a *hypothetical* reaction, if the rate were proportional to the second power of the concentration of the substrate, then the rate law would contain the concentration of the substrate *squared:*

$$\text{Rate} = k_2[\text{R--L}]^2$$

number of points. The line may be drawn *with bias*, purposely giving more "weight" to the majority of points that will lie close to the line and ignoring or giving less credence to those points that appear to be in error. The slope of this line ("the best straight line")[12] is the rate constant k_1 if equation (25) is used (natural logrithms) or $k_1/2.303$ if equation (26) is used. In the latter case the slope must be multiplied by 2.303 to obtain k_1.

It should be noted that equations (25) and (26) contain only the *ratio* of the concentrations of the reactants, so that the *units* of concentration cancel. Hence, one does not need to express the concentrations in the customary units, but may instead use any units so long as both C_o and C_t are given in the same units. In fact, one may use *total amounts*, *i.e.*, grams or moles, of reactants at time zero and at time t (A_o and A_t, respectively) to give the following equation:

$$2.303 \log (A_o/A_t) = k_1 t \qquad (27)$$

The usefulness of this equation will be demonstrated later.

Both S_N1 and S_N2 reactions can be studied from a kinetic standpoint, but bimolecular reactions are generally more difficult to examine experimentally and the calculations for them are more tedious. Therefore, we shall limit our studies to a detailed examination of the kinetics of S_N1 reactions, the rate law for which we have just discussed. The experiments presented here are designed to illustrate methods of studying chemical kinetics and measuring the effects of structure on reactivity, as exemplified by the solvolysis of tertiary alkyl halides.

Solvolysis refers to a substitution reaction in which the solvent (SOH) functions as the nucleophile [equation (28)]. In principle, solvolysis reactions can be carried out in

$$R—X + SOH \rightarrow R—OS + H^{\oplus} + X^{\ominus} \qquad (28)$$

any nucleophilic solvent, such as water (hydrolysis), alcohol (alcoholysis), carboxylic acids (acetolysis with acetic acid and formolysis with formic acid) or liquid ammonia (ammonolysis). A major limitation in choosing a solvent is dictated by the solubility of the substrate in the solvent. It is important that the reaction mixture be homogeneous, for if it is not, surface effects at the interface of the phases will make the kinetic results difficult to interpret and, probably nonreproducible, as well. In the experiments described here, the solvolyses will be carried out in mixed solvents consisting of isopropyl alcohol and water.

In order to determine the rate constants (k_1) for these solvolysis reactions [equation (28)], one must make use of the rate expression given in equation (26). Recall that the quantities which must be obtained in order to use equation (26) are C_o and C_t. Of these, C_t is the more difficult to obtain. One possible method would be to withdraw an aliquot (a measured sample of the reaction mixture, *e.g.*, 5 ml) at time t and to

---

[12] An alternative method of obtaining the best straight line is the application of a *least-squares treatment* to the data. This mathematical treatment will also yield a "biased" line, but will do so without the subjective wishful-thinking that may contaminate human judgment. The student who wishes to apply this method may consult the following reference for details: G. H. Brown and C. M. Sallee, *Quantitative Chemistry*, Prentice-Hall, Inc., Englewood Cliffs, N. J., 1963, p. 123 ff.

isolate the remaining alkyl halide contained therein. This could be a very inaccurate procedure, for it would be difficult to isolate remaining starting material quantitatively from such a mixture. The best procedure is based upon the fact that for every molecule of alkyl halide that reacts, one molecule of HX is produced [equation (28)]. Thus the progress of the reaction may be followed by determining the concentration of hydrogen ion, [H$^+$], produced as a function of time. Note, however, that [H$^+$] gives the amount of alkyl halide which has *reacted* and not the amount which remains, *i.e.*, [H$^+$] = C$_o$ − C$_t$. C$_t$ is then determined by subtracting [H$^+$] at time t from C$_o$ [equation (29)].

$$C_t = C_o - [H^+] \tag{29}$$

Experimentally, [H$^+$] at time t is determined by withdrawing an aliquot from the reaction mixture with a pipet (the volume must be accurately measured) and adding it to a sufficient quantity of 98% isopropyl alcohol sufficient to "quench" the reaction so that it no longer proceeds at a measureable rate. The elapsed time is noted, and the quenched sample is titrated with standardized dilute sodium hydroxide solution.

The reaction is started by dissolving a weighed quantity of alkyl halide in an appropriate amount of the solvent to be used. If a temperature other than room temperature is to be used, the solvent and halide should be separately adjusted to this temperature before mixing. If a volumetric flask is available, the solution should be prepared in this flask and enough additional solvent added to bring the total volume to the mark on the flask. Thus the exact volume is known, as well as the weight of sample, so that the initial concentration, C$_o$, may be calculated. The time, t_o, is taken as that time when the final adjustment of volume is made.

An easier and more reliable method of determining C$_o$ is to allow the reaction to go to completion (the "infinity point," the point at which essentially all of the alkyl halide has been converted into product) and to titrate an aliquot of this final reaction mixture. The [H$^+$] at this point should be equal to the initial concentration of alkyl halide. The solvolysis of *t*-butyl chloride will reach 99.5% completion in about 12 hr with 60% isopropyl alcohol, or in about 4 hr with 50% isopropyl alcohol.

If the latter method is used to obtain C$_o$, equation (27) may be used to determine k_1, where A$_o$ = (ml of NaOH from final titration), and A$_t$ = (ml of NaOH from final titration) − (ml of NaOH from titration at time t). As noted before, the concentration units in equation (26) cancel giving rise to equation (27). To apply equation (27) in this way, the aliquot size must be the same for each titration, including the final titration.

Factors which influence the rate of S$_N$1 reactions. Some of the particular factors which influence the rate of S$_N$1 reactions are described in the following paragraphs; experiments are provided which will allow some of them to be investigated.

1. Solvent composition.   The nature of the solvent influences greatly the rate of S$_N$1 reactions. The solvent effect may be considered as a dependence of the rate on the *polarity* of the solvent. The polarity influences the rate of the initial ionization [equation (6a)], the slow step of the S$_N$1 reaction. The more polar the solvent, the more rapid is the ionization. This is because of the greater ability of the more polar solvent to stabilize charged species through solvation.

As mentioned before, the experiments make use of mixed solvents consisting of isopropyl alcohol and water. The effects of solvent composition on rate can be studied using mixtures containing 60:40, 55:45, and 50:50 isopropyl alcohol:water.

2. *Effect of concentration.* Equation (23) indicates that the rate of solvolysis is proportional to the concentration of alkyl halide. This equation assumes that the rate constant is independent of the concentration of alkyl halide.

3. *Effect of temperature.* Temperature changes also influence the rate of a chemical reaction. A *rough* "rule of thumb" is that the rate doubles for each 10° rise in temperature. Thus, in going from 25° to 45° (a 20° rise in temperature), the rate of reaction increases by a factor of *about four*. To examine the effects of temperature on solvolysis reactions, there are two temperatures which can be utilized conveniently in the laboratory; one of these is room temperature, and the other is 0°, which can be obtained with the use of an ice-water bath.

4. *Effect of alkyl group.* In solvolysis reactions, tertiary halides may be used with complete confidence that the substitutions are proceeding by the S_N1 mechanism. Two compounds which are suitable for the study are *t*-butyl chloride and *t*-pentyl chloride, whose preparations are described in Section 12.2. A determination of the rate of solvolysis of these compounds will provide data which can be used to illustrate the reactivity difference between two tertiary alkyl groups, the *t*-butyl group, $(CH_3)_3C-$, and the *t*-pentyl group, $(CH_3)_2(C_2H_5)C-$.

5. *Effect of leaving group.* Two common leaving groups are chloride and bromide; although iodide and fluoride could be used, compounds containing them are generally difficult to obtain and expensive. Also, tertiary alkyl iodides are notably unstable. *t*-Butyl chloride and *t*-butyl bromide are used to illustrate the effect of the leaving group on reactivity. If all other factors are kept constant, then the relative abilities of chloride and bromide to leave can be measured by comparing the rates of solvolysis of these two compounds.

Products from S_N1 solvolysis reactions. Equation (28) implies that the product of a solvolysis reaction is only R—OS, but the reaction is not as simple as it might appear. Recall that the intermediate in an S_N1 reaction is a carbonium ion [equation (6)], which may undergo several reactions. Equation (30) indicates the major products that occur when *t*-butyl chloride is solvolyzed in an isopropyl alcohol-water solvent mixture. It should be noted that the carbonium ion **9** [equation (30a)] undergoes reactions which are typical of most carbonium ions: (1) loss of a proton to give an alkene [equation (30b)], (2) reaction with water to give an alcohol [equation (30c)], and (3) reaction with isopropyl alcohol to give an ether [equation (30d)].

The products which are obtained in solvolysis reactions are highly dependent upon the nature and composition of the solvent. In the present experiment, the relative amounts of alcohol, ether and alkene produced are different for the different solvent mixtures and conditions used. The polarity of the solvent plays an important role in dictating the product distribution, and there are no generalities which can be used to predict this distribution. The reaction temperature and nature of the leaving group also play a role in determining the product ratio.

$$\underset{\underset{CH_3}{|}}{\overset{\overset{CH_3}{|}}{CH_3-C-Cl}} \xrightarrow{\text{slow}} \underset{CH_3}{\overset{\oplus}{CH_3-C}}\overset{CH_3}{\diagup} + Cl^{\ominus} \tag{30a}$$

$$\mathbf{9}$$

$$\underset{CH_3}{\overset{\oplus}{CH_3-C}}\overset{CH_3}{\diagup} \xrightarrow{\text{fast}} \underset{CH_3}{CH_3-C}\overset{CH_2}{\diagdown\!\!\!\!\diagup} + H^{\oplus} \tag{30b}$$

$$\underset{CH_3}{\overset{\oplus}{CH_3-C}}\overset{CH_3}{\diagup} + H-OH \xrightarrow{\text{fast}} \underset{\underset{CH_3}{|}}{\overset{\overset{CH_3}{|}}{CH_3-C-OH}} + H^{\oplus} \tag{30c}$$

$$\underset{CH_3}{\overset{\oplus}{CH_3-C}}\overset{CH_3}{\diagup} + (CH_3)_2CHOH \xrightarrow{\text{fast}} \underset{\underset{CH_3}{|}}{\overset{\overset{CH_3}{|}}{CH_3-C-OCH(CH_3)_2}} + H^{\oplus} \tag{30d}$$

Although no directions are given here for determining the composition of the mixture obtained from the solvolysis reactions, gas-liquid partition chromatography (glpc) is frequently used.

EXPERIMENTAL PROCEDURE

Successful results in the present experiment depend greatly upon being systematic and organized. The best way to insure being systematic is to prepare a neat table in the notebook beforehand, and to check regularly during the course of the kinetic run to see that all necessary entries have been recorded. Provision should be made for recording the following data: (1) the volume of solvent used in the reaction mixture; (2) the exact weight, or volume, of alkyl halide added; (3) the time, t_o, at which the kinetic run is initiated; (4) the temperature of the reaction solution; (5) the results of a "blank" titrimetric determination; (6) a series of times at which aliquots are withdrawn; and (7) the titrimetric data for each aliquot.

This experiment involves a series of quantitative measurements carried out in a relatively short time. Therefore the student should be prepared to work *rapidly*, although *carefully* so that a high standard of accuracy can be maintained. Buret readings should be made to the nearest 0.02 ml if possible, although precision within 0.05 ml will normally be satisfactory. If a rate measurement with *t*-butyl bromide has been assigned, read part B completely *before* beginning the experiment.

A. GENERAL KINETIC PROCEDURE

Throughout this experiment, use the same pipet and buret as well as standardized base solution.

Measure out 100 ml of the appropriate solvent accurately with a graduated cylinder, transferring the solvent into a 250-ml Erlenmeyer flask equipped with a well-fitting rubber stopper. Obtain 80 ml of 98%[13] isopropyl alcohol to use for quenching purposes.

Obtain 125–150 ml of approximately 0.04 *N standardized* sodium hydroxide solution. Stopper the flask with a well-fitting solid rubber stopper. *Note and record the concentration of the standard base solution.* Set up a buret, rinse it with a small amount of the base, fill it, see that all air bubbles are out of the tip, and cover the top of the buret with a test tube or small beaker to minimize absorption of carbon dioxide from the air.

Put about 2 ml of phenolphthalein indicator solution in a test tube and have it available, with a dropper, for use in the different titrations. Also connect a short length of rubber tubing to the nearest aspirator or vacuum line for use in drawing air through the pipet for a minute or two after each sampling in order to dry it before taking the next sample. Have available a watch with a dial marked in minutes. *Record the laboratory temperature.*

To initiate a kinetic run, add the alkyl halide to the solvent mixture. (A sample size of approximately 1 g is satisfactory; the sample must be measured *accurately* by means of a pipet, or by weighing.) Swirl the solution gently to obtain homogeneity. Note the time of addition to the nearest minute. Keep the flask well stoppered to avoid evaporation.

While waiting to make the first measurement, determine a "blank" correction for the solvent. Using a graduated cylinder, measure a separate 10-ml portion of the isopropyl alcohol-water mixture being used into a 125-ml Erlenmeyer flask. (*Note:* Do *not* use the mixture containing the alkyl halide.) Next, add 10 ml of 98% isopropyl alcohol and 4–5 drops of phenolphthalein to the "blank," and titrate with base to a faint pink color that persists for 30 sec. In all titrations, use a white background (paper or a towel) below the titration flask, and accentuate the lower edge of the meniscus in the buret by holding dark paper or some other dark object just below it to make it easier to read. The blank correction will probably be no more than 0.05–0.15 ml.

At regular intervals, take a 10-ml sample from the reaction mixture with a 10-ml pipet and add it to 10 ml (measured with a graduated cylinder) of 98% isopropyl alcohol contained in a 125-ml flask. Be sure to note the time of addition of the aliquot, probably best taken as the time at which half of the aliquot has been added to the alcohol used to "quench" the reaction. Titrate with base to the phenolphtholein end-point as in the blank determination above.

The suggested *approximate* times of taking aliquots using various solvents under various conditions are listed below:

(a) 50% isopropyl alcohol-water and *t*-butyl chloride at room temperature: 10, 20, 35, 50, 75, and 100 min.

(b) 60% isopropyl alcohol-water and *t*-butyl chloride at room temperature: 20, 40, 70, 100, 130, and 170 min.

(c) 55% isopropyl alcohol-water and *t*-butyl chloride at room temperature: 15, 30, 50, 75, 100, and 135 min.

(d) 50% isopropyl alcohol-water and *t*-pentyl chloride at room temperature: 10, 20, 30, 40, 50, and 60 min.

---

[13]Commercial isopropyl alcohol is 98% pure and is satisfactory for quenching the solvolysis reactions. The term "98% isopropyl alcohol" is used throughout to indicate the commercial grade.

(e) 60% isopropyl alcohol-water and *t*-pentyl chloride at room temperature: 20, 40, 60, 80, 100, and 120 min.

(f) 55% isopropyl alcohol-water and *t*-pentyl chloride at room temperature: 15, 30, 45, 60, 80, 110, and 140 min.

The "infinity" titrations (after the reaction is *complete*) for each of these sets of conditions should be done during the next laboratory period. Be sure to leave the flask containing the reaction mixture *tightly* stoppered.

B. SOLVOLYSIS OF *t*-BUTYL BROMIDE

The kinetics of solvolysis of *t*-butyl bromide may conveniently be studied at room temperature and at 0°. Since the reactions are quite fast it will be advantageous for students to work in pairs, with one student drawing and quenching samples while the other carries out the necessary titrations.

For the room temperature measurements, the 10-ml aliquot is withdrawn at the suggested time interval and added to an Erlenmeyer flask containing 20 ml of 98% isopropyl alcohol which has previously been cooled to 0°. The titration is carried out at 0° by keeping the flask in an ice-water bath during the titration. These precautions minimize the extent of solvolysis which may occur during the time necessary for titration.

For the kinetics measurements at 0°, proceed as follows. Add the designated solvent mixture to a dry 250-ml Erlenmeyer flask and place the flask in a pan. Add enough ice and water so that the level of the water in the pan is about the same as that in the flask. Be sure that the flask is fitted with a rubber stopper. Note and record the temperature of the bath, and with a *dry* thermometer, determine the temperature of the contents *within* the flask after it has come to equilibrium with the ice water. Proceed according to the directions given in part A by adding an accurately known amount of *t*-butyl bromide (*ca.* 1 g) to the solvent. More ice will have to be added from time to time and some water will have to be removed periodically to maintain a constant level in the pan. It may be helpful to hold the flask with a clamp; however, be sure to mix the contents of the flask well after adding the alkyl bromide.

Infinity point measurements should be taken during the next laboratory period. Be sure to leave the flask containing the reaction mixture *tightly* stoppered.

Suggested time intervals for sampling are given below:

(g) 50% isopropyl alcohol-water and *t*-butyl bromide at room temperature: 1, 2, 3, 4, 5, 6, 8, and 10 min.

(h) 55% isopropyl alcohol-water and *t*-butyl bromide at room temperature: 2, 4, 6, 8, 10, 12, and 15 min.

(i) 60% isopropyl alcohol-water and *t*-butyl bromide at room temperature: 2, 5, 7, 10, 15, 20, and 25 min.

(j) 50% isopropyl alcohol-water and *t*-butyl bromide at 0°: 2, 5, 8, 11, 15, 20, and 28 min.

(k) 55% isopropyl alcohol-water and *t*-butyl bromide at 0°: 4, 9, 15, 20, 30, 40, and 50 min.

(l) 60% isopropyl alcohol-water and *t*-butyl bromide at 0°: 4, 12, 18, 25, 35, 50, and 70 min.

C. TREATMENT OF DATA

1. If an infinity titration was made, proceed as follows. From the titration volumes, determine C_o and C_t for each time t. If the same pipet, buret, and standardized base solution were used for each titration, "ml of NaOH" may be used as A_o and A_t. For each kinetic point, compute the logarithm of C_o/C_t or A_o/A_t, and then multiply this value by 2.303 [see equation (26) or (27)]. Plot the resulting number vertically *vs.* t (in hr) for each kinetic "point." Draw the best straight line through the points which were plotted. Determine the slope of the line; the slope is the rate constant, k_1. It should be about 0.2 hr^{-1} for 60% isopropyl alcohol and t-butyl chloride at room temperature and about 0.8 hr^{-1} for 50% isopropyl alcohol and t-butyl chloride at room temperature.

When this method is used, it is not necessary to apply the blank correction since it is a constant which does not affect the slope of the line (see Exercise 3).

2. If an infinity titration was *not* made, proceed as follows. Using the known weight of the alkyl halide, determine the initial concentration (C_o) of the alkyl halide in the original reaction mixture. (If only the volume of the starting alkyl halide was known, its weight may be determined from its density.) Using the normality of the standard base solution and its volume, as well as the volume of each sample which was titrated, determine C_t for each kinetic point. With this method, it will be necessary to apply the blank correction to C_o and to each value of C_t. Determine the values of $2.303 \log (C_o/C_t)$ for each time t, and plot these data as described in 1 (above). Determine the value of k_1 from the slope of the line.

3. Take the same data and calculate the rate constant separately for *each* kinetic point by putting the data directly into equation (26) and solving for k_1 (in hours^{-1}).

4. Compare each of the values obtained in 3 (above) with the average of those values and with the value obtained from the graph in 1 or 2 (above). Which method allows one to spot an error more easily?

5. From equation (26), it can be seen that the half-life ($t_{1/2}$, that time necessary for half of the original alkyl chloride to react) is given by the relation:

$$t_{1/2} = \frac{0.69}{k_1} \tag{31}$$

Calculate the half-life using the value of k_1 obtained from the graph. After the half-life has been calculated, go back and examine the experimental data. If about one-half of the total volume of NaOH used in the final ("infinity") titration would not have been consumed in a titration at this time, an error has been made in the calculations. Recheck them carefully.

6. Consider the magnitudes and types of errors involved in the various measurements made in this experiment. (Most general chemistry laboratory manuals include a discussion of errors.) Using the unavoidable errors in measurement which would have been expected in volume, time, and titration measurements, calculate the *maximum* error in k_1 which can be expected (*i.e.,* $k_1 = 0.80$ hour$^{-1} \pm 0.05$ hour^{-1}).

7. If you have performed rate determinations for t-butyl bromide at room temperature *and* at 0°, the two rate constants obtained in the above calculations provide sufficient data for the determination of the energy of activation for the solvolysis

in the solvent system used. Calculate the energy of activation, E_{act}, using the integrated form of the Arhenius equation:

$$E_{act} = 2.303 \frac{RT_2T_1}{T_2 - T_1} \log \frac{k_2}{k_1} \tag{32}$$

Use for R, the ideal gas constant, the value of 0.0198 kcal/mole deg in order to obtain the energy of activation in kcal/mole. T_2 and k_2 are the temperature and rate constant at room temperature, and T_1 and k_1 are the temperature and rate constant at 0°; temperature is expressed in degrees Kelvin.

EXERCISES

1. Give as many advantages as possible for allowing the reaction to proceed essentially to completion, allowing for a final titration, as opposed to adding a known volume of alkyl halide and *not* making a final titration.
2. Explain clearly why "ml of NaOH" can be used [equation (27)] instead of using C_o and C_t in moles/liter in equation (26). Is the normality of the NaOH needed? Why or why not?
3. In the plot of 2.303 log (C_o/C_t) *vs.* t, why doesn't the line go through 2.303 log $(C_o/C_t) = 0$ when $t = 0$, *i.e.*, why is C_o not equal to C_t at $t = 0$?
4. Show how equation (31) can be derived from equation (26).
5. Suppose that the flask had not been stoppered until the next laboratory period when the final titration was taken. Would the rate constant be expected to be larger or smaller than it would have been with the flask stoppered?
6. Why does the titration endpoint fade after 30–60 seconds?
7. Using the chemical equations involved for the hydrolysis and for the titration of the acid formed with base, show that in order to get C_t one must subtract [H^+] from C_o. Explain the reasoning. (C_t is the concentration of alkyl halide which remains at any time t.)
8. In a titration done at 0°, the reaction mixture is stored at room temperature until the next laboratory period before carrying out the final titration. Since the rest of the titrations were done at 0°, why is the value of the final titration valid?
9. List the errors involved in the determinations of rate constants by the procedure used. State the relative importance of each.

chapter thirteen

carbenes and arynes

Highly Reactive Intermediates

For a more complete understanding of any organic reaction, recognition and characterization of possible intermediates formed during the conversion of starting materials to products is required. The types of intermediates encountered in organic chemistry are varied, both in their relative reactivities and in their electronic structures. Charged intermediates, such as carbonium ions and carbanions, are usually very reactive, and their preparation and isolation commonly requires special experimental precautions such as the use of inert atmospheres and the rigorous exclusion of moisture from the reaction. For instance, the preparation of a Grignard reagent, RMgX (X = halogen), an organometallic substance having the properties anticipated for a carbanion, is successful only if the reagents and apparatus used are dry. Some charged intermediates, however, are sufficiently stable so that they can be isolated as salts without the use of special experimental techniques, as is the case in the preparation of triphenylmethyl fluoborate, $(C_6H_5)_3C^+BF_4^-$, a carbonium ion salt (Chapter 20).

Uncharged intermediates quite often are stable molecules which can be readily isolated by use of appropriate reaction conditions. As an example, benzophenone (1), a stable ketone, undoubtedly intervenes as an intermediate in the preparation of triphenylmethanol by reaction of methyl benzoate with two moles of phenylmagnesium bromide [equation (1)]. Although benzophenone is not isolated in the procedure

$$C_6H_5CO_2CH_3 + C_6H_5MgBr \longrightarrow \underset{\mathbf{1}}{C_6H_5COC_6H_5} \xrightarrow[\text{2) } H_3O^\oplus]{\text{1) } C_6H_5MgBr} (C_6H_5)_3COH \qquad (1)$$

described in Chapter 14, it could be if the conditions of the reaction were modified. Many neutral intermediates cannot be easily isolated, however, because of their high reactivity. Typical of these are free radicals such as those generated in the free radical-chain halogenation of alkanes (Chapter 5).

In this chapter, we shall consider the generation and reactions of two very reactive neutral intermediates which owe their high reactivity to two quite different molecular properties. The first of these substances has the general formula **2** (R = H, alkyl, aryl, halogen, cyano, etc.) and is called a *carbene*. The presence of only six rather than eight electrons in the outer valence shell imparts high reactivity to such intermediates. The experiment described in Section 13.1 will involve the generation and chemistry of *dichlorocarbene* (**2**, R = Cl).

The other intermediate to be considered in this chapter is *benzyne* (**3**), a member of a class of compounds designated as *arynes*. The high reactivity of benzyne, or dehydrobenzene as this molecule is sometimes called, is considered to be due to the large amount of strain resulting from introduction of a triple bond into a six-membered ring.

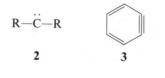

$$R—\overset{..}{C}—R$$

2   **3**

13.1   Carbenes

Carbenes can be generated in a variety of ways, two of which are given below. One method involves decomposition of a diazo compound (**4**) by irradiation or heat, or with a metallic catalyst [equation (2)]. This method is generally much too hazardous to

$$\left[R_2C{=}N{=}N \longleftrightarrow R_2\overset{\ominus}{C}{-}\overset{\oplus}{N}{\equiv}N: \right] \xrightarrow[\text{or metallic catalyst}]{\text{light or heat}} R_2C: + :N{\equiv}N: \qquad (2)$$

$$\qquad\qquad\qquad\qquad\qquad \mathbf{4} \qquad\qquad\qquad\qquad\qquad\qquad\qquad\qquad \mathbf{2}$$

be attempted in the introductory organic laboratory because the diazo compounds are often explosive.

An alternative approach to producing carbenes is that termed α-elimination. This technique fundamentally involves initial formation of a carbanion of the type **5**, followed by decomposition of the carbanion to a carbene by loss of an anion such as halide ion [equation (3), M = metal, X = halogen]. This route to carbenes is quite safe

$$R—\overset{\displaystyle R}{\underset{\displaystyle X}{\overset{|}{\underset{|}{C}}}}:^{\ominus}M^{\oplus} \longrightarrow R_2C: + X^{\ominus} + M^{\oplus} \qquad (3)$$

$$\qquad\qquad \mathbf{5} \qquad\qquad\qquad \mathbf{2}$$

experimentally but does have the disadvantage that yields of products are lower than when diazo compounds are utilized as sources of carbenes.[1]

Two types of reactions which are characteristic of carbenes are their *insertion* into

[1] It should be noted here that current investigations indicate that α-elimination may not lead to a "free" carbene, as does photochemical or thermal decomposition of a diazo compound, but rather that the reactive intermediate in the former procedure is either **5** or **2** complexed with the salt MX. Whatever the true nature of the active intermediate resulting from α-elimination, it is observed to react in much the same way in the addition reaction as a carbene generated from a diazo compound.

carbon-hydrogen bonds [equation (4)] and their *addition* to carbon-carbon double and triple bonds [equation (5)]. The insertion reaction is of considerable interest mechanistically, but has little synthetic utility. The addition reaction, however, has received wide application in the synthesis of molecules containing three-membered rings. Of importance in terms of the mechanism of the reaction is the observation that the addition of carbenes usually occurs in a stereospecifically *cis* fashion. As an example, the reac-

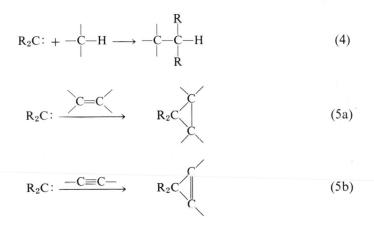

$$R_2C: + \underset{|}{\overset{|}{-C}}-H \longrightarrow \underset{|}{\overset{R}{-C}}-\underset{|}{\overset{|}{C}}-H \tag{4}$$

$$R_2C: \xrightarrow{\diagdown C=C\diagup} R_2C\diagdown \overset{C}{\underset{C}{\diagup}} \tag{5a}$$

$$R_2C: \xrightarrow{-C\equiv C-} R_2C\diagdown \overset{C}{\underset{C}{\diagup}} \tag{5b}$$

tion of *trans*- and *cis*-2-butene with a carbene, $R_2C:$, produces the two different adducts **6** and **7**, respectively.

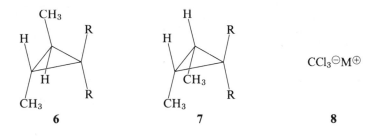

6   **7**   **8**

Precursors **5**, required for generation of carbenes by the method of α-elimination are commonly prepared by the reaction of a strong base with a halogenated substrate. Thus, treatment of chloroform with potassium *t*-butoxide yields **8** (M = K), which can subsequently decompose, with loss of potassium chloride, to dichlorocarbene, (**2**, R = Cl). Alternatively, the salt **8** (M = Na), can be produced by the thermal decomposition of sodium trichloroacetate [equation (6)], which is the procedure to be followed in this experiment.

$$Cl_3C\!\!-\!\!C\!\!-\!\!O^{\ominus}Na^{\oplus} \xrightarrow[\text{heat}]{-CO_2} Cl_3C^{\ominus}Na^{\oplus} \xrightarrow{-NaCl} Cl_2C: \tag{6}$$

8

The decomposition of sodium trichloroacetate will be performed in a mixture of tetrachloroethylene and diglyme, $(CH_3OCH_2CH_2)_2O$, as solvent. This mixture has the advantages that its boiling point, approximately 110°, is high enough to produce a suitable rate of decomposition of the sodium trichloroacetate, and that the acetate is only slightly soluble in the medium so that its concentration in solution is rather low. This low concentration minimizes the importance of side reactions, such as that shown in equation (7), which act to decrease the yield of the desired addition product.

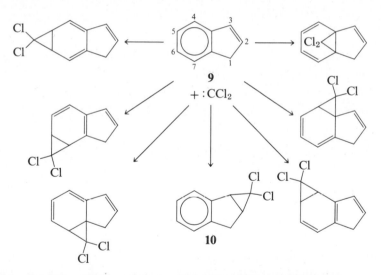

$$CCl_3^{\ominus}Na^{\oplus} + :CCl_2 \longrightarrow Cl_2\overset{Cl}{\underset{}{C}}\overset{Na^{\oplus}}{\underset{}{\overset{\ominus}{C}Cl_2}} \longrightarrow \overset{Cl}{\underset{Cl}{}}C=C\overset{Cl}{\underset{Cl}{}} + NaCl \qquad (7)$$

Indene (9) will be used as the unsaturated substrate with which the dichlorocarbene is to react. In principle a number of products resulting from both the insertion and addition reactions is possible. Fortunately, the reaction mixture is not as complex as it might be since carbenes formed by α-elimination preferentially give products resulting from the addition rather than from the insertion of the carbene. However, even if the reaction is limited to addition of dichlorocarbene to indene, there is still possible a variety of products, all of which are shown in Chart 13.1.

CHART 13.1 Possible addition products of dichlorocarbene and indene.

A prediction concerning which of the possible products might be favored can be made on the basis that a fundamental difference exists between compound 10, which bears the imposing name, 1,1-dichloro-1a,6a-dihydrocycloprop[a]indene, and all the other potential products of addition. The formation of 10, by attack of dichlorocarbene on the double bond between carbon atoms 2 and 3 of indene, permits preservation of the aromaticity of the benzene ring, whereas the production of any of the other adducts results in the loss of the aromaticity in the molecule. Therefore a preference for addition of dichlorocarbene across the 2,3-double bond might be anticipated.

In fact, this preference is realized experimentally, since generation of dichlorocarbene in the presence of indene, followed by work-up of the reaction mixture under mild conditions, affords **10** in 70% yield as the only isolable product.

Although the attack of dichlorocarbene on indene turns out to be selective, a possible further complication should be noted. Three-membered rings bearing *geminal* halo-

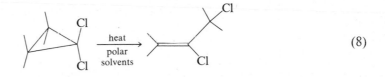

$$(8)$$

gens, such as that in **10**, are relatively unstable and rearrange in polar media and/or with heating to ring-opened isomers [equation (8)]. In the case of **10**, rearrangement is known to occur readily at temperatures greater than 50°. Compound **10**, by analogy to the general reaction given in equation (8), would be expected to rearrange to a mixture of **11** and **12** [equation (9)], but it has been observed experimentally that upon treatment with an ethanolic solution of potassium hydroxide, **10** yields 2-chloronaphthalene (**13**) in high yield. Compounds **11** and **12** may serve as intermediates in the formation of **13**, but they readily lose the elements of hydrogen chloride to produce the more highly stable naphthalene system.

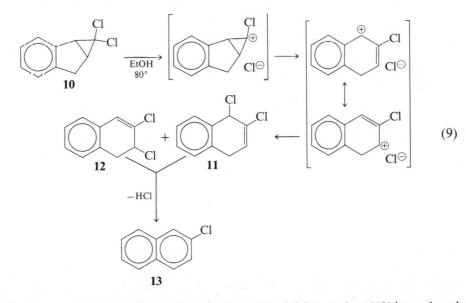

$$(9)$$

In the experimental procedure described below, the addition product (**10**) is produced by heating indene and sodium trichloroacetate at a temperature greater than 100° in a polar medium. As a consequence **13**, the rearrangement product from **10**, is isolated in this experiment.[2] The overall equation for the reaction is therefore the following:

[2] Students interested in preparing compound **10** should consult reference 2 given at the end of this section.

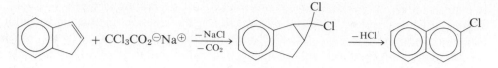

It is important to note that this experiment demonstrates the marked effect that reaction conditions and purification techniques may sometimes have on the structures of products ultimately isolated from organic reactions. Minor modifications in an experimental procedure may be the difference between obtaining a desired product in high yield or producing major amounts of undesired compounds.

EXPERIMENTAL PROCEDURE

Equip a 250-ml round-bottomed flask with a reflux condenser fitted with a thermo-meter adapter, or 1-hole rubber stopper, through which is passed a glass tube. Attach to this tube a length of rubber tubing bearing a short piece of glass tubing. Place 0.081 mole of sodium trichloroacetate,[3,4] which has previously been dried for at least one hour at 90°, 20 ml of tetrachloroethylene, 5 ml of diglyme, and 0.086 mole of indene [5] in the flask. Bring this mixture to a gentle reflux with the aid of a microburner, swirling the flask occasionally to insure good mixing. Continue to heat until the evolution of carbon dioxide has ended, as indicated by the absence of bubbling when the tube lead-ing from the top of the condenser is inserted into a few milliliters of tetrachloro-ethylene contained in a test tube. At this time, the granular sodium trichloroacetate should have disappeared, and finely divided sodium chloride should have separated.★

Before removing the burner from beneath the flask, disconnect the bubbler in order to prevent suction of tetrachloroethylene into the reaction flask. Detach the flask from the condenser, add 25 ml of water to the reaction mixture, and attach the flask to an apparatus set for steam distillation using an external source of steam (see Figure 2.11).

Steam-distil the reaction mixture as rapidly as the ability of the condenser to cool the distillate will permit, collecting the distillate in a graduated cylinder. When approxi-

---

[3] If commercial sodium trichloroacetate (Columbia Organic Chemicals) is used, take into account the fact that the product is of 95% purity when making the calculation of the number of grams re-quired.

[4] If desired, sodium trichloroacetate can be prepared in the following manner. Prepare a solution of 0.090 mole of sodium hydroxide in 10 ml of water and cool this solution thoroughly in an ice-water bath. Place 0.086 mole of trichloroacetic acid in a 125-ml suction flask and, while swirling this flask in the ice-water bath, add to it about 9 ml of the cold solution of base. Now add one drop of 0.04% Bromocresol Green solution so that a yellow color is visible. Complete the titration of trichloroacetic acid by adding the hydroxide solution dropwise to a point where a single drop produces a color change from yellow to blue. If the end-point is overshot, add some additional acid and repeat the titration.★ Stopper the flask and attach it to a water aspirator by means of suction tubing. Place the flask *within* the rings of a steam cone and concentrate the solution by evacuating the flask and heating with steam. After the solution has been evaporated to dryness,★ which will require about 20 minutes, transfer the solid to a watch glass, break up any large lumps, and dry the sodium trichloroacetate in an oven at 90°.

[5] Indene is not stable for long periods of time and therefore should be purified immediately before use by means of a simple distillation.

mately 26 ml of an organic layer has been obtained, which will require that a total volume of 60 to 65 ml of distillate be collected, replace the graduated cylinder with a 250-ml Erlenmeyer flask and discard the liquid in the graduated cylinder.* Continue the distillation until no more oily material can be detected in the distillate. Collection of about 200 ml of distillate will be required to obtain all of the crude 2-chloronaphthalene.* During the course of the distillation, check to make sure that solid does not collect in the condenser and block it. If solid does begin to form, shut off the flow of water through the condenser and allow the uncondensed steam to sweep the solid out of the condenser.

With the aid of a water bath, cool the distillate to room temperature and then transfer it to a separatory funnel. Some crystalline 2-chloronaphthalene may remain in the Erlenmeyer flask, but it will be recovered later. Extract the aqueous solution in the funnel three times with separate 25-ml portions of either dichloromethane or technical ether and combine the extracts in the 250-ml Erlenmeyer flask used to collect the steam distillate.*

Remove the solvent from the crude product by heating the solution on a steam bath (preferably in the hood).* Alternatively, the solvent can be removed by performing a simple distillation (*No flames* if ether was used!). Transfer the crude 2-chloronaphthalene to a smaller Erlenmeyer flask, rinse the larger flask with 15 ml of 95% ethanol, and add this rinse along with some decolorizing carbon to the product. Boil this solution for a few minutes on the steam cone and then filter to remove the carbon.* Adjust the temperature of the filtrate to about 45° and, while maintaining this temperature, add water dropwise until the solution becomes turbid. At this point, allow the solution to stand undisturbed until it cools to room temperature.* If the 2-chloronaphthalene fails to crystallize but rather begins to oil out, a problem which is characteristic of many low-melting solids, scratch the walls of the flask at the interface between the solution and the air in an attempt to induce crystallization. In some cases, it may be necessary to evaporate a small sample of the solution to obtain the oily crude product which, upon scratching, may provide a few seed crystals that can be added to the solution to encourage crystallization.

When no more solid crystallizes at room temperature, cool the flask in an ice-water bath to complete the crystallization, and collect the product. A second crop of product can be obtained if the filtrate is reheated to 45° and water added dropwise until the solution again becomes turbid. Cooling and scratching as before should provide additional 2-chloronaphthalene of somewhat lower purity than the first crop.

Pure 2-chloronaphthalene is an irridescent white solid [6] (mp 58–59°). The yield of product in this experiment should be 20–30%.

EXERCISES

1. Why is tetrachloroethylene rather than some other liquid such as water used to test for the evolution of CO_2?

---

[6] For a discussion of the reason for the blue color often observed in the crude 2-chloronaphthalene and even in the recrystallized product obtained in this experiment, see Exercise 6.

2. Predict which double bond in phenanthrene would be most reactive toward a carbene and give an explanation for your choice.

Phenanthrene

3. What properties of the diglyme-tetrachloroethylene mixture make it suitable as a solvent in the reaction?

4. What organic compounds distil in the first stage of the steam distillation?

5. When attempting a mixed-solvent recrystallization of the type performed in this experiment, the solvent in which the solute is sparingly soluble is normally added to a *boiling* solution of the solute and the other solvent. What special circumstances prohibit the addition of water to a boiling solution of 2-chloronaphthalene in 95% ethanol in order to effect saturation? Why is it more efficient to saturate the ethanolic solution at about 45° than at, for example, room temperature?

6. The 2-chloronaphthalene isolated by the procedure described in this experiment sometimes has a bluish tinge caused by the presence of a small amount of a chloroazulene. The chlorine substituent is in the seven- rather than the five-membered ring but its exact position is unknown. Propose a mechanism for the formation of the chloroazulene from indene and dichlorocarbene. Suggest a reason why it is difficult to remove the azulene from 2-chloronaphthalene even after repeated recrystallization.

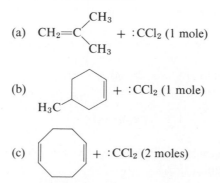

7. Write out the expected products of the reaction between dichlorocarbene and the following alkenes under mild conditions.

(a) $CH_2{=}C\overset{\displaystyle CH_3}{\underset{\displaystyle CH_3}{\big\langle}}$ + :CCl_2 (1 mole)

(b)   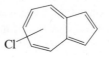   + :CCl_2 (1 mole)

(c)   + :CCl_2 (2 moles)

8. Suggest why sodium trichloroacetate decarboxylates much more readily than sodium acetate.

REFERENCES

1. W. Kirmse, *Carbene Chemistry,* 2nd Ed., Academic Press, Inc., New York, 1971.
2. W. E. Parham, H. E. Rieff, and P. Swartzentruber, *Journal of the American Chemical Society,* **78,** 1437 (1956).

SPECTRA OF STARTING MATERIAL AND PRODUCT

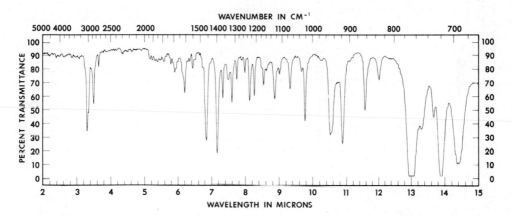

FIGURE 13.1 IR spectrum of indene.

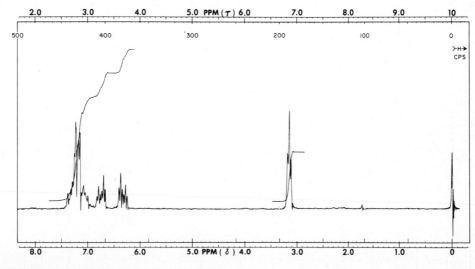

FIGURE 13.2 NMR spectrum of indene.

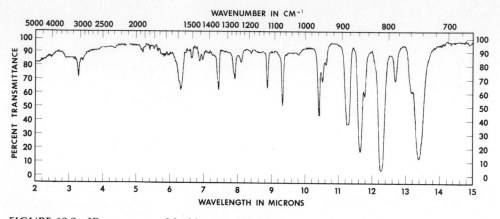

FIGURE 13.3   IR spectrum of 2-chloronaphthalene.

13.2   Arynes

Arynes, which can be considered to be cyclohexadienynes, have been recognized as reaction intermediates only since the early 1950s. Evidence which prompts postulation of their existence includes the observations that treatment of **14** with amide ion produces an equimolar mixture of **15** and **16** [equation (10)], that reaction of **17** with hydroxide ion generates **18** and **19** [equation (11)] and that pyrolysis of **20** in the

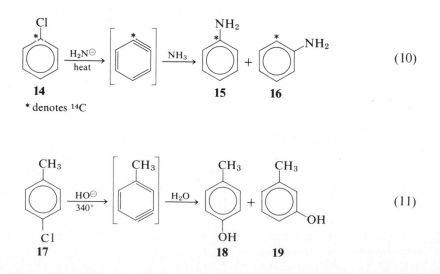

presence of anthracene yields triptycene **(23)** [equation (12)]:

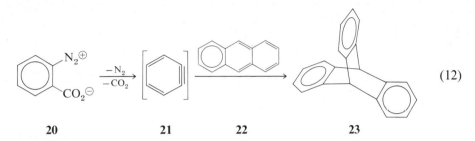

$$20 \qquad\qquad 21 \qquad 22 \qquad\qquad 23 \tag{12}$$

Equations (10) and (11) show the production of an aryne by an elimination reaction followed by reaction of the aryne with a nucleophile. Equation (12) exemplifies the generation of the aryne, benzyne **(21),** by thermal decomposition of an unstable "zwitterionic" species, **20,** with subsequent reaction of the benzyne, acting as a dienophile, with a diene.

Although benzyne itself is often generated by reaction of an aryl halide with a strong base [equations (10) and (11)], a more convenient procedure involving synthesis and decomposition of benzenediazonium-2-carboxylate **(20)** will be used here. As noted above, **20** is unstable and can be violently explosive, particularly when dry, if it is subjected to shock. The isolation of **20,** therefore, represents a hazardous and, at the least, an unwise endeavor. It is possible to circumvent the hazards in this instance, and similarly in many other cases where preparation and reaction of dangerous intermediates are involved, by generation *and* decomposition of **20** *in situ,* the goal being never to have a very high concentration of benzene-diazonium-2-carboxylate **(20)** present.

Because it is the desire to decompose **20** to benzyne about as fast as **20** is produced, it is necessary to have any reagent that is to react with the benzyne present from the very start of the reaction; ideally, then, the reagent for trapping the benzyne should not itself react with the reagents involved in the production of **20.** This is the case in the procedure outlined below; anthracene **(22),** an aromatic hydrocarbon, does not react to any significant extent with either isoamyl nitrite **(24)** or anthranilic acid **(25)** (Chart 13.2) under the reaction conditions used.

Benzenediazonium-2-carboxylate **(20)** is the product of a reaction between isoamyl nitrite **(24),** a diazotizing reagent, and anthranilic acid **(25).** A possible reaction mechanism is outlined in Chart 13.2. Intermediate **20** is prepared at a temperature sufficiently high to induce its decomposition to benzyne, which is subsequently trapped by anthracene **(22)** to provide the Diels-Alder adduct triptycene **(23)** [equation (12)]. Anthracene is functioning as a diene in this reaction.

The formation of triptycene indicates that benzyne adds preferentially across the 9- and 10-positions of anthracene. The fundamental reason for this selectivity is that **23** has the greatest amount of aromatic stabilization of all the possible products.[7]

---

[7] A similar explanation was used to rationalize the preferred attack of dichlorocarbene on the 2,3-carbon-carbon double bond of indene (Section 13.1).

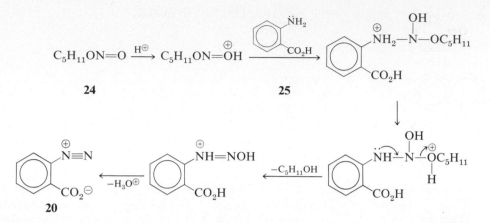

CHART 13.2   The preparation of benzenediazonium-2-carboxylate.

The choice of solvents for this reaction is crucial. Clearly, it is important that all starting reagents and the intermediate **20** be soluble in the medium. It is particularly crucial that **20** not be insoluble because if it were it would precipitate from the reaction mixture and accumulate, an occurrence that is to be avoided. Furthermore, the boiling point of the solvent must be sufficiently high so that **20** is formed and decomposed at a reasonable rate. A final requirement is that the solvent be *aprotic* in order to minimize side reactions involving nucleophilic attack of solvent on benzyne; for example, *protic* solvents such as alcohols would react with benzyne by adding across the highly reactive "triple bond" [equation (13)]. All of these

$$\text{⟨benzyne⟩} + \text{RO–H} \longrightarrow \text{⟨ArOR/H⟩} \tag{13}$$

requirements are met by 1,2-dichloroethane, an aprotic solvent of modest polarity. The anthranilic acid is added to the reaction mixture as a solution in diethylene glycol diethyl ether, another aprotic solvent of moderate polarity. The high boiling point and water solubility of this ether facilitate isolation of the final product (see Exercise 2).

An important technical problem that is encountered in the purification of the reaction mixture is the removal of unreacted anthracene which, as might be anticipated on structural grounds, has solubility characteristics similar to those of triptycene. This problem is solved by preparing a derivative of anthracene which is soluble in aqueous base. Thus, treatment of the crude reaction mixture with maleic anhydride will result in the destruction of excess anthracene by the formation of the Diels-Alder adduct, **26,** which, on hydrolysis with aqueous potassium hydroxide, affords the water-soluble salt **28.**

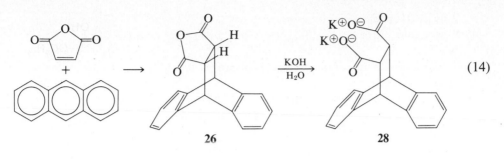

$$\text{26} \qquad\qquad\qquad \text{28} \qquad\qquad (14)$$

EXPERIMENTAL PROCEDURE

Place 5.4 g (0.030 mole) of anthracene, 2.7 g (0.023 mole) of isoamyl nitrite, and 40 ml of 1,2-dichloroethane in a 100-ml round-bottomed flask equipped with a Claisen connecting tube bearing an addition funnel and a reflux condenser; the funnel should be on the straight arm of the connecting tube. Heat the mixture to reflux over a steam bath. After the mixture has started to reflux, remove the apparatus from the steam bath. While the mixture is still warm, add a solution of 3.0 g (0.022 mole) of anthranilic acid in 15 ml of diethylene glycol diethyl ether dropwise from the addition funnel over a period of about 10 min. Agitate the reaction mixture frequently to ensure good mixing. The exothermicity of the reaction should maintain a moderate rate of reflux. After the addition is complete, heat the mixture at reflux over a steam bath for 15 min. Refit the flask for simple distillation and distil until a head temperature of 150° is reached. Allow the reaction mixture to cool somewhat, add 3.0 g (0.030 mole) of maleic anhydride, and heat this mixture at reflux for 5 min.★ Cool the flask in an ice-water bath and slowly add with stirring a solution of 6.0 g of potassium hydroxide in 20 ml of methanol and 10 ml of water.★ Collect the crude triptycene by suction filtration of the cold mixture and wash it with 4 : 1 methanol-water solution until the washings are colorless (15–20 ml).★

Purify the crude product by dissolving it in butanone (10 ml per gram of triptycene), heating the solution on a steam bath, adding a small amount of decolorizing carbon, and filtering the solution.★ Concentrate the filtrate to about two-thirds of its original volume, add an equal amount of methanol, and cool the solution in an ice-water bath. Collect the product by suction filtration and wash it with a small amount of cold methanol. The reported melting point of triptycene is 254–255.5°.

EXERCISES

1. What substances are removed by the simple distillation?
2. Outline the reasons why the high-boiling point and the water solubility of the particular ether chosen as one of the solvents in this experiment facilitate isolation of the triptycene formed in the reaction.

3. Why is the simple distillation necessary?
4. What is the purpose of adding methanol during the recrystallization of triptycene?
5. Predict the structure of the product resulting from addition of 1 mole of bromine to anthracene.
6. Explain, on the basis of stabilization (delocalization) energies, why the product of reaction between benzyne and anthracene is triptycene, rather than the compound shown below.

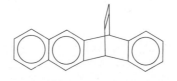

REFERENCES

1. L. Friedman and F. Logullo, *Journal of Organic Chemistry,* **34,** 3091 (1969).
2. H. Heaney, *Chemical Reviews,* **62,** 81 (1962).
3. R. W. Hoffman, *Dehydrobenzene and Cycloalkynes,* Academic Press, Inc., New York, 1967.

SPECTRUM OF STARTING MATERIAL AND PRODUCT

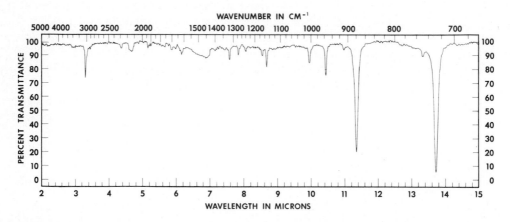

FIGURE 13.4   IR spectrum of anthracene.

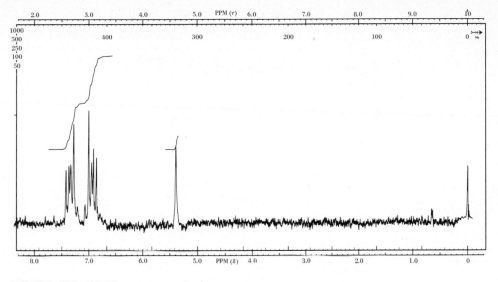

FIGURE 13.5 NMR spectrum of triptycene.

chapter fourteen

organometallic chemistry

Organometallic reagents represent perhaps one of the single most useful synthetic tools for laboratory reactions. This chapter is devoted to discussion and experimental application of the Grignard and the organolithium reagents to organic syntheses.

14.1 The Grignard Reagent: Its Preparation and Reactions

The preparation of the Grignard reagent, $RMgX$, is represented in equation (1),

$$RX + Mg \xrightarrow[\text{(solvent)}]{\text{dry ether}} RMgX \qquad (1)$$

where RX represents an organic halide. The exact nature of the Grignard reagent in solution is not known. It is believed to be a mixture of numerous species, but some of the principal species are undoubtedly R_2Mg and MgX_2 in equilibrium with $2\,RMgX$. These species are highly solvated by ether and are complexed with one another. It is customary to represent the Grignard reagent by the formula $RMgX$ when writing chemical equations, but it should be kept in mind that the species in solution are of a much more complex nature.

The ether solvent is an essential part of the Grignard reagent, for it is known to form a complex with the magnesium which is present in the reagent. Several cases are known where Grignard reagents have been prepared in the absence of ether, but the yields are not good. Satisfactory yields are usually obtained when ether is present. The most common ether solvent is ethyl ether, $(C_2H_5)_2O$, due to its low cost and ease of removal (its boiling point is 36°). Other ethers, for example n-butyl ether, bp 142°, have been used successfully when higher-boiling solvents are required. Some cyclic ethers, such as tetrahydrofuran, have been used with excellent success in the preparation of certain Grignard reagents (see below).

The organic halide may, in general, be of any organic moiety (alkyl or aryl) and the halide may be bromide, chloride, or iodide. Alkyl iodides, bromides, or chlorides and aryl iodides or bromides can be converted to the corresponding Grignard reagents using ethyl ether as the solvent. Aryl chlorides, on the other hand, do not react with magnesium in ethyl ether, but they react readily when tetrahydrofuran is used as the solvent.

The initiation of the reaction between the halide and magnesium is sometimes difficult, with the iodides being more reactive than the bromides, which are in turn more reactive than chlorides. Aryl halides are less reactive than their alkyl counterparts, and it has been found that the aryl bromides and the alkyl chlorides react about equally well. In cases where one is dealing with a particularly unreactive halide, the initiation of the reaction may sometimes be accomplished by the addition of a small amount of iodine to the reaction mixture. This is believed to convert a small amount of the halide to the iodide, which is more reactive.

The preparation of the Grignard reagent *must* be carried out under anhydrous conditions and if possible in the absence of oxygen. It is exceedingly important to maintain completely dry conditions throughout, for the presence of water inhibits the initiation of the reaction as well as destroys the reagent once it is formed.

The reaction that occurs when the Grignard reagent comes in contact with water is shown in equation (2). The reagent is a strong base, as one of the carbon atoms bears

$$RMgX + H_2O \longrightarrow RH + HOMgX \tag{2}$$

substantial negative charge ($R^-Mg^{++}X^-$). The Grignard reagent, acting as a base, removes a proton from water, which acts as an acid. The overall effect is the destruction of the reagent, with the formation of a hydrocarbon (RH) and a basic magnesium salt. The hydrocarbon is inert, so that the net effect is the production of hydroxide ion in solution. In fact, it is possible to standardize a solution of a Grignard reagent by adding a known volume of it to a known volume of standardized mineral acid and backtitrating with standardized base.

In addition to the reaction with water [equation (2)], there are other side reactions that may occur during formation of the Grignard reagent, as shown in equations (3–5).

REACTION WITH OXYGEN:

$$2\,RMgX + O_2 \rightarrow 2\,ROMgX \tag{3}$$

REACTION WITH CARBON DIOXIDE:

$$RMgX + CO_2 \rightarrow RCO_2MgX \tag{4}$$

COUPLING:

$$RMgX + RX \rightarrow R\!-\!R + MgX_2 \tag{5}$$

It is possible to minimize these reactions by taking certain precautions when carrying out the experimental work. The reaction with oxygen and carbon dioxide may be avoided by carrying out the reaction under an inert atmosphere (such as nitrogen or helium gas). In research work, this is routinely done; however, when ethyl ether is used as a solvent, it excludes a certain amount of air from the reaction vessel due to its very high vapor pressure. The coupling reaction [equation (5)] is an example of a Wurtz reaction. It is not possible to eliminate this coupling reaction completely, but it may be minimized by using dilute solutions so as to avoid localized high concentrations of halide. This can be done by using very efficient stirring and by slowly adding the halide to the magnesium in ether. Normally the rate of addition of halide (dissolved in ether) and the rate of reflux (when ethyl ether is used) should be adjusted so that they are about equal. Alkyl iodides are much more prone to coupling reactions than are the bromides and chlorides, so that the latter are preferable for preparing Grignard reagents even though they are less reactive.

The experiments which appear in this chapter involve the preparation of two Grignard reagents from organic bromides and three typical reactions of these reagents, to produce a tertiary alcohol, a carboxylic acid, and a secondary alcohol.

Preparation of phenylmagnesium bromide and its reaction with methyl benzoate. The equation for the preparation of phenylmagnesium bromide indicates that 1 mole of magnesium must be added for each mole of bromobenzene which is used; the main reaction is

$$C_6H_5Br + Mg \xrightarrow{(C_2H_5)_2O} C_6H_5MgBr \tag{6}$$

The most important side reaction is the coupling process

$$C_6H_5MgBr + C_6H_5Br \rightarrow C_6H_5-C_6H_5 + MgBr_2 \tag{7}$$

but it is not a significant problem. Once the reagent is prepared, it is used directly in subsequent reactions; it is not necessary to separate the small amount of the by-product biphenyl from the reagent before using it.

When phenylmagnesium bromide is allowed to react with an ester such as methyl or ethyl benzoate, the chief product of the reaction is triphenylmethanol. The formation of this compound can be envisioned as arising from the two-step reaction shown by equations (8) and (9). The intermediate benzophenone, whose formation is shown in equation (8) as arising from the reaction of 1 mole of methyl benzoate with 1 mole of phenylmagnesium bromide, then reacts with 1 more mole of phenylmagnesium bromide to give the salt of triphenylmethanol, as shown in equation (9). Subsequent hydrolysis of the salt gives triphenylmethanol. In order to avoid precipitation of a basic magnesium salt (HOMgX), the hydrolysis is carried out with an acid solution (H_2SO_4 or NH_4Cl), which keeps the magnesium in solution in the form of normal salts. The reaction for the overall process becomes

$$C_6H_5\underset{\underset{O}{\|}}{C}-OCH_3 + 2\ C_6H_5MgBr + H_2SO_4 \longrightarrow$$
$$(C_6H_5)_3COH + CH_3OH + MgBr_2 + MgSO_4 \quad (10)$$

The synthesis of triphenylmethanol is accomplished experimentally by first preparing the Grignard reagent and then adding the methyl benzoate to it with stirring and cooling (the reaction is highly exothermic). In working up the reaction mixture, water is added to hydrolyze the magnesium salts to the final organic products; a little mineral acid is also commonly added to dissolve the basic magnesium salt which is formed. If this is not done, the magnesium hydroxide will produce an unworkable emulsion. The organic products (triphenylmethanol, benzophenone, biphenyl, unchanged methyl benzoate, and benzene—the benzene coming from action of water upon unchanged phenylmagnesium bromide) all remain in the organic layer (consisting primarily of ethyl ether). The principal organic products present are triphenyl-methanol and biphenyl, with the former being present in the major amount. It is possible to separate these two compounds from one another, owing to the solubility of the hydrocarbon product (biphenyl) in hydrocarbon solvents (such as petroleum ether, hexane, etc.). Recrystallization of the crude product mixture from these solvents leaves pure triphenylmethanol as the solid. This is because the polar, hydroxylic triphenylmethanol is much less soluble in hydrocarbon solvents than is biphenyl.

Reaction of phenylmagnesium bromide with carbon dioxide. The reaction that occurs when phenylmagnesium bromide is allowed to react with carbon dioxide and the resulting magnesium salt hydrolyzed is shown in equation (11); the major product is a carboxylic acid (benzoic acid). The principal side reactions which are encountered in

$$C_6H_5:{}^{\ominus}MgBr^{\oplus} + O{=}C{=}O \longrightarrow C_6H_5-\underset{\underset{O}{\|}}{C}-O^{\ominus}MgBr^{\oplus} \xrightarrow{\ H_2O\ } C_6H_5\underset{\underset{O}{\|}}{C}-OH \quad (11)$$
$$+\ HOMgBr$$

this process result in the formation of a ketone or a tertiary alcohol, as shown by the steps in equation (12). These side reactions can, however, be minimized in several ways.

$$C_6H_5CO_2MgBr + C_6H_5MgBr \longrightarrow (C_6H_5)_2CO + MgBr_2 + MgO$$
$$\downarrow {\scriptstyle C_6H_5MgBr}$$
$$(C_6H_5)_3COMgBr \xrightarrow{\ H_2O\ } (C_6H_5)_3COH \quad (12)$$
$$+\ HOMgBr$$

First of all, the bromomagnesium salt of the carboxylic acid, because it is moderately insoluble in ether, precipitates from solution and is thus effectively prevented from undergoing further reaction. Secondly, dry ice is used as the source of carbon dioxide, so that the temperature is maintained at $-78°$ throughout. Finally, a large excess of carbon dioxide is always used thus enhancing the probability that phenylmagnesium bromide will react with carbon dioxide rather than with the magnesium salt (to produce the ketone) or with the ketone (to yield the magnesium salt of the alcohol), both of which are present in lower concentrations than carbon dioxide and are also less reactive toward the Grignard reagent.

Experimentally, this reaction is carried out by preparing the Grignard reagent, which is then poured slowly and with swirling onto an excess of finely powdered dry ice. The excess dry ice is then allowed to evaporate, and on treatment of the resulting mixture (which contains ether as the solvent) with dilute mineral acid, the carboxylic acid is liberated and dissolves in the ether. All of the possible organic compounds, which include benzoic acid, benzophenone, triphenylmethanol, benzene, and biphenyl, are extracted into the ether layer. However the desired product, benzoic acid, is separated quite readily from the by-products by extracting it into a dilute solution of sodium hydroxide. The acid is converted to its soluble sodium salt, whereas the other compounds remain in the organic layer. Careful acidification of the aqueous layer regenerates the benzoic acid, which can be purified further by recrystallization.

Preparation of n-butylmagnesium bromide and reaction with 2-methylpropanal. *n*-Butylmagnesium bromide is prepared from *n*-butyl bromide by a procedure entirely analogous to the preparation of phenylmagnesium bromide. It is then allowed to react with 2-methylpropanal to give the salt of 2-methyl-3-heptanol as shown in equation (13), and the salt is hydrolyzed with dilute sulfuric acid.

$$n\text{-}C_4H_9MgBr + CH_3\overset{\overset{\displaystyle O}{\|}}{C}H\underset{\underset{\displaystyle CH_3}{|}}{C}H \longrightarrow CH_3CH_2CH_2CH_2\underset{\underset{\displaystyle CH_3}{|}}{C}H\overset{\overset{\displaystyle O^{\ominus}MgBr^{\oplus}}{|}}{C}HCHCH_3 \tag{13}$$

$$\Big\downarrow H_2O \;|\; H_2SO_4$$

$$CH_3CH_2CH_2CH_2\underset{\underset{\displaystyle CH_3}{|}}{C}H\overset{\overset{\displaystyle OH}{|}}{C}HCHCH_3$$

It should be pointed out that the experiments described here represent two possible ways in which organometallic compounds can be allowed to react with substrates. In the preparation of triphenylmethanol and 2-methyl-3-heptanol, the substrates (methyl benzoate and 2-methylpropanal, respectively) are added to the Grignard reagent; this mode of addition is designated *normal addition*. On the other hand, in the preparation of benzoic acid, the Grignard reagent is added to the substrate (carbon dioxide); this procedure is designated *inverse addition*. The method of addition is dictated by the chemical reactions involved. For example, in the preparation of benzoic acid, if one bubbled carbon dioxide gas into the Grignard reagent (normal addition), the reaction illustrated in equation (12) would be the predominant one. There would be an excess of the Grignard reagent present, which would react further with the bromomagnesium salt of benzoic acid once it formed. To obtain a good yield of benzoic acid, it is necessary to minimize this reaction, so the Grignard reagent should be added to a large excess of carbon dioxide. On the other hand, in the preparation of triphenylmethanol, one could equally well add the Grignard reagent to the methyl benzoate. However, to do this would involve transferring the Grignard reagent to a dropping funnel and adding it to a clean flask containing the substrate.

EXPERIMENTAL PROCEDURE

Phenylmagnesium bromide.   Equip a 250-ml round-bottomed flask with a Claisen connecting tube to which are attached an efficient reflux condenser and a dropping funnel. (If a drying oven is available, dry the equipment before assembling.) Place 0.1 mole of magnesium turnings in the flask, and dry the apparatus by flaming it with a burner; do not have water in the condenser during the drying. After heating, immediately attach a calcium chloride drying tube to the condenser; use a drying tube which will also fit the flask. *Allow the flask to cool to room temperature before continuing.* (At this point, start water through the condenser.)

Note: *From this point on, it is imperative that all flames be extinguished. Care should be taken to insure that there are no open flames (yours or your neighbor's) present within several feet.*

As a precaution for controlling the reaction, prepare an ice bath and have it ready for immediate use if needed. Add 0.115 mole of bromobenzene to 15 ml of *anhydrous* ether [1] contained in the dropping funnel and swirl until the mixture is homogeneous.

Add about 25 ml of anhydrous ether to the flask containing the magnesium turnings, and then add a portion (*ca.* 5 ml) of the bromobenzene–ether mixture. Shake the flask gently. A change in the appearance of the reaction mixture, as evidenced by the presence of a slightly cloudy solution and by the formation of bubbles at the surface of the magnesium turnings, indicates that the reaction has commenced.[2] Once the reaction has started, the ether will be observed to reflux and the flask will become slightly warm. At

---

[1] Commercial anhydrous ether is normally dry enough that it can be used as is. Anhydrous ether may be prepared by drying technical grade ether over phosphorus pentoxide, but the hazards and cost of this procedure far exceed the cost of commercial anhydrous ether.

[2] If there is no apparent change after the portion of the solution of bromobenzene has been added, remove the condenser and mash a piece of magnesium firmly against the bottom of the flask with a *dry stirring rod.* (Be careful not to produce a hole in the flask!) This has the effect of providing a clean, metallic surface on magnesium, which often accelerates initiation of the reaction. If this does not work, the following remedies might be tried, in the order listed: (a) Warm the flask on a steam bath, with swirling. If boiling continues after the heat is removed, reaction has probably started. (b) Crush another piece of magnesium metal with a dry stirring rod. (c) Add a small crystal of iodine. Mix thoroughly, and if reaction does not appear to start, warm over a steam bath again. If iodine is added, then the ether solution containing the final product must be washed with a dilute solution of sodium bisulfite or sodium thiosulfate to remove the iodine. (d) Start the reaction over again; use greater care to insure complete drying of the equipment and reagents.

this time add 10–15 ml of ether to the reaction mixture through the condenser. The rest of the bromobenzene–ether mixture should be added over a period of *ca.* 15–20 min by dropwise addition. If the reaction becomes too vigorous, cool the flask in an ice bath. If the spontaneous refluxing of the reaction becomes slow or if it is slow, mount the flask on a steam bath and heat gently at reflux; be sure that the condenser and drying tube are attached. Continue heating at reflux until most of the magnesium has reacted; during this time, the solution normally acquires a tan to brown as well as a cloudy appearance. The completion of the reaction is indicated by the nearly complete disappearance of the magnesium turnings; usually residual bits of metal remain.

Use the Grignard reagent as soon as possible after its preparation. Two typical reaction procedures making use of it are given below. It is possible to divide the solution and do both preparations; consult the instructor about this possibility. *If you are to do both experiments, be sure to decrease the amount of the other starting materials used in each part by a factor of two.*

Triphenylmethanol.   Dissolve 0.045 mole of methyl (or ethyl) benzoate in about 20 ml of anhydrous ether, and add this mixture to the dropping funnel used in the preparation of the Grignard reagent. Cool the flask containing the phenylmagnesium bromide solution with an ice-water bath (note that the *same* apparatus which was used for the preparation of the Grignard reagent is being used for this synthesis and that the phenylmagnesium bromide has not been removed from the flask). Add the ether solution slowly to the Grignard reagent, cooling the reaction mixture as needed with an ice bath in order to control the exothermic reaction which occurs on mixing. Gentle refluxing of the ether may occur. Quite often a white solid forms as the reaction proceeds; this is a further sign that the reaction is progressing normally. After the addition is complete, remove the dropping funnel and gently swirl the flask until its contents warm to room temperature and the exothermic reaction has subsided. To complete the reaction, either of the following may be done: (1) Heat the mixture at reflux for 30 min using the steam bath, or (2) stopper the flask and allow the mixture to stand until the next laboratory period (no reflux required).★

To a 250-ml Erlenmeyer flask add about 50 ml of 6 *N* sulfuric acid and about 25-30 g of ice; pour the entire reaction mixture into this flask with swirling. Swirl until the heterogeneous mixture is completely free of undissolved solids. It may be necessary to add more technical ether in order to dissolve all of the organic material. Transfer the entire mixture to a separatory funnel, shake vigorously but carefully (be sure to vent the funnel often), and remove the aqueous layer.★ Wash the organic layer with dilute sulfuric acid and then with saturated salt solution. Repeat the washing with salt solution until the resulting aqueous solution is no longer acidic. Dry the organic layer by filtering it through anhydrous sodium sulfate into an Erlenmeyer flask of suitable size. Remove the ether either in the hood or by attaching a funnel to an aspirator line and holding the funnel upside down over the Erlenmeyer flask while the ether is boiled off (use a steam bath for heating). The crude residue which is left as a solid should weigh about 11.5 g and may melt over a wide range. Determine its melting point.★

The crude solid may be purified by fractional crystallization from a 2:1 mixture of petroleum ether–ethanol.[3] Use a minimum amount of the solvent and carry out this

[3] Refer to footnote 13, p. 45 in Chapter 2.

operation either in the hood or with a funnel inverted over the flask and attached to an aspirator. Once all of the material is in solution, evaporate the solvent slowly until small crystals of triphenylmethanol start to form. Remove the heat and allow crystallization to continue at room temperature; complete the crystallization by cooling to 0° until no further crystals appear to form. Isolate the crystalline triphenylmethanol by filtration; it should be colorless.[4] Determine the melting point and yield of the product. The yield should be about 5.5–6.0 g. The mother liquor may be concentrated to yield a second crop, although this will give only a very small amount of additional product. If the melting point is lower than 160° or if the melting range is wide, recrystallize the product once more from a minimum amount of 2:1 petroleum ether – ethanol.

Benzoic acid.   Place 25–30 g of solid carbon dioxide (dry ice), which has been coarsely crushed and which has been protected from moisture as much as possible, in a 500-ml Erlenmeyer flask. Add slowly, with gentle swirling, the phenylmagnesium bromide solution. The mixture becomes thick, and an additional 50–75 ml of technical ether should be added after the addition is complete. The excess carbon dioxide is allowed to escape. This is perhaps most easily done by allowing the flask to stand overnight and to come to room temperature on its own accord; it would be preferable to leave the flask in the hood, properly labeled with your name, rather than to leave it in a locker.[5] ★

After the excess dry ice is gone, treat the mixture with about 50 ml of 6 N sulfuric acid which has been mixed with 25–30 g of ice. The addition of the acid to the crude reaction mixture should be done with care in order to avoid foaming; if the ether has evaporated appreciably, more technical ether may be added so that the total volume of ether is about 100 ml. Transfer the entire mixture to a separatory funnel after mixing it thoroughly by swirling. Rinse the flask with a small portion of technical ether and add the rinse to the separatory funnel. Shake the funnel cautiously (with venting) and separate the layers. Extract the aqueous layer two additional times with 15-ml portions of ether and combine all of the ether extracts.

Extract the ethereal solution with three 20-ml portions of a 1 N solution of sodium hydroxide. Treat the alkaline extract with decolorizing carbon and filter. Add 6 M hydrochloric acid until precipitation is complete and filter the solid.★ The solid (benzoic acid) may be recrystallized from water if needed. After drying the product, determine the melting point and yield; the yield should be about 6 g.

---

[4] An alternative method of recovering the triphenylmethanol from the crude mixture containing it and biphenyl is as follows. Add about 150 ml of petroleum ether and about 50 ml of technical grade ether to the crude product. Heat the mixture on a steam bath (preferably in the hood) until the crystals dissolve; add more ether if needed. Then boil off the solvent *until* the first crystals appear, and then stop heating. Allow the flask to cool and collect the crystals which form. Proceed to determine the melting point of the crystals, and recrystallize if needed.

[5] The process of removing the excess CO_2 can be expedited in one of several ways: (1) Shake or swirl the flask while warming it *very slightly* in a warm water bath; (2) stir the viscous mixture that forms during the reaction; or (3) add small amounts of warm water to the reaction mixture. All of these methods may cause a sudden loss of CO_2 gas, and great care should be taken in hastening this process. At no time should the flask be stoppered. Any warming which is done also causes the ether to evaporate, and care should be taken to replace the ether before going to the next step.

n-Butylmagnesium bromide.   Following the procedure for the preparation of phenylmagnesium bromide, prepare the Grignard reagent from 0.21 mole of *n*-butyl bromide. Use a 500-ml flask and double the amounts of magnesium and ether used in the preparation of phenylmagnesium bromide.

2-Methyl-3-heptanol.   Dilute 0.20 mole of 2-methylpropanal with 25 ml of anhydrous ether and place this solution in the dropping funnel attached to the flask in which the *n*-butylmagnesium bromide was prepared. Begin dropwise addition of the 2-methylpropanal solution to the *n*-butylmagnesium bromide solution (which should be at room temperature), and be prepared to cool the reaction flask with an ice-water bath if the addition leads to uncontrolled refluxing. The addition may require as much as 30 min. After the addition is complete, allow the reaction mixture to stand about 15 min.★

Place about 150 ml of crushed ice in a 500-ml beaker and add 9 ml of concentrated sulfuric acid. Pour the reaction mixture slowly and with stirring into the ice-acid mixture. After the addition is complete, transfer the cold mixture, which may contain some precipitate, to a separatory funnel and shake it gently. The precipitate should dissolve. Separate the layers and extract the aqueous layer with two 25-ml portions of ether; add these extracts to the main ether layer. Wash the combined ether solutions (in the separatory funnel) with 30 ml of saturated sodium bisulfite solution, venting the funnel to relieve pressure, then with two 30-ml portions of saturated sodium bicarbonate solution, and finally with 30 ml of saturated sodium chloride solution. Dry the ether solution over anhydrous magnesium sulfate.★

Filter the solution from the drying agent into a 250-ml round-bottomed flask and remove most of the ether from the product by simple distillation using a steam bath, taking the usual precautions against the flammability of ether. Transfer the residue (the product with some remaining ether) to a 50-ml round-bottomed flask and prepare for distillation.

Insulate the top of the distilling flask and the still head with glass wool or aluminum foil in order to insure steady distillation of the rather high-boiling product. Collect separately any forerun boiling below 165°, and the product, 2-methyl-3-heptanol, boiling at 165–168°. Weigh the product and calculate the yield.

EXERCISES

1. Comment on the use of steam distillation as a possible alternative procedure for the purification of the crude triphenylmethanol. Consider the possible unchanged starting materials and the products which are formed, and indicate which of these would steam distil and which would not. Explain in detail how this method of purification would yield pure triphenylmethanol (if it would at all).
2. Ethyl alcohol quite often is present in technical grade ether. If this grade were used, what effect, if any, would the ethyl alcohol have on the formation of the Grignard reagent? Explain.
3. In the work-up of the reaction mixture from *n*-butylmagnesium bromide and 2-methylpropanal, successive washes with solutions of sodium bisulfite, sodium

bicarbonate, and sodium chloride are used. What is the purpose of each of these steps?

4. The infrared spectrum of 2-methyl-3-heptanol prepared by the procedure of this chapter sometimes shows an absorption at 1720 cm^{-1}. Offer an explanation for this absorption.

5. Give a plausible structure for $C_6H_5MgBr \cdot 2(C_2H_5)_2O$. How do you think the ether molecules are bound to C_6H_5MgBr?

6. Arrange the following compounds in increasing order of reactivity toward a Grignard reagent: methyl benzoate, benzoic acid, benzaldehyde, acetophenone, benzoyl chloride. Explain the basis for your decision, making use of mechanisms where needed. What are the product(s) of reaction of each of the above carbonyl-containing compounds with *excess* phenylmagnesium bromide?

7. How might primary, secondary, and tertiary alcohols be prepared from an aldehyde or a ketone and an organometallic reagent? Suggest suitable carbonyl-containing compounds, as well as suitable organometallic reagents. Write chemical reactions for these preparations, and indicate stoichiometry where important.

SPECTRA OF STARTING MATERIALS AND PRODUCTS

The spectra of 2-methylpropanal are given in Figures 15.3 and 15.4.

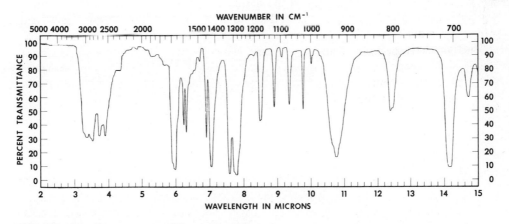

FIGURE 14.1   IR spectrum of benzoic acid.

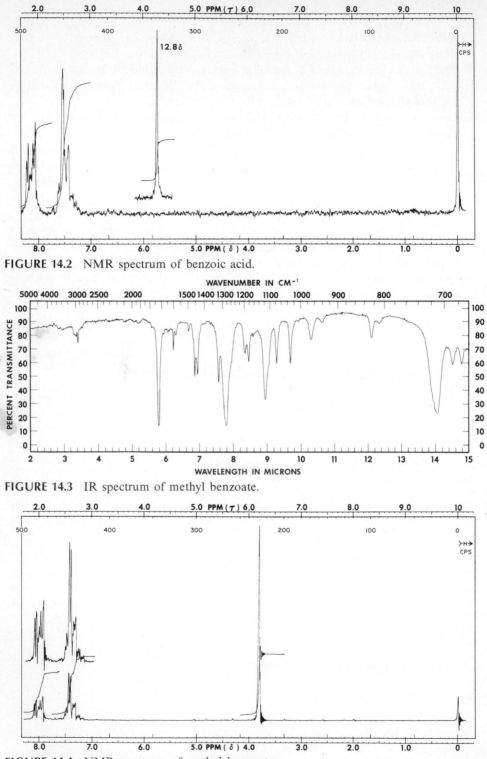

FIGURE 14.2 NMR spectrum of benzoic acid.

FIGURE 14.3 IR spectrum of methyl benzoate.

FIGURE 14.4 NMR spectrum of methyl benzoate.

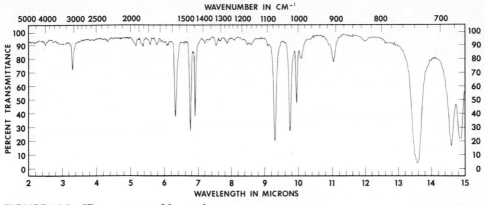

FIGURE 14.5 IR spectrum of bromobenzene.

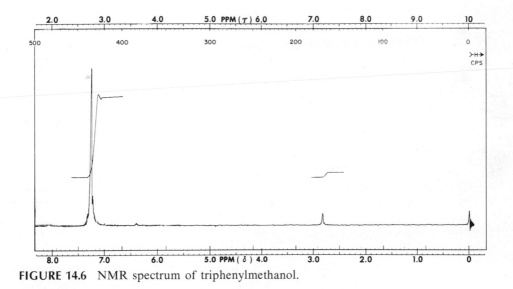

FIGURE 14.6 NMR spectrum of triphenylmethanol.

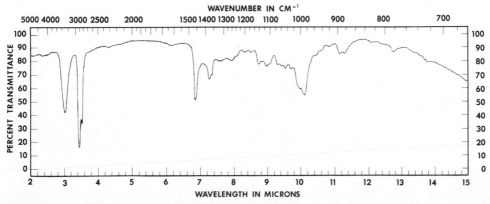

FIGURE 14.7 IR spectrum of 2-methyl-3-heptanol.

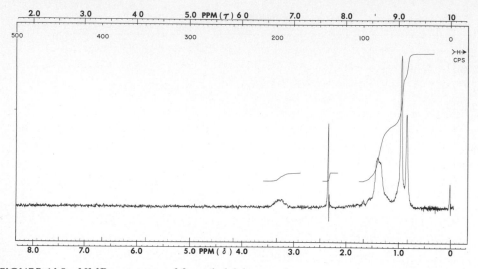

FIGURE 14.8   NMR spectrum of 2-methyl-3-heptanol.

14.2   The Organolithium Reagent

Organolithium reagents play an important role in organic synthesis, and because of their increasingly extensive use, it is desirable to discuss briefly the essential differences in the preparation and reactions of these from those of the Grignard reagent. These reagents give similar types of reactions, but owing to difficulties of conducting experiments involving highly reactive lithium metal in the undergraduate organic laboratory, no experimental procedures for the preparation and reactions of an organolithium reagent are included here.

The preparation of the lithium reagent, RLi, is indicated in equation (14), where RX represents an alkyl or aryl halide. The solvent does not appear to play as important a

$$RX + 2\,Li \xrightarrow[\substack{\text{or} \\ \text{hydrocarbon solvent}}]{\text{dry ether solvent}} RLi + LiX \tag{14}$$

role in the preparation of organolithium reagents as it does in the preparation of Grignard reagents. This has been shown by the fact that it is possible to prepare organolithium reagents in both hydrocarbon solvents (such as pentane and hexane) and ether solvents. However, there is a difference in reactivity of the reagents toward the solvent. Ethereal solutions of the Grignard reagent may be stored under anhydrous conditions for long periods of time, but organolithium reagents may be kept only when a hydrocarbon solvent is used. The latter cleave ethers, even though one can prepare the reagent in ether solution. If this is done, the reagent must be used immediately after its preparation and may not be stored.

It should be noted that two moles of lithium are required for each mole of organic halide, in contrast to the one to one mole ratio of magnesium to halide. Pure lithium metal can be used effectively in preparing the lithium reagent, although it has been found that the use of lithium containing 0.8% by weight of sodium metal often increases the ease with which the reaction starts.

The reactivity of the alkyl and aryl halides toward lithium metal is the same as that for the Grignard reagent. The major exception is the necessity of using tetrahydrofuran as the solvent for the preparation of the Grignard reagent from aryl chlorides. Most alkyl halides form lithium reagents by reaction with lithium metal in either ethyl ether or in a hydrocarbon solvent.

The side reactions arising from the preparation of the lithium reagent are the same as those which are found for the Grignard reagents [see equations (3), (4), and (5)]. The same conditions of extreme dryness must be utilized, and the side reactions may be minimized by use of an inert atmosphere and dilute solutions.

The lithium reagents usually react in a manner similar to Grignard reagents. In some instances, lithium reagents are preferable due to the fact that they are generally more reactive than their halomagnesium counterparts. Furthermore, once the lithium reagent has reacted with a substrate, the resulting salt which forms is generally more soluble in the solvent than the corresponding salt resulting from the Grignard reagent.

An advantage of the greater reactivity of the lithium reagent is realized in *metalation reactions*, wherein the lithium reagent, RLi, abstracts a proton from a hydrocarbon, R'H, as shown in equation (15).

$$\text{RLi} + \text{R'H} \longrightarrow \text{RH} + \text{R'Li} \tag{15}$$

This reaction has been used to determine the acidity of hydrocarbons and can also be used to introduce substituents into various aromatic compounds. *n*-Butyllithium [equation (16)] has been used extensively for this purpose, and some typical examples of metalation reactions are shown in equations (17)-(19).

$$\text{n-C}_4\text{H}_9\text{Br} + 2\,\text{Li} \xrightarrow[-10°\ 85\%\ \text{yield}]{\text{ether}} \text{n-C}_4\text{H}_9\text{Li} + \text{LiBr} \tag{16}$$

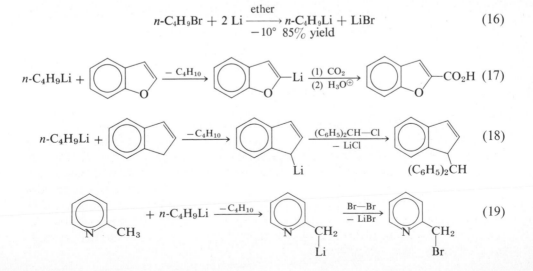

chapter fifteen

oxidation reactions of alcohols and carbonyl compounds

15.1   Preparation of Aldehydes and Ketones by Oxidation of Alcohols

The considerable utility of aldehydes and ketones in organic synthesis makes their efficient preparation of great importance. These substances can be synthesized from alkynes by acid-catalyzed hydration [equation (1)] or by hydroboration followed by

$$R-C\equiv C-R(H) \xrightarrow[\substack{H_2SO_4 \\ H_2O}]{HgSO_4} R-\overset{\overset{\displaystyle O}{\|}}{C}-CH_2-R(H) \tag{1}$$

oxidation [equation (2)], and from carboxylic acids or their derivatives by reaction with organometallics or a variety of reducing agents [equations (3)–(6)]. One of the most common synthetic schemes, however, is the oxidation of primary and secondary

$$R-C\equiv C-R(H) \xrightarrow[\substack{2)\ H_2O_2/HO^\ominus}]{1)\ B_2H_6} R-CH_2-\overset{\overset{\displaystyle O}{\|}}{C}-R(H) \tag{2}$$

$$R-\overset{\overset{\displaystyle O}{\|}}{C}-OH \xrightarrow[\substack{2)\ H_3O^\oplus}]{1)\ R'Li} R-\overset{\overset{\displaystyle O}{\|}}{C}-R' \tag{3}$$

$$R-\overset{\overset{\displaystyle O}{\|}}{C}-Cl + R_2'Cd \longrightarrow R-\overset{\overset{\displaystyle O}{\|}}{C}-R' \tag{4}$$

$$R-C\equiv N \xrightarrow[\substack{2)\ H_3O^\oplus}]{1)\ R'MgX} R-\overset{\overset{\displaystyle O}{\|}}{C}-R' \tag{5}$$

$$R-C\equiv N \xrightarrow[\substack{2)\ H_3O^\oplus}]{1)\ LiAlH_4/-80^\circ} R-\overset{\overset{\displaystyle O}{\|}}{C}-H \tag{6}$$

alcohols with chromic acid, H_2CrO_4, or with potassium permanganate [equation (7)]. A description of the use of the former oxidizing agent for conversion of alcohols to aldehydes and ketones follows; procedures involving the use of both reagents are included in the experimental section.

$$\underset{\overset{|}{R-CH-R'(H)}}{\overset{OH}{}} \xrightarrow[\text{or KMnO}_4]{\text{H}_2\text{CrO}_4} \underset{\overset{\|}{R-C-R'(H)}}{\overset{O}{}} \qquad (7)$$

Chromic acid is not stable for long periods of time and is therefore produced when required by reaction of sodium or potassium dichromate with an excess of an acid such as sulfuric or acetic [equation (8)] or by dissolution of chromic anhydride in water [equation (9)]. In the latter preparation, either sulfuric or acetic acid is also added since

$$Na_2Cr_2O_7 + 2\ H_2SO_4 \longrightarrow [H_2Cr_2O_7] \xrightarrow{H_2O} 2\ H_2CrO_4 + 2\ NaHSO_4 \qquad (8)$$

$$CrO_3 + H_2O \longrightarrow H_2CrO_4 \qquad (9)$$

the rate of oxidation of alcohols by chromic acid is much greater in acidic solutions. For preparation or oxidation of substances which would decompose under strongly acidic conditions, either chromic anhydride dissolved in pyridine or basic potassium permanganate can be used as the oxidizing agent.

The stoichiometry of the chromic acid oxidation of alcohols is such that two equivalents of acid oxidize three equivalents of an alcohol to the corresponding carbonyl compound [equation (10)]. A consideration of the mechanism of the oxidation will illustrate some of the individual steps which result in the overall stoichiometry given in equation (10).

$$\underset{\overset{|}{3\ RCHR'(H)}}{\overset{OH}{}} + 2\ H_2CrO_4 + 3\ H_2SO_4 \longrightarrow \underset{\overset{\|}{3\ RCR'(H)}}{\overset{O}{}} + Cr_2(SO_4)_3 + 8\ H_2O \quad (10)$$

In the presence of chromic acid, alcohols form esters, just as they do when allowed to react with carboxylic acids (*cf.* esterification, Chapter 17.2). Subsequent decomposition of the chromate ester, **1**, yields the carbonyl compound [equation (11)].

$$\underset{R}{\overset{R'(H)}{}} \!\!\diagdown\!\! \underset{}{CH-OH} + HO\overset{\overset{O}{\|}}{C}rOH \xrightarrow{-H_2O} \underset{R}{\overset{R'(H)}{}} \!\!\diagdown\!\! CH-O\overset{\overset{O}{\|}}{C}rOH \longrightarrow \underset{R}{\overset{R'(H)}{}} \!\!\diagdown\!\! C=O + H_2CrO_3 \qquad (11)$$

1

During the course of the oxidation of the alcohol, chromium undergoes reduction from the $+6$ to the unstable $+4$ valence state. A disproportionation rapidly occurs between chromium (VI) and chromium (IV) to produce chromium (V), a new oxidizing agent which we shall write as HCr^VO_3 [equation (12)]. The two moles of chromium (V) will oxidize two more moles of alcohol to the carbonyl compound, and chromium (III), a stable oxidation state of this metal, will be generated [equation (13)]. The sum of equations (11)–(13) corresponds to equation (10).

$$H_2Cr^{IV}O_3 + H_2Cr^{VI}O_4 \longrightarrow 2\ HCr^{V}O_3 + H_2O \tag{12}$$

$$2\ HCrO_3 + 2\ R{-}\overset{\overset{\displaystyle R'(H)}{|}}{C}HOH + 3\ H_2SO_4 \longrightarrow 2\ \underset{R}{\overset{R'(H)}{\diagup}}C{=}O + Cr_2(SO_4)_3 + 6\ H_2O \tag{13}$$

There are some important side reactions which complicate the oxidation of a primary alcohol to an aldehyde. Perhaps the most important of these is the ready oxidation of aldehydes to carboxylic acids by chromic acid [equation (14)]. This further, undesired, oxidation can be minimized by adding the chromic acid *to* the primary alcohol, so that an excess of the oxidizing agent is *not* present in the reaction mixture, and by distilling the aldehyde from the reaction mixture as it is formed. As a consequence, the desired aldehyde must be quite volatile, *i.e.*, boiling point less than about 150°, if it is to be prepared in high yield by chromic acid oxidation.

$$3\ R{-}\overset{\overset{\displaystyle O}{\|}}{C}{-}H + 2\ H_2CrO_4 + 3\ H_2SO_4 \longrightarrow 3\ R{-}\overset{\overset{\displaystyle O}{\|}}{C}{-}OH + Cr_2(SO_4)_3 + 5\ H_2O \tag{14}$$

Even in cases in which the aldehyde is volatile, a poor yield of product is sometimes obtained. This is due to the facile formation from the aldehyde of the hemiacetal, **2**, which is subsequently oxidized to the ester of a carboxylic acid [equation (15)]. Occasionally, this "side" reaction is turned into a useful synthesis of the ester as is the case in the preparation of *n*-butyl butyrate by the chromic acid oxidation of 1-butanol.

$$RCH_2OH \xrightarrow{\ H_2CrO_4\ } R{-}\overset{\overset{\displaystyle O}{\|}}{C}{-}H \xrightarrow[RCH_2OH]{\ H^{\oplus}\ } R{-}\overset{\overset{\displaystyle OH}{|}}{\underset{\underset{\displaystyle OCH_2R}{|}}{C}}{-}H \xrightarrow{\ H_2CrO_4\ } R{-}\overset{\overset{\displaystyle O}{\|}}{C}{-}OCH_2R \tag{15}$$

2

Ketones are much more stable toward oxidizing agents in mildly acidic media than are aldehydes, so that the side reactions mentioned in the oxidation of primary alcohols do not occur to a significant extent in the converson of secondary alcohols to ketones. However, under alkaline or *strongly* acidic conditions, enolizable ketones will undergo oxidation with cleavage to give two carbonyl fragments. For example, cyclohexanone, which can be obtained in good yield by the chromic acid oxidation of cyclohexanol, can be converted to adipic acid by treatment with potassium permanganate under mild alkaline conditions [equation (16)]. The reaction undoubtedly requires initial conversion of the ketone to the enolate ion, **3**, which is then oxidized by the permanganate [equation (17)].

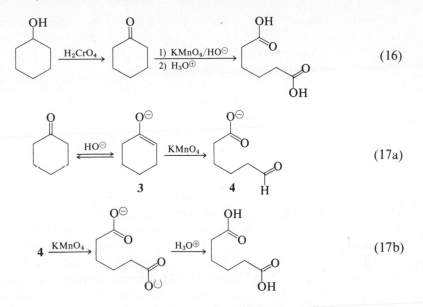

$$(16)$$

$$(17a)$$

$$(17b)$$

Cyclohexanone is a symmetrical ketone and can give only the single enolate ion, **3**. If a ketone is not symmetrical, it can produce two different enolate ions, each of which will be oxidized by permanganate to different products [equation (18)]. The formation of complex mixtures of products when unsymmetrical ketones are oxidized is a complication which detracts from the synthetic utility of this reaction.

$$RCH_2\overset{O}{\overset{\|}{-}}C\overset{}{-}CH_2R' \overset{HO^{\ominus}}{\rightleftharpoons} RCH{=}\overset{O^{\ominus}}{\overset{|}{C}}{-}CH_2R' + R{-}CH_2\overset{O^{\ominus}}{\overset{|}{-}}C{-}CHR' \qquad (18)$$

EXPERIMENTAL PROCEDURE

A. PREPARATION OF 2-METHYLPROPANAL

Prepare a solution of chromic acid by dissolving 0.20 mole of sodium dichromate dihydrate in 290 ml of water and then *slowly* adding with swirling 77 g of concentrated sulfuric acid. While this solution is cooling to room temperature, fit a 500-ml round-bottomed flask with a Claisen adapter equipped with a dropping funnel and a Hempel column bearing a still head with a thermometer. The still head is connected to a water-cooled condenser which leads to a 100-ml graduated cylinder. Place 0.60 mole of isobutyl alcohol and 45 ml of water in the flask along with a boiling stone and heat the mixture with a Bunsen burner until gentle boiling begins. Remove the burner and immediately begin to add the red-orange solution of chromic acid from the dropping funnel to the hot mixture of alcohol and water, at a rate such that all of the acid will have been added in *ca*. 15–25 min. This rate of addition will cause the reaction mixture to boil vigorously, and a mixture of alcohol, aldehyde and water will steam distil, giving a head temperature of 80–85°. This temperature should be maintained by appropriate adjustment of the rate of addition of acid. After all the chromic acid has been added, replace the Bunsen burner and distil the dark green [chromium (III)] reaction mixture for an additional 15 min.* Note the amount of water contained in the two-phase distillate, transfer the mixture to a separatory funnel, and add 1 g of sodium carbonate. Shake the funnel thoroughly, with venting to relieve any pressure, and then saturate the aqueous layer by adding about 0.2 g of sodium chloride for each milliliter of water in order to salt out any dissolved product. Shake the funnel again to effect solution of the salt, remove the organic layer and dry it over anhydrous sodium sulfate.* Filter the crude organic product into a 100-ml distilling flask and isolate the 2-methyl-propanal, bp 63–66°, by fractional distillation. If the fractionation is continued after the aldehyde distils, unchanged alcohol, bp 107–108°, and some isobutyl isobutyrate, bp 148–149°, can be isolated.

Prepare the 2,4-dinitrophenylhydrazone of 2-methylpropanal, mp 186–188°, by the procedure given in Chapter 16.1. Apply the chromic acid in acetone test described in Section 15.2 to the 2-methylpropanal.

The preparation of the sodium bisulfite addition product of the aldehyde can be accomplished by adding 6 ml of saturated sodium bisulfite solution to 1 ml of the aldehyde contained in a 125-ml Erlenmeyer flask. Swirl the flask and then allow the mixture to stand for 10 min. Reaction will be indicated by warming of the solution. Add 25 ml of 95% ethanol, swirl well and then cool the solution in an ice-salt bath. Collect the sodium bisulfite addition product by suction filtration, washing the filter cake once with ethanol and once with ether. You should *not* attempt to determine the melting point of the solid. The 2-methylpropanal can be regenerated by adding 5–10 ml of either 10% sodium carbonate solution or dilute hydrochloric acid to the addition compound and gently warming the mixture.

Although sodium bisulfite addition compounds are not, in general, good derivatives for characterizing carbonyl compounds, they often are used to purify aldehydes, methyl ketones, and cyclic ketones such as cyclohexanone and cyclopentanone.

B. PREPARATION OF CYCLOHEXANONE

Prepare a solution of chromic acid by dissolving 0.081 mole of sodium dichromate dihydrate in 125 ml of water contained in a 250-ml Erlenmeyer flask and *carefully* adding, with swirling, 19 ml of concentrated sulfuric acid. Cool the resulting orange-red solution of chromic acid to room temperature and then, in one portion, *add it to* a mixture of 0.20 mole of cyclohexanol and 75 ml of water in a 500-ml Erlenmeyer flask. Throughly mix the solutions by swirling and determine the temperature of the reaction mixture. The mixture should quickly become warm; when it reaches a temperature of 55°, cool the flask in an ice-water bath or a pan of cold water so that a temperature of 55–60° is maintained. When the temperature of the solution no longer exceeds 60° upon removal of the flask from the cooling bath, allow the flask to stand for one hour with occasional swirling.*

Transfer the reaction mixture to a 500-ml round-bottomed flask, add 100 ml of water and a boiling stone, and attach the flask to an apparatus set for fractional distillation. Distil the mixture until approximately 100 ml of distillate, which consists of an aqueous and an organic layer, has been obtained.* Place the distillate in a separatory funnel, saturate the aqueous layer by adding sodium chloride (about 0.2 g of salt per ml of water will be required) and swirling to effect solution. Separate the layers and extract the aqueous layer with 15 ml of either dichloromethane or technical ether. Combine this extract with the organic layer and dry the solution over anhydrous magnesium sulfate.* Filter the solution into a 50- or 100-ml flask and attach the flask to an apparatus set for simple distillation. Remove the flammable ether by heating with a water bath and then distil the remaining cyclohexanone with the aid of a Bunsen burner. If dichloromethane was used, the burner can be used throughout the distillation. Cyclohexanone is a colorless liquid, bp 152–155°.

Prepare the 2,4-dinitrophenylhydrazone of cyclohexanone, mp 162–163°, by the procedure outlined in Chapter 16.1. Also apply the chromic acid in acetone test described in Section 15.2 to cyclohexanone.

C. OXIDATION OF CYCLOHEXANONE TO ADIPIC ACID

Combine 0.10 mole of cyclohexanone and a solution of 0.20 mole of potassium permanganate in 250 ml of water in a 500-ml Erlenmeyer flask. To this mixture add 2 ml of 3 N sodium hydroxide solution and note the temperature of the reaction mixture. The mixture will warm to 45° and should be held at that temperature by intermittent cooling with an ice- or cold-water bath. When the temperature no longer rises above 45° upon removal of the flask from the cooling bath, allow the flask to stand for an additional 5–10 min (a drop in the temperature should occur), and then complete the reaction by heating the mixture at gentle reflux for a few minutes with a Bunsen burner. Test for the presence of permanganate by placing a drop of the reaction mixture on a piece of filter paper; any unchanged permanganate will appear as a purple ring around the brown manganese dioxide. If permanganate remains, add *small* portions of *solid* sodium bisulfite to the reaction mixture until the spot test is negative.* Filter the mixture by suction through a pad of filter-aid,[1] thoroughly wash the brown filter cake with water * and then concentrate the filtrate to a volume of about 65 ml by heating over a flame.* If the concentrate is colored, add decolorizing carbon, reheat the solution to boiling for a few minutes and then filter. Carefully add concentrated hydrochloric acid to the filtrate until the solution tests acid to litmus paper and then add an additional 15 ml of acid.* Allow the solution to cool to room temperature and isolate the precipitated adipic acid by suction filtration. Adipic acid is a colorless solid, mp 152–153°, and can be recrystallized from ethanol–water if necessary.

EXERCISES

1. What is the purpose of washing the steam distillate with sodium carbonate in the preparation of 2-methylpropanal?
2. Write the structure of the sodium bisulfite addition product of 2-methylpropanal.
3. Why is the chromic acid solution added *to* the alcohol, rather than the reverse, when the preparation of an aldehyde is being attempted?
4. Why are the two-phase distillates obtained in the oxidation of isobutyl alcohol

---

[1] The pad can be prepared by first making a slurry of 0.5–1 g of filter-aid in a few milliliters of water. The slurry is poured onto a filter paper contained in a Büchner funnel attached to a suction flask; suction is then slowly applied to remove the water from the slurry and leave a thin, even pad of filter-aid.

and of cyclohexanol first saturated with sodium chloride before separation of the organic layer?

5. Point out what modifications might be employed in the oxidation of isobutyl alcohol with chromic acid in order to *maximize* the yield of isobutyl isobutyrate.

6. Write out the balanced equation for the oxidation of cyclohexanol to cyclohexanone by sodium dichromate and aqueous sulfuric acid.

7. The costs of sodium dichromate dihydrate and potassium dichromate are $3.65/lb and $3.24/lb, respectively. Determine which reagent would be more economical to use for oxidation of 1 mole of cyclohexanol to cyclohexanone.

8. Why is the reaction mixture made alkaline in the oxidation of cyclohexanone to adipic acid?

9. Write the balanced equation for the oxidation of cyclohexanone to adipic acid.

10. Give the products to be expected on oxidation of 2-methylcyclohexanone with alkaline permanganate.

REFERENCES

1. *Oxidation in Organic Chemistry,* K. B. Wiberg, editor, Academic Press, Inc., New York, 1965, Vol. 5-A, Chapters 1 and 2.
2. *Oxidation,* R. L. Augustine, editor, M. Dekker, Inc., New York, 1969, Vol. 1, Chapters 1 and 2.

SPECTRA OF STARTING MATERIAL AND PRODUCTS

The ir and nmr spectra of cyclohexanone are given in Figures 11.5 and 11.6, respectively; those of cyclohexanol are given in Figures 6.14 and 6.15.

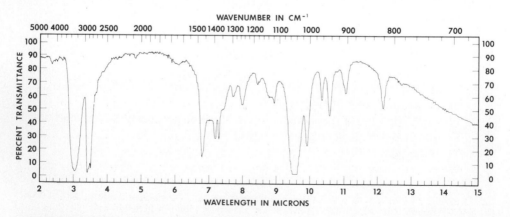

FIGURE 15.1   IR spectrum of isobutyl alcohol.

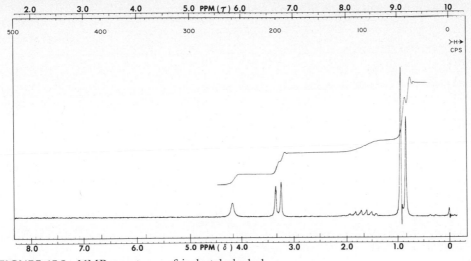

FIGURE 15.2 NMR spectrum of isobutyl alcohol.

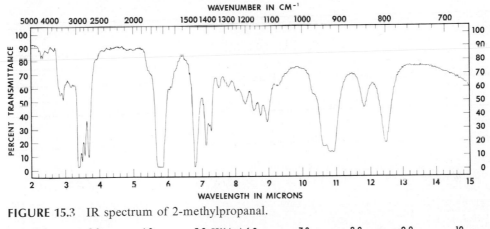

FIGURE 15.3 IR spectrum of 2-methylpropanal.

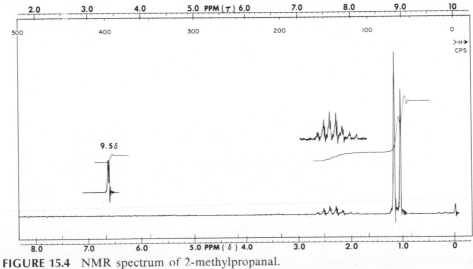

FIGURE 15.4 NMR spectrum of 2-methylpropanal.

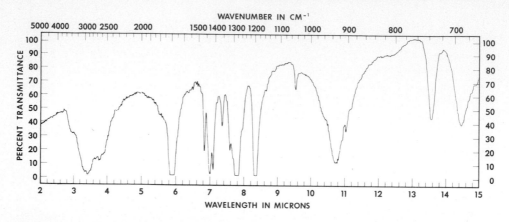

WAVENUMBER IN CM⁻¹

FIGURE 15.5 IR spectrum of adipic acid.

15.2 Classification of Alcohols and Carbonyl Compounds by Means of Chromic Acid

The greater ease of oxidation of aldehydes than of ketones (Section 15.1) is the basis of several classical qualitative tests. Fehling's and Benedict's reagents contain complex salts (tartrate and citrate, respectively) of cupric ion as oxidizing agents, and Tollens' reagent consists of a silver diammine salt. The first two have been used as tests for aliphatic aldehydes (although they are more often used to detect reducing sugars), and the Tollens' test has been widely employed to distinguish between aldehydes (both aliphatic and aromatic) and ketones.

Tertiary alcohols are much more resistant toward oxidation than primary and secondary alcohols. Potassium permanganate and certain other oxidizing agents have been used to distinguish tertiary alcohols from those of the other two types.

There are some disadvantages to all of these tests. A newer reagent which gives quick and definitive results is chromic acid, prepared by dissolving chromic anhydride in sulfuric acid (Section 15.1), and used in acetone solution. This reagent oxidizes primary and secondary alcohols and all aldehydes with a distinctive color change, and it gives no visible reaction with tertiary alcohols and ketones under the conditions of the test.

$$RCH_2OH \xrightarrow{H_2CrO_4} RCHO \xrightarrow{H_2CrO_4} RCO_2H$$

$$R_2CHOH \xrightarrow{H_2CrO_4} R_2CO \xrightarrow[/\!/]{H_2CrO_4} \text{no visible reaction}$$

$$R_3COH \xrightarrow[/\!/]{H_2CrO_4} \text{no visible reaction}$$

Primary and secondary alcohols and aliphatic aldehydes give a positive test within five seconds. Aromatic aldehydes require 30 seconds to one minute. Tertiary alcohols and ketones produce no visible change in several minutes. Phenols and aromatic amines give dark precipitates, as do aromatic aldehydes having hydroxyl or amino groups on the aromatic ring.

Thus, the chromic acid reagent gives a clear-cut distinction between primary and secondary alcohols and aldehydes, on the one hand, and tertiary alcohols and ketones, on the other. Aldehydes may be distinguished from primary and secondary alcohols by means of Tollens', Benedict's, or Fehling's tests, and lower-molecular-weight primary and secondary alcohols may be differentiated on the basis of their rates of reaction with concentrated hydrochloric acid containing zinc chloride—the Lucas reagent (Chapter 25, Section A.2 under "Alcohols").

EXPERIMENTAL PROCEDURE

If the chromic acid reagent is not supplied to you, prepare it as follows: Add 25 g of chromic anhydride (CrO_3) to 25 ml of concentrated sulfuric acid and stir until a smooth paste is obtained. Dilute the paste (*cautiously*) with 75 ml of distilled water and stir until a clear orange solution is obtained.

Place 1 ml of acetone (reagent grade, or solvent grade which has been distilled from potassium permanganate) in a test tube and dissolve one drop of a liquid or *ca.* 10 mg of a solid alcohol or carbonyl compound in it. Add one drop of the acidic chromic anhydride reagent to the acetone solution and shake the tube to mix the contents. A positive oxidation reaction is indicated by disappearance of the orange color of the reagent and the formation of a green or blue-green precipitate or emulsion.

Test a number of compounds of each type chosen from the following classes of compounds: primary alcohols, secondary alcohols, tertiary alcohols, aldehydes, ketones.

EXERCISES

1. Write the formula of the expected oxidation product from each alcohol or aldehyde tested.
2. What is the green precipitate formed in a positive test?
3. What structural feature do tertiary alcohols and ketones have in common, and what is the relationship of this feature to their resistance toward oxidation?
4. Why can the Lucas reagent *not* be used to distinguish between 1-octanol and 2-octanol? (Refer to Chapter 25, Section A.2 under "Alcohols.")

REFERENCES

1. F. G. Bordwell and K. M. Wellman, *Journal of Chemical Education*, **39**, 308 (1962).
2. J. D. Morrison, *ibid.*, **42**, 554 (1965).

15.3   Base-Catalyzed Oxidation-Reduction of Aldehydes: The Cannizzaro Reaction

Aldehydes which have no hydrogen atoms on the carbon atom adjacent to the carbonyl group (the α-carbon atom) undergo mutual oxidation and reduction in the presence of strong alkali. Those which *have* hydrogens on the α-carbon atom undergo other reactions preferentially, as described in Chapter 16.2. Since aldehydes are intermediate in oxidation state between alcohols and carboxylic acids, it is not too surprising to find that this reaction, called the Cannizzaro reaction, takes place. The mechanism, which has been supported by considerable evidence, follows quite logically in view of the well-known ease of addition of nucleophiles to the carbonyl group. The first step explains the function of the strong basic catalyst:

$$
R-C\underset{H}{\overset{O}{\Big\backslash}} + HO^{\ominus} \longrightarrow R-\underset{H}{\overset{O^{\ominus}}{\underset{|}{C}}}-OH \tag{19}
$$

$$
R-\underset{H}{\overset{O^{\ominus}}{\underset{|}{C}}}-OH + R-C\overset{O}{\underset{H}{\Big\backslash}} \longrightarrow R-C\underset{OH}{\overset{O}{\Big\backslash}} + R-\underset{H}{\overset{O^{\ominus}}{\underset{|}{C}}}-H \tag{20}
$$

$$
R-C\underset{OH}{\overset{O}{\Big\backslash}} + HO^{\ominus} \longrightarrow R-C\underset{O^{\ominus}}{\overset{O}{\Big\backslash}} + HOH \tag{21}
$$

$$
R-\underset{H}{\overset{O^{\ominus}}{\underset{|}{C}}}-H + HOH \longrightarrow R-\underset{H}{\overset{OH}{\underset{|}{C}}}-H + HO^{\ominus} \tag{22}
$$

Summation of equations (19) through (22), gives the equation for the overall reaction:

$$
2\ RCHO + HO^{\ominus} \longrightarrow RCO_2^{\ominus} + RCH_2OH \tag{23}
$$

Aromatic aldehydes are the most common type which undergo the Cannizzaro reaction, but formaldehyde and α, α, α-trisubstituted acetaldehydes also react in this way. Reaction between two *different* aldehydes of these types may occur, producing the acids and alcohols corresponding to each of the two aldehydes. Such reactions are referred to as *crossed-Cannizzaro reactions*.

EXPERIMENTAL PROCEDURE

Dissolve 2 g of solid potassium hydroxide in 2 ml of distilled water by swirling in a 100-ml beaker; cool the mixture to room temperature in a water bath. Put 2 ml of benzaldehyde in an 18 × 150-mm test tube and add the concentrated potassium hydroxide solution to it. Cork the tube securely and shake the mixture vigorously until an emulsion is formed. Allow the stoppered tube to stand in your desk until the next laboratory period.

Add about 1 ml of water to the mixture, stopper the tube and shake it. If all of the crystals obtained do not dissolve, add a little water, break up the solid mass with a glass rod, stopper, and shake again. Repeat this procedure until all the solid is in solution. Pour the solution into a separatory funnel and extract it three times with 10-ml portions of ether (shake the mixture *gently* to avoid forming an emulsion).

The ether solution may be dried over anhydrous magnesium sulfate and examined by gas chromatography [2] to determine the proportion of benzyl alcohol and unchanged benzaldehyde present. If the reaction is carried out on a larger scale, the ether solution may be shaken with aqueous sodium bisulfite solution to remove the benzaldehyde, the ether solution washed, dried, and distilled to yield benzyl alcohol as a high-boiling fraction, bp 205°. Alternatively, a crystalline derivative of benzyl alcohol may be prepared. The phenyl urethan (mp 78°) or 3,5-dinitrobenzoate (mp 113°) are suitable (see Chapter 25, Sections B.1 and B.2 under "Alcohols").

Following its extraction with ether, pour the alkaline aqueous solution into a mixture of 5 ml of concentrated hydrochloric acid, 4 ml of water, and *ca.* 5 g of crushed ice, with vigorous stirring. Cool the resultant mixture by placing its container in an ice bath and then collect the crystals of benzoic acid.* Dry the crystals and determine their melting point. Save some of this product for comparison with a product of an experiment in Chapter 16, p. 284.

EXERCISES

1. When the Cannizzaro reaction is carried out in D_2O solution, no deuterium becomes attached to carbon in the alcohol produced. How does this support the mechanism given above?
2. By what means would aqueous sodium bisulfite remove unchanged benzaldehyde from the reaction mixture?
3. Write an equation for the reaction of benzaldehyde and formaldehyde with concentrated potassium hydroxide solution. Show all products.
4. How would propionaldehyde react with potassium hydroxide solution under the conditions of this experiment? Answer the same question for the reaction of trimethylacetaldehyde.

SPECTRA OF STARTING MATERIAL AND PRODUCTS

The ir and nmr spectra of benzoic acid are given in Figures 14.1 and 14.2.

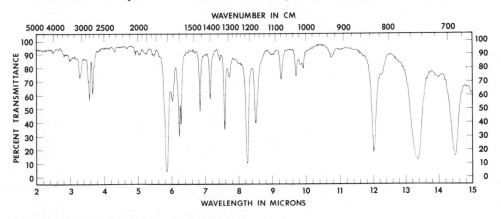

FIGURE 15.6   IR spectrum of benzaldehyde.

[2] A 5-ft column with 5% silicone gum rubber as the stationary phase is satisfactory. In one experiment, about 20% of the original benzaldehyde was found present after 24 hr at room temperature.

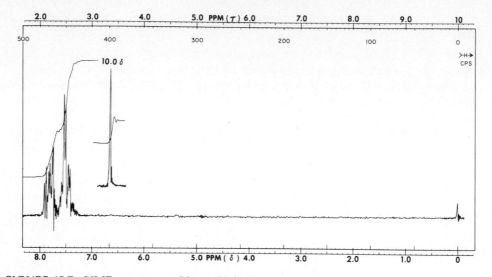

FIGURE 15.7   NMR spectrum of benzaldehyde.

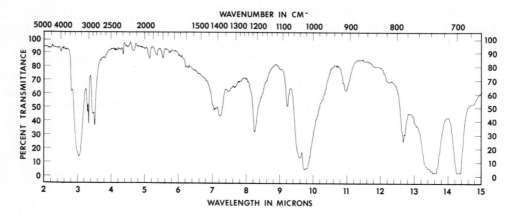

FIGURE 15.8   IR spectrum of benzyl alcohol.

chapter sixteen

some typical reactions of carbonyl compounds

16.1 Reactions involving Nucleophilic Addition to the Carbonyl Group

The addition of a nucleophilic reagent (Nu—E) to a carbonyl compound may be written in the general form

$$
\begin{array}{c}
R \\
\diagdown \\
C{=}O + Nu^{\ominus} \\
\diagup \\
R
\end{array}
\longrightarrow
\begin{array}{c}
R \quad O^{\ominus} \\
\diagdown \; \diagup \\
C \\
\diagup \; \diagdown \\
R \quad Nu
\end{array}
\xrightarrow{\;E^{\oplus}\;}
\begin{array}{c}
R \quad O \quad E \\
\diagdown \; \diagup \\
C \\
\diagup \; \diagdown \\
R \quad Nu
\end{array}
\qquad (1)
$$

where Nu is a nucleophile and E is an electrophile. One of the most important examples of this reaction is the addition of Grignard reagents in which Nu is an alkyl group and E is MgX. In Chapter 14.1 experiments are described in which phenylmagnesium bromide adds to carbon dioxide and methyl benzoate; the reactions of Grignard reagents with aldehydes, ketones, and various other carbonyl compounds also provide extremely valuable synthetic procedures.

Addition of derivatives of ammonia. Nitrogen compounds that may be considered to be derivatives of ammonia constitute other important types of nucleophilic reagents that add to carbonyl compounds [equation (2)]. In the case of *primary amines* (R =

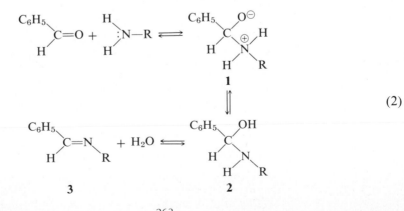

$$ (2) $$

alkyl or aryl) the addition product, **2**, readily loses a molecule of water to produce an *imine*, **3**. Although aliphatic amines give imines that are unstable and polymerize, aromatic amines give stable imines, also called Schiff bases, that can be isolated. These imines are readily hydrolyzed to regenerate the carbonyl compound; that is, the reaction of equation (2) is reversible.

Imines may undergo addition of a mole of hydrogen in the presence of a nickel catalyst to produce secondary amines [equation (3)] in a process which is analogous to hydrogenation of alkenes to alkanes and of carbonyl compounds to alcohols. It is

$$\underset{\textbf{3}}{\underset{H}{\overset{C_6H_5}{\diagdown}}C=N\underset{R}{\diagup}} + H_2 \xrightarrow{\text{Ni}} \underset{\textbf{4}}{C_6H_5-\overset{H}{\underset{|}{C}}H-\overset{H}{\underset{|}{N}}-R} \qquad (3)$$

not necessary to isolate the imine, and, in fact, reaction mixtures from an aldehyde or ketone and an amine (either aromatic or aliphatic) or even ammonia may be subjected directly to catalytic hydrogenation to produce primary or secondary amines in good yields. Presumably imines are transitory intermediates in these reactions.

One of the limitations to this procedure for the synthesis of amines is that any other functional group in the imine that is sensitive to catalytic hydrogenation, such as C=C, N=O (as in NO$_2$), or C—X (X = Cl, Br), may be reduced. This limitation has been removed by the discovery of the alkali metal hydrides, which are more selective reducing agents. For example, N-cinnamyl-*m*-nitroaniline (**9**) may be synthesized starting with cinnamaldehyde (**5**) and *m*-nitroaniline (**6**) using sodium borohydride as the reducing agent. Thus the carbon-nitrogen double bond of the intermediate imine (**7**) is reduced, but neither the carbon-carbon double bond nor the nitro group is affected [equations (4)–(6)].

The reaction of sodium borohydride with the imine [equation (5)] is analogous to the addition of a Grignard reagent to a carbonyl compound. A hydride ion (H:⁻) is trans-

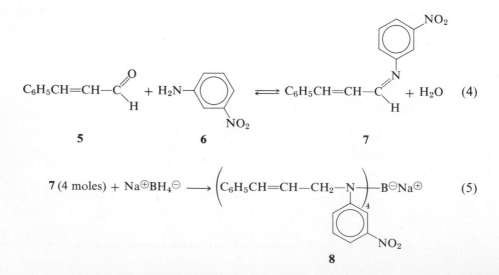

$$8 + 3 H_2O \longrightarrow 4 C_6H_5CH{=}CH{-}CH_2{-}NH + NaH_2BO_3 \qquad (6)$$

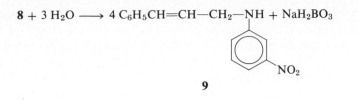

9

ferred from the borohydride anion (BH_4^-) to the electrophilic carbon of the carbon-nitrogen double bond, and the electron-deficient boron adds to nitrogen. All four hydrogens of the borohydride anion are transferred to imine carbons in this way, producing an organoboron anion, **8**, which is subsequently decomposed with water to yield the secondary amine[**9**, equation (6)].

Although the formation of the imine is a reversible reaction [equation (4)], it can be brought effectively to completion by removing the water from the reaction mixture as it is produced.[1] A convenient way to do this is to heat the carbonyl compound and primary amine in benzene solution, allowing the benzene to distil continuously. The benzene forms a minimum-boiling azeotrope with the water and removes it as it is formed.

Sodium borohydride is conveniently used in methanol solution. Although it reacts slowly with the solvent, its rate of reaction with the imine is much faster. In small scale reactions, it is more convenient to use a sufficient excess of the borohydride to allow for its partial decomposition by reaction with the solvent than to use a less reactive solvent in which the borohydride is less soluble.

Other useful derivatives of ammonia that react with carbonyl compounds according to equation (2) are hydroxylamine (R = OH), semicarbazide (R = $NHCONH_2$), and various arylhydrazines (R = NHAr). The reaction of semicarbazide with carbonyl compounds to produce semicarbazones was presented in Chapter 11 to illustrate the principle of kinetic and equilibrium control. However, the main importance of the reaction of carbonyl compounds with semicarbazide, hydroxylamine, and the arylhydrazines lies in the fact that the products (semicarbazones, oximes, and arylhydrazones) are almost invariably crystalline solids, whereas the carbonyl compounds from which they are derived are often liquids. Arylhydrazines which are commonly employed in making crystalline derivatives of carbonyl compounds include phenylhydrazine, *p*-nitrophenylhydrazine, and 2,4-dinitrophenylhydrazine. The formation of a 2,4-dinitrophenylhydrazone, **10**, is represented by equation (7).

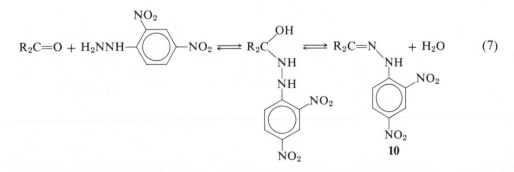

(7)

10

---

[1] Compare with the similar problem encountered in esterification reactions, which are also equilibrium processes, Chapter 17.2.

As the oximes, semicarbazones, and arylhydrazones are generally very insoluble in the reaction medium, they precipitate from solution, and the reversibility of the reaction therefore does not prevent their isolation in high yields.[2]

Directions for the preparation of semicarbazones, oximes, and 2,4-dinitrophenyl-hydrazones are given in the Experimental Procedure section. The directions for semi-carbazones are more generally applicable than those given in Chapter 11.

Addition of Wittig reagents. In 1953 G. Wittig discovered that certain organophos-phorus compounds, called ylides, **13**, add to carbonyl compounds to form unstable intermediates called betaines, **14**, which decompose to produce an alkene and a phosphine oxide, **15**. The highly reactive ylide is generated in the presence of the carbonyl compound by the action of a strong base, usually phenyllithium, on a phospho-nium halide, **12**. The latter compound is produced most commonly by reaction of a primary or secondary organic halide with triphenylphosphine [**11**, equation (8); R may be alkyl, aryl, CN, $CO_2C_2H_5$, etc.].

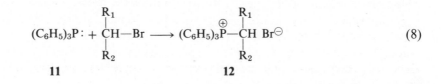

$$(C_6H_5)_3P: \;+\; \underset{\underset{R_2}{|}}{\overset{\overset{R_1}{|}}{CH}}{-}Br \;\longrightarrow\; (C_6H_5)_3\overset{\oplus}{P}{-}\underset{\underset{R_2}{|}}{\overset{\overset{R_1}{|}}{CH}}\;Br^{\ominus} \qquad (8)$$

$$\underset{\mathbf{11}}{} \qquad\qquad\qquad \underset{\mathbf{12}}{}$$

$$\mathbf{12} \;\xrightarrow{C_6H_5Li}\; \left[(C_6H_5)_3P{=}\underset{\underset{R_2}{}}{\overset{\overset{R_1}{}}{C}} \;\longleftrightarrow\; C_6H_5)_3\overset{\oplus}{P}{-}\overset{\ominus}{\underset{\underset{R_2}{}}{\overset{\overset{R_1}{}}{C}}} \right] \;+\; C_6H_6 \;+\; LiBr \qquad (9)$$

$$\underset{\mathbf{13}}{}$$

$$\mathbf{13} \;+\; O{=}\underset{\underset{R_4}{}}{\overset{\overset{R_3}{}}{C}} \;\longrightarrow\; \begin{matrix} & R_1 & \\ (C_6H_5)_3\overset{\oplus}{P}{-}\overset{|}{C}{-}R_2 \\ \overset{\ominus}{O}{-}\underset{\underset{R_4}{|}}{\overset{|}{C}}{-}R_3 \end{matrix} \;\longrightarrow\; \begin{matrix}(C_6H_5)_3P \\ \| \\ O\end{matrix} \;+\; \underset{R_4}{\overset{R_1}{}}{}\underset{R_3}{\overset{R_2}{}}{C{=}C} \qquad (10)$$

$$\underset{\mathbf{14}}{} \qquad\qquad\qquad \underset{\mathbf{15}}{}$$

As may be judged from the polarized resonance structure of the ylide, the electron-rich carbon is nucleophilic, and it adds to a carbonyl group in the expected way to give the betaine. The decomposition of the betaine is thought to involve a four-center transi-tion state; as the second C—C bond and the P—O bond are forming, the P—C and O—C bonds are breaking.

The overall effect of reactions (8)–(10) is conversion of a carbon-oxygen double bond to a carbon-carbon double bond, *i.e.*, C=O → C=C. The Wittig synthesis thus

---

[2] These reactions exhibit an interesting and important dependence of rate on the pH of the medium. Consult Chapter 11 for a discussion of this subject.

represents a very general method of preparation of alkenes, one which has two important advantages over older procedures: (1) The carbonyl group is replaced *specifically* by a carbon-carbon double bond, without the formation of isomeric alkenes, and (2) the reactions are carried out in basic rather than in acidic media under very mild conditions.

Various modifications of the original Wittig synthesis have been made. One of these that is convenient for an introductory experiment (because it does not require an inert atmosphere and the use of the somewhat dangerous organolithium reagents) involves the use of triethyl phosphite (**16**) in place of triphenylphosphine. Activated organic halides (such as benzyl and allyl chlorides and α-bromo esters) will react satisfactorily

$$(C_2H_5O)_3P + C_6H_5CH_2{-}Cl \longrightarrow (C_2H_5O)_3\overset{\oplus}{P}{-}CH_2C_6H_5 \quad Cl^{\ominus} \tag{11}$$
$$\mathbf{16} \qquad\qquad\qquad\qquad\qquad\quad \mathbf{17}$$

$$\mathbf{17} \xrightarrow[\text{heat}]{} (C_2H_5O)_2\overset{\overset{O}{\|}}{P}{-}CH_2C_6H_5 + C_2H_5Cl \tag{12}$$
$$\mathbf{18}$$

$$\mathbf{18} + CH_3O^{\ominus}Na^{\oplus} \longrightarrow (C_2H_5O)_2\overset{\overset{O}{\|}}{P}{-}\overset{\ominus}{C}HC_6H_5 \; Na^{\oplus} + CH_3OH \tag{13}$$
$$\mathbf{19}$$

$$\mathbf{19} + O{=}CHC_6H_5 \longrightarrow (C_2H_5O)_2\overset{\overset{O}{\|}}{P}{-}CHC_6H_5 \; Na^{\oplus} \tag{14}$$
$$\overset{|}{{}^{\ominus}O{-}CHC_6H_5}$$

$$\downarrow$$

$$(C_2H_5O)_2\overset{\overset{O}{\|}}{P}O^{\ominus}Na^{\oplus} + \overset{\overset{CHC_6H_5}{\|}}{C}HC_6H_5$$
$$\mathbf{21} \qquad\qquad\qquad \mathbf{20}$$

with esters of phosphorous acid [equation (11)]. The phosphonium salt, **17**, produced is unstable toward heat; for example, the product from benzyl chloride evolves ethyl chloride and yields the phosphonate ester, **18**. Treatment of this ester with sodium methoxide produces a nucleophile analogous to an ylide and, in the presence of benz-aldehyde, the reactions of equations (13) and (14) take place to yield an alkene, **20** (stilbene) and sodium diethylphosphate (**21**). The latter is water-soluble and hence easily separated from the alkene.

There are two geometric isomers of stilbene: The *trans* isomer is a solid at room temperature, mp 126–127°, whereas the *cis* isomer is a liquid, mp 6°. The phosphonate ester modification of the Wittig reaction is reported to give the pure *trans* isomer, even though the original Wittig procedure (using triphenylphosphine as starting material) is reported to give a mixture of 30% *cis*-, 70% *trans*-stilbene. The infrared spectra of the two isomers are distinctly different and may be used in both qualitative and quantitative analysis of mixtures.

Optional or additional experiments. Other aldehydes may be used in place of benzaldehyde in the phosphonate modification of the Wittig reaction. With *trans*-cinnamaldehyde **(22)**, the product is *trans,trans*-1,4-diphenyl-1,3-butadiene **(23)**. If some

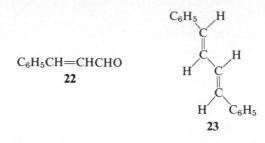

C$_6$H$_5$CH=CHCHO
22

23

students prepare this diene, it will be interesting to have the *ultraviolet* absorption spectrum of this product compared with that of the stilbene product to see the effect of an additional conjugated double bond. If it is not practical to record the spectra of the products in your laboratory, the literature may be consulted (see references 3 and 4).

When 2-furaldehyde **(24)** is substituted for benzaldehyde, the product is *trans*-2-styrylfuran **(25)**.

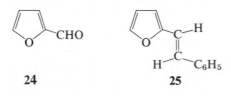

24   **25**

EXPERIMENTAL PROCEDURE

A. ADDITION OF DERIVATIVES OF AMMONIA

1. Addition of primary amines to produce imines. Synthesis of secondary amines by sodium borohydride reduction of imines.

a. Cinnamaldehyde and m-nitroaniline. In a 100-ml round-bottomed flask, place 0.022 mole of cinnamaldehyde, 0.020 mole of *m*-nitroaniline, 25 ml of benzene, and several boiling stones. Set up apparatus for simple distillation, using a graduated cylinder as a receiver, and heat the reaction mixture on a steam cone or in a boiling water bath. (If a burner is used to heat a water bath, take proper precautions against the flammability of the benzene.) Distil until almost all of the benzene has been removed; 22–24 ml of distillate should be obtained in about 25–30 min.

During the distillation, prepare a solution of 0.02 mole of sodium borohydride in 15 ml of methanol.[3] When the distillation has been completed, remove the flask from

[3] The solution should be prepared no more than 5 or 10 min before it is used, because the sodium borohydride reacts with the methanol. Do not stopper the flask containing the solution, as pressure will build up from hydrogen gas produced by the reaction:

$$NaBH_4 + 4\,CH_3OH \rightarrow 4\,H_2 + NaOCH_3 + B(OCH_3)_3$$

the steam cone and either pour out about 0.5 ml of the residual liquid or, preferably, insert a micropipet and take a 0.5-ml sample of the liquid. Add 3–4 ml of methanol to this sample in a test tube, swirl to dissolve it, and then place the test tube in ice. (Go to the directions of the next paragraph while the solution is cooling.) Collect any crystals which separate, dry them, and determine their melting point. The reported melting point of the imine, N-cinnamylidene-*m*-nitroaniline, is 92–93°. The imine may be purified by recrystallization from methanol.

To the remainder of the residue from the distillation (crude imine) add 20 ml of methanol. Attach a Claisen connecting tube equipped with a water-cooled reflux condenser and an addition funnel.[4] Pour the methanolic solution of sodium borohydride into the addition funnel (stopcock closed), and then add this solution to the imine solution dropwise, or in several portions, at a rate such that the addition is completed within 5 min. Swirl the reaction mixture while the addition is being made. After all of the borohydride solution has been added, heat the reaction mixture at reflux for 15 min.

Cool the reaction mixture to room temperature and pour it into 50 ml of water. Stir the mixture and allow it to stand with occasional stirring for 10–15 min. Collect the orange crystals by suction filtration and wash them with water.★ Weigh the dry product and calculate the yield. Determine the melting point of the product, N-cinnamyl-*m*-nitroaniline. It may be recrystallized from 95% ethanol; the melting point of the pure secondary amine is 106–107°.

b. Cinnamaldehyde and aniline. Aniline may be substituted for *m*-nitroaniline and the above procedure followed, with the following modification: After the reduction step has been completed, pour the reaction mixture into 50 ml of 3 *N hydrochloric acid* instead of water. The free secondary amine, N-cinnamylaniline, is a liquid at room temperature; for ease of handling, the product is converted by hydrochloric acid to the salt, which is only slightly soluble in cold water.

The melting point of the intermediate imine, N-cinnamylideneaniline (which may be isolated as described above for the corresponding imine from *m*-nitroaniline) is reported to be 109° (yellow leaflets from ethanol). The reported melting point of N-cinnamylanilinium chloride is 185°.

The picric acid salt (picrate) of N-cinnamylaniline (mp 137°) may be prepared as follows: Add 0.3 g of N-cinnamylanilinium chloride to 3 ml of 1.5 *N* sodium hydroxide solution and 2 ml of chloroform in a test tube. Shake the mixture, allow it to settle, and then remove the aqueous layer with a pipet. Add 2 ml of water, shake, and again remove the aqueous layer. Add a small amount of anhydrous magnesium sulfate to the chloroform solution, shake the mixture intermittently during 5–10 min, and then filter the solution from the magnesium sulfate. ★ Evaporate the chloroform solution and dissolve the residue (the free secondary amine) in 2 ml of ethanol. Add 3 ml of a cold, saturated solution of picric acid, prepared by shaking excess picric acid with ethanol at 0° and filtering the mixture. Warm the ethanolic solution in a water bath at 100°

---

[4] If a Claisen adapter is not available, the condenser may be attached directly to the flask and an addition funnel inserted in the top of the condenser through a *slotted* cork or rubber stopper. (The slot is to prevent a closed system.)

for 5 min and then cool it in an ice bath. Collect the yellow crystals and determine their melting point (reported, 137°).

c. Benzaldehyde and m-nitroaniline. The imine and the amine may be prepared from these reactants by the procedure used with cinnamaldehyde and *m*-nitroaniline. The melting point of the imine, N-benzylidene-*m*-nitroaniline (crystallized from methanol) is 70°. The melting point of N-benzyl-*m*-nitroaniline is 106°.

d. Benzaldehyde and aniline. The imine and the amine–hydrochloric acid salt may be prepared from these reactants by the procedure used with cinnamaldehyde and aniline. The imine, N-benzylideneaniline, is more soluble in ethanol than the corresponding *m*-nitroaniline derivative, and its melting point is lower (54°), so it may be more difficult to crystallize. The reported melting point of N-benzylanilinium chloride is 214–216°.

2. The preparation of crystalline derivatives of carbonyl compounds.

a. Semicarbazones. Dissolve 0.5 g of semicarbazide hydrochloride[5] and 0.8 g of sodium acetate in 5 ml of water in a test tube and then add about 0.5 ml of the carbonyl compound. Stopper and shake the tube vigorously, remove the stopper, and place the test tube in a beaker of boiling water. Discontinue heating the water and allow the test tube to cool to room temperature in the beaker of water. Remove the test tube to an ice bath and scratch the side of the tube with a glass rod at the interface between the liquid and air. The semicarbazone may be recrystallized from water or aqueous ethanol.

If the carbonyl compound is insoluble in water, dissolve it in 5 ml of ethanol. Add water until the solution becomes turbid, then add a little ethanol until the turbidity disappears. Add the semicarbazide hydrochloride and sodium acetate and continue as above from this point.

b. Oximes. Dissolve 0.5 g of hydroxylamine hydrochloride in 5 ml of water and 3 ml of 3 *N* sodium hydroxide solution and then add 0.5 g of the aldehyde or ketone. If the carbonyl compound is insoluble in water, add just enough ethanol to give a clear solution. Warm the mixture on a steam bath (or boiling water bath) for 10 min and then cool it in an ice bath. If crystals do not form immediately, scratch the side of the container at and below the liquid level with a glass rod. The oxime may be recrystallized from water or aqueous ethanol.

In some cases the use of 3 ml of pyridine and 3 ml of absolute ethanol in place of the 3 ml of 3 *N* sodium hydroxide solution and 5 ml of water will be found to be more effective. A longer heating period is often necessary. After the heating is finished, pour the mixture into an evaporating dish and remove the solvent with a current of air in a hood. Triturate (grind) the solid residue with 3–4 ml of cold water and filter the mixture. Recrystallize the oxime from water or aqueous ethanol.

c. 2,4-Dinitrophenylhydrazones. If the reagent is not supplied, it is prepared by dissolving 1 g of 2,4-dinitrophenylhydrazine in 5 ml of concentrated sulfuric acid. This

---

[5] Neither semicarbazide nor hydroxylamine is stable as the free base, so they are usually stored in the form of the hydrochloric acid salt, or "hydrochloride."

solution is then added, with stirring, to 7 ml of water and 25 ml of 95% ethanol. After stirring vigorously, the solution is filtered from any undissolved solid.

Dissolve one or two drops of a liquid (or about 100 mg of a solid) in 2 ml of 95% ethanol and add this solution to 2 ml of the 2,4-dinitrophenylhydrazine reagent. Shake the mixture vigorously; if a precipitate does not form immediately, let the solution stand for 15 min.

If more crystals are desired for a melting point, dissolve 200–500 mg of the carbonyl compound in 20 ml of 95% ethanol and add this solution to 15 ml of the reagent. The product may be recrystallized from aqueous ethanol.

B. ADDITION OF WITTIG REAGENTS
Caution: Organic phosphorus compounds are toxic, and benzyl chloride is irritating and lachrymatory. Avoid contact of these substances with the skin and inhalation of their vapor.

1. *trans*-Stilbene. Weigh 0.045 mole of triethyl phosphite into a 25- or 50-ml round-bottomed flask and add 0.045 mole of benzyl chloride and several small boiling stones. Connect a water-cooled condenser to the flask and heat the mixture at gentle reflux for 1 hr. While this reaction mixture is cooling to room temperature, place 0.045 mole of sodium methoxide in a 125-ml Erlenmeyer flask and immediately add 20 ml of N,N-dimethylformamide. Pour the *cool* phosphonate reaction mixture into the Erlenmeyer flask and rinse the round-bottomed flask with 20 ml of N,N-dimethylformamide, adding the rinse solution to the rest of the reaction mixture. Cool the Erlenmeyer flask and its contents in an ice-water bath, stirring with a thermometer until the temperature of the solution is below 20°. While continuing to stir the cooled solution, slowly add 0.045 mole of benzaldehyde, cooling as necessary so that the temperature of the mixture does not rise above 35°. Remove the flask from the cooling bath and allow it to stand at room temperature for about 10 min. Add 15 ml of water with stirring, collect the crystals on a Büchner funnel, and wash them with cold 1:1 methanol-water. Determine the weight and the melting point of the crystalline product. A mixture of ethanol and ethyl acetate may be used to recrystallize impure stilbene. Submit a sample of the product in carbon disulfide or carbon tetrachloride solution for infrared analysis and compare the spectrum with those of Figures 16.13 and 16.14.

2. 1,4-Diphenyl-1,3-butadiene. This diene may be prepared by following the procedure above up to the point of addition of benzaldehyde. Use 0.04 mole of cinnamaldehyde in place of benzaldehyde, and continue to follow the procedure up to the point of collecting the crystals. Wash the crystals first with water and then with methanol until the filtrate is colorless. Weigh the product and determine its melting point. The melting point of *trans, trans*-1,4-diphenyl-1,3-butadiene is reported to be 150–151°.

3. *trans*-2-Styrylfuran. Follow the procedure for *trans*-stilbene exactly except substitute an equimolar amount of freshly distilled 2-furaldehyde for benzaldehyde. The melting point of *trans*-2-styrylfuran is reported to be 54–55°. This product may be recrystallized from methanol.

EXERCISES

A. ADDITION OF DERIVATIVES OF AMMONIA

1. What causes the turbidity in the distillate collected during the heating of the carbonyl compound and primary amine in benzene solution?
2. Why was the use of *dry* benzene not specified?
3. Although 95% ethanol is a satisfactory solvent for recrystallization of N-cinnamyl-*m*-nitroaniline, it is not suitable for recrystallization of the corresponding imine. Why?
4. Note the molar proportion of $NaBH_4$ to imine used in your experiment. How does this compare with the stoichiometric requirement? Account for the proportion used.
5. Sodium borohydride may be "stabilized" toward reaction with solvent methanol by addition of sodium hydroxide. Explain.
6. In Figures 16.3 and 16.4, assign as many of the ir absorption peaks to structural components of the molecules as you can. Which peaks give evidence of the conversion of $-\overset{|}{C}=N-$ to $-\overset{|}{\underset{H}{C}}-\overset{|}{\underset{H}{N}}-$ and of retention of the carbon-carbon double bond in the secondary amine?
7. Can you identify the ir absorptions of the NO_2 group by comparing Figures 16.3 and 16.4 with Figure 16.6? Explain the basis of your identification.
8. By comparing the nmr spectra of Figures 16.5 and 16.11, identify the vinyl hydrogen $(-CH=CH-)$ signals in Figure 16.5 and the methylene $(-CH_2-)$ signals in both Figures 16.5 and 16.11.
9. By calculating the integrations in Figures 16.5, 16.8, and 16.11, and, particularly, by considering the signals of Figure 16.8, see if you can discover the "hiding place" of the elusive N—H signal.
10. There are two NH_2 groups in the semicarbazide molecule, yet only one of them is nucleophilic. Note which one this is, and explain.
11. The melting points of 2,4-dinitrophenylhydrazones are generally higher than those of the corresponding phenylhydrazones. Suggest an explanation for this behavior.
12. How does scratching the side of a container with a glass rod help to induce crystallization?

B. ADDITION OF WITTIG REAGENTS

1. Compare the mechanism of the addition of Grignard and Wittig reagents to the carbonyl group, pointing out similarities and differences.
2. Write equations for the preparation of the following alkenes by the original Wittig synthesis from triphenylphosphine and the necessary organic halides and carbonyl compounds:
   (a) $C_6H_5CH=C(CH_3)_2$ (b) $C_6H_5C(CH_3)=CHCH_3$ (c) $CH_3CH_2C(CH_3)=CH_2$
3. Write equations for Wittig syntheses of the compounds of Exercise 2 using alternative organic halides and carbonyl compounds.

4. Write equations for the preparation of the following alkenes by the phosphonate ester modification (starting with triethyl phosphite):

(a) $C_6H_5C(CH_3)\!\!=\!\!C(CH_3)C_6H_5$ (b) $CH_2\!\!=\!\!CH\!\!-\!\!CH\!\!=\!\!CH\!\!-\!\!C_6H_5$

(c) $(CH_3)_2C\!\!=\!\!CH\!\!-\!\!CO_2C_2H_5$

5. Would you expect an ylide, **13**, in which R = CN to be more or less stable than one in which R = alkyl?

6. Why should the sodium methoxide not be exposed to the atmosphere for more than a few minutes?

7. Why should the aldehydes used as starting materials in Wittig syntheses be free of carboxylic acids?

8. What peaks in the ir spectra of the stilbenes (Figures 16.13 and 16.14) would be most useful for quantitative analysis of a mixture of *cis* and *trans* isomers?

REFERENCES

1. E. J. Seus and C. V. Wilson, *Journal of Organic Chemistry*, **26**, 5243 (1961).
2. G. Wittig and U. Schollkopf, *Chemische Berichte*, **87**, 1318 (1954).
3. J. L. Bills and C. R. Noller, *Journal of the American Chemical Society*, **70**, 958 (1948).
4. J. H. Pinckard, B. Willie, and L. Zechmeister, *ibid.*, **70**, 1939 (1948).
5. A. Maercker, in *Organic Reactions*, A. C. Cope, editor, John Wiley & Sons, Inc., New York, 1965, Vol. 14, Chap. 3. A good general review.

SPECTRA OF STARTING MATERIALS AND PRODUCTS

The ir and nmr spectra of 2-furaldehyde are given in Figures 11.3 and 11.4. The ir and nmr spectra of benzaldehyde are given in Figures 15.6 and 15.7. The ir and nmr spectra of aniline are given in Figures 17.3 and 17.4.

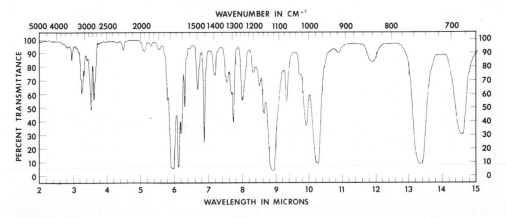

FIGURE 16.1   IR spectrum of cinnamaldehyde.

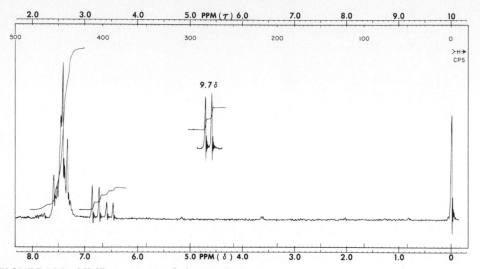

FIGURE 16.2   NMR spectrum of cinnamaldehyde.

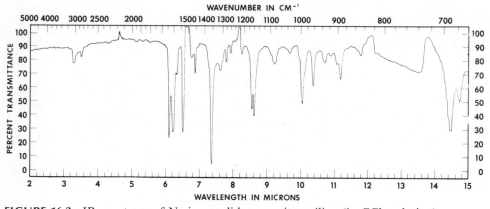

FIGURE 16.3   IR spectrum of N-cinnamylidene-*m*-nitroaniline (in CCl_4 solution).

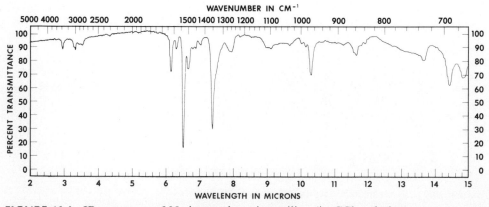

FIGURE 16.4   IR spectrum of N-cinnamyl-*m*-nitroaniline (in CCl_4 solution).

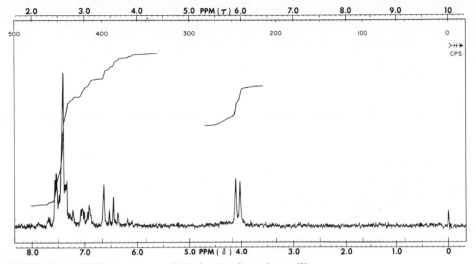

FIGURE 16.5   NMR spectrum of N-cinnamyl-*m*-nitroaniline.

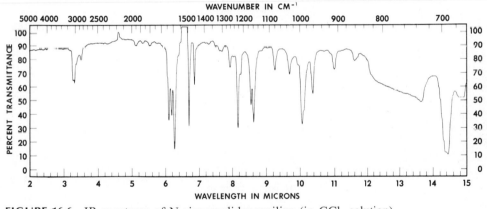

FIGURE 16.6   IR spectrum of N-cinnamylideneaniline (in CCl_4 solution).

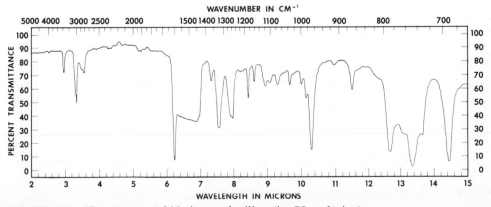

FIGURE 16.7   IR spectrum of N-cinnamylaniline (in CS_2 solution).

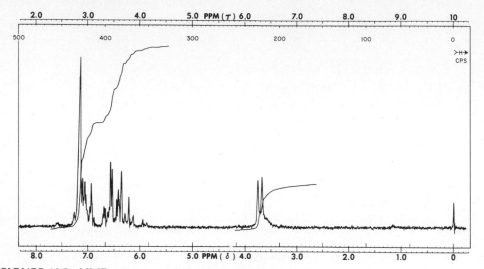

FIGURE 16.8   NMR spectrum of N-cinnamylaniline.

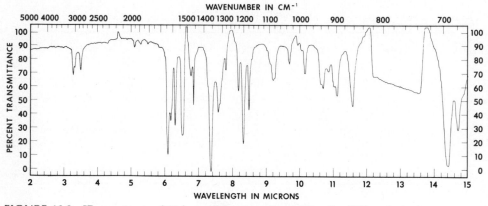

FIGURE 16.9   IR spectrum of N-benzylidene-*m*-nitroaniline (in CCl$_4$ solution).

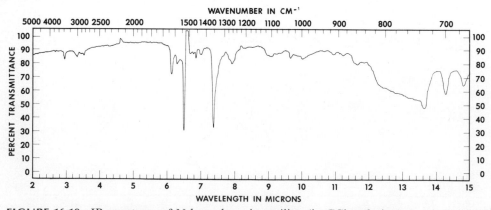

FIGURE 16.10   IR spectrum of N-benzyl-*m*-nitroaniline (in CCl$_4$ solution).

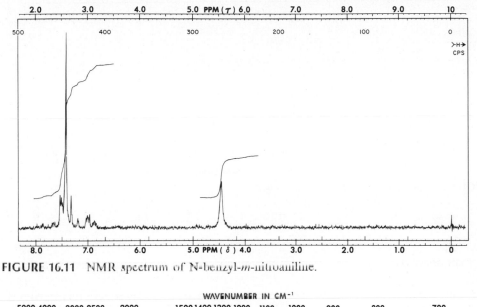

FIGURE 16.11   NMR spectrum of N-benzyl-*m*-nitroaniline.

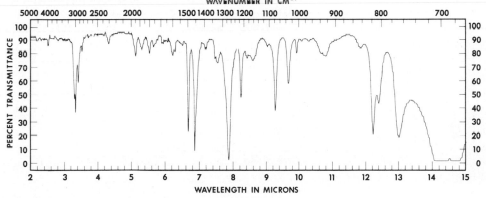

FIGURE 16.12   IR spectrum of benzyl chloride.

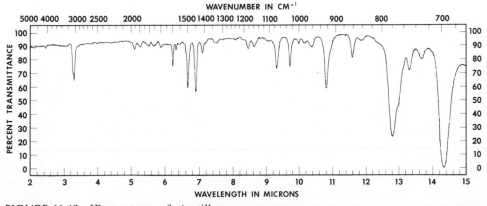

FIGURE 16.13   IR spectrum of *cis*-stilbene.

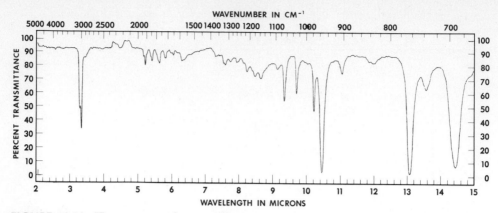

FIGURE 16.14 IR spectrum of *trans*-stilbene.

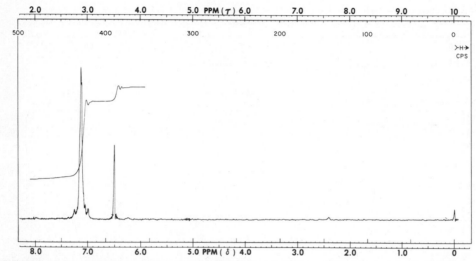

FIGURE 16.15 NMR spectrum of *cis*-stilbene.

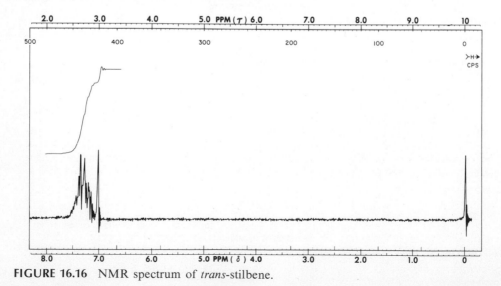

FIGURE 16.16 NMR spectrum of *trans*-stilbene.

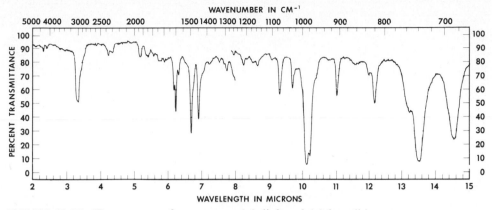

FIGURE 16.17   IR spectrum of *trans, trans*-1,4-diphenyl-1,3-butadiene.

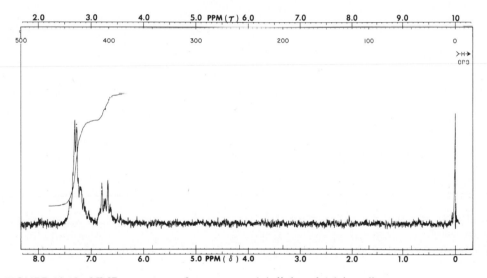

FIGURE 16.18   NMR spectrum of *trans, trans*-1,4-diphenyl-1,3-butadiene.

16.2   Reactions of Carbonyl Compounds involving the α- and β-Carbon Atoms

The haloform reaction.   The reactions studied in Chapter 15 and in the preceding sections of this Chapter all take place at the carbonyl group, *i.e.*, oxidation to a carboxyl group, reduction to an alcohol group, or addition of various reactants across the carbon-oxygen π-bond. The nature of the carbonyl group also affects its neighboring α- and β-carbon atoms. The partial positive charge on the carbonyl carbon tends to make the hydrogens on the α-carbons (the "α-hydrogens") easily removed as protons, and the resulting carbanion is stabilized by resonance with the carbonyl π-electron system to produce what is termed an enolate anion [**26**, equation (15)].

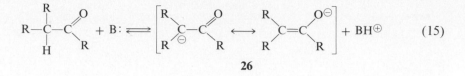

$$\text{26} + X_2 \longrightarrow R{-}\overset{\overset{\displaystyle R}{|}}{\underset{\underset{\displaystyle X}{|}}{C}}{-}C\overset{\displaystyle O}{\underset{\displaystyle R}{\diagdown}} + X^{\ominus} \tag{16}$$

case of acetaldehyde or a methyl ketone, all three of the α-hydrogens of the methyl group are replaced by halogen to give **27** [equation (17), R = H or alkyl].

$$H{-}\overset{\overset{\displaystyle H}{|}}{\underset{\underset{\displaystyle H}{|}}{C}}{-}\overset{\overset{\displaystyle O}{\|}}{C}{-}R + 3\,\text{NaOH} + 3\,X_2 \longrightarrow X{-}\overset{\overset{\displaystyle X}{|}}{\underset{\underset{\displaystyle X}{|}}{C}}{-}\overset{\overset{\displaystyle O}{\|}}{C}{-}R + 3\,\text{NaX} + 3\,H_2O \tag{17}$$

27

When the first α-hydrogen is replaced by halogen, the remaining hydrogens attached to this α-carbon atom become more acidic because of the inductive effect of the halogen, so that further substitution by halogen at this site occurs more rapidly than at the other α-carbon atom. The inductive effect of the three halogens makes the carbon atom of the carbonyl group particularly susceptible to nucleophilic addition of hydroxyl. The intermediate adduct **28** readily undergoes cleavage as shown in equation (18) and the fragments **29** and **30** are immediately converted to the observed products, a haloform, **31**, and a carboxylic acid salt, **32**, as shown in equations (19) and (20). (**31** and **32** may also be formed simply by transfer of a proton from **30** to

$$X{-}\overset{\overset{\displaystyle X}{|}}{\underset{\underset{\displaystyle X}{|}}{C}}{-}\overset{\overset{\displaystyle O}{\|}}{C}{-}R + HO^{\ominus} \longrightarrow X{-}\overset{\overset{\displaystyle X}{|}}{\underset{\underset{\displaystyle X}{|}}{C}}{-}\overset{\overset{\displaystyle O^{\ominus}}{|}}{\underset{\underset{\displaystyle OH}{|}}{C}}{-}R \longrightarrow X_3C^{\ominus} + HO{-}\overset{\overset{\displaystyle O}{\|}}{C}{-}R \tag{18}$$

28   **29**   **30**

$$X_3C^{\ominus} + H_2O \longrightarrow HCX_3 + HO^{\ominus} \tag{19}$$

31

$$R{-}\overset{\overset{\displaystyle O}{\|}}{C}{-}OH + HO^{\ominus} \longrightarrow R{-}\overset{\overset{\displaystyle O}{\|}}{C}{-}O^{\ominus} + H_2O \tag{20}$$

32

29). The overall equation for the reaction of a methyl ketone, or acetaldehyde (R = H), is given by equation (21).

$$\underset{\substack{\| \\ O}}{CH_3CR} + 4\ NaOH + 3\ X_2 \longrightarrow HCX_3 + \underset{\substack{\| \\ O}}{NaOCR} + 3\ NaX + 3\ H_2O \qquad (21)$$

Although all carbonyl compounds having hydrogens on the α-carbons undergo halogenation at these positions, only those with *methyl* groups adjacent to the carbonyl group undergo carbon-carbon cleavage, presumably because *three* halogens attached to one carbon atom are required to weaken the bond to the point of rupture. Thus, although nearly all carbonyl compounds react with halogens in basic media, only those with α-methyl groups produce haloforms and carboxylic acid salts. This fact is exploited in two important ways: (1) Since iodoform (HCI_3) is a highly insoluble crystalline yellow solid with a characteristic odor, it is readily detected, and its formation is used as a *qualitative test* for the structural moieties shown below (R = H, alkyl, or

$$\underset{\substack{\| \\ O}}{CH_3-C-R} \qquad \underset{\substack{| \\ OH}}{CH_3-CH-R}$$

aryl).[6] (2) The conversion of a methyl ketone to a carboxylic acid with one less carbon atom is often useful in synthesis. In this case chlorine is the halogen of choice because it is the least expensive and most readily available.

$$\underset{\substack{\| \\ O}}{R-C-CH_3} \xrightarrow[\text{(2) } H_3O^{\oplus}]{\text{(1) } Cl_2,\ NaOH} \underset{\substack{\| \\ O}}{R-C-OH} \qquad (22)$$

Aldol additions and condensations.   Another important and general reaction of enolate anions (besides reaction with halogens) is addition to the carbonyl group of the aldehyde or ketone from which the enolate anion was derived. In this way a dimeric anion, **33**, is produced [equation (23)], which stabilizes itself by abstracting a proton from the solvent (water or alcohol), as shown in equation (24). The β-hydroxy

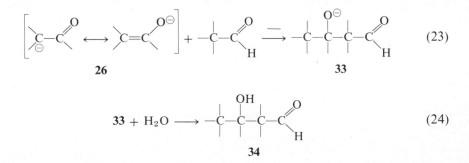

$$\qquad\qquad (23)$$

26   **33**

$$33 + H_2O \longrightarrow \underset{\textbf{34}}{-\overset{|}{\underset{|}{C}}-\overset{\overset{\displaystyle OH}{|}}{\underset{|}{C}}-\overset{|}{\underset{|}{C}}-C\underset{H}{\overset{O}{\diagup}}} \qquad (24)$$

carbonyl compound of type **34** is called an "aldol," since it is both an aldehyde and an alcohol. The term "aldol *addition*" is also applied generally to the base-catalyzed self-addition of ketones as well as of aldehydes. The overall change for the reaction of

---

[6] Alcohols of this structural type respond to the test because of the oxidizing power of the reagent, which converts them partially to the corresponding carbonyl compounds.

acetaldehyde is given by equation (25). Most β-hydroxy aldehydes and ketones undergo dehydration readily to α, β-unsaturated aldehydes and ketones [equation (26)]. In this

$$2\ CH_3CHO \xrightarrow{HO^{\ominus}} CH_3\overset{\displaystyle OH}{\underset{\displaystyle |}{CH}}-CH_2CHO \tag{25}$$

$$CH_3\overset{\displaystyle OH}{\underset{\displaystyle |}{CH}}-CH_2CHO \longrightarrow CH_3CH{=}CHCHO + H_2O \tag{26}$$

case the overall reaction may be referred to as an "aldol *condensation*," since a molecule of water is eliminated from the adduct.

In general, ketones do not undergo self-addition as readily as aldehydes; in fact, special conditions must usually be employed to obtain good yields in such reactions. (Refer to a textbook for a discussion of the reaction of acetone, for example). "Mixed (or crossed) aldol condensations" of two different aldehydes or of an aldehyde with a ketone are possible. Such mixed condensations are practical for synthesis in the case of two aldehydes *if* one of the aldehydes has no α-hydrogens (and hence only serves as a carbonyl acceptor of the enolate anion derived from the other aldehyde) and in the case of a ketone and aldehyde under conditions such that the ketone undergoes no appreciable self-condensation. A good example of the latter case is the reaction of benzaldehyde with acetophenone in the presence of dilute sodium hydroxide solution [equations (27)–(30)]. The product, **38,** is called benzalacetophenone. Under the

$$C_6H_5\overset{\displaystyle O}{\overset{\|}{C}}-CH_3 + HO^{\ominus} \xrightarrow{-H_2O} \left[C_6H_5\overset{\displaystyle O}{\overset{\|}{C}}-\overset{\ominus}{C}H_2 \longleftrightarrow C_6H_5\overset{\displaystyle O^{\ominus}}{\overset{|}{C}}{=}CH_2 \right] \tag{27}$$
$$\underset{\textbf{35}}{}$$

$$\textbf{35} + C_6H_5\overset{\displaystyle O}{\overset{\|}{C}}H \longrightarrow C_6H_5\overset{\displaystyle O^{\ominus}}{\underset{}{\overset{|}{C}}}H-CH_2\overset{\displaystyle O}{\overset{\|}{C}}C_6H_5 \tag{28}$$
$$\underset{\textbf{36}}{}$$

$$\textbf{36} + H_2O \longrightarrow C_6H_5\overset{\displaystyle OH}{\underset{}{\overset{|}{C}}}H-CH_2\overset{\displaystyle O}{\overset{\|}{C}}C_6H_5 + HO^{\ominus} \tag{29}$$
$$\underset{\textbf{37}}{}$$

$$\textbf{37} \longrightarrow C_6H_5CH{=}CH\overset{\displaystyle O}{\overset{\|}{C}}C_6H_5 + H_2O \tag{30}$$
$$\underset{\textbf{38}}{}$$

conditions of the experiment, the dehydration of the "aldol," **37,** is spontaneous [equation (30)].

Reactions of benzalacetophenone. As expected, benzalacetophenone gives reactions typical of both the ketone function and the alkenic double bond. For example, a semicarbazone and a 2,4-dinitrophenylhydrazone may be obtained by the procedures of Section 16.1.

Bromine adds readily to the carbon-carbon double bond. Since two asymmetric carbon atoms are produced in this reaction, there should be four possible stereoisomers of the dibromide, two enantiomeric pairs (see Chapter 21.1).

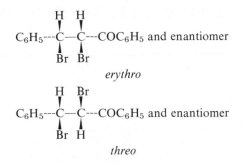

$$C_6H_5\text{---}\overset{\overset{\displaystyle H}{|}}{\underset{\underset{\displaystyle Br}{|}}{C}}\text{---}\overset{\overset{\displaystyle H}{|}}{\underset{\underset{\displaystyle Br}{|}}{C}}\text{---}COC_6H_5 \text{ and enantiomer}$$

erythro

$$C_6H_5\text{---}\overset{\overset{\displaystyle H}{|}}{\underset{\underset{\displaystyle Br}{|}}{C}}\text{---}\overset{\overset{\displaystyle Br}{|}}{\underset{\underset{\displaystyle H}{|}}{C}}\text{---}COC_6H_5 \text{ and enantiomer}$$

threo

The higher-melting pair (mp 159–160°) is thought to have the *erythro* configuration, and the lower-melting pair (mp 123–124°) the *threo* configuration. It should be noted that benzalacetophenone is capable of geometric (*cis*-*trans*) isomerism; the mixed aldol condensation gives predominantly the more stable *trans* isomer. If the addition of bromine to benzalacetophenone were stereospecific, only one of the pairs of enantiomers should result from each geometric isomer (see Chapter 6.2); however, the higher melting dibromide is the major product of addition of bromine to both *cis*- and *trans*-benzalacetophenone.

As mentioned in Chapter 6.2, nucleophiles will add to compounds having carbon-carbon double bonds conjugated with carbonyl groups, although they do not add to simple alkenes. The addition of aniline to benzalacetophenone represents a good example of the addition of a nucleophile to an α,β-unsaturated ketone [equation (31)].

$$C_6H_5CH{=}CH\overset{\overset{\displaystyle O}{\|}}{C}C_6H_5 + C_6H_5NH_2 \longrightarrow C_6H_5\underset{\underset{\displaystyle NHC_6H_5}{|}}{C}H\text{---}CH_2\overset{\overset{\displaystyle O}{\|}}{C}C_6H_5 \qquad (31)$$

$$\textbf{38} \hspace{5cm} \textbf{40}$$

The driving force for this addition is the formation of the resonance-stabilized intermediate, **39** (or the anion formed from it by loss of one of the protons from the nitrogen

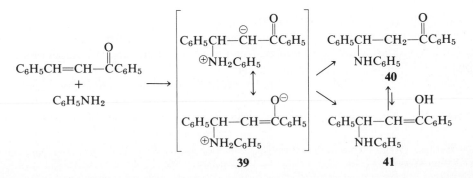

atom). The final product, β-anilino-β-phenylpropiophenone (**40**), may be formed directly from **39** or via the 1,4-addition product **41**, the less stable enol form of **40**.

EXPERIMENTAL PROCEDURE

A. THE HALOFORM REACTION

1. Sodium hypoiodite—the "iodoform test." Apply the following procedure to small samples of isopropyl alcohol, acetophenone, cyclohexanone, and pinacolone.

If the substance is water-soluble, dissolve 2–3 drops of a liquid or an estimated 50 mg of a solid in 2 ml of water in a small test tube, add 2 ml of 3 *N* sodium hydroxide, and then slowly add 3 ml of iodine solution.[7] In a positive test the brown color disappears, and yellow iodoform separates. If the substance tested is insoluble in water, dissolve it in 2 ml of dioxane, proceed as above, and at the end dilute with 10 ml of water.

Iodoform can be recognized by its odor and yellow color and, more definitely, by its melting point, 119°. The substance can be isolated by suction filtration of the test mixture or by adding 2 ml of chloroform, shaking the stoppered test tube to extract the iodoform into the small lower layer, withdrawing the clear part of this layer with a capillary dropping tube, and evaporating it in a small tube on the steam bath. The crude solid is recrystallized from methanol-water.

2. Sodium hypochlorite—preparation of benzoic acid. To 40 ml of 5% sodium hypochlorite solution ("Clorox," "Purex," etc.) in a 150-ml beaker add 1 ml of acetophenone. Swirl the mixture vigorously and notice whether any heat of reaction may be detected. Place the beaker on the steam cone and heat gently, swirling from time to time, for about 10 min to volatilize the chloroform. Add about 10 drops of acetone to the warm solution to destroy any excess hypochlorite.

Add a small amount of decolorizing carbon to the warm solution, heat it to boiling while swirling gently but continuously, and filter the hot mixture. To the hot filtrate add 5 ml of concentrated hydrochloric acid with stirring, allow the mixture to cool to room temperature, and finally cool it in an ice bath. Collect the crystals, wash them with a little cold water, dry them, and determine their melting point.

Determine the melting point of a mixture of this product with the one from the Cannizzaro reaction (Chapter 15.3).

B. THE ALDOL CONDENSATION

1. Preparation of *trans*-benzalacetophenone. In a 125-ml Erlenmeyer flask mix 25 ml of 3 *N* sodium hydroxide solution, 15 ml of 95% ethanol, and 0.05 mole of acetophenone. Cool the mixture in an ice bath and, while swirling, add an equimolar amount of benzaldehyde. Allow the mixture to warm to room temperature and occasionally shake it (stoppered) vigorously during 1 or 2 hr. When a yellow oil separates, it will be helpful to induce crystallization by adding a few seed crystals

---

[7] The iodine reagent is prepared by dissolving 25 g of iodine in a solution of 50 g of potassium iodide in 200 ml of water.

of benzalacetophenone and/or scratching the side of the flask with a stirring rod. (*Caution:* Do not allow benzalacetophenone to come in contact with the skin as it is a skin irritant.)

Cool the reaction mixture in an ice bath until crystallization appears to be complete and collect the product by vacuum filtration. Wash the crystals with cold water and then with a little ice-cold 95% ethanol. Reserve a few crystals to be used for seeding and recrystallize the crude product from 95% ethanol, using 4–5 ml of solvent per gram of crystals. Determine the melting point of the recrystallized material. The melting point of pure *trans*-benzalacetophenone is reported to be 58–59°.

2. Reactions of benzalacetophenone. Benzalacetophenone dibromide. Dissolve 1 g of dry *trans*-benzalacetophenone in 4 ml of carbon tetrachloride. Cool the solution in an ice bath and add dropwise, while swirling, 4 ml of a solution of bromine in carbon tetrachloride, prepared by dissolving 1 g of bromine in 6 ml of carbon tetrachloride. Allow the mixture to stand for 5 or 10 min and collect the crystals, wash them with a little cold 95% ethanol, dry them, and determine their melting point.

Addition of aniline to benzalacetophenone. Dissolve 1 g of benzalacetophenone in 20 ml of 95% ethanol and add an equimolar amount of aniline. Stir or shake until the solution is homogeneous and allow it to stand (stoppered) overnight. If crystals have not separated, scratch the side of the tube or flask with a glass rod, and cool the mixture in an ice bath. Collect the crystals, dry them, and determine their melting point (the reported melting point of β-anilino-β-phenylpropiophenone is 175°).

EXERCISES

A. THE HALOFORM REACTION

1. Ethanol gives a clearly positive "iodoform test," yet when sodium hypoiodite is used to oxidize ethanol to acetaldehyde, the yield is found to be less than 10%. Explain how one is able to obtain a good qualitative test in spite of this fact. (*Hint:* Calculate the molecular weight of iodoform.)

2. Suggest an explanation for the failure of ethyl acetoacetate, $CH_3COCH_2CO_2C_2H_5$, to give iodoform on treatment with sodium hypoiodite. (*Hint:* Consider the mechanism of the base-catalyzed Claisen condensation, which is a reversible reaction.)

3. Dibenzoylmethane, $C_6H_5COCH_2COC_6H_5$, gives iodoform when treated with sodium hypoiodite, even though the compound is not a methyl ketone. Explain this behavior, using equations.

B. THE ALDOL CONDENSATION

1. Why is the mixed aldol condensation, which produces benzalacetophenone, the main reaction that occurs in this experiment rather than the two possible side reactions, self-condensation of acetophenone and Cannizzaro reaction of benzaldehyde?

2. Write equations to show the major products from reaction of the following carbonyl compounds and bases:
   (a) propionaldehyde, formaldehyde, dilute sodium hydroxide
   (b) benzaldehyde, concentrated potassium hydroxide

    (c) benzaldehyde, ethyl acetate, sodium ethoxide in ethanol

    (d) acetone, benzaldehyde (2 moles), dilute sodium hydroxide

3. Suggest a reason for the fact that the *trans* isomer of benzalacetophenone predominates in the reaction mixture from the aldol condensation.

4. How may *cis*-benzalacetophenone be isomerized to the *trans* isomer?

SPECTRA OF STARTING MATERIALS AND PRODUCT

The spectra of benzaldehyde are given in Figures 15.6 and 15.7.

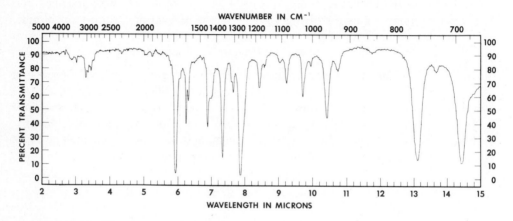

FIGURE 16.19 IR spectrum of acetophenone.

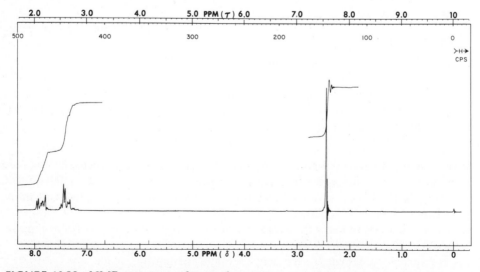

FIGURE 16.20 NMR spectrum of acetophenone.

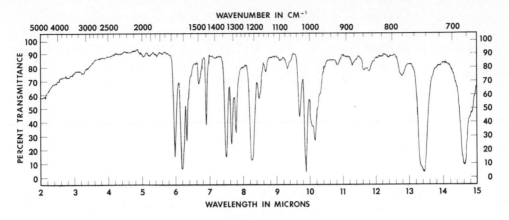

FIGURE 16.21 IR spectrum of *trans*-benzalacetophenone.

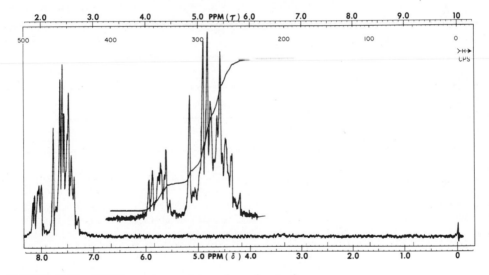

FIGURE 16.22 NMR spectrum of *trans*-benzalacetophenone.

16.3 Identification of an Alcohol or Carbonyl Compound by Means of Three Reagents

The reactions of alcohols, aldehydes, and ketones with the reagents, chromic acid in acetone, 2,4-dinitrophenylhydrazine, and sodium hypoiodite, have been discussed in Sections 15.2, 16.1, and 16.2. By use of only these three reagents, with proper interpretation of the results of their reactions, it is possible to identify an unknown alcohol or carbonyl compound, if there is a known limitation on the number of possibilities for the unknown.

As an illustration, it is possible to decide on the identity of any one of the compounds in the following sets of five by using only the three reagents. In some instances only two tests may be required. Your instructor will provide you with an unknown compound labeled as belonging to one of the sets. Identify the compound by using a *minimum* number of tests.

Set A	*Set B*	*Set C*
1-pentanol	acetone	1-octanol
2-butanol	2-methyl-1-propanol	cinnamaldehyde
2-butanone	2-methylpropanal	acetophenone
benzaldehyde	2-propanol	4-methyl-2-pentanol
2-methyl-2-butanol	2-methyl-2-propanol	cyclohexanone

EXERCISE

Write out balanced equations for all positive reactions which you observe and give a careful explanation of their meaning in terms of the structure of your unknown compound.

chapter seventeen

carboxylic acids and their derivatives

Organic compounds which possess the general structural formula, RCOX, in which R is an alkyl or aryl substituent and X represents a functional group having an atom other than carbon or hydrogen bonded to the carbonyl carbon, are classified as carboxylic acids or their derivatives. Thus, X is OH in a carboxylic acid, OR' in an ester, OCOR' in an anhydride and NR_2 in an amide. Nitriles, RCN, are also sometimes classified as derivatives of carboxylic acids because their hydrolysis produces these acids.

It is of interest that many naturally occurring products are carboxylic acids or their derivatives. For example, fats are esters of straight-chain carboxylic acids containing from twelve to eighteen carbon atoms and glycerol (1,2,3-trihydroxypropane), and proteins are basically polyamides. In addition, volatile esters are partially responsible for the pleasant odors of many fruits, whereas carboxylic acids are often the cause of unpleasant odors such as those of rancid butter and of goats. Soaps are the sodium or potassium salts of the same carboxylic acids contained in fats.

The chemistry of carboxylic acids and their derivatives in many ways parallels that of aldehydes and ketones (see Chapters 15 and 16) in that reactions at both the carbonyl carbon and the α-carbon are known. An example of attack at the α-carbon is the Hell-Volhard-Zelinsky reaction in which a carboxylic acid is converted to an α-haloacid [equation (1), X = Cl or Br]. The Claisen ester condensation exemplifies reaction at the α-carbon of one molecule of an ester and at the carbonyl carbon of a second [equation (2)]. These two conversions are analogous to the base-catalyzed halo-

$$\underset{\overset{|}{R'}}{\overset{\overset{H}{|}}{R-C}}-COOH + X_2 \xrightarrow[\text{or PX}_3 \text{ (cat.)}]{P \text{ (cat.)}} \underset{\overset{|}{R'}}{\overset{\overset{X}{|}}{R-C}}-COOH + HX \qquad (1)$$

$$2\ RCH_2\overset{\overset{O}{\parallel}}{C}OR' \xrightarrow{R'O^\ominus} RCH_2\overset{\overset{O}{\parallel}}{C}-\underset{\overset{|}{R}}{\overset{\overset{O}{\parallel}}{C}}HCOR' + R'OH \qquad (2)$$

genation of aldehydes and ketones and to the base-catalyzed aldol condensation, respectively. Of course, a third site of reaction, not present in aldehydes and ketones, is the functional group X, present in carboxylic acids and their derivatives.

The types of reactions to be considered in this chapter will be limited to those occurring at the carbonyl carbon or the substituent X. Although the specific experiments to be described all involve benzoic acid or its derivatives, C_6H_5COX, the student should bear in mind that other carboxylic acids and their derivatives are prepared or react in a similar manner. As a rule, however, aliphatic derivatives of acids are more reactive than their aryl counterparts, undergoing hydrolysis much more readily, for example.

17.1   Carboxylic Acids

Carboxylic acids can be synthesized in a variety of ways, which include the hydrolysis of nitriles or of any derivative of the acid, and oxidation of an aldehyde or ketone. The carbonation of a Grignard reagent, described in Chapter 14.1, is another useful synthetic route to these acids.

A special method available for the preparation of aryl carboxylic acids involves the oxidation of the side chain R of an alkyl-substituted benzene with alkaline permanganate or with chromic acid solution [equation (3)]. The carbon atom bonded to

$$ArCH_2R \xrightarrow[\text{or } H_2CrO_4]{\text{KMnO}_4/\text{base}} ArCOOH \tag{3}$$

the aromatic ring must bear *at least one* hydrogen for this method of preparation to be efficient. If more than one alkyl side chain is present, each will be oxidized to a carboxyl group. Identification of the resulting acid will allow determination of the positions of attachment of the side chains on the aromatic ring, if this is not known.

In the preparation of a carboxylic acid described in this experiment, ethylbenzene, which is readily available from the Friedel-Crafts alkylation of benzene, is oxidized with alkaline permanganate to benzoic acid [equation (4)]. The mechanism of this conversion is not well understood but probably involves the removal of the benzylic hydrogen in a free-radical process initiated by permanganate ion; subsequent reactions lead first to the alcohol and then to the ketone, which is oxidized further to the carboxylic acid. The mechanism of this final oxidation step parallels that given for the oxidation of cyclohexanone to adipic acid (Chapter 15.1).

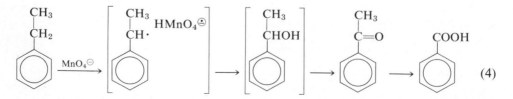

As a group, carboxylic acids are considered weak acids as compared to mineral acids, such as hydrochloric, and have K_a's of the order of 10^{-5}. Nevertheless, carboxylic acids are sufficiently acidic to give a test with litmus paper and readily to undergo re-action with bases to form salts.

Use is often made of the acidity of the carboxylic acids in their removal from a re-action mixture by extraction of the mixture with base. The organic acid can be re-generated from the basic solution by acidification with mineral acid. A good way to differentiate carboxylic acids from phenols (ArOH, another type of weak organic acid) depends on the ability of carboxylic acids to react with the weak base bicarbonate to produce the carboxylate salt and carbon dioxide, which bubbles out of the solution. Most phenols, being weaker acids (K_a's $\approx 10^{-10}$), will not react with this base.

The relatively high acidity of carboxylic acids enables ready determination of the *equivalent weight* (sometimes called "neutralization equivalent") of the acid by titra-tion with standard base. The equivalent weight of an acid is that weight, in grams, of acid which reacts with one equivalent of base. As an example, suppose that 0.1000 g of an unknown acid required 16.90 ml of 0.1000 N sodium hydroxide solution to be titrated to a phenolphthalein endpoint. This means that 0.1000 g of the acid corre-sponds to (16.90 ml) (0.1000 equivalent/1000 ml) or 0.00169 equivalent of the acid, or that one equivalent of the acid weighs 0.1000/0.00169 or 59.2 grams. Thus, the follow-ing expression applies:

$$\text{Equivalent weight} = \frac{\text{grams of acid}}{(\text{volume of base consumed in liters})(N)}$$

where N is the normality of the standard base.

Because each carboxylic acid function in a molecule will be titrated with base, the equivalent weight corresponds to the molecular weight of the acid divided by n, where n is the number of acid functions present in the molecule. Thus, for the example above, the molecular weight may be 59.2, if a single acid function is present, 118.4 if two are present, and 177.6 if three are present. If the molecular formula of an unknown com-pound is known, then the number of acid groups in the molecule can be calculated by dividing the molecular weight by the equivalent weight. Hence, if the molecular formula of the unknown compound is $C_4H_6O_4$ and its equivalent weight is 59.2, the un-known must have two titratable acid functions.

17.2 Carboxylic Acid Esters

The esters of carboxylic acids are most commonly obtained by allowing the acid and an alcohol to react in the presence of a mineral acid [equation (5)]. The role of the catalyst undoubtedly is to promote attack of the nucleophilic oxygen of the alcohol on the carbonyl carbon of the acid by protonation of the carbonyl oxygen to give **1**. The process is one of equilibria between a variety of compounds, and, in order that the esterification be a reaction of high yield, the equilibria must be shifted toward the products, the desired ester and water. This can be accomplished either by removing one or more of the products from the reaction mixture as formed or by employing a large excess of one of the starting reagents.

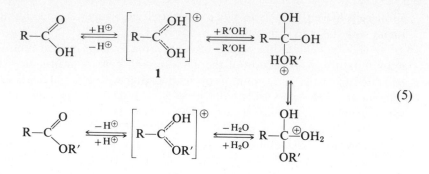

$$(5)$$

The effect of the latter approach can be understood by consideration of the mass law relating starting materials and products [equation (6)]. Increasing the amount of either the alcohol or the carboxylic acid will result in an increase in the amount of products formed since the equilibrium constant, K, for the reaction must remain constant, at a given temperature, no matter what quantity of either reagent is used.

$$RCOOH + R'OH \overset{K}{\rightleftharpoons} RCOOR' + H_2O$$

$$K = \frac{[RCOOR'] \, [H_2O]}{[RCOOH] \, [R'OH]} \tag{6}$$

Occasionally it is economically unfeasible to use a large excess of either reagent, and other methods must be employed. These may involve removal of water by azeotropic distillation with the esterifying alcohol, effective removal of the water from the reaction mixture by use of a large excess of acid catalyst (which converts water to its conjugate acid, H_3O^+) or by use of a solvent such as ethylene chloride in which the ester is soluble, but water is not. The procedure adopted in the esterification of benzoic acid with methanol [equation (6), $R = C_6H_5$, $R' = CH_3$], a reaction described in the experimental section of this chapter, is to use a large excess of the alcohol, which is quite inexpensive.

If the carboxylic acid has bulky substituents on the carbons α and/or β to the carboxyl group, the esterification either may be very slow or may not occur at all because of steric hindrance to the attack of the alcohol on the intermediate 1 [equation (5)]. An alternative method for the preparation of esters of such hindered acids involves initial conversion of the acid to its acid chloride (see below) and subsequent reaction of this compound with an alcohol to produce the ester [equation (7)].

$$R-\overset{\overset{\displaystyle O}{\|}}{C}-OH \longrightarrow R-\overset{\overset{\displaystyle O}{\|}}{C}-Cl \xrightarrow{R'OH} R-\overset{\overset{\displaystyle O}{\|}}{C}-OR' + HCl \tag{7}$$

Hydrolysis of esters to carboxylic acids, the reverse of esterification, can be accomplished by heating the ester with aqueous acid or base, or, in the case of esters which are insoluble in water, by heating with methanolic or ethanolic potassium hydroxide [equation (8)]. Catalysis by base rather than acid is generally preferred because the rate of hydrolysis of an ester is greater in basic medium. Moreover, if at least a molar equivalent of base is used, the acid will be converted to its salt, which makes the reaction irreversible.

$$R-\overset{\overset{\displaystyle O}{\|}}{C}-OR' + H_2O \xrightarrow{\text{H}_3\text{O}^\oplus \text{ or HO}^\ominus} R-\overset{\overset{\displaystyle O}{\|}}{C}-OH + R'OH \qquad (8)$$

It is possible to carry out the hydrolysis of an ester with alkali in a quantitative manner so that a value, termed the *saponification equivalent*, can be derived. This value is analogous to the equivalent weight of an acid in that it is the molecular weight of the ester divided by the number of ester functions in the molecule. Therefore, the saponification equivalent is the number of grams of ester required to react with one gram-equivalent of alkali. This equivalent is determined by hydrolyzing a weighed amount of the ester with an excess of standardized alkali and then titrating the excess alkali to a phenolphthalein end-point with standardized hydrochloric acid. The saponification equivalent is then given as follows:

$$\text{Saponification Equivalent} = \frac{\text{grams of ester}}{\text{equivalents of alkali consumed}}$$

$$-\frac{\text{grams of ester}}{(\text{volume of alkali in liters})\,(N) - (\text{volume of acid in liters})\,(N')}$$

where N is the normality of the standard base and N' is the normality of the standard acid.

17.3   Carboxylic Acid Halides (Acyl Halides) and Anhydrides

Acid halides, substances in which the hydroxyl function of the carboxylic acid has been replaced by a halogen, are exceedingly useful and reactive compounds. Some of the more important reactions of acid halides are those with various nucleophiles, and a few of these are shown in Chart 17.1. Thus, acid halides are converted to carboxylic acids upon hydrolysis, to esters upon reaction with alcohols, to amides when treated with ammonia or amines, and to anhydrides when allowed to react with carboxylic acids or their salts.

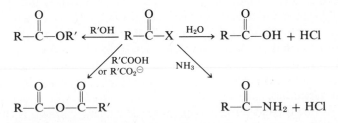

CHART 17.1   Some reactions of acid halides.

A generalized mechanism, which is common to almost all reactions of derivatives of carboxylic acids with nucleophiles, can be applied to the conversions shown in Chart 17.1. In this mechanism, the nucleophile, NuH (or Nu⁻), adds to the carbonyl carbon of the derivative, RCOX, to produce the tetrahedral intermediate, **2**, which then decomposes with elimination of HX (or X⁻) to give the product [equation (9)]. Although the overall reaction is one of *substitution* of Nu for X, the mechanism of the conversion involves an addition step and an elimination step, and is therefore markedly different from the familiar substitutions of the S_N1 and S_N2 types. The success of such a

$$
R-\overset{\overset{O}{\parallel}}{C}-X \; + \; :NuH \; \longrightarrow \; R-\overset{\overset{O^{\ominus}}{\mid}}{\underset{\overset{\mid}{\oplus NuH}}{C}}-X \; \longrightarrow \; R-\overset{\overset{O}{\parallel}}{C}-Nu \; + \; HX \tag{9}
$$

$$\mathbf{2}$$

substitution reaction might be expected to depend to a great extent upon the relative basicities of the reagent, NuH (or Nu⁻), and the leaving group, XH (or X⁻), and such a dependence is observed experimentally. Thus, it is impossible to convert an ester (X = OR) to an acid chloride (X = Cl) by reaction of the ester with chloride ion because chloride ion is a much weaker base than is alkoxide, RO⁻.

The rate at which the substitution reaction occurs with acid halides is enhanced compared to other acid derivatives because the halogen is electron-withdrawing and therefore increases the electrophilicity of the carbonyl carbon. In addition, halide ions, being very weak bases, are good leaving groups.

Acid chlorides can be prepared by treatment of the carboxylic acid with thionyl chloride [equation (10)], phosphorus trichloride [equation (11)] or phosphorus penta-

$$
RCOOH + SOCl_2 \rightarrow RCOCl + HCl + SO_2 \tag{10}
$$

$$
3\,RCOOH + PCl_3 \rightarrow 3\,RCOCl + H_3PO_3 \tag{11}
$$

$$
RCOOH + PCl_5 \rightarrow RCOCl + POCl_3 + HCl \tag{12}
$$

chloride [equation (12)]. Acid bromides may be produced by using the corresponding bromine-containing reagents but seldom are because, in sharp contrast to the relative reactivities of *alkyl* bromides and chlorides in S_N1 and S_N2 reactions, acid bromides are *less* reactive than the corresponding chlorides.

The phosphorus pentahalides are more reactive halogenating agents than the others but are more difficult to handle in the laboratory. Consequently they are usually employed only in instances in which the other reagents fail, as is the case in the conversion of some aryl carboxylic acids and sterically hindered acids to their acid halides. It should be noted that the most complete utilization of the halogens present in the halogenating agent is accomplished in the reaction of the carboxylic acid with phosphorus trichloride, but use of this reagent is sometimes precluded because of possible experimental difficulties.

Thionyl chloride is used for the preparation of benzoyl chloride from benzoic acid [equation (10), R = C_6H_5] as described in the Experimental Procedure. It is used in preference to phosphorus trichloride because the latter reagent, although somewhat less expensive, leads to reaction mixtures containing phosphorous acid, H_3PO_3, from which it is difficult to separate the benzoyl chloride. In contrast, the by-products from the reaction of thionyl chloride with benzoic acid are gases (HCl and SO_2) and therefore are readily removed from the reaction mixture. Furthermore, thionyl chloride has a low boiling point (79°) so that it may be separated from the product by simple distillation.

Methods are also given for the conversion of benzoyl chloride to esters, amides and an anhydride. Although the reactions are typical of acid chlorides in general, it is important to note, once again, that the reaction conditions are somewhat more vigorous than those required for aliphatic acid chlorides. Thus, hydrolysis of benzoyl chloride requires gentle heating, whereas acetyl chloride, CH_3COCl, reacts almost violently with cold water.

The preparation of benzoic anhydride from benzoyl chloride deserves some special comment. The reaction of benzoic acid with benzoyl chloride, using pyridine, C_5H_5N, as a catalyst, produces a high yield of the anhydride [equation (13)]. The role of the

$$C_6H_5\overset{\displaystyle O}{\overset{\|}{C}}-Cl \longrightarrow C_6H_5\overset{\displaystyle O}{\overset{\|}{C}}-\overset{\oplus}{N} \underset{Cl^{\ominus}}{} \quad\underset{\mathbf{3}}{} \xrightarrow{C_6H_5CO_2^{\ominus} \ \overset{\oplus}{HN}} C_6H_5\overset{\displaystyle O}{\overset{\|}{C}}-OCOC_6H_5 \qquad (13)$$

catalyst is twofold: (1) It activates the acid chloride by formation of a pyridinium salt, 3, which is more susceptible to nucleophilic attack than is the acid chloride itself, and (2) it converts benzoic acid to the more nucleophilic benzoate ion. It is possible to obtain a still higher yield of benzoic anhydride if the benzoic acid is generated *in situ* by reaction of the acid chloride-pyridine complex, 3, with 0.5 equivalent of water. The benzoic acid resulting from hydrolysis of the complex reacts immediately with a second mole of the complex to produce the desired anhydride.

Because of their greater ease of handling in the laboratory, anhydrides are often used in place of the more reactive acid halides for the acylation of alcohols and of amines to produce esters and amides, respectively [equations (14) and (15)]. Use of acetic anhydride to acetylate an amine is described in Chapter 19.2 and 19.3.

$$R\overset{\displaystyle O}{\overset{\|}{C}}-O-\overset{\displaystyle O}{\overset{\|}{C}}-R + R'OH \longrightarrow R\overset{\displaystyle O}{\overset{\|}{C}}-OR' + R\overset{\displaystyle O}{\overset{\|}{C}}-OH \qquad (14)$$

$$R\overset{\displaystyle O}{\overset{\|}{C}}-O-\overset{\displaystyle O}{\overset{\|}{C}}-R + R'NH_2 \longrightarrow R\overset{\displaystyle O}{\overset{\|}{C}}-NHR' + R\overset{\displaystyle O}{\overset{\|}{C}}-OH \qquad (15)$$

17.4 Carboxylic Acid Amides and Nitriles

The amides of carboxylic acids can be prepared by reaction of an acid halide or anhydride with either anhydrous or aqueous ammonia, if an unsubstituted amide is desired, or with an amine, if an N-alkylated amide is required. Thermal decomposition of the ammonium salt of a carboxylic acid also generates the corresponding amide [equation (16)].

$$RCOOH + NH_3 \longrightarrow RCOO^{\ominus}NH_4^{\oplus} \xrightarrow{\text{heat}} R-\overset{\overset{\displaystyle O}{\|}}{C}-NH_2 + H_2O \qquad (16)$$

Amides can also be prepared by treatment of esters with amines [equation (17)]. This conversion generally requires heat and, in the case of volatile amines or ammonia, must be carried out under pressure.

$$R-\overset{\overset{\displaystyle O}{\|}}{C}-OR' \xrightarrow{HNR_2} R-\overset{\overset{\displaystyle O}{\|}}{C}-NR_2 \; + R'OH \qquad (17)$$

Although amides undergo a variety of reactions, two of the more important are their reduction to amines with lithium aluminum hydride [equation (18)] and their conversion to nitriles with dehydrating agents such as phosphorus oxychloride, POCl$_3$, phosphorus pentoxide, P$_2$O$_5$, or thionyl chloride [equation (19)]. The fact that amides

$$R-\overset{\overset{\displaystyle O}{\|}}{C}-NH_2 \xrightarrow[\text{2) } H_3O^{\oplus}]{\text{1) LiAlH}_4} R-CH_2-NH_2 \qquad (18)$$

$$R-\overset{\overset{\displaystyle O}{\|}}{C}-NH_2 \xrightarrow[\text{or SOCl}_2]{\substack{\text{POCl}_3 \\ \text{or P}_2\text{O}_5}} R-C\equiv N \qquad (19)$$

are reduced to amines is somewhat unexpected since all other derivatives of carboxylic acids and, in fact, the acids themselves are reduced to primary alcohols on treatment with lithium aluminum hydride [equation (20)]. In short, the functional group X is retained on reduction of amides but is lost in the case of the other derivatives of acids.

$$R-\overset{\overset{\displaystyle O}{\|}}{C}-X \xrightarrow[\text{2) } H_3O^{\oplus}]{\text{1) LiAlH}_4} R-CH_2-OH \qquad (20)$$
$$(X \neq NR_2)$$

The mechanism of dehydration of an amide to a nitrile is interesting. As an example, the conversion of benzamide to benzonitrile with thionyl chloride probably involves attack of the amide oxygen on the sulfur atom of thionyl chloride to produce an intermediate salt, **4**, which then decomposes to the nitrile by loss of sulfur dioxide and two moles of hydrogen chloride [equation (21), R = C$_6$H$_5$].

4

Nitriles can also be prepared by the S_N2 reaction between cyanide ion and a primary or secondary alkyl halide. These substances can be hydrolyzed to amides, which are isolable in some cases, and then to the corresponding carboxylic acids by treatment with aqueous acid or base [equation (22)]. They can also be converted into esters by heating with an alcohol in the presence of one molar equivalent of water and a catalytic amount of mineral acid [equation (23)].

$$RC\equiv N \xrightarrow[\text{H}_2\text{O}]{\text{H}^\oplus \text{ or HO}^\ominus} \left[\underset{\displaystyle \text{RC}}{\overset{\displaystyle \text{O}}{\|}} -NH_2 \right] \longrightarrow \underset{\displaystyle \text{RC}}{\overset{\displaystyle \text{O}}{\|}} -OH \qquad (22)$$

$$RC\equiv N \xrightarrow[\text{R'OH}]{\text{HCl}} \underset{\displaystyle \text{RC}}{\overset{\displaystyle \text{O}}{\|}} -OR' \qquad (23)$$

EXPERIMENTAL PROCEDURE

A. OXIDATION OF AN AROMATIC SIDE CHAIN: PREPARATION OF BENZOIC ACID FROM ETHYLBENZENE

Place 0.071 mole of potassium permanganate, 120 ml of water, 1.5 ml of 3 N aqueous sodium hydroxide, and 0.016 mole of ethylbenzene in a 500-ml round-bottomed flask fitted with a reflux condenser, and gently heat this mixture at reflux for 2 hr. ★ Severe bumping will occur if strong heating is used. Test the hot solution for unchanged permanganate by placing a drop of the reaction mixture on a piece of filter paper; if a purple ring appears around the brown spot of manganese dioxide, permanganate remains. Destroy any excess permanganate by adding small amounts of solid sodium bisulfite to the mixture until the spot test is negative. Do not add a large excess of bisulfite. Add 2 g of filter-aid to the hot mixture to promote its rapid filtration; manganese dioxide is so finely dispersed that it would tend to clog the pores of the filter paper otherwise. Filter the mixture by suction and rinse the reaction flask and filter cake with two 10-ml portions of hot water.★ Concentrate the filtrate to about 30 ml by performing a simple distillation; if the concentrate is turbid, clarify it by a gravity filtration.★ Acidify the aqueous residue with concentrated hydrochloric acid, adding acid until no more benzoic acid separates (about 5–10 ml of acid will be required).★ Cool this mixture in an ice-water bath and isolate the benzoic acid by suction filtration. The crude acid can be purified by recrystallization from hot water or by sublimation. If the latter procedure is to be used, the crude acid must be *dry*. Determine the melting point and yield of benzoic acid obtained.

B. REACTIONS OF BENZOIC ACID

1. Acidity. Saturate 1 ml of hot water with benzoic acid and test the solution with litmus paper, noting the result. Add a small amount of solid sodium bicarbonate or a few drops of 10% aqueous sodium bicarbonate to the saturated solution and note the result. Acidify the resulting solution with concentrated hydrochloric acid and note the result. Compare the solubility of benzoic acid both in cold water and in cold 10% aqueous sodium hydroxide.

2. Esterification; preparation of methyl benzoate.   Place 0.082 mole of benzoic acid and 0.62 mole of methanol in a 100-ml round-bottomed flask and carefully pour 3 ml of concentrated sulfuric acid down the walls of the flask. Swirl the flask to mix the components thoroughly, attach a reflux condenser and gently heat the mixture at reflux for 1 hr.★ Cool the solution and transfer it to a separatory funnel containing 50 ml of water. Rinse the flask with 40 ml of ether and add the rinsings to the funnel. Shake the funnel thoroughly to facilitate extraction of methyl benzoate into the ether layer, venting the funnel occasionally. Drain off the aqueous layer and wash the organic layer with a second 25-ml portion of water. Separate the layers, add 25 ml of 5% aqueous sodium bicarbonate to the funnel (*Caution! Foaming may occur.*) and shake the mixture, *frequently* venting the funnel. Drain off the aqueous washes and test to see that they are basic. If not, repeat the washing of the organic layer with aqueous bicarbonate until basic washes are obtained. After a final wash with saturated sodium chloride solution, dry the ether solution thoroughly over anhydrous magnesium sulfate.★

Filter the solution and remove the ether by distillation on the steam bath (*No flames!*). Then decant the methyl benzoate into a 100-ml round-bottomed flask, and attach the flask to an apparatus set up for simple distillation. Distil the ester using an air-cooled rather than a water-cooled condenser (which might crack because of the high boiling point, 199°, of the ester), collecting the material boiling above 190° in a tared Erlenmeyer flask.

Recover unchanged benzoic acid by acidifying the basic washes with concentrated hydrochloric acid and collecting any precipitate that results. Dry and weigh the precipitate so that recovered starting material can be allowed for when calculating the percentage yield in this reaction. Ascertain that the precipitate is benzoic acid by determining its melting point.

3. Preparation of benzoyl chloride.   (Carry out this reaction in a hood if possible. If not, use the gas trap called for.)[2] Equip a dry 100-ml round-bottomed flask with a water-cooled condenser connected to a gas trap like that shown in Figure 5.1, and place 0.20 mole of dry benzoic acid in the flask. Measure out 0.70 mole of thionyl chloride *in the hood,* pour it into the flask, and quickly attach a reflux condenser. Bring the mixture to a gentle reflux and continue heating until the vigorous evolution of HCl and SO_2 ceases (about 30 min).★ Quickly remove the flask from the condenser, attach it to an apparatus set for simple distillation and add a fresh boiling stone. Distil the excess thionyl chloride using a burner, collecting the distillate in a receiver *attached to a vacuum adapter* carrying a calcium chloride tube. This tube should be connected to a gas trap if the distillation is not performed in a hood. In this way, the distillation can be carried out in an open system without exposure of the distillate to atmospheric moisture. After no more thionyl chloride distils, drain the water from the condenser and continue the distillation. Material boiling between 85–192° should be collected in a second flask, and the product, bp 192–198°, in a third. Return the low boiling fraction to a bottle labeled "Recovered Thionyl Chloride," and pour the intermediate fraction down a sink in the hood, flushing

---

[2] Acid halides generally are lachrymators, and exposure to their vapors should consequently be avoided. Care must also be taken to keep these compounds off the skin since they can produce burns.

with copious amounts of water. Store the benzoyl chloride (*Caution:* Lachrymatory) in an appropriately labeled *glass stoppered* bottle. Determine the yield of product.

C. REACTIONS OF BENZOYL CHLORIDE

Caution: Benzoyl chloride is a strong lachrymator. Perform the reactions in a hood if possible.

1. Hydrolysis. Add 5 drops of benzoyl chloride to 5 ml of water contained in an 18×150-mm test tube and warm the mixture gently, with occasional shaking, until it is clear. Cool the solution and collect the precipitated benzoic acid. Compare the reactivity of this acid chloride to that of acetyl chloride by *cautiously* adding 2–3 drops of the latter to 5 ml of ice-cold water contained in a test tube, carefully shaking the tube, and observing the result.

2. Ammonolysis. Add 1 ml of benzoyl chloride dropwise to 5 ml of cold concentrated ammonium hydroxide in an 18×150-mm test tube, stopper the tube and shake for *ca.* 2–3 min, venting the test tube periodically to relieve any pressure. Pour the mixture into 10 ml of water contained in an Erlenmeyer flask, and collect the precipitated benzamide. Determine the yield and melting point of this product. If you are to prepare enough benzamide to use in the preparation of benzonitrile described below, the amounts of reagents used in the ammonolysis must be increased by a factor of 15.

Benzanilide, $C_6H_5CONHC_6H_5$, can be prepared by combining 0.027 mole of aniline and 0.017 mole of benzoyl chloride in a 25-ml Erlenmeyer flask and adding, in portions and with vigorous shaking, 12 ml of 3 N aqueous sodium hydroxide. Isolate the product by suction filtration and recrystallize the crude benzanilide from 95% ethanol. The reported melting point of this compound is 163°.

3. Alcoholysis. Methyl benzoate can be prepared by adding 0.01 mole of benzoyl chloride to 0.05 mole of methanol contained in an 18×150-mm test tube. Warm the tube gently on a steam cone for 2–3 min and then add 2 ml of saturated sodium chloride solution. Note the odor of the layer which separates.

To prepare phenyl benzoate, add 1.5 ml of 90% aqueous phenol to 10 ml of $3N$ aqueous sodium hydroxide contained in a 25-ml Erlenmeyer flask and to this solution add 1 ml of benzoyl chloride. Shake the mixture for 5–10 min, cool it in an ice-water bath, and isolate the product by suction filtration. Phenyl benzoate can be recrystallized from 95% ethanol-water and has a reported melting point of 72°.

4. Conversion to benzoic anhydride. Add 0.056 mole of benzoyl chloride to 15 ml of dry dioxane contained in a 50-ml Erlenmeyer flask, stopper the flask, and cool the solution to about 5°. Add 0.060 mole of pyridine, which has been dried over potassium hydroxide, and swirl the mixture for a few minutes, keeping the temperature below 10°. Introduce 0.028 mole of water with vigorous swirling and then allow the solution to stand with occasional swirling in the ice-water bath for *ca.* 10 min. Pour the reaction mixture into a well-stirred mixture of 25 ml of concentrated hydrochloric acid, 40 g of cracked ice, and 100 ml of water and collect the solid benzoic anhydride by suction filtration. (Should the product "oil out" rather than crystallize, cool the solution in an ice-water bath and scratch to induce crystallization.) Shake the solid with 15 ml of cold 5% aqueous sodium bicarbonate for a minute to remove any residual acid and then filter the mixture. Wash the filter cake with cold water. The crude benzoic anhydride can be recrystallized by dissolving it in 10–15 ml of 95% ethanol, adjusting the temperature of the solution to 40°, and adding water until the mixture just becomes turbid. Allow the solution to cool to room temperature and scratch to induce crystallization. It may be necessary to add a seed crystal and cool the solution in an ice-water bath to obtain the product. Complete the recrystallization by cooling the solution thoroughly in an ice-water bath and isolate the product. Pure benzoic anhydride is colorless, mp 42–43°.

D. PREPARATION OF BENZONITRILE:
DEHYDRATION OF AN AMIDE

Dissolve 0.083 mole of finely pulverized benzamide in 50 ml of benzene contained in a 100-ml round-bottomed flask. Equip the flask with a reflux condenser and connect the latter to a gas trap. Add 0.084 mole of thionyl chloride. Observe the resulting solution to see if some white crystals can be seen suspended in it. Bring the mixture to a gentle reflux and continue the reflux until the evolution of hydrogen chloride and sulfur dioxide ceases; about an hour will be needed.* Attach the flask to an apparatus set for simple distillation, and distil the benzene using a microburner (*Caution: Benzene is flammable. All connections in the apparatus must be tight*).* Transfer the residue to a 25-ml flask, rinsing the larger flask with a little benzene and adding the rinse to the 25-ml flask, and continue the distillation. After all the benzene has distilled, drain the water from the condenser and distil the benzonitrile (bp 188–192°). Record the yield of product.

EXERCISES

1. Write a balanced equation for conversion of ethylbenzene and permanganate to benzoic acid and manganese dioxide in the presence of hydroxide ion.
2. Write possible structures for a dicarboxylic acid which has the molecular formula, $C_4H_6O_4$.
3. Calculate the equivalent weight of an acid, 0.2370 g of which could be titrated to a phenolphthalein end-point with 27.50 ml of 0.0750 N sodium hydroxide solution.
4. Write a mechanism for the conversion of a carboxylic acid to an acid chloride with thionyl chloride.
5. Would it be feasible to attempt the preparation of benzaldehyde by permanganate oxidation of toluene? Why or why not?
6. Would you expect a carboxylic acid to be more or less soluble in water than in water containing mineral acids? Explain.
7. Write a balanced equation for reaction of benzoic acid with sodium bicarbonate.
8. Suggest a reason why phenols are less acidic than carboxylic acids.

9. Assuming that the equilibrium constant for esterification of benzoic acid with methanol is 3, calculate the yield of methyl benzoate expected using the molar amounts employed in this experiment.

10. Show how the expression for the saponification equivalent is derived.

11. Calculate the saponification equivalent that would be expected for dimethyl succinate, $CH_3O_2CCH_2CH_2CO_2CH_3$.

12. Give a mechanism for the hydrolysis of an ester in acidic and in basic media.

13. Why is it important that all apparatus used in the preparation of an acid chloride be as dry as possible?

14. Why is it possible to prepare unsubstituted amides by reaction of an acid chloride with ammonium hydroxide, which is nothing more than an *aqueous* solution of ammonia?

15. Suggest why acetyl chloride is significantly more reactive than benzoyl chloride.

16. Why must one carry out the reaction of benzoyl chloride with phenol in basic solution when no base is required for reaction of the acid halide with methanol?

17. Explain why anhydrides are less susceptible to nucleophilic attack than are acid chlorides. What evidence is there from experiments described in this chapter that this may be true? (*Hint:* Consider the method of purification of benzoic anhydride.)

18. Why might benzene be less suitable than dioxane as the solvent for the preparation of benzoic anhydride described in this chapter?

19. What might the crystals be that can sometimes be observed when benzamide and thionyl chloride are combined?

SPECTRA OF STARTING MATERIALS AND PRODUCTS

For spectra of benzoic acid and of methyl benzoate, see Figures 14.1, 14.2, 14.3, and 14.4.

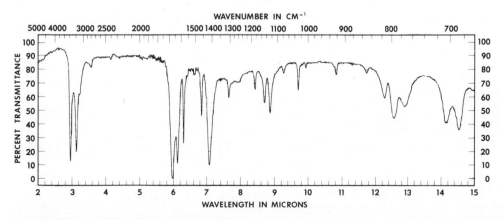

FIGURE 17.1 IR spectrum of benzamide.

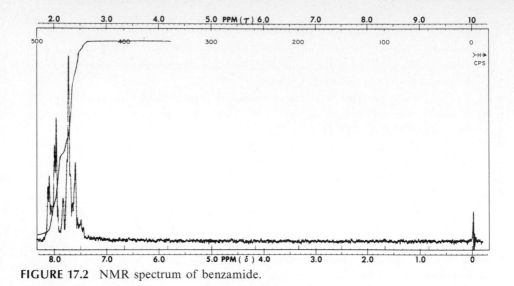

FIGURE 17.2 NMR spectrum of benzamide.

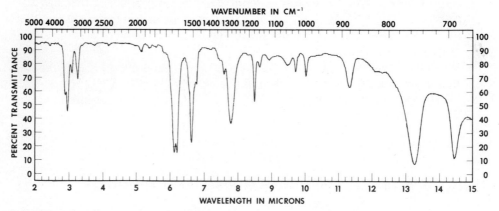

FIGURE 17.3 IR spectrum of aniline.

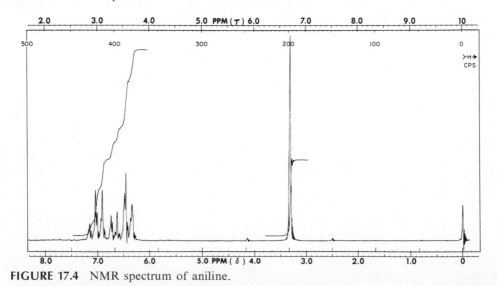

FIGURE 17.4 NMR spectrum of aniline.

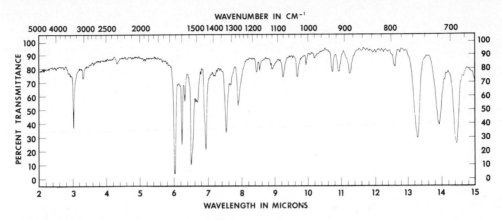

FIGURE 17.5   IR spectrum of benzanilide.

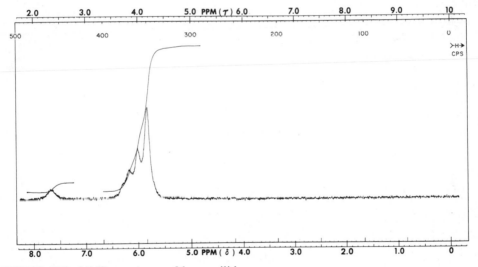

FIGURE 17.6   NMR spectrum of benzanilide.

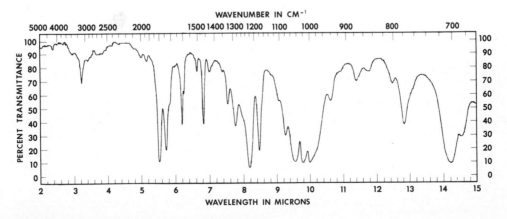

FIGURE 17.7   IR spectrum of benzoic anhydride.

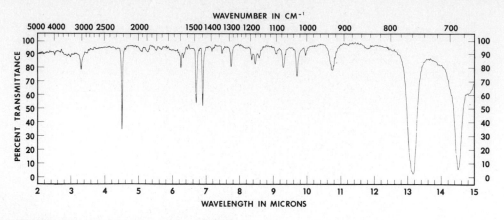

FIGURE 17.8 IR spectrum of benzonitrile.

chapter eighteen

heterocyclic synthesis

Heterocyclic compounds (heterocycles) are those cyclic organic compounds which contain, in addition to carbon atoms, one or more atoms such as oxygen, nitrogen, or sulfur as part of a ring. These three heteroatoms are by far the most common. The importance of heterocycles in organic chemistry is emphasized by the fact that about one-third of the pages of *Beilstein* (see Chapter 26) are devoted to them, and about one-fourth of the research articles in organic chemistry being published currently involve them. If we include the organic compounds in biochemical publications, the proportion of heterocyclic compounds is even much higher. This is because heterocyclic compounds occur widely in nature and many others are synthesized because of the broad spectrum of biological activity they exhibit.

Oxygen-containing heterocycles are found in carbohydrates, in flower pigments, and in marijuana; nitrogen heterocycles occur in proteins, alkaloids, nucleic acids, vitamins, and coenzymes. Some of these latter compounds also contain sulfur heterocycles. Almost all medicinal compounds—whether natural or synthetic—contain one or more heterocyclic rings.

Heterocyclic compounds may usefully be divided into two classes in the same way as carbocyclic compounds, *i.e.*, aliphatic and aromatic. The chemistry of the aliphatic heterocyclic compounds requires little special attention because it is generally predictable on the basis of a knowledge of noncyclic compounds. Exceptions to this rule are the unusual properties of those heterocycles with very small rings (3- and 4-members) and certain transannular reactions of medium-ring compounds.

Although heterocyclic compounds classed as aromatic have properties in common with aromatic carbocyclic compounds as might be expected, they may exhibit substantial differences. These differences may generally be attributed to the presence of polarized bonds resulting from differences in electronegativity between heteroatoms and carbon atoms or to the nonbonding electron pairs that the heteroatoms possess. Formally, aromatic heterocycles are related to aromatic hydrocarbons by (1) replacement of one or more carbon atoms by isoelectric heteroatoms (those providing an equal number of electrons for covalent bonding) and (2) replacement of a carbon-carbon double bond of a Kekulé structure by a heteroatom with at least one pair of unshared electrons. Some heterocycles contain both types of replacements. Examples of common parent ring systems which illustrate these relationships are as follows:

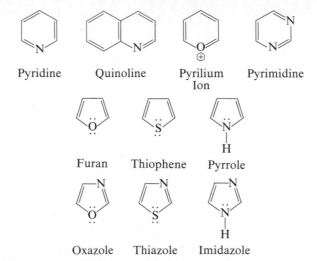

Pyridine   Quinoline   Pyrilium Ion   Pyrimidine

Furan   Thiophene   Pyrrole

Oxazole   Thiazole   Imidazole

Four heterocyclic bases are involved in the celebrated genetic sequence code of DNA. Two of these (cytosine and thymine) are derivatives of pyrimidine and the other two (adenine and guanine) are derivatives of purine, which may be considered to be a fusion of pyrimidine and imidazole.

Purine

In this chapter we shall not be concerned primarily with the chemical properties of heterocyclic compounds, but rather with the synthesis of two typical aromatic heterocycles. These experiments are illustrative of many similar synthetic procedures used to produce a wide variety of biologically active compounds.

18.1   2-Aminothiazole

The thiazole ring system is found in various naturally occurring compounds, the most important of which is vitamin B_1, or thiamine. This "vital amine" is one of the most important vitamins, and indeed the name *vitamin* resulted from this fact. It functions as a coenzyme in critical biochemical decarboxylations; its complete absence in the diet leads to beriberi in humans and to death in a week or two in experimental animals.

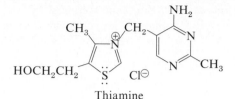

Thiamine

Sulfathiazole has been one of the more successful antibacterial drugs. A late step in its synthesis consists of reaction of 2-aminothiazole with *p*-acetamidobenzenesulfonyl chloride, with the ultimate formation of a heterocyclic derivative of sulfanilamide [see equation (28) in Chapter 19]. In this chapter we are concerned with the preparation of the 2-aminothiazole used in this synthesis.

2-Aminothiazole

Because the thiazole ring system is not nearly as stable as the analogous carbon ring system, cyclopentadiene, it is not possible to substitute directly on the thiazole ring without destroying the molecule. Consequently, most thiazole systems are made by simple condensation reactions. The preparation of 2-aminothiazole is typical of the general synthetic route to this type of compound, namely, condensation of an α-haloketone or aldehyde with a thioamide.

The main reaction leading to the preparation of 2-aminothiazole is shown in equation (1). Chloroacetaldehyde is condensed with thiourea by heating an aqueous solution of the two at reflux.

$$\underset{\text{O}}{\overset{\text{O}}{ClCH_2\overset{\parallel}{C}H}} + H_2N\overset{\overset{\text{S}}{\parallel}}{C}NH_2 \longrightarrow \text{(thiazole ring)} \; NH_2 + HCl + H_2O \qquad (1)$$

Chloroacetaldehyde is extensively hydrated in aqueous solution and can be obtained commercially as a 30–40% solution in water. Alternatively, it can be prepared *in situ* from acid hydrolysis of either an α,β-dichloroethyl alkyl ether

$$ClCH_2\overset{OR}{\underset{|}{C}H}—Cl \xrightarrow{H_3O^{\oplus}} ClCH_2\overset{O}{\overset{\parallel}{C}H} + HCl + ROH \qquad (2)$$

or of a dialkyl chloroacetal

$$ClCH_2\overset{OR}{\underset{|}{C}H}—OR \xrightarrow{H_3O^{\oplus}} ClCH_2\overset{O}{\overset{\parallel}{C}H} + 2\,ROH \qquad (3)$$

Chloroacetaldehyde is believed to condense with thiourea through the enol form of the latter (perhaps more properly called an "iminethiol"). Nucleophilic addition of the imino group to the carbonyl function, followed by the loss of a molecule of water, gives 2-aminothiazole.[1] A plausible mechanism for these steps is

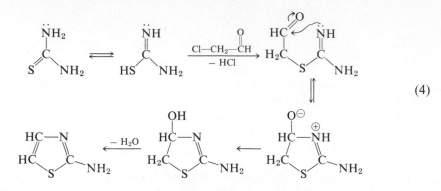

$$(4)$$

Important side reactions include the polymerization of chloroacetaldehyde, which produces materials that are water soluble and removed easily during filtration.

EXPERIMENTAL PROCEDURE

Prepare a solution of 0.2 mole of thiourea and 90 ml of water in a 500-ml round-bottomed flask; the water may need to be warmed to effect complete solution. Add 0.2 mole of chloroacetaldehyde dimethyl acetal and 1.5 ml of 85% phosphoric acid,[2] and equip the flask with an efficient reflux condenser. Heat the mixture under reflux by means of a water bath for 2.5 hr. Fit the flask for simple distillation and slowly collect and then discard 60 ml of distillate. Cool the reaction mixture in an ice-water bath and make it slightly basic to litmus by slow addition of 12 M sodium hydroxide solution. Avoid a large excess of base. After cooling for an additional 15 min, collect the precipitated solid by suction filtration. Concentrate the mother liquor to about two-thirds of its initial volume, readjust the pH by adding more base until just basic to litmus, and recool. Collect the second crop of crystals and combine it with the first crop. Wash the product twice with 20-ml portions of ice-cold half-saturated sodium bisulfite solution, and press as dry as possible.★

For purification of this crude product, dissolve it in hot toluene (10 ml per gram of product), treat with decolorizing carbon, and filter by gravity while hot (*Caution: Extinguish flames!*); filter-aid will aid in the complete removal of the decolorizing carbon. Collect the filtrate in a distilling flask, and distil about one-half of the toluene from the mixture. If the pressing of the initial precipitate has been thorough,

[1] Hydrogen chloride is produced as the condensation reaction [equation (4)] proceeds and converts the 2-aminothiazole into its hydrochloric acid salt.

[2] Alternative ways of obtaining chloroacetaldehyde are discussed in the text; if one of them is to be used, *substitute* the following reagents for the chloroacetaldehyde dimethyl acetal and phosphoric acid.

1. *Use of α,β-dichloroethyl ethyl ether:* Substitute 0.2 mole of α,β-dichloroethyl ethyl ether for the above reagents.

2. *Use of* 40% *aqueous chloroacetaldehyde:* Dissolve the thiourea in 44 ml of water and add 36 ml of a 40% aqueous solution of chloroacetaldehyde.

the small amount of water retained is removed at this point by azeotropic distillation with toluene. When the remaining toluene solution is cooled, 2-aminothiazole (mp 89–90°) is obtained. Dry the solid, and determine its melting point and the yield.

EXERCISES

1. Offer an explanation for the fact that chloroacetaldehyde forms a much more stable hydrate than acetaldehyde.
2. What is the purpose of washing the crude 2-aminothiazole with sodium bisulfite solution?

18.2   4-Aminoquinolines

Quinine, which is obtained from the bark of the cinchona tree, has long been used as an antimalarial drug.

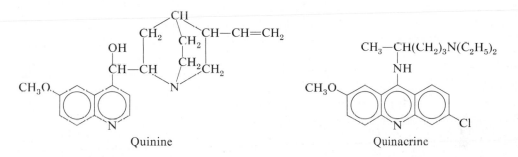

Quinine   Quinacrine

One of the first successful synthetic antimalarial drugs was quinacrine (or Atebrin), first produced in Germany. During and after World War II, thousands of heterocyclic compounds related structurally to quinine and quinacrine were synthesized by organic chemists and tested for antimalarial effectiveness. Many of these were derivatives of 4-aminoquinoline. One of the most successful was chloroquine **(9a),** which may be seen to be related structurally to both quinine and quinacrine.

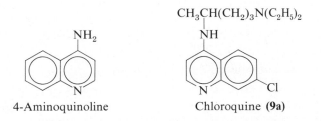

4-Aminoquinoline   Chloroquine **(9a)**

A practical scheme for the synthesis of chloroquine, and of other 4-aminoquinoline derivatives is outlined in Chart 18.1. This synthesis is typical of the majority of quinoline syntheses in that the cyclization is carried out to produce the pyridine

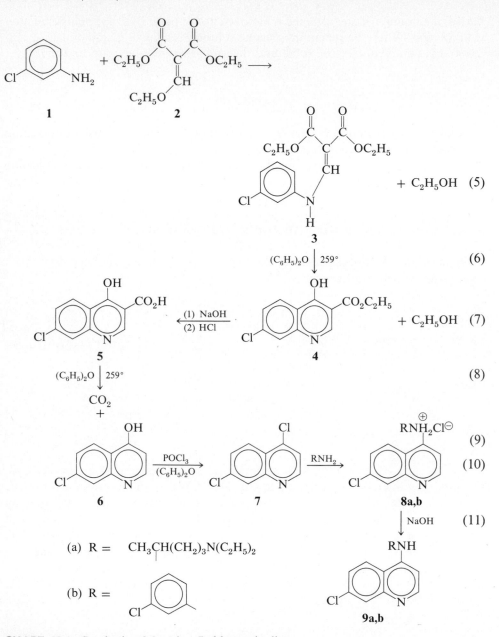

CHART 18.1 Synthesis of 4-amino-7-chloroquinolines.

ring rather than the benzene ring. Besides *m*-chloroaniline, many other substituted anilines may be used to produce quinolines having other substituents in the benzene ring, and the amine side chain may be varied by the reaction of different primary amines with 4,7-dichloroquinoline (**7**) in the last step.

In our experiment, the primary amine used in the last step is *m*-chloroaniline, for several reasons: (1) It is relatively inexpensive, (2) it is the same compound needed for the first step of the synthesis, (3) it does not have the extremely disagreeable odor of the diamine required to produce chloroquine, and (4) the product, 4-*m*-chloroanilino-7-chloroquinoline **(9b),** is a crystalline compound which is easily purified by recrystallization. It should be noted, however, that this product **(9b)** is *not* an effective antimalarial drug.

The first step in the synthesis is the condensation of *m*-chloroaniline **(1)** with diethyl ethoxymethylenemalonate **(2)** to produce **3,** with the elimination of a molecule of ethanol [equation (5)]. A high-boiling solvent (phenyl ether, bp 259°) is added to the crude product **3,** which need not be isolated before carrying out the cyclization step [equation (6)]. The phenyl ether solution is heated to reflux, whereupon the cyclization reaction begins, as is evidenced by the separation of crystals of 3-carbethoxy-7-chloro-4-hydroxyquinoline **(4)** from the boiling solution.

This cyclization step may be rationalized as involving an orbital symmetry allowed electrocyclic reaction, as shown in equation (12); see reference 7 in Chapter 10. The

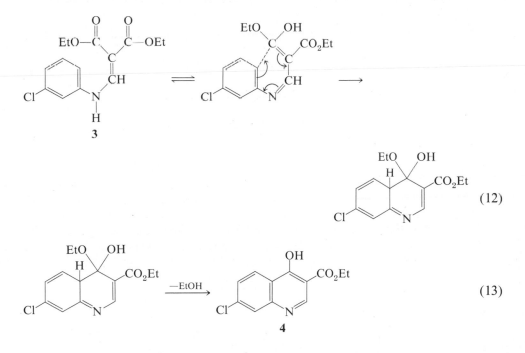

(12)

(13)

second carbethoxy group may appear to have no function in this step, and it is removed in subsequent steps. However, it apparently serves a useful purpose in influencing the *direction* of the cyclization. If ring closure took place *ortho* to the chlorine atom of intermediate **3,** 5-chloroquinoline derivatives would result. The 5-chloroquinoline isomer of chloroquine is not an effective antimalarial; therefore, any cyclization in this direction represents an undesirable side reaction. Fortunately, little of this occurs; the 5-chloroquinoline isomers have only been detected when the synthesis was carried out on a very large scale. It is interesting to speculate on the explanation for the selectivity of the ring closure; see Exercise 1 at the end of this chapter.

The third step of the synthesis, hydrolysis of the ester **4,** is carried out by adding aqueous sodium hydroxide solution and heating to vigorous reflux. As hydrolysis takes place, the 7-chloro-4-hydroxyquinoline-3-carboxylic acid **(5)** produced goes into solution as the sodium salt. After cooling, the organic acid **(5)** is precipitated by the addition of hydrochloric acid.

The next step of the synthesis is the decarboxylation of **5**. This may be done successfully without drying or purification of the crude acid by heating it in the high-boiling solvent used in the previous steps, phenyl ether. A clue to the ease of this decarboxylation may be gained from a consideration of a tautomeric form of the acid, **(5a)**, which may be seen to be a β-keto acid, a structure generally sensitive to decarboxylation.

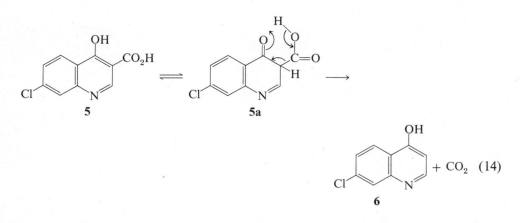

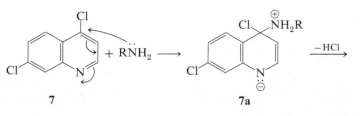

$+ CO_2$ (14)

Subsequently, reaction with phosphorus oxychloride replaces the phenolic hydroxyl group of **6** with chlorine, producing 4,7-dichloroquinoline **(7)**. The success of the final step depends on the wide difference in the ease of nucleophilic substitution of the two chlorine atoms in **7**. The chlorine in the 7-position has the inactivity characteristic of chlorobenzene, whereas the chlorine in the 4-position is quite reactive toward nucleophilic aromatic substitution by a primary amine. This is owing to the activating effect of the nitrogen in the same ring.

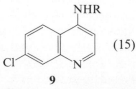

(15)

The nitrogen is able to stabilize the intermediate ion **7a** (one of several resonance structures) by bearing the negative charge. Although the 7-chlorine may be considered also to be in conjugation with the nitrogen, the negative charge cannot be transmitted to nitrogen without disrupting the aromatic resonance of both rings,

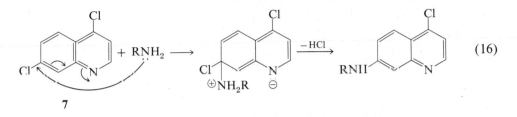

(16)

7

whereas this is not true in the case of reaction at the 4-position.

The last reaction is best carried out in the presence of a little hydrochloric acid, which acts as a catalyst,

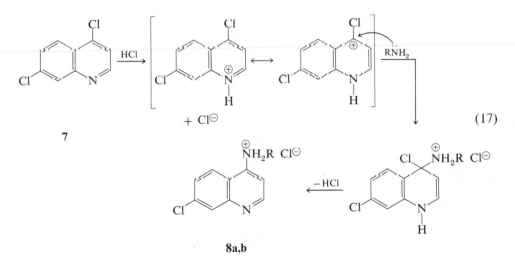

(17)

7

8a,b

Protonation by the acid of the basic nitrogen of the ring further enhances the reactivity of the ring toward nucleophilic aromatic substitution.

EXPERIMENTAL PROCEDURE

A. 4,7-DICHLOROQUINOLINE

Equip a 250-ml round-bottomed flask with a still head to which a capillary ebullition tube[3] is attached by means of a Neoprene fitting. Connect a vacuum adapter to the still head and attach a small round-bottomed receiving flask to the vacuum adapter. In the 250-ml flask, place 0.050 mole of *m*-chloroaniline and 0.051 mole of diethyl ethoxymethylenemalonate; connect the vacuum adapter to a

[3]See Figure 2.8, page 36.

water aspirator and heat the reaction mixture on a steam cone under vacuum for 1 hr. During this time, the ethanol produced according to equation (5) will be removed and pass into the water stream of the aspirator before condensing. Disconnect the aspirator hose from the adapter and then remove the still head from the reaction flask.[4]★

To the reaction mixture add 50 ml of phenyl ether and attach a reflux condenser *without water hoses connected* and heat the mixture to reflux using a flame diffused by a wire gauze. Continue heating for 15 min after reflux temperature has been attained. During this period crystals of 3-carbethoxy-7-hydroxyquinoline (4) will separate from the boiling solution. Remove the flame and allow the reaction mixture to cool to room temperature. Collect the crystalline ester (4) by suction filtration; press the filter cake with a clean cork or stopper to remove the solvent. Transfer the crystals to a 250-ml beaker and triturate them well with about 75 ml of petroleum ether (bp 60–80°); collect the crystals again on a filter and wash them with a little more petroleum ether. Resuspend the crystals once more in *ca.* 50 ml of petroleum ether, stir, and filter again.[5]★

Place the crystals of 4 (which should be almost free of solvent) in a 250-ml round-bottomed flask and add 50 ml of 3N sodium hydroxide solution and a few boiling stones. Attach a water-cooled reflux condenser and heat the mixture to vigorous reflux for 1 hr.★ Discontinue heating the reaction mixture and allow it to cool for a few minutes, disconnect the condenser, and add about 0.5 g of decolorizing charcoal. Replace the condenser and heat to reflux again for 5 min, swirling the flask if bumping occurs.★

Allow the reaction mixture to cool almost to room temperature and filter it, using a fluted filter. Wash the filter with 20 ml of hot water, adding the wash to the main filtrate. Acidify the filtrate to pH 4 with 3 N hydrochloric acid. Collect the 7-chloro-4-hydroxyquinoline-3-carboxylic acid (5) by suction filtration and wash it on the filter with a little water. Take a small amount of the acid for purification by recrystallization from a large volume of ethanol. The reported melting point of the pure acid is 273–274°, dec. (see *Caution* in footnote 5).

Place the main part of the 7-chloro-4-hydroxyquinoline-3-carboxylic acid in a 250-ml round-bottomed flask with 50 ml of phenyl ether and a few boiling stones. Attach a reflux condenser (without water cooling) and heat the mixture to reflux for 1 hr; the acid should dissolve in the hot solvent as the decarboxylation proceeds.

Allow the reaction mixture to cool to room temperature.[6]★ Add 5 ml of phosphorus oxychloride and some fresh boiling stones. Attach a water-cooled reflux condenser

---

[4] If desired, a small sample of 3 may be removed by pipet. Upon cooling in ice it should crystallize; it may be purified by recrystallization from petroleum ether (bp 60–80°). The melting point of pure 3 is 55–56°.

[5] If desired, a sample of 3-carbethoxy-7-chloro-4-hydroxyquinoline (4) may be purified by recrystallization from pyridine. The reported mp is 295–297°. (*Caution:* Do not attempt to take a mp over 200° with a liquid bath apparatus; a metal block apparatus is appropriate for high-melting substances.)

[6] If isolation of pure 7-chloro-4-hydroxyquinoline is desired, pour out about 10 ml of the reaction mixture into a flask containing 10 ml of petroleum ether (bp 60–80°). Stir the mixture and then collect the crystals and wash them with more petroleum ether. The crude product may be recrystallized from a large volume of water, using charcoal if necessary. The melting point of pure 7-chloro-4-hydroxyquinoline is 270–272°, sintering from 260°. (See *Caution* in footnote 5.)

and suspend a 360°-thermometer from the top of the condenser by means of a copper wire so that the bulb of the thermometer is in the liquid reaction mixture. Heat the reaction mixture slowly to 135–140° and swirl the flask from time to time during 1 hr of heating.★

When the reaction mixture has cooled to room temperature, transfer it to a separatory funnel with the aid of about an equal volume of ethyl ether and extract with four 50-ml portions of 3 N hydrochloric acid.[7] Add a little crushed ice to the combined acid extracts and neutralize with 3 N sodium hydroxide. Collect the precipitated 4,7-dichloroquinoline (**7**) on a filter. (*Caution: 4,7-dichloroquinoline is lachrymatory and irritating to the skin. Take care not to inhale its fumes or get any on the skin.*) Resuspend the crystals in water, stir well, and collect them again on a filter. The product may be air-dried in a fume hood or, better, in a vacuum desiccator. The melting point of pure 4,7-dichloroquinoline is 83.5–84.5°.

B.   4-*m*-Chloroanilino-7-chloroquinoline.   Place 2.0 g of 4,7-dichloroquinoline and 1.3 g of *m*-chloroaniline in 100 ml of water and add a few drops of concentrated hydrochloric acid. Attach a reflux condenser and heat the mixture to a gentle boil. Within a few minutes the hydrochloric acid salt of the product will begin to precipitate. Continue heating for an additional 10 min, allow the mixture to cool to room temperature, and collect the crystals by suction filtration. Resuspend the crystalline hydrochloride in about 20 ml of 3 N sodium hydroxide solution; heat the mixture to boiling while stirring for about 5 min. Collect the solid product and recrystallize it from ethanol. Shining white plates of 4-*m*-chloroanilino-7-chloroquinoline should be obtained, mp 223–225° (see *Caution* in footnote 5).

EXERCISES

1. 7-Chloro-4-hydroxyquinoline has also been synthesized starting with *m*-chloroaniline and diethyl oxaloacetate, which condense to give compound **10**. The sequence of reactions by which **10** is converted into 7-chloro-4-hydroxyquino-

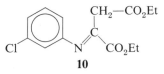

10

line is parallel to that in the present synthesis. However, the decarboxylation of 7-chloro-4-hydroxyquinoline-2-carboxylic acid is more difficult than that of the isomer, **5**, and a significant amount of 5-chloro-4-hydroxyquinoline is found as well as the desired product, **6**.

   Outline all of the steps in the synthesis of **6** from **10**, and suggest an explanation for the more difficult decarboxylation and for the production of 5-chloro-4-hydroxyquinoline.

---

[7] Take care to vent the separatory funnel immediately after each shaking operation, as some heat may be produced in the extraction.

2. Why is water *not* circulated through the reflux condenser during some of the heating periods in the procedures?

3. In taking the melting point of 7-chloro-4-hydroxyquinoline-3-carboxylic acid, a student noticed that the sample began to froth and bubble as it melted in the capillary tube. Explain.

4. (a) Identify the three major nmr absorptions in the spectrum of diethyl ethoxy-methylenemalonate with the responsible hydrogen atoms. (b) Rationalize the very low-field position of the absorption of 7.5 δ. (c) What does the nmr spectrum tell you about the equivalence or nonequivalence of the three ethyl moieties in this molecule?

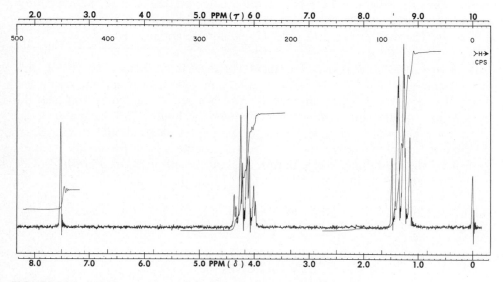

FIGURE 18.1   NMR spectrum of diethyl ethoxymethylenemalonate.

chapter nineteen

multi-step organic syntheses

The synthesis of complex molecules from simpler starting materials is one of the most important aspects of organic chemistry. Some syntheses require as many as 20 or 30 sequential reactions or "steps." Multi-step syntheses of such lengths are usually economically feasible only for end-products which are biologically active, and which cannot be obtained readily from natural sources (*e.g.*, cortisone). This is because of the cumulative effect on overall yield of less-than-quantitative yields of individual steps, when there are many such steps. For example, even a five-step synthesis in which each step takes place in 80% yield gives an overall yield of only 33% [$(0.8)^5 \times 100 = 32.8$]. Although thirty-step syntheses are rare, five-step syntheses are common in both academic and industrial laboratories; because of the cumulative effect of multiple steps on overall yield, the organic chemist searches for reaction sequences capable of high yields, and he seeks to develop optimum experimental techniques to minimize losses in each step.

In this chapter we present three examples of multi-step syntheses. These will provide experience in using the product of one reaction in subsequent steps, an experience which will emphasize the importance of good experimental technique and will give some insight into the excitement of producing a complex organic substance from simpler starting materials.

In some of the procedures there are alternatives of either isolating intermediate products in pure form, or using crude reaction mixtures in the next step. This is the sort of choice the synthetic organic chemist faces constantly. It will be instructive for some students in a laboratory section to be assigned to one procedure and other students to the other; the two groups will then have an opportunity to compare the results of the alternative procedures.

19.1   Polystyrene

In this section a synthesis of polystyrene from benzene in four steps is described. An outline of the synthesis is presented in Chart 19.1.

The main industrial method of producing polystyrene, outlined in Chart 19.2, in-

317

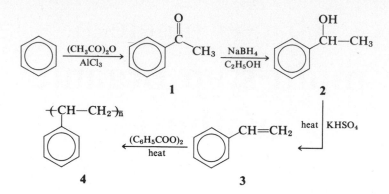

CHART 19.1   Laboratory synthesis of polystyrene.

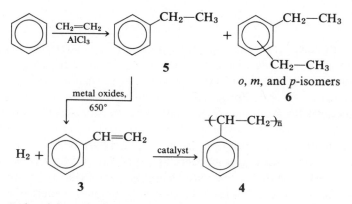

CHART 19.2   Industrial synthesis of polystyrene.

volves one less step than this laboratory synthesis, but it is not easily adapted for a student experiment for several reasons. It utilizes ethylene and benzene as starting materials and, since ethylbenzene (**5**), the initial product, is even more easily alkylated than benzene, a large ratio of benzene to ethylene must be used to minimize the formation of diethylbenzenes, **6**. Even under these conditions, some diethylbenzenes form and must be separated from the ethylbenzene. The catalytic dehydrogenation step also requires the difficult separation of the product, styrene (**3**), from unchanged ethylbenzene.

   Acetophenone.   Instead of the Friedel-Crafts *alkylation* of benzene employed in the industrial method, the laboratory synthesis utilizes *acylation* in the first step. The acylation of benzene is accomplished with acetic anhydride and aluminum chloride and is an example of a Friedel-Crafts ketone synthesis. The quantity of aluminum chloride used in the acylation, which in this case may be referred to as an acetylation, is much more than a catalytic amount such as was used in the alkylations of Chapter 8.2. This is because some aluminum chloride is "used up" in reaction with the acetate anion produced in the first step [equation (1)], which generates the active electrophile, the acylium (or "acylonium") cation, **7**.

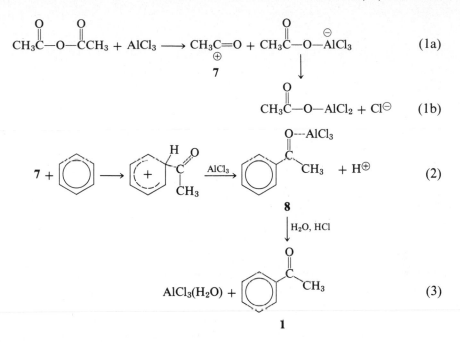

$$CH_3\overset{\overset{O}{\|}}{C}-O-\overset{\overset{O}{\|}}{C}CH_3 + AlCl_3 \longrightarrow CH_3\overset{\oplus}{C}=O + CH_3\overset{\overset{O}{\|}}{C}-O-\overset{\ominus}{AlCl_3} \qquad (1a)$$

$$\mathbf{7}$$

$$CH_3\overset{\overset{O}{\|}}{C}-O-AlCl_2 + Cl^{\ominus} \qquad (1b)$$

$$\mathbf{8}$$

Furthermore, the acetophenone, as it is formed, interacts with a mole of aluminum chloride to form a strong complex [**8**, equation (2)]. Thus, more than two moles of aluminum chloride per mole of acetic anhydride are used in order to insure an adequate supply of active catalyst. The acetophenone is set free from the complex at the end of the reaction by the addition of water [equation (3)]; precipitation of basic aluminum salts is prevented by the addition of concentrated hydrochloric acid.

Acylations are free of two undesirable features of alkylations: (1) The tendency toward polysubstitution, and (2) rearrangement of the alkylating agent. The introduction of an acyl group into an aromatic ring deactivates the ring toward further electrophilic substitution. The lack of rearrangement of acyl moieties makes acylation, followed by reduction, one of the most reliable synthetic procedures for the preparation of alkylbenzenes of authentic structures; *e.g., n*-propylbenzene [**9**, equation (4)].

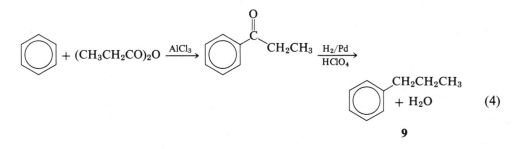

$$\mathbf{9}$$

1-Phenylethanol. The second step of the synthesis illustrates the use of one of the versatile metal hydrides as a reducing agent [equation (5)]. Both lithium aluminum

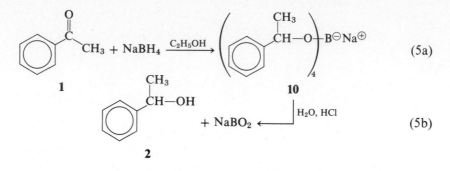

(5a)

(5b)

2

hydride and sodium borohydride reduce aldehydes and ketones to alcohols; the virtue of the latter (milder) reagent is that it may be used in alcoholic and even aqueous solution, in contrast to lithium aluminum hydride. Although sodium borohydride reacts slowly with alcohols and water, it reacts so much more rapidly with carbonyl compounds that its reaction with these solvents is not troublesome.

The sodium borohydride transfers a hydride ion to the carbonyl carbon in the first step of the mechanism [equation (6)]. All of the hydrogens attached to boron are

$$Na^{\oplus}\left[\begin{matrix}H\\H-B(-H\\H\end{matrix}\right]^{\ominus}\quad\begin{matrix}R\\C=O\\R\end{matrix}\longrightarrow Na^{\oplus}\left[\begin{matrix}R&H\\H-C-O-B-H\\R&H\end{matrix}\right]^{\ominus}\qquad(6)$$

transferred in this way to produce the intermediate borate salt [**10**, equation (5a)], which is decomposed upon the addition of water and acid to yield 1-phenylethanol [**2**, equation (5b)]. The 1-phenylethanol must then be separated from water, ethanol, inorganic acids, and salts. This is accomplished by evaporating the ethanol on a steam bath and then extracting the 1-phenylethanol with ether. The ether solution is dried by a dessicant and the ether is evaporated. The 1-phenylethanol may be isolated by distillation under reduced pressure; the next step (dehydration) may be carried out on the crude product, however.

Styrene. The dehydration of 1-phenylethanol is accomplished by heating the alcohol with a mild dehydrating agent, potassium hydrogen sulfate, and collecting the water and styrene (**3**) as they distil at atmospheric pressure [equation (7)]. The crude styrene is

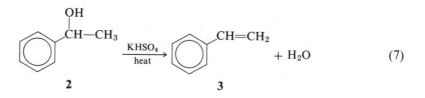

2   **3**

then either extracted into ether, washed and distilled under reduced pressure, or else is diluted with toluene and the solution is dried by azeotropic distillation. The choice depends on whether one wishes to carry out the fourth step, polymerization, on neat styrene or on a toluene solution of styrene.

Polystyrene. The polymerization of styrene is represented most simply by equation (8). The mechanism of this free radical-catalyzed polymerization should be compared

$$
n \underset{}{\overset{CH=CH_2}{\bigcirc}} \xrightarrow[\text{heat}]{(C_6H_5COO)_2} \overset{-(CH-CH_2)_{\overline{n}}}{\bigcirc} \tag{8}
$$

with that of chlorination of hydrocarbons discussed in Chapter 5.1. The reaction is initiated by thermal decomposition of benzoyl peroxide; the free radicals so produced (In·) *initiate* the polymerization as shown in equation (9).

$$
\text{In} \cdot + \underset{C_6H_5}{CH_2{=}CH} \longrightarrow \underset{C_6H_5}{In{-}CH_2{-}CH} \cdot \tag{9}
$$

$$
\underset{C_6H_5}{In{-}CH_2{-}CH} \cdot + n\,\underset{C_6H_5}{CH_2{=}CH} \longrightarrow \underset{C_6H_5}{In{-}CH_2{-}CH}\underset{C_6H_5}{(CH_2{-}CH)_{\overline{n}}} \cdot \tag{10}
$$

$$
\mathbf{11}
$$

$$
\mathbf{11} \longrightarrow \underset{C_6H_5}{In{-}CH_2{-}CH}\underset{C_6H_5}{(CH_2{-}CH)_{n-1}}\underset{C_6H_5}{CH_2}\;\;\underset{}{CH_2}
$$

$$
\mathbf{12}
$$

$$
+ \underset{C_6H_5}{In{-}CH_2{-}CH}\underset{C_6H_5}{(CH_2{-}CH)_{\overline{n}-1}}\underset{C_6H_5}{CH{=}CH} \tag{11}
$$

$$
\mathbf{13}
$$

$$
\mathbf{11} + R \cdot \longrightarrow \underset{C_6H_5}{In{-}CH_2{-}CH}\underset{C_6H_5}{(CH_2{-}CH)_{\overline{n}}}R \tag{12}
$$

Equation (10) represents the *propagation* of the growing polymer chain. Equations (11) and (12) show possible *termination* processes. In equation (11) the free radical end of one growing polymer chain abstracts a hydrogen atom, with its electron, from the carbon atom next to the end of another polymer radical to produce one polymer molecule which is saturated at the end and one polymer molecule which is unsaturated at the end, a process termed *disproportionation*. In equation (12), an R· may be one of the initiating radicals, In·, or another growing polymer chain.

In the procedure below directions are given for producing the polymer in the form of an amorphous solid, a film, and a clear "glass."

EXPERIMENTAL PROCEDURE

A. ACETOPHENONE
Connect a reflux condenser to a 500-ml round-bottomed flask and attach a Claisen adapter and separatory funnel at the top of the condenser, as shown in Figure 19.1. By means of a simple adapter and a rubber tube, connect the sidearm of the Claisen adapter to a gas trap. Place a few boiling stones in the reaction flask.

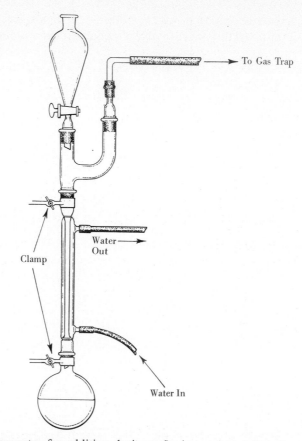

To Gas Trap

Water
Out

Clamp

Water In

FIGURE 19.1 Apparatus for addition during refluxing and trapping of a water-soluble gas.

When the apparatus has been set up and adjusted correctly, weigh out 0.41 mole of anhydrous powdered aluminum chloride (quickly so as to avoid atmospheric moisture) and introduce it into the reaction flask, having temporarily disconnected the condenser. A good technique to prevent the solid from sticking in the neck of the flask, particularly if a greased ground-jointed flask is being used, is to roll a piece of paper into a tube, insert it in the neck of the flask, and then to pour the aluminum chloride through this paper tube. Remove the paper and immediately pour 70 ml of dry benzene [1] into the reaction flask so as to cover the aluminum chloride as quickly as is possible. Now re-connect the condenser and the rest of the apparatus. Mix 0.20 mole of acetic anhydride and 10 ml of dry benzene in the separatory funnel. Begin dropwise addition of the acetic anhydride solution to the benzene–aluminum chloride mixture; loosen the lower clamp holder so that by moving the clamp gently the mixture can be swirled continuously during the addition operation. Hydrogen chloride will be evolved (fumes will be observed in the trap) and the reaction mixture will become hot. If the benzene begins to boil, reduce the rate of addition and, if necessary, bring a cold-water bath up around the reaction flask. The addition should be completed in about 15 to 20 min.

[1] Benzene may be dried satisfactorily by azeotropic distillation as described in footnote 2, Chapter 8.2, page 155.

After all of the acetic anhydride solution has been added and the vigor of the reaction has decreased, move the apparatus carefully to a steam bath and heat the flask for *ca.* 20 min. Then remove the apparatus from the steam bath and place a pan of cold water around the flask. While the reaction mixture is cooling, place 120 g of crushed ice in a 600-ml beaker and add 100 ml of concentrated hydrochloric acid with stirring. Now pour the cool reaction mixture slowly into the ice-acid mixture. If any basic aluminum salts remain undissolved after the mixture has been stirred well, add enough hydrochloric acid to dissolve them.

Transfer the two liquid layers to a large separatory funnel and rinse the beaker with 25 ml of ether, adding the ether to the mixture in the separatory funnel. Shake the mixture gently and then remove the organic layer. Extract the aqueous layer with 25 ml of ether and combine the ether solution with the main organic layer.★ Wash the organic layer with 35 ml of 3 *N* sodium hydroxide solution and then with 25 ml of water. Separate the organic solution and dry it with two 5-g portions of anhydrous magnesium sulfate, decanting from the drying agent the first time and filtering the second time.★ Remove the solvent by distillation, using a steam bath, and pour the residue into a 50-ml round-bottomed flask. Add boiling stones and then distil, taking care to keep flames away from the receiver while the residual ether and benzene distil. After these solvents have been collected and the distillation temperature has begun to rise, change the receiver and drain the water from the condenser. Collect the acetophenone boiling at 195–200°. The yield should be 65–85%. Weigh the product and calculate the percentage yield on this step.

B. 1-Phenylethanol

In a 150-ml beaker, dissolve 0.032 mole of sodium borohydride (*Caution: Do not allow this strongly caustic reagent to come in contact with your skin*) in 25 ml of 95% ethanol and add 0.10 mole of acetophenone dropwise. Stir the mixture with a thermometer during the addition, and keep the temperature below 50° by decreasing the rate of addition and by external cooling with an ice bath, if necessary. After the addition of the acetophenone is complete, allow the reaction mixture (which contains a white precipitate) to stand for 15 min at room temperature. Add *ca.* 10 ml of 3 *N* hydrochloric acid solution dropwise and with stirring; hydrogen will be evolved and most of the white solid will dissolve. Place the beaker containing the reaction mixture on a steam bath and concentrate the solution by evaporation of the ethanol until two liquid layers separate. Add 20 ml of ether and transfer the mixture to a 125-ml separatory funnel. Shake the mixture gently, allow the layers to separate, and remove the ether layer.★ Extract the aqueous layer with a 10-ml portion of ether and add this to the other ether solution. Dry the combined ether solutions over anhydrous magnesium sulfate; briefly swirl the solution with the drying agent and then decant the solution into a dry flask. Add another portion of drying agent and repeat the swirling; filter the ether solution into a tared 100-ml round-bottomed flask.★

If pure 1-phenylethanol is to be isolated, add about 1 g of anhydrous potassium carbonate and arrange for vacuum distillation.[2] (The potassium carbonate present in

---

[2] See Chapter 2.5 and references at the end of Chapter 2.

the distillation pot helps to avoid acid-catalyzed dehydration during the distillation.) Remove the ether at atmospheric pressure (steam bath) and then reduce the pressure to below 20 mm, if possible, and distil using an oil bath; the boiling point of 1-phenyl-ethanol is 102.5–103.5° (19 mm), 97° (13 mm).

If pure 1-phenylethanol is not to be isolated by vacuum distillation, proceed in the following way. Connect the flask to an aspirator by means of an adapter or a one-hole rubber stopper fitted with a short glass tube. (The aspirator must be connected through a safety trap with a pressure release such as that shown in Figure 2.16.) Apply gentle suction to the flask by closing the screwclamp partially or completely, but momentarily; swirl the flask containing the ether solution so that the ether boils, but does not bump. As the ether boils away, the pressure may be gradually lowered until the full aspirator vacuum is applied. Warm the flask with your hand or with a water bath at room temperature. Calculate the yield of crude 1-phenylethanol.

C. STYRENE

At this stage, two or more of the students should combine their yields of 1-phenyl-ethanol so that a total of at least 0.09 mole may be used in the dehydration step. This may be done by pouring the crude products together in one 100-ml round-bottomed flask with the aid of a little dry ether, which is then removed as described above.

To *ca.* 0.09 mole of crude 1-phenylethanol, add 1.0 g of anhydrous potassium hydrogen sulfate and *ca.* 0.1 g of copper powder.[3] Attach a distilling head and a condenser to the 100-ml flask and distil the mixture by heating in an oil bath whose temperature is gradually (during about 10 min) raised to 200–220°. The temperature of the distilling vapor should not exceed 130°. Collect the distillate and calculate the approximate yield of crude styrene from the volume of the organic layer. If the layers do not separate well, add a little sodium chloride to saturate the water layer.★

If you are to isolate pure styrene, add the mixture of styrene and water obtained as distillate above to 30 ml of ether in a 125-ml separatory funnel. Shake gently, allow the layers to separate, and drain off the aqueous layer. Wash the ether solution first with 15 ml of saturated sodium carbonate solution and then with 10 ml of water. Dry the ether solution by swirling with *ca.* 1 g of anhydrous calcium chloride.★ Decant the ether solution into a 50-ml round-bottomed flask, wash the drying agent with 5 ml of ether, and combine the wash with the main ether solution. Attach a still head and condenser, add some boiling stones, and distil using a steam bath. Remove the ether at atmospheric pressure, then arrange for vacuum distillation, using an aspirator.[2] The boiling point of styrene is 55° (30 mm), 48° (20 mm). Collect the styrene in a weighed flask and calculate the yield, based on 1-phenylethanol (one step), on acetophenone (two steps), and on acetic anhydride (three steps), considering the actual amount of material used in each step.

D. POLYSTYRENE

1. Polymerization of pure styrene. Place about 5 g of pure, freshly distilled styrene in an 18 × 150-mm soft glass test tube and add *ca.* 0.1 g of benzoyl peroxide. (*Caution: Benzoyl peroxide is sensitive to shock and to static electricity. It should be*

---

[3] Copper powder has been reported to reduce the tendency of the styrene (produced in this step) to polymerize during the distillation.

dispensed from cardboard containers and transferred by means of a folded 3 × 5-inch card. Do not use a metal spatula.) Stopper the tube *loosely* and heat it on a steam bath for several hours; allow the tube to stand at room temperature for 24 hr, or until the next laboratory period. After the viscosity has increased sufficiently, one may suspend a small object (such as a coin) in the viscous liquid and allow the polymerization to proceed until a hard, transparent "glass" is obtained. The test tube may then be cracked away from the polystyrene.

2. Solution polymerization of crude styrene.   Add the mixture of styrene and water collected in the step above to three times its volume of toluene or xylene (commercial mixture of isomers) and drain off the small water layer in a 125-ml separatory funnel. Distil the toluene or xylene solution until the distillate dropping from the end of the condenser is clear (all the water will have been removed by azeotropic distillation). Add 0.1 g of benzoyl peroxide (see *Caution* in part 1) for each 5 g of styrene (the amount estimated in the crude distillate from the dehydration reaction above) in the solution, connect a reflux condenser, and heat to reflux with a burner for 1 to 3 hr. (Since the boiling temperature of xylene is considerably higher than that of toluene, a comparable degree of polymerization will be obtained in a shorter time in xylene solution than in toluene solution.) Cool the solution to room temperature and then pour it into about 150 ml of methanol. Collect the white precipitate of polystyrene by decantation, or by suction filtration if decantation is not practical. Resuspend the polystyrene in fresh methanol and stir it vigorously; then collect it on a filter and allow it to dry.

Redissolve a portion of the polystyrene in toluene (the *minimum* amount required to dissolve it when heated on a steam bath) and pour the solution on a watch glass or the bottom of a large inverted beaker and allow it to evaporate. A clear film of polystyrene should result.

EXERCISES

1. Why is the water drained from the condenser after the solvent has been removed from the acetophenone and before the acetophenone distils?
2. Suggest a formula for the white precipitate formed in the reaction of acetophenone with sodium borohydride, and write an equation for its reaction with water and hydrochloric acid.
3. When 1-phenylethanol is purified by distillation, reduced pressure is always used. Why?
4. Styrene is distilled under reduced pressure. Why?
5. Calculate the overall percentage yield of cortisone, an important drug, produced in a 33-step synthesis, assuming an average yield of 90% on each step.
6. Suggest an explanation for the fact that acylation takes place without rearrangement, whereas rearrangement often accompanies alkylation; for example, compare reaction of CH_3CH_2COCl and $CH_3CH_2CH_2Cl$ with benzene and aluminum chloride.
7. What other reagents (besides $NaBH_4$ and $LiAlH_4$) will reduce acetophenone to 1-phenylethanol?

8. What reagents will reduce acetophenone to ethylbenzene?
9. Suppose that in the first step of this synthesis an appreciable amount (*ca.* 10%) of diacetylation had occurred, and the other steps had been carried out without purification of the intermediates. What difference in the physical properties of the resultant polystyrene might result?

REFERENCE

S. H. Wilen, C. B. Kremer, and I. Waltcher, *Journal of Chemical Education,* **38,** 304 (1961).

SPECTRA OF STARTING MATERIALS AND PRODUCTS

The ir and nmr spectra of acetophenone are given in Figures 16.19 and 16.20, respectively.

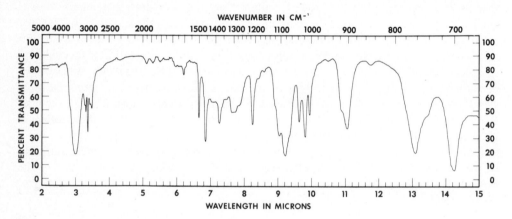

FIGURE 19.2   IR spectrum of 1-phenylethanol.

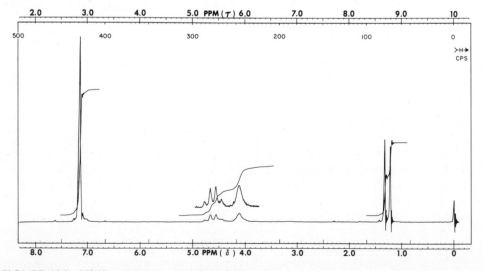

FIGURE 19.3   NMR spectrum of 1-phenylethanol.

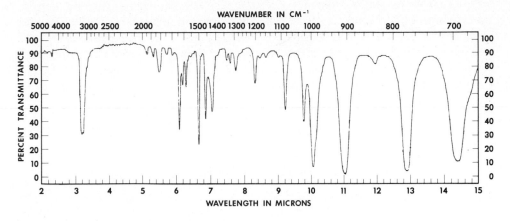

FIGURE 19.4 IR spectrum of styrene.

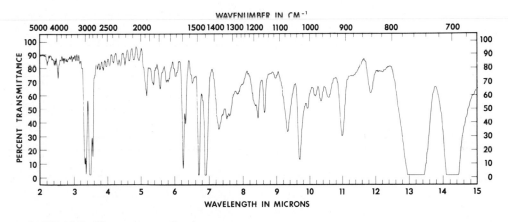

FIGURE 19.5 IR spectrum of polystyrene.

19.2 The Synthesis of Sulfathiazole

Some of the most effective drugs belong to a general class called "sulfa" drugs, the members of which have a common structural feature in that they are derivatives of the parent compound, p-aminobenzenesulfonamide (14), often called sulfanilamide. It should be noted that this compound contains two different $-NH_2$ moieties, one as the amino group and the other as the sulfonamido ($-SO_2NH_2$) group. Different substituents on the sulfonamido nitrogen give rise to the family of sulfa drugs.

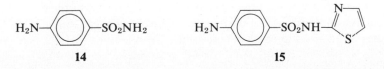

One of the more familiar examples of a sulfa drug is sulfathiazole (**15**), and the current experiment presents its total synthesis. This preparation can be started, if desired, from benzene and simple aliphatic compounds; the steps of the total synthesis are outlined in Chart 19.3. The preparations of several interesting intermediate compounds, such as sulfanilamide itself, are included, and many of them can be isolated if desired or used in subsequent steps without extensive purification. A discussion of the reactions involved in each step is included.

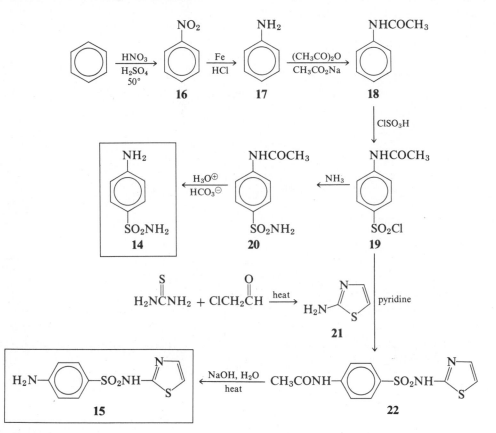

CHART 19.3 Laboratory synthesis of sulfanilamide (**14**) and sulfathiazole (**15**).

Aromatic nitration: nitrobenzene (16). Another typical electrophilic aromatic substitution reaction (Chapter 8) is found in nitration, where a nitro group ($-NO_2$) replaces a hydrogen atom on an aromatic ring. The exact conditions required for nitration depend on the aromatic compound. In extremely reactive compounds, such as phenols, dilute nitric acid can be used. In other less reactive compounds, such as benzene, concentrated nitric acid must be used in the presence of concentrated sulfuric acid. In very unreactive compounds, such as nitrobenzene, fuming sulfuric acid and fuming nitric acid must be used in order to effect a good yield of the desired product. Not only can the concentration of the nitrating agent be changed, but also the temperature may be varied so as to assist in controlling the reaction.

In the present work, the nitration reaction will be applied to benzene itself. The main reactions for the preparation of nitrobenzene from benzene are illustrated in equations (13) and (14). The first step is believed to be an equilibrium reaction, in which the concentration of nitronium ion, NO_2^+ (23), is increased by the presence of sulfuric acid. This reaction is followed by a slow, rate-determining step in which NO_2^+ attacks the benzene ring; this attack is followed by a fast reaction, which results in the loss of a proton to give nitrobenzene.

EQUILIBRIUM:

$$HO-NO_2 + 2\,H_2SO_4 \rightleftarrows NO_2^{\oplus} + H_3O^{\oplus} + 2\,HSO_4^{\ominus} \qquad (13)$$

23

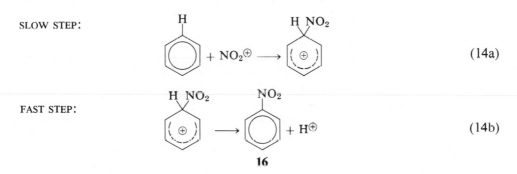

SLOW STEP: (14a)

FAST STEP: (14b)

16

The principal side reaction is the formation of *m*-dinitrobenzene [equation (15)], but the formation of this compound is very slow under the reaction conditions. This is because of the fact that the nitro group is strongly deactivating, a trend characteristic of groups which orient a new group *meta* to the first group. It is possible to isolate pure nitrobenzene by separating the organic layer, washing to remove any acid, and fractionally distilling.

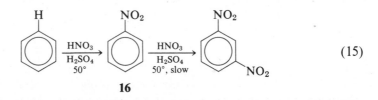

(15)

16

Reduction of aromatic nitro compounds: aniline (17). A variety of methods is available for the reduction of an aromatic nitro compound to the corresponding amine. It should be noted that reduction of aliphatic nitro compounds is seldom found, owing to the difficulty in obtaining the starting compounds; on the other hand, aromatic nitro compounds are easily obtained. The general methods of reduction include catalytic hydrogenation and electrolytic and chemical reduction. In the laboratory, the most commonly used method is chemical reduction, in which various metals are used in acidic solution; the most important commercial method is probably catalytic hydrogenation.

The mechanisms of these reductions are not well understood. The fact that various stable intermediate compounds have been isolated suggests that it is a stepwise reaction. Some of the intermediates for the reduction are shown in equation (16). Differences in rates of reduction of the various intermediates may be important, but the nature of the reducing agent dictates the final product. For example, the reduction of nitrobenzene with tin metal and hydrochloric acid gives only aniline, whereas the use of zinc metal and ammonium chloride gives only N-phenylhydroxylamine (25). Note that in equation (16), [H] means reduction by any of the possible reducing agents.

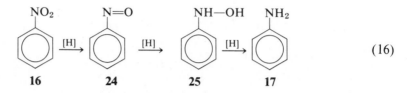

$$\text{(16)}$$

In this experiment, iron powder with a small amount of hydrochloric acid will be used to reduce nitrobenzene to aniline. The main reaction is shown in equation (17), in unbalanced form. This is a typical oxidation-reduction equation which can perhaps

$$\text{(17)}$$

most easily be balanced by considering the two half-reactions given in equations (17a) and (17b).

$$\text{(17a)}$$

$$3\ H_2O + Fe = Fe(OH)_3 + 3\ H^{\oplus} + 3\ e^{\ominus} \qquad \text{(17b)}$$

Several possible side reactions take place but only to a minor extent under the reaction conditions used. These side reactions [equation (18)] occur due to the presence of some of the intermediate products [equation (16)], which then react with one another. Nitrosobenzene (24) can react with aniline to give azobenzene (26), which can be reduced to hydrazobenzene (27) by iron. Hydrazobenzene can also undergo rearrangement in the presence of the acid catalyst to yield benzidine (28); this rearrangement is often referred to as the **benzidine rearrangement**. N-Phenylhydroxylamine (25) can undergo acid-catalyzed rearrangement to p-aminophenol (29). Thus, some of the anticipated side products are compounds 25, 26, 27, 28, and 29; their separation from aniline is discussed in the following paragraph.

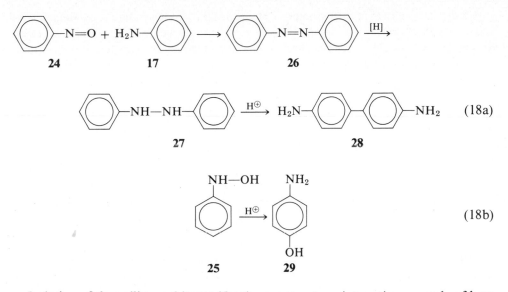

Isolation of the aniline and its purification represents an interesting example of how solubility differences and steam distillation can be used. It is necessary to remove the aniline from its principal impurities, which are unchanged nitrobenzene, benzidine, and *p*-aminophenol. When the reaction is completed, some of the aniline will be present as its hydrochloric acid salt. The reaction mixture is made distinctly basic and then steam distilled. The phenol, which will have been converted to its water-soluble, non-volatile salt, will remain in the aqueous residue along with the high-molecular-weight benzidine. The aniline and nitrobenzene will steam distil together. To separate these two from one another, it is necessary to convert the amine to its soluble hydrochloric acid salt, and steam distil again. The nitrobenzene will now steam distil, whereas the salt will remain in the aqueous residue. The aniline may be recovered by making the acidic residue basic and extracting the amine into an organic solvent such as dichloromethane.

Alternatively, one could isolate the aniline from the unchanged nitrobenzene by treating the mixture of the two (obtained from steam distillation) with dilute acid and extracting the neutral nitrobenzene (the aniline will have been converted to the salt and will be in the aqueous layer) with an organic solvent, such as dichloromethane. Making the aqueous layer basic, followed by extraction with dichloromethane will render the aniline in the second dichloromethane layer.

Acetanilide (18). Aniline **(17)** is acetylated in aqueous solution, using a mixture of acetic anhydride and sodium acetate [equation (19)]. The water-insoluble aniline is first converted to the water-soluble hydrochloric acid salt; to this salt in water is added acetic anhydride, followed by the addition of sodium acetate to free the aniline from its salt. The free amine is then rapidly acetylated by the acetic anhydride. This provides an easy method of acetylation, and the overall yields are high. Under the conditions used, the acetanilide is produced as a solid, which can be filtered and recrystallized from water.

A side reaction which might occur is diacetylation, shown in equation (20). It appears that this is minimized by the general method described above, whereas acetylation in pure acetic anhydride (using the free amine) frequently leads to the production of the diacetyl compound.

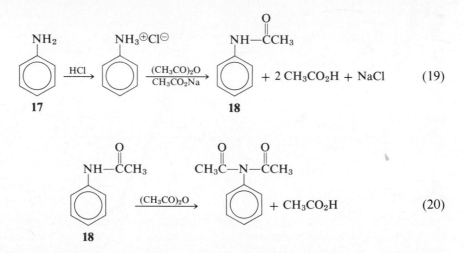

$$+ \; 2 \; CH_3CO_2H + NaCl \qquad (19)$$

$$\xrightarrow{(CH_3CO)_2O} \qquad + \; CH_3CO_2H \qquad (20)$$

Chlorosulfonation reactions: *p*-**Acetamidobenzenesulfonyl chloride (19).** A simple, one-step reaction can be used to introduce the sulfonyl chloride group, $-SO_2Cl$, onto an aromatic ring. This reaction, known as chlorosulfonation, involves the use of chlorosulfonic acid ($ClSO_3H$) and an aromatic compound. As applied to acetanilide (**18**), the major product is *p*-acetamidobenzenesulfonyl chloride [**19**, equation (21)]. Two moles of the chlorosulfonic acid are needed. The reaction is known to proceed through the sulfonic acid, which is converted to the sulfonyl chloride on further reaction with chlorosulfonic acid [(equation (22)].

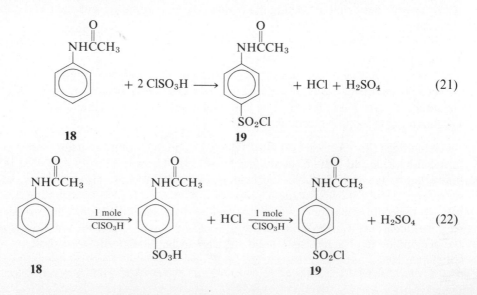

$$+ \; 2 \; ClSO_3H \longrightarrow \qquad + \; HCl + H_2SO_4 \qquad (21)$$

$$\xrightarrow[ClSO_3H]{1 \; mole} \qquad + \; HCl \xrightarrow[ClSO_3H]{1 \; mole} \qquad + \; H_2SO_4 \qquad (22)$$

The sulfonation of acetanilide is a typical electrophilic aromatic substitution reaction, in which SO_3 is probably the electrophile. The SO_3 may arise from chlorosulfonic acid according to the equilibrium shown in equation (23). It should be noted that the

$$ClSO_3H \rightleftharpoons SO_3 + HCl \tag{23}$$

acetamido group orients an incoming group predominantly *para*, and indeed virtually none of the *ortho* isomer is observed. Presumably this is due to the size of the acetamido group.

To isolate the product, the reaction mixture is poured into ice water, and the product is obtained as a precipitate. The water serves to hydrolyze the excess chlorosulfonic acid [equation (24)].

$$ClSO_3H + H_2O \rightarrow HCl + H_2SO_4 \tag{24}$$

The preparation of sulfanilamide may be carried out without drying or purifying the p-acetamidobenzenesulfonyl chloride, since the latter will be treated with aqueous ammonia. However, if the sulfonyl chloride is to be converted to sulfathiazole, it must be dried, since this reaction requires anhydrous conditions. In general, sulfonyl chlorides are much less reactive towards water than are acid chlorides, but they do hydrolyze slowly to give the corresponding sulfonic acid [equation (25)]. In order to purify

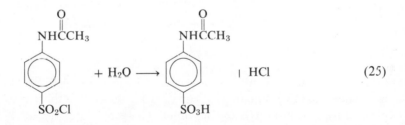

p-acetamidobenzenesulfonyl chloride, the wet product is dissolved in chloroform; the small amount of water which is present forms as a second layer owing to its very low solubility in chloroform. The organic layer containing the desired product is then drawn off and the product is recrystallized from a suitable solvent. This represents a generally useful technique for purifying solids which contain small amounts of water.

Sulfanilamide [p-aminobenzenesulfonamide (14)].   The preparation of sulfanilamide starts with p-acetamidobenzenesulfonyl chloride **(19)**. The moist product can be used without purification, but then it must be used *immediately*. The amide is produced by treatment of the sulfonyl chloride with an excess of aqueous ammonia, followed by removal of the acetyl group. This latter process can be carried out in acidic or basic solution without affecting the sulfonamido group, which hydrolyzes much less rapidly than the acetamido group. Acid hydrolysis will be used here, and the resulting amine forms a soluble hydrochloric acid salt under these conditions. In order to obtain free sulfanilamide, the acidic solution must be neutralized with dilute base (sodium bicarbonate). These reactions are shown in equation (26). The sulfanilamide is purified further

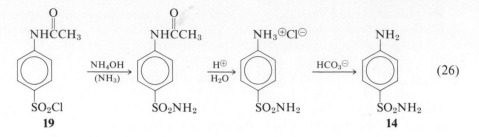

by recrystallization from water. A possible side reaction, though one which is not important under the conditions of the experiment, is the hydrolysis of the sulfonyl chloride during sulfonamide formation.

An important synthetic procedure has been introduced in the preparation of sulfanilamide—the use of a protective group (the acetyl group) which is subsequently removed at a later stage. It would appear that one reasonable method for preparing sulfanilamides would be to allow p-aminobenzenesulfonyl chloride (30) to react with ammonia or with some substituted amine, as is the case in the preparation of sulfathiazole. However, it is not practical to generate a sulfonyl chloride group in the presence of an amino group contained in the same molecule. The amino group of one molecule would react with the sulfonyl chloride group of another molecule and give a polymeric material containing sulfonamide linkages, as illustrated in equation (27). Thus, in order

$$n\ H_2N\text{—}\bigcirc\text{—}SO_2Cl \xrightarrow{\ -n\ HCl\ } \text{~~~NH—}\bigcirc\text{—}SO_2NH\text{—}\bigcirc\text{—}SO_2\text{~~~}\Big]_n \qquad (27)$$

30

to prepare the sulfanilamides, it is necessary first to "protect" the free amino group so as to allow the other functional group to be introduced. In the present synthesis, this is most easily done by acetylating the amine. This general practice must be followed when a molecule contains functional groups which are reactive toward one another. Of course, the free amine can be regenerated by removing the acetyl group (called either the "protective group" or the "blocking group") after the sulfonamido group has been introduced. In using such a technique, care must be taken to insure that the blocking group can be removed without affecting the second functional group; in the present case, the acetyl group can easily be removed without hydrolyzing the sulfonamido group.

Sulfathiazole (15).   The final reaction in the synthesis of sulfathiazole involves the condensation of p-acetamidobenzenesulfonyl chloride (19) with 2-aminothiazole (21), which occurs with the liberation of hydrogen chloride [see equation (28)]. (The preparation of 2-aminothiazole is described in Chapter 18.) As the reaction proceeds, the hydrogen chloride which is given off would react with 2-aminothiazole and convert it to its unreactive hydrochloric acid salt. Thus, it is important to remove the hydrogen chloride as it is produced. Pyridine is most often used for this purpose because it is inexpensive and easily removed from the reaction mixture. In this preparation, pyridine will be used both to remove hydrogen chloride and

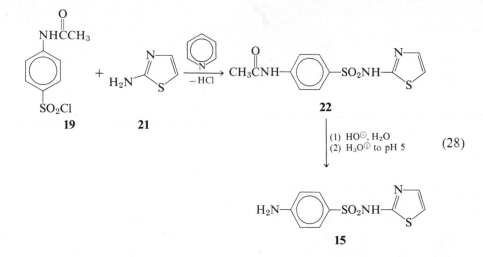

$$(28)$$

to serve as a solvent for the reaction. Precautions must be taken to insure that all equipment and reagents are dry to minimize hydrolysis of the sulfonyl chloride.

Another important principle of organic synthesis is involved in this preparation, this being the conservation of the more (or most) valuable reagent. 2-Aminothiazole is more expensive than is p-acetamidobenzenesulfonyl chloride,[4] so the latter is used in slight excess to maximize conversion of the more expensive reagent to product.

Important side reactions occur when the sulfonyl chloride reacts with 2-amino-thiazole in alternative ways, and some of these reactions are shown in equation (29). None of the side products so obtained have any hydrogen atoms on a sulfonamide nitrogen, and they will therefore be insoluble in the alkaline solution which is used in the work-up procedure (see the Hinsberg reaction, Chapter 25, Section A.1 under "Amines").

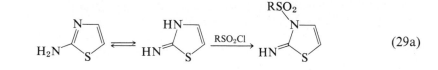

$$(29a)$$

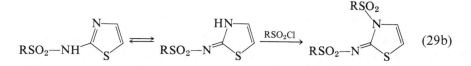

$$(29b)$$

Purification of the sulfathiazole is accomplished by treating the reaction mixture with water, collecting the solid which forms, and treating the solid with excess base and filtering. Compound 22 is soluble in base, whereas the side products indicated by equa-

---

[4] This may not seem to be a valid statement to the student who started the preparation with benzene. However, acetanilide is available at very low cost, so that it is only one simple step to convert acetanilide to p-acetamidobenzenesulfonyl chloride. On the other hand, chloroacetaldehyde, which is needed for the preparation of 2-aminothiazole, is very expensive.

tion (29) are not. After removal of the side products, the basic solution is heated to effect hydrolysis of the amide. Under the reaction conditions, the sulfonamide group is not affected. The basic solution is acidified and then treated with sodium acetate until the pH of the solution is about 5. The acetate reacts with the mineral acid present to produce acetic acid, which in the presence of acetate ion yields a buffer solution. In strongly acidic solution, sulfathiazole is protonated and is water soluble, but after the solution is adjusted to pH 5, the sulfathiazole is liberated.

EXPERIMENTAL PROCEDURE

The experimental procedures start with benzene and carry through to sulfathiazole. The instructor will indicate at which point to start, and how far along the synthesis to proceed. Remember to adjust quantities of reagents according to the amount of starting material which is available. Beginning with benzene, for example, some of the reactions may give better yields than those indicated, whereas others may give poorer yields. In going to the next step, this must be taken into account; however, do not change the reaction times unless told to do so. It may be convenient to use different sized flasks.

A. NITROBENZENE[5]

Place 28 ml of concentrated nitric acid in a 250-ml Erlenmeyer flask and cool the flask in an ice bath. To the cool acid, add 36 ml of concentrated sulfuric acid, with continued cooling and shaking. After the exothermic reaction subsides and the mixture has cooled to room temperature, transfer it to a 250-ml separatory funnel. Place 0.30 mole of benzene in a 500-ml Erlenmeyer flask and then add about 7 ml of the sulfuric-nitric acid mixture. Insert a thermometer in the reaction mixture and leave it there during the duration of the nitration. Gently swirl the flask, and maintain the temperature of the reaction mixture between 50 and 60° by means of a water bath. As the exothermic reaction ceases, add the acid mixture in small portions until it has all been added, being careful to control the temperature between 50 and 60° throughout the additions.[6] After the addition is complete, gently swirl the mixture until the temperature drops below 40°. Cool the flask and transfer the mixture to a separatory funnel. Remove the organic layer and wash it with about 50 ml of water, 50 ml of 0.5 M sodium hydroxide, and another 50 ml of water. Discard the aqueous washings and transfer the nitrobenzene to a clean Erlenmeyer flask. Add about 5 g of anhydrous calcium chloride and warm the flask with a water bath until the cloudiness of the organic layer disappears.★ Filter the crude nitrobenzene into a distilling flask, and fractionally distil. Collect the product boiling 205–212°.[7] The presence of a small amount of water, as evidenced by cloudiness in the nitrobenzene fraction, is a result of incomplete drying. Such a cloudy fraction should be collected separately, redried, and redistilled. *When distilling, do not heat the flask to dryness.*

[5] Nitrobenzene is toxic. Care should be taken to avoid inhalation of its vapors.

[6] The yield of dinitrobenzene increases if the reaction temperature is allowed to rise above 60°.

[7] Unless special distillation columns are used, high-boiling materials, such as nitrobenzene, are normally collected over a wide range. This range is caused by air cooling of the column, and the boiling points observed under these conditions are normally slightly lower than the "true" boiling point.

The residue, which contains mostly *m*-dinitrobenzene, may decompose if heated above the boiling point of nitrobenzene. Only a small amount of dinitrobenzene forms, but it still may be enough to cause a small explosion. The yield of nitrobenzene should be 83–85%.

B. Aniline

Place 30 g of iron powder[8] in a 500-ml round-bottomed flask and add to it 0.24 mole of nitrobenzene and 100 ml of water. Attach a reflux condenser and add about 0.5 ml of concentrated hydrochloric acid through the top of the condenser. Shake the reaction mixture vigorously, and if the reaction does not start soon (as evidenced by the production of heat), heat it gently with a small flame (be prepared to cool the flask in a pan of water should it appear that the reaction is going to get out of control). After the reaction has started, add another 0.5 ml of concentrated hydrochloric acid and shake the flask vigorously. When the reaction has subsided, bring the mixture to gentle reflux. After 15 min, add 1 ml of concentrated hydrochloric acid and then continue the heating under reflux for an additional 45 min.[9]

When the reflux period is complete, add 5 ml of 6 *M* sodium hydroxide solution directly to the reaction mixture and equip the flask for steam distillation (Chapter 2.6 and Figures 2.11, 2.12, and 2.13). Steam distil the mixture until the distillate dropping from the condenser no longer contains any visible amount of organic product. (The distillate may still be slightly cloudy, but if there is no visible amount of oil in it, it may be assumed that the distillation is complete.) ★ Add 20 ml of concentrated hydrochloric acid to the distillate, and steam distil this mixture until the *residue* in the distilling flask is clear and free from oily material.★

Make the acidic distillation residue basic with a minimum volume of 12 *M* sodium hydroxide solution. (*Use care:* Heat is evolved.) Saturate the basic solution with sodium chloride (roughly 25 g of salt per 100 ml of solution), cool the mixture and transfer it to a separatory funnel. Extract the organic product with two 50-ml portions of technical ether using the first portion to rinse the flask in which the neutralization was done. Separate the aqueous layer from the organic layer as thoroughly as possible; transfer the combined organic extracts to a small Erlenmeyer flask and dry with several sodium hydroxide pellets until the solution is clear.★ Transfer the solution by decantation into a distilling flask and distil it. Collect three fractions, one boiling between 35 and 90°, the second boiling between 90 and 180° and the third boiling between 180 and 185°. Discard the first fraction and, if the second fraction is of significant volume, redistil it to obtain more product. Pure aniline is colorless but may darken immediately following distillation owing to air oxidation. Aniline is often redistilled just before use to remove colored oxidation products. The yield of aniline should be 85–90%.

[8] Iron powder should be finely divided and should be free of an oxide coating. The reduction reaction is a heterogeneous surface reaction, so that a large area of metal surface is needed. Iron metal obtained by reduction of the oxide with hydrogen is termed "reduced with hydrogen" and is most suitable for this reduction reaction. Tin powder may be used in place of iron.

[9] If the reaction required heating with a burner for initiation, heat under reflux for at least 90 additional min.

C. ACETANILIDE

Dissolve 0.215 mole of aniline in 500 ml of water to which 18 ml of concentrated hydrochloric acid has been added; use a 1000-ml Erlenmeyer flask. Swirl the mixture to aid in dissolving the aniline. (If the solution is dark-colored, add about 1 g of decolorizing carbon to it, swirl and filter.) Prepare a solution containing 0.24 mole of sodium acetate trihydrate dissolved in 100 ml of water, and measure out 24 ml of acetic anhydride. Warm the solution containing the dissolved aniline to 50°, and add the acetic anhydride; swirl the flask to dissolve the anhydride. Then add the sodium acetate solution *immediately and in one portion.* Cool the reaction mixture in an ice bath with stirring while the product crystallizes. Collect the acetanilide by suction filtration, wash it with a small portion of cold water, and dry.★ Determine the melting point and yield of product; the yield should be between 65 and 75%. If impure or slightly colored acetanilide is obtained, it may be recrystallized from a minimum volume of hot water, using decolorizing carbon if necessary to give a colorless product.

D. *p*-ACETAMIDOBENZENESULFONYL CHLORIDE

If possible, either carry out the following reaction in the hood or use the gas-removal apparatus described below to prevent hydrogen chloride or sulfur dioxide gases from being vented in the laboratory room. Equip a dry 250-ml round-bottomed flask with a Claisen connecting tube. Place a vacuum adapter filled with 4–6 mesh calcium chloride on the side arm of the Claisen tube and connect the vacuum adapter to an aspirator trap via a length of tubing. Grease all joints carefully because an air-tight seal is required at all connections. Place the 0.155 mole of *dry* acetanilide in the flask. *In the hood,* measure 0.92 mole of chlorosulfonic acid into a 125-ml separatory funnel (be certain that the stopcock of the funnel is firmly seated). *Use care in handling chlorosulfonic acid. Be careful to avoid contact with the skin and with moisture. Chlorosulfonic acid reacts vigorously with water; care should be used in washing any equipment that has contained the acid.* Stopper the funnel, and place it on the straight arm of the Claisen tube so that the chlorosulfonic acid will drop directly onto the acetanilide contained in the flask.

Cool the flask to 10–15° in a water bath containing a little ice, but do not cool below 10°. Turn on the water aspirator so that a slight flow of air occurs through the vacuum adapter; it is not necessary to turn the aspirator on full force. Open wide the stopcock of the funnel so that the chlorosulfonic acid is added as rapidly as possible to the flask; it may be necessary to lift the stopper on the funnel to equalize the pressure in the system should the flow of acid become slow. After completion of the addition, *gently* swirl the apparatus from time to time to speed the rate of dissolution of the acetanilide; maintain the temperature of the mixture below 20°. After most of the solid has dissolved, allow the reaction mixture to warm to room temperature and then heat it on the steam cone until moderate agitation of the apparatus produces no increase in the rate of gas evolution; 10–20 min of heating will be required.

Cool the reaction mixture to room temperature (or slightly lower), using an ice bath. Working in the hood, place 600 g of cracked ice and 100 ml of water in a large beaker or Erlenmeyer flask and pour the reaction mixture slowly and with stirring onto the ice. *Use care in this step; do not add the mixture too quickly and avoid splattering of the chlorosulfonic acid.* After the addition is complete, rinse the flask with a little cold water and transfer this to the beaker. The precipitate that forms in the beaker is crude *p*-acetamidobenzenesulfonyl chloride, which may be white to pink in color. It soon may become a hard mass, and any lumps that form should be broken up with a stirring rod. Collect the crude material by suction filtration. Wash the solid with a small amount of cold water, and press it as dry as possible with a cork.

If sulfanilamide is to be prepared from the sulfonyl chloride, the crude product does not need to be purified, but it should be used immediately. *If sulfathiazole is to be made, the sulfonyl chloride must be purified.*

For purification, dissolve the crude product in a minimum amount of boiling chloroform (the approximate volume required is 135 to 185 ml). Preheat a separatory funnel on the steam bath and transfer the boiling chloroform mixture to it. Remove the organic layer from the water layer as quickly as possible; be very careful in separating the layers. Rinse the funnel with an additional 40-ml portion of hot chloroform, and combine this rinse with the main organic layer. On cooling, colorless crystals of pure product are obtained.★ The yield should be 77–90%; the reported melting point of pure *p*-acetamidobenzenesulfonyl chloride is 149°. Air-dry the product and store it in a closed container.

E. Sulfanilamide

Transfer the crude *p*-acetamidobenzenesulfonyl chloride to a 500-ml Erlenmeyer flask and add 75 ml of concentrated ammonium hydroxide (28%). A very rapid exothermic reaction should occur. Break up any lumps of solid which may remain with a stirring rod; the reaction mixture should be thick but homogeneous insofar as is possible. Heat the mixture on a steam bath for about 30 min. Because some ammonia vapors will be released into the air, heating preferably should be done in a hood; alternatively, invert a funnel over the flask and attach the funnel to a water aspirator. Cool the flask in an ice bath and add 6 *M* sulfuric acid to the cool reaction mixture until it is acidic to Congo red paper. Cool the reaction mixture again in an ice bath, and collect the product by suction filtration. Wash the crystals with cold water and dry.★ The yield should be 89–96%. The product is pure enough for the hydrolysis reaction, but if desired it may be purified as follows. Dissolve the crude *p*-acetamidobenzenesulfonamide in a minimum amount of hot water. If needed, use decolorizing carbon. After filtering, cool the aqueous solution in an ice bath and collect and dry the crystals.★ The melting point of the pure product is 220°.

Weigh the *p*-acetamidobenzenesulfonamide and transfer it to a 250-ml round-bottomed flask. Prepare a solution of dilute hydrochloric acid by mixing equal volumes of concentrated hydrochloric acid and water. Add to the amide an amount of dilute hydrochloric acid solution *twice* the weight of the amide. Attach a reflux condenser to the flask and heat at a gentle reflux for 30 min. Boil gently and swirl the reaction mixture when first heating so as to aid in the dissolution of the organic material. To the homogeneous reaction mixture, add an equal quantity of water, and transfer the new mixture to a 600-ml beaker. Neutralize the excess acid which is present by the addition of small quantities of solid sodium carbonate; continue addition of the base until the solution is just alkaline to litmus paper. (*Caution:* Add the sodium carbonate in small quantities, for foaming will occur!) A precipitate should form during neutralization. After making the solution just basic, cool the mixture in an ice bath to complete the precipitation of the product. Collect the crystals by suction filtration, wash with a small amount of cold water, and allow the product to air-dry.★ Purify the crude product by recrystallization from hot water (12 to 15 ml of hot water per gram of compound will be required). Decolorize the hot solution, if necessary, filter and cool the filtrate in an ice bath. It may be necessary to preheat the filter funnel so that the product will not crystallize in the funnel. Sulfanilamide will form on cooling to give long, white needles; collect the product by suction filtration. Determine the melting point (reported 163°) and the yield, which should be about 0.7 g per gram of starting *p*-acetamidobenzene-sulfonamide.

Test the solubility of sulfanilamide in 1.5 *N* hydrochloric acid solution and in 1.5 *N* sodium hydroxide solution.

F. SULFATHIAZOLE

Dry about 40 ml of reagent grade pyridine over 2–3 g of potassium hydroxide pellets.[10] To a 250-ml round-bottomed flask equipped with a Claisen connecting tube, add 0.075 mole of 2-aminothiazole (Chapter 18) and about 30 ml of dry pyridine. Transfer the dry pyridine by use of a graduated pipet. Equip the flask with a calcium chloride drying tube and a thermometer. Add 0.082 mole of purified *p*-acetamido-benzenesulfonyl chloride to the reaction flask in small portions; swirl gently to insure mixing, and make the addition at such a rate that the temperature inside the flask does not exceed 40°. Do not cool the reaction mixture at any time during the addition. After the addition is complete, heat the mixture on a steam bath for *ca.* 30 min. Then cool the mixture and pour it into about 200 ml of warm water while stirring vigorously with a stirring rod. The oil which forms initially should solidify if a stirring rod is used to induce crystallization. Remove the solid by filtration,★ wash it with a small amount of water, and dissolve it in 38 ml of 2 *M* sodium hydroxide solution; warm on a steam bath, if needed, to bring the solid into solution. If any solid remains, filter it off and discard it. Make the resulting filtrate acidic to litmus by the addition of glacial acetic acid, collect the solid which forms by suction filtration, and press as dry as possible.★ Weigh the solid, add 10 times its weight of 2 *M* sodium hydroxide solution, and heat the mixture at reflux for one hour. After heating, cool the mixture and slowly add concentrated hydrochloric acid; at first a solid will form, which will then redissolve as excess acid is added. Add only enough acid so that the solid redissolves; do not add excess acid. After decolorizing and filtering the solution, neutralize it with 2 *M* sodium hydroxide, and then add solid sodium acetate in small portions until the solution is basic to litmus paper. Bring the heterogeneous mixture to boiling and then cool in an ice bath; collect the solid by suction filtration ★ and recrystallize it from a minimum amount of boiling water. Collect the sulfathiazole by suction filtration and air-dry it. Determine the melting point (reported 201–202°) and the yield.

---

[10] This should be done during the laboratory period before the experiment is to be performed. Add potassium hydroxide pellets to the pyridine and store in the laboratory desk until ready to use.

EXERCISES

1. Outline a possible synthesis for the compound shown below, using benzene as the only source of an aromatic ring. Use any needed aliphatic or inorganic reagents.

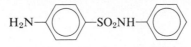

$$H_2N-\text{⟨◯⟩}-SO_2NH-\text{⟨◯⟩}$$

2. Outline the procedure for the purification of aniline in flow diagram form. Indicate the importance of each step in the procedure, and give reasons for doing the steam distillation first with a basic solution and then with an acidic solution. Write equation(s) for reactions which occur when base and then acid are added.

3. In the purification of aniline, sodium hydroxide pellets are used as the drying agent. Why is this compound used rather than magnesium sulfate, calcium chloride or other common drying agents? Would potassium hydroxide be an acceptable substitute for sodium hydroxide?

4. In the preparation of sulfanilamide from *p*-acetamidobenzenesulfonamide, only the acetamido group is hydrolyzed. Provide an explanation for this difference in reactivity of acetamido and sulfonamido groups.

5. Explain the results which were obtained when the solubility of sulfanilamide was determined in 1.5 *N* hydrochloric acid and in 1.5 *N* sodium hydroxide. Write equations for any reaction(s) occurring.

6. What would be observed if *p*-acetamidobenzenesulfonamide were subjected to vigorous hydrolysis conditions (*i.e.*, concentrated hydrochloric acid and heat for a long period of time)? Write an equation for the reaction which would occur.

7. With the aid of chemical equations, explain the behavior of sulfathiazole towards acid and base. How do these reagents affect its solubility in water?

8. Outline the workup of sulfathiazole in flow diagram form. Indicate the importance of each step in the procedure, and give reasons for each of them. Write equation(s) for the reactions which occur in the purification process.

SPECTRA OF STARTING MATERIALS AND PRODUCTS

The ir and nmr spectra of aniline are given in Figures 17.3 and 17.4, respectively.

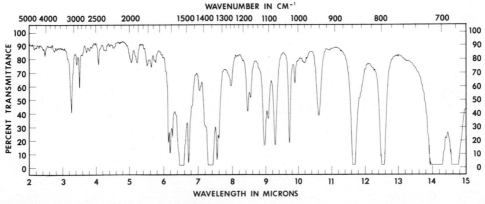

FIGURE 19.6   IR spectrum of nitrobenzene.

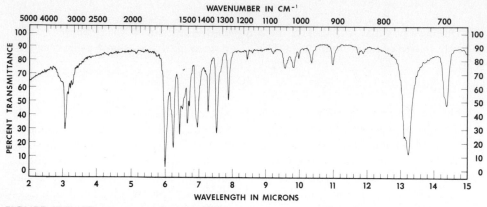

FIGURE 19.7   IR spectrum of acetanilide.

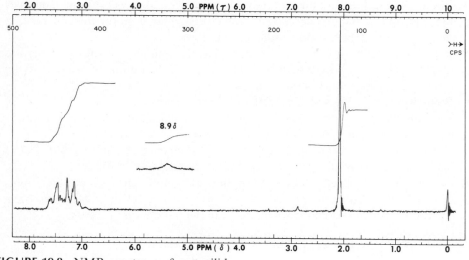

FIGURE 19.8   NMR spectrum of acetanilide.

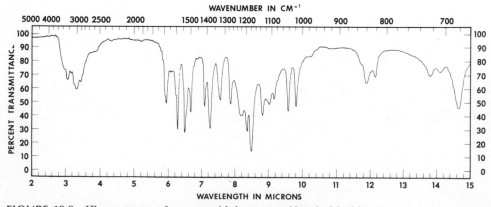

FIGURE 19.9   IR spectrum of *p*-acetamidobenzenesulfonyl chloride.

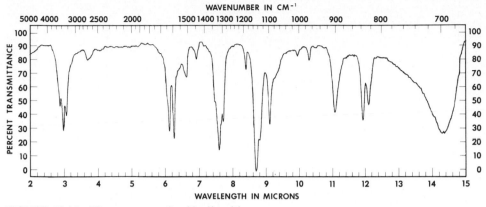

FIGURE 19.10 IR spectrum of sulfanilamide.

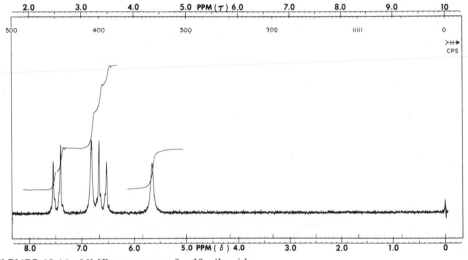

FIGURE 19.11 NMR spectrum of sulfanilamide.

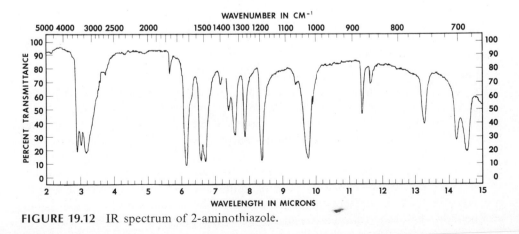

FIGURE 19.12 IR spectrum of 2-aminothiazole.

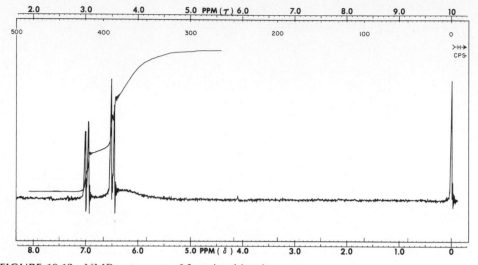

FIGURE 19.13 NMR spectrum of 2-aminothiazole.

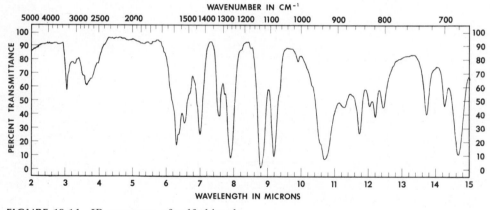

FIGURE 19.14 IR spectrum of sulfathiazole.

19.3 The Synthesis of 1-Bromo-3-chloro-5-iodobenzene

Another interesting multi-step synthesis is the preparation of 1-bromo-3-chloro-5-iodobenzene **(31)**, which can be initiated from benzene. The complete sequence of this eight-step synthesis is outlined in Chart 19.4. Most of the reactions involve electrophilic aromatic substitution. Two important stratagems of aromatic chemistry are incorporated in the synthesis: (1) An amino group is "protected" by conversion to the amide and, after several electrophilic substitution reactions are performed on the molecule, the free amine is regenerated, and (2) an amino group, which was introduced into the molecule initially because of its directing influence and activating properties, is removed in the final step to yield the desired product. The usefulness of the amino group will be indicated as the various steps of the reaction sequence are discussed. It should be noted that the yields of each individual step range from 60 to 96%, which is always

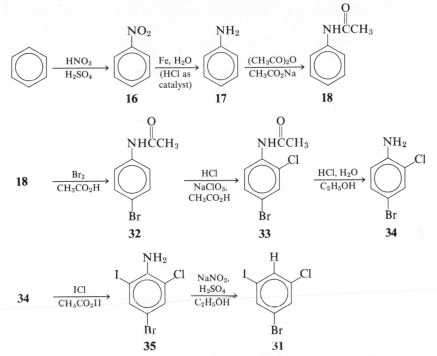

CHART 19.4   Laboratory synthesis of 1-bromo-3-chloro-5-iodobenzene (**31**).

an important consideration in multi-step syntheses. In general, the reaction times are short enough so that one or more steps can be carried out in a single laboratory period.

The synthesis of **31** may be started anywhere along the reaction sequence and, if desired, may be started from benzene. The nitration of benzene, reduction of nitrobenzene, and conversion of aniline to acetanilide have already been discussed under Section 19.2 (sulfathiazole), and the experimental details are provided therein. The discussion here will start with the synthesis of 4-bromoacetanilide (**32**) from acetanilide.

4-Bromoacetanilide (32).   The first halogen to be introduced into the molecule is bromine. The amino group activates the aromatic ring so greatly that any bromination conditions lead to trisubstitution on aniline, and indeed an aqueous solution of bromine is sufficient to do this [equation (30)].

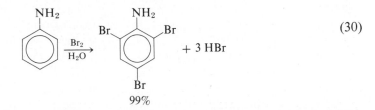

$$+ \; 3 \; HBr \tag{30}$$

In order to decrease the reactivity of aniline so that monosubstitution can be realized, aniline is acetylated to give acetanilide. Under mild conditions (bromine in acetic acid), acetanilide can be monobrominated easily (note that no Lewis acid catalyst such as $FeBr_3$ is required, for acetanilide is still quite reactive toward bromine).

As the acetamido group, CH_3CONH-, is *ortho,para*-directing, both 2-bromo- and 4-bromoacetanilide are produced; however, under the conditions of the reaction, 95% of the 4-bromo isomer is obtained [equation (31)]. 4-Bromoacetanilide can be separated from the 2-bromo isomer by a single recrystallization of the crude product from methanol.

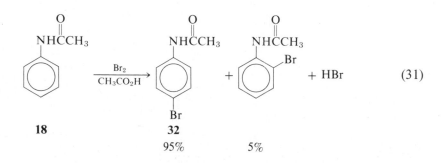

$$\text{18} \qquad \text{32} \qquad \qquad (31)$$
$$95\% \qquad\qquad 5\%$$

In general, the separation and purification of *ortho* and *para* isomers from a mixture containing both of them is fairly simple, as illustrated by the ease with which 4-bromoacetanilide is purified. Even though different functional groups may be attached to an aromatic ring, the *para* isomer is more "symmetrical" than is the *ortho* isomer, and one generally observes the former to possess a much higher melting point than the latter. Also, the *para* isomer is normally less soluble in a given solvent than is the *ortho* isomer, so that fractional crystallization can be used to separate them from one another. For example, in this experiment, the melting point of 4-bromoacetanilide is 167° and that of 2-bromoacetanilide is 99°. The 4-bromo isomer is much less soluble in methanol than is the 2-bromo isomer, so that recrystallization of a mixture of these compounds from methanol gives 4-bromoacetanilide as a solid whereas the 2-bromo isomer remains in solution.

2-Chloro-4-bromoacetanilide (33). The next halogen to be introduced into the molecule is chlorine. The bromine atom in 4-bromoacetanilide deactivates the ring slightly towards further electrophilic substitution (as compared to acetanilide), but the compound is still reactive enough so that it can be chlorinated without using a catalyst. Chlorine in acetic acid will monochlorinate 4-bromoacetanilide to produce 2-chloro-4-bromoacetanilide [**33**, equation (32)]. Although it is possible to use chlorine gas

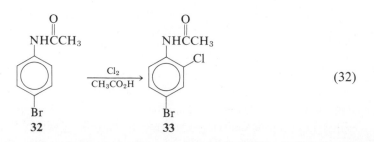

$$\text{32} \qquad\qquad\qquad \text{33} \qquad\qquad\qquad (32)$$

dissolved in acetic acid, the difficulties and hazards of using this gas with large numbers of students make it desirable to generate chlorine gas *in situ*. One convenient method of *in situ* preparation of chlorine is to allow hydrochloric acid and sodium chlorate ($NaClO_3$) to react; an oxidation-reduction reaction occurs, in which chloride ion is oxidized to chlorine and chlorate ion is reduced to chlorine as shown in equation (33).

OXIDATION: $2\,Cl^\ominus = Cl_2 + 2\,e^\ominus$

REDUCTION: $12\,H^\oplus + 2\,ClO_3{}^\ominus + 10\,e^\ominus = Cl_2 + 6\,H_2O$ (33)

OVERALL: $5\,Cl^\ominus + ClO_3{}^\ominus + 6\,H^\oplus = 3\,Cl_2 + 3\,H_2O$

In addition to ease and safety in handling, just the right amount of chlorine gas can be prepared by this method. The reaction between chlorine and 4-bromoacetanilide occurs very rapidly, and little, if any, dichlorination product is obtained. 4-Bromo-2-chloro-acetanilide is purified by fractional crystallization from methanol.

2-Chloro-4-bromoaniline (34). 2-Chloro-4-bromoacetanilide (**33**) can be converted to 2-chloro-4-bromoaniline in excellent yield using acid-catalyzed hydrolysis of the amide [equation (34)]. Concentrated hydrochloric acid in ethanol is employed, with the ethanol being used to increase the solubility of the amide.

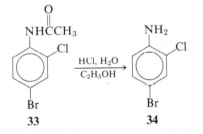

(34)

2-Chloro-4-bromo-6-iodoaniline (35). The 2-chloro-4-bromoaniline (**34**), which was obtained in the previous reaction, is highly activated due to the presence of the amino group. Of the positions *ortho* and *para* to the NH_2 group, only the 6-position is unsubstituted. Thus, iodine may be added in that position by starting with the free amine and without need for the deactivating effect of the acetamido group. Any electrophilic substitution reaction which might be carried out on **34** would result in monosubstitution at the free 6-position.

Electrophilic substitution reactions involving iodine are generally more difficult to carry out than are chlorination and bromination reactions. One general method uses iodine and potassium carbonate, with the base being present to absorb the hydroiodic acid which forms. Another good method makes use of iodine monochloride, ICl, as the iodinating agent. Of the two halogens present in ICl, chlorine is more *electronegative* than is iodine. Thus, when the molecule dissociates, the bonding electrons leave with chlorine to give Cl^- and I^+ [equation (35)]. I^+ serves as an electrophile which attacks the aromatic ring and replaces H^+ at the 6-position [equation (36)].

$$I—Cl \xrightarrow{\text{CH}_3\text{CO}_2\text{H}} I^{\oplus} + Cl^{\ominus} \tag{35}$$

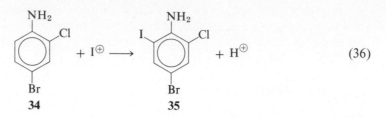

$$\tag{36}$$

It should be pointed out that I^+ is a very weak electrophile compared to Br^+ or Cl^+. One of the main reasons for introducing iodine in the final step of the substitution sequence is that a very activated aromatic ring must be used in order to realize a good yield in any iodination substitution reaction. Thus, bromine and chlorine are introduced with the acetamido group present, since electrophilic substitution with these elements requires a less activated ring than does iodine. Also, with the acetamido group present, the reaction can be controlled to give monosubstitution. Once the chlorine and bromine atoms are in the desired positions, the highly activating free amino group is regenerated and iodination is carried out with ease and without the possibility of disubstitution.

There are no important side reactions in this preparation; unchanged starting material can be separated from the product by fractional crystallization.

1-Bromo-3-chloro-5-iodobenzene (31). The final step in the synthesis is the removal of the amino group from 2-chloro-4-bromo-6-iodoaniline **(35)**. After the amino group has been diazotized with nitrous acid at 0°, the diazo group can be removed by treatment with hypophosphorous acid (H_3PO_2) or by treatment with absolute ethanol. The former is the method most often encountered in textbooks, and indeed it is the best general reagent for replacing the diazo group by a hydrogen. When hypophosphorous acid is used, it is oxidized to phosphorous acid as the diazonium salt is reduced.

The use of ethanol has been found to be successful with certain types of aromatic compounds, in particular those which contain halogen. As this method is easier to use in the laboratory, it will be employed here. Presumably an oxidation–reduction reaction occurs with ethanol as well; in this case ethanol is oxidized to acetaldehyde or acetic acid as the diazo compound is reduced [equation (37)].

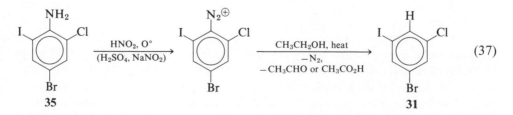

$$\tag{37}$$

There is one possibly important side reaction which might occur when this procedure is used. The diazonium salt might dissociate to give a carbonium ion and nitrogen. The carbonium ion could then attack ethanol to give an aryl alkyl ether as a side product [equation (38)]. It is well known that this type of reaction occurs with certain types of

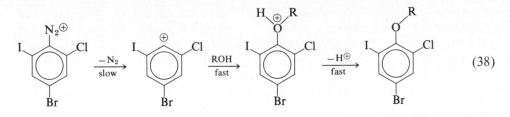

$$(38)$$

aromatic amines; for example, *o*- and *m*-toluidine (methylanilines) give high yields of the corresponding ethers when diazotized and treated with an alcohol. It appears, however, that ether formation is not important in the present reduction, and if any ether is produced, it is removed in the recrystallization of the product.

In diazotization reactions involving aromatic halo compounds, hydrohalic acids containing a *different* halogen should never be used. There are numerous examples of halogen interchange between the ring and the mineral acid during diazotization and subsequent reduction of the diazo group. This interchange occurs only with the *ortho* and *para* halogens (of which there are three in compound **35**), but not with the *meta* halogens. When a halogen is present in the ring, it is best to produce the nitrous acid from sodium nitrite and *sulfuric acid* as is done here.

EXPERIMENTAL PROCEDURE

It is possible to start the synthesis anywhere along the sequence, but this normally will not extend beyond 4-bromoacetanilide, which is the last compound that is commercially available at a reasonable cost.

In each of the procedures except the final step, the crude product from the previous reaction can be used without further purification. However, methods of purification for each intermediate compound are given, should the student wish to purify the starting materials or to stop somewhere along the reaction sequence. The melting points reported are those for once recrystallized material.

In carrying out a multi-step synthesis, it *may* be necessary to increase or decrease the scale of a subsequent reaction, as dictated by the yield of a previous product. However, do not change reaction *times* unless told to do so by the instructor.

A. Nitrobenzene, aniline, and acetanilide
See Section 19.2 for experimental procedures.

B. 4-Bromoacetanilide
Dissolve 0.06 mole of acetanilide in 30 ml of glacial acetic acid in a 250-ml round-bottomed flask, and add to this mixture slowly and with stirring a solution of 0.061 mole of bromine dissolved in 6 ml of glacial acetic acid. *Use special care when handling bromine; wear gloves and avoid contact with the skin. Consult the instructor before using bromine.* After addition of the bromine-acetic acid solution is complete, stir for several minutes and then add slowly, with stirring, 200 ml of water. On addition of water, a solid will form. Prepare a saturated solution of sodium bisulfite in water, and add just enough of it to discharge the yellow color of the solution. Collect the product by suction filtration; wash the solid well with water

and allow it to air dry.★ The yield of crude product should be about 96%. Recrystal-lization of the crude product from methanol (about 3–4 ml of methanol per gram of compound) gives 4-bromoacetanilide, mp 171–172°.

C. 2-CHLORO-4-BROMOACETANILIDE

Note: It would be preferable to carry out this reaction in the hood or to invert a funnel over the reaction flask and attach the funnel to the water aspirator.

Suspend 0.05 mole of 4-bromoacetanilide in a mixture of 23 ml of concentrated hydrochloric acid and 28 ml of glacial acetic acid, using a 250-ml flask. Heat the mixture gently on a steam bath until it becomes homogeneous, and then cool the solution to 0°. To the cold mixture, add 0.026 mole of sodium chlorate dissolved in about 7 ml of water. During the addition of the sodium chlorate solution, some chlorine gas is evolved. As the addition is carried out, a yellow precipitate forms and the solution turns yellow. Allow the reaction mixture to stand at room temperature for one hour, and collect the precipitate by suction filtration.★ The material which is obtained is crude 2-chloro-4-bromoacetanilide; the yield should be about 97%. The crude product can be recrystallized from methanol (7–8 ml of methanol per gram of crude product) to give pure 2-chloro-4-bromoacetanilide, mp 153–154°.

D. 2-CHLORO-4-BROMOANILINE

Mix 0.045 mole of crude 2-chloro-4-bromoacetanilide with 20 ml of 95% ethanol and 13 ml of concentrated hydrochloric acid in a 250-ml Erlenmeyer flask. Heat the mixture on the steam bath for *ca.* 30 min; during the heating, the yellow precipitate should dissolve and be replaced by a white precipitate. At the end of the heating, add 90 ml of hot water. Swirl the flask to dissolve the white solid completely and pour the solution onto 50 g of ice. Add to the resulting mixture 12 ml of 14 *N* sodium hydroxide solution, stirring well during the addition and mixing process. Light brown crystals should precipitate during the addition of the base. Collect these by suction filtration and dry them as thoroughly as possible.★ The yield should be about 91%. Recrystallize the crude product from 30–60° petroleum ether (3–4 ml per gram of product) to give 2-chloro-4-bromoaniline, mp 65–66°.

E. 2-CHLORO-4-BROMO-6-IODOANILINE

Dissolve 0.024 mole of recrystallized 2-chloro-4-bromoaniline in 80 ml of glacial acetic acid, and add about 20 ml of water to the mixture. Prepare a solution of 0.03 mole of technical iodine monochloride in 20 ml of glacial acetic acid in an Erlenmeyer flask and add this solution to the reaction mixture over a period of 8 min. Heat the resulting black mixture on the steam bath until its temperature is 90°, and then add just enough saturated sodium bisulfite solution to turn the color of the mixture bright yellow; note the volume of sodium bisulfite solution added. Dilute the reaction mixture with enough extra water such that the volume of the sodium bisulfite solution used *plus* the volume of added water is about 25 ml. Cool the reaction mixture in an ice bath; 2-chloro-4-bromo-6-iodoaniline will separate as light brown crystals. Collect the solid by suction filtration and wash the crystals with a small amount of 33% acetic acid and then with water.★

It would be advisable to purify the product before going on to the final step of the synthesis. The crude product may be recrystallized from acetic acid-water as follows. Mix the product with glacial acetic acid in the ratio of about 20 ml of glacial acetic acid per gram of product. Heat the mixture on the steam bath and slowly add 5 ml of water per gram of product to the solution as it is heating. On slow cooling, long colorless crystals of 2-chloro-4-bromo-6-iodoaniline will form; filter and dry the pure product. The recovery should be about 80% (based on crude product) and the melting point should be 96–98°.

F. 1-BROMO-3-CHLORO-5-IODOBENZENE

Suspend 0.006 mole of 2-chloro-4-bromo-6-iodoaniline in about 10 ml of absolute ethanol in a 250-ml round-bottomed flask. While stirring the mixture, add 4.0 ml of concentrated sulfuric acid dropwise. Equip the flask with a condenser and add 0.01 mole of powdered sodium nitrite in small portions through the condenser. When the addition is complete, heat the mixture on a steam bath for about 10 min. Add 50 ml of hot water to the flask through the condenser, and steam distil. Collect about 80 ml of distillate (which should be clear and not contain any organic product). The desired product will form as a solid in the condenser, so that care should be taken to insure that the condenser does not plug up completely during the distillation.[11] It is easiest to remove the solid product from the condenser by pouring ether through it; this dissolves and removes the product.* Dry the ether solution over anhydrous magnesium sulfate, filter, and distil the ether. (*Caution:* Use a steam bath and be sure that no flames are close by.) Recrystallize the residue from about 10 ml of methanol[12] to give 1-bromo-3-chloro-5-iodobenzene in about 40% yield. The long, nearly colorless needles should melt between 82 and 84°.

EXERCISES

1. Outline synthetic procedures, starting from benzene, for each of the following compounds:

    (a) 1,3,5-tribromobenzene; (b) 2-bromo-4-chloro-6-iodophenol; (c) 2-bromo-4,6-dichloroaniline.

2. In the bromination of acetanilide, using bromine in acetic acid, the major product is 4-bromoacetanilide. Suggest a reason for this. Give another example of an electrophilic substitution reaction which gives the *para* isomer as the predominant product.

3. Give the complete, step-wise mechanism for the reaction which occurs between iodine monochloride and 2-chloro-4-bromoaniline. Suppose that this same reaction were carried out using bromine monochloride, BrCl. What electrophilic substitution reaction might occur when this compound is allowed to react with 2-chloro-4-bromoaniline? Briefly explain.

---

[11] Should the condenser plug up during the distillation, it can be opened up by running hot water or a slow stream of steam through the condenser jacket.

[12] It may be necessary to heat the filter funnel before filtering the solution, because the solid will crystallize in the stem of a cold funnel before the filtrate reaches the flask. See the discussion in Chapter 2.7 under "Hot Filtration" (page 46) and Figure 2.14.

4. Hydroxylic solvents, such as water or low-molecular-weight alcohols, are often used to purify amides. Briefly explain why these are used rather than hydrocarbon solvents, such as petroleum ether.

REFERENCE

A. Ault and R. Kraig, *Journal of Chemical Education*, **43**, 213 (1966).

SPECTRA OF STARTING MATERIALS AND PRODUCTS

The ir and nmr spectra of aniline are given in Figures 17.3 and 17.4, respectively, and the corresponding spectra of acetanilide in Figures 19.7 and 19.8.

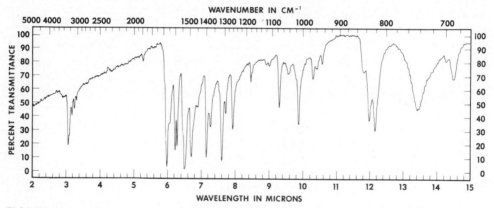

FIGURE 19.15   IR spectrum of 4-bromoacetanilide.

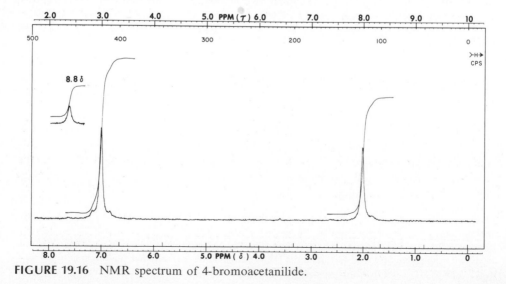

FIGURE 19.16   NMR spectrum of 4-bromoacetanilide.

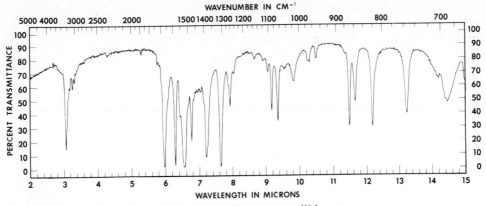

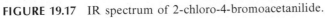

FIGURE 19.17   IR spectrum of 2-chloro-4-bromoacetanilide.

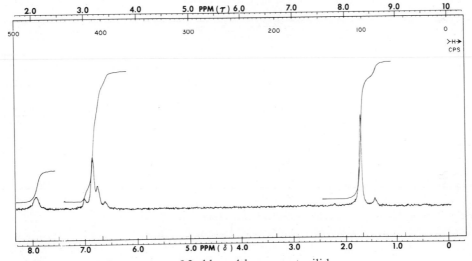

FIGURE 19.18   NMR spectrum of 2-chloro-4-bromoacetanilide.

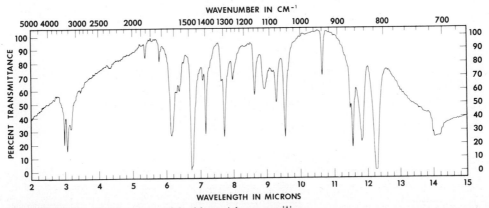

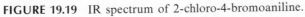

FIGURE 19.19   IR spectrum of 2-chloro-4-bromoaniline.

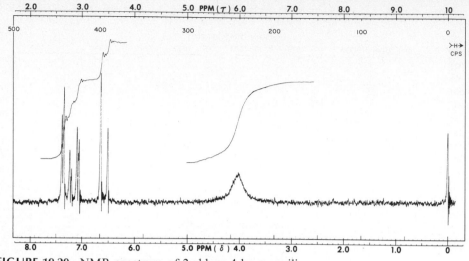
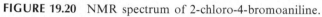

FIGURE 19.20 NMR spectrum of 2-chloro-4-bromoaniline.

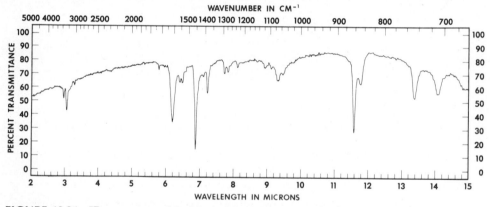

FIGURE 19.21 IR spectrum of 2-chloro-4-bromo-6-iodoaniline.

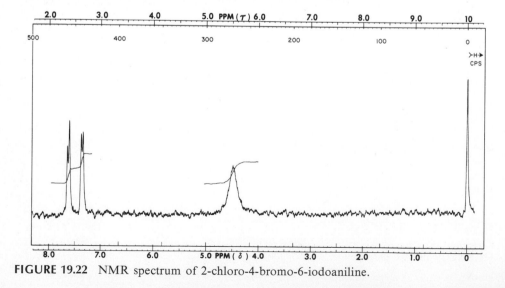
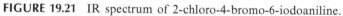

FIGURE 19.22 NMR spectrum of 2-chloro-4-bromo-6-iodoaniline.

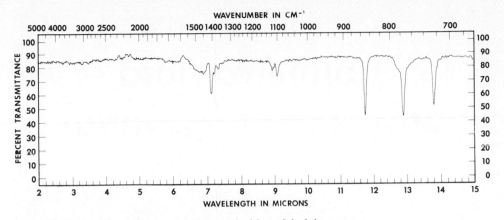

FIGURE 19.23   IR spectrum of 1-bromo-3-chloro-5-iodobenzene.

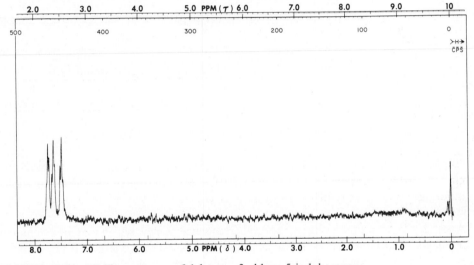

FIGURE 19.24   NMR spectrum of 1-bromo-3-chloro-5-iodobenzene.

chapter twenty
nonbenzenoid aromatic compounds

20.1 Aromaticity

In recent years many fascinating examples of compounds exhibiting chemical properties typical of benzene, yet having no apparent molecular similarity to it, have been discovered. These substances are said to have aromatic character, or "aromaticity," although as yet there is not complete agreement among chemists as to an exact definition of this term.

The problem revolves about which property or properties of benzene are considered to be most typical. The older concept of aromaticity involved resistance toward addition reactions and ease of substitution. This has been updated in terms of mechanisms; both ionic addition and free radical substitution are considered typical of aliphatic compounds, and electrophilic substitution is considered typical of aromatic compounds.

Another criterion that has been applied is *resonance energy*, which may be calculated from heats of combustion or hydrogenation. The resonance energy of an aromatic compound is a measure of the unusual stability of the molecule compared to some aliphatic model and is attributed to the cyclic conjugation of the aromatic compound. Naphthalene (1), phenanthrene (2), and other polycyclic aromatic hydrocarbons and their derivatives are obviously "benzenoid," since they essentially consist of two or more benzene rings fused together. The nitrogen-containing analogs of benzene, such as pyridine (3) and pyrimidine (4), are also usually considered benzenoid.

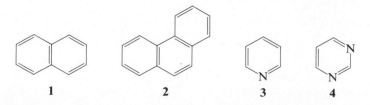

1   2   3   4

A different situation is encountered in azulene (5), which contains a cyclic conjugated system in fused five- and seven-membered rings. Azulene, then, must be considered nonbenzenoid. It is an isomer of naphthalene, to which it is converted quantitatively by heating above 350° in the absence of air [equation (1)]. Although less stable than naphthalene, azulene undergoes certain typical electrophilic substitution reactions.[1]

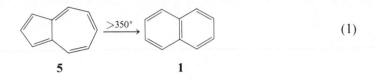

$$>350°$$

(1)

5   **1**

The heterocyclic compounds with five-membered rings, such as furan (**6**), pyrrole (**7**), and thiophene (**8**), constitute another class of aromatic compounds. Strictly speaking

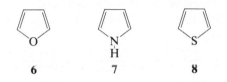

6   **7**   **8**

they are not benzenoid, since they do not have six-membered rings. However, they are closely related to benzene and pyridine in that they also have an "aromatic sextet" of electrons. The hetero atoms provide a pair of electrons which enter into conjugation with the four π-electrons of the carbon-carbon bonds, as shown in the resonance structures **6**, **9**, and **10** (there are two additional resonance structures equivalent to **9** and **10**).

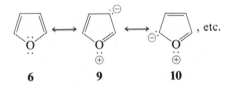

, etc.

6   **9**   **10**

A hydrocarbon system closely related to these heterocyclic compounds is the cyclopentadienyl anion, represented by resonance structures **12**, **13**, and **14**, etc.,

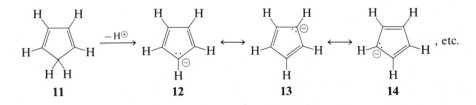

, etc.

11   **12**   **13**   **14**

derived from cyclopentadiene (**11**). This compound is unusually acidic for a hydrocarbon (pK$_a$ = *ca.* 20), presumably owing to stabilization of its anion by delocalization of negative charge as represented by resonance structures **12**, **13** and **14**, etc.

One might infer from these examples that any cyclic conjugated system for which a

[1] In connection with this discussion of azulene, see Exercise 6 in Chapter 13.1.

number of reasonable contributing resonance structures may be written will have aromatic character. That this is not true is demonstrated by the fact that the cyclopentadienyl *cation* (**15**) has never been prepared, even though the same number of contributing resonance structures (**15**, **16**, and **17**, etc.) can be written as for the corre-

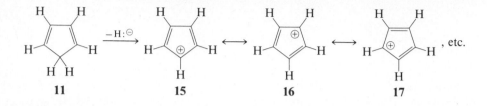

11   **15**   **16**   **17**

sponding anion. Thus, it may be seen that simple *valence bond theory* is inadequate to explain the difference in stability of cyclopentadienyl cation and anion, and certain other aspects of aromaticity.

A *molecular orbital theory* proposed by the German chemist E. Hückel in 1931 has provided an explanation for such problems. Hückel's theory, which was based on quantum mechanics, stated that conjugated cyclic systems in which the number of delocalized π-electrons is $4n + 2$, where $n = 0, 1, 2, 3$, etc., will exhibit special stability, *i.e.*, aromaticity. Benzene, pyridine, furan, and the cyclopentadienyl anion all fit "Hückel's Rule" for $n = 1$; that is, the number of π-electrons is six ($4 \times 1 + 2 = 6$). The cyclopentadienyl *cation* system involves *four* π-electrons, so according to Hückel's Rule, it should not have special stability.

Hückel's Rule also offers an explanation for the fact that cyclooctatetraene (**18**), with eight π-electrons, and cyclobutadiene (**19**), with four, are nonaromatic. On the

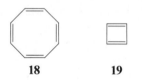

18   **19**

other hand, cyclopropenyl cations (**20**, $n = 0$), and cycloheptatrienyl cation (**21**, $n = 1$), are predicted to be unusually stable.[2] One of the most significant examples of

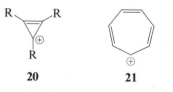

20   **21**

the success of prediction based on theory in organic chemistry is the realization of the preparation of these cations and the demonstration of their remarkable stability.

---

[2] It is interesting to note that the cycloheptatrienyl anion, which has eight π-electrons, is extremely unstable, even though one can write the same number of equivalent contributing resonance structures for the cycloheptatrienyl cation and anion.

A modern criterion of aromaticity may be described in connection with these cations. A carbonium ion may react with water reversibly to form an equilibrium mixture containing the corresponding alcohol and hydronium ion [equation (2)]. According to

$$R^{\oplus} + 2\ HOH \rightleftharpoons ROH + H_3O^{\oplus} \tag{2}$$

$$K_R{}^{\oplus} = \frac{[ROH][H_3O^{\oplus}]}{[R^{\oplus}]} \tag{3}$$

$$\text{When}\quad \frac{[ROH]}{[R^{\oplus}]} = 1,\ \text{then}\quad K_R{}^{\oplus} = [H_3O^{\oplus}] \tag{4}$$

$$\text{and therefore}\quad pK_R{}^{\oplus} = \log \frac{1}{[H_3O^{\oplus}]} = pH \tag{5}$$

equation (5), the $pK_R{}^{\oplus}$ of a carbonium ion is equal to the pH of a solution in which the carbonium ion is half converted to the corresponding alcohol. An unconjugated carbonium ion is extremely reactive toward water, so that the $pK_R{}^{\oplus}$ is a large negative number. The $pK_R{}^{\oplus}$ of an alkyl carbonium ion, such as the *t*-butyl cation, cannot be measured, and even the triphenylmethyl cation has a $pK_R{}^{\oplus}$ of -6.63. By contrast, the cycloheptatrienyl cation (**21**) has a $pK_R{}^{\oplus}$ of $+4.7$ and the cyclopropenyl cation, (**20**, R $= n$-C_3H_7) has a $pK_R{}^{\oplus}$ of $+7.2$. The latter cation is thus stable even in neutral water solution; it is the most stable hydrocarbon cation yet known. Since the $pK_R{}^{\oplus}$ of a cation is a measure of its stability, it may also be used as a criterion of aromaticity in appropriate systems.

Another modern criterion of aromaticity is a *ring current* which is induced in an aromatic compound when it is placed in a magnetic field. An nmr spectrometer makes use of a strong external magnetic field, and evidence for the ring current is provided by the nmr spectrum of an aromatic compound.[3] The ring current attributable to the circulation of π-electrons in benzene results in the deshielding of the six hydrogens (which are *outside* the aromatic ring) so that their nmr signals are shifted downfield (to larger δ values). According to the ring current theory, hydrogen atoms on the *inside* of an aromatic ring should experience a shielding effect, resulting in an upfield shift of their nmr signal. This theory has been confirmed by the nmr spectrum of the interesting hydrocarbon **22**, which is called [18]annulene. This compound, which has 18 π-

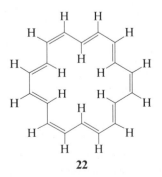

22

[3] Refer to Chapter 4.2, p. 91.

electrons and thus should be aromatic according to Hückel's Rule ($n = 4$), has 6 hydrogen atoms inside the conjugated ring system and 12 outside. The two nmr signals occur at 1.9 ppm *upfield* from TMS (integration = 6H) and 8.8 ppm *downfield* from TMS (integration = 12H); thus, the inside hydrogen atoms are strongly shielded, and the outside hydrogen atoms are deshielded, as expected for a ring current effect.

20.2   Ferrocene

Ferrocene **(23)** was discovered accidentally in 1951. It is an orange, crystalline solid whose structure consists of two cyclopentadienyl anions bonded to a ferrous cation. The bonding between the metal and the organic anions involves the π-electrons of the two rings in such a way that all of the carbon atoms are bonded equally to the central ferrous ion. Several other metals form "sandwich-type" compounds with cyclopentadienyl anions, called *metallocenes*. Among the most stable of these, besides ferrocene, are those from the dipositive ions of ruthenium and osmium, since in these compounds, as well as in ferrocene, the metal ion achieves the electronic structure of an inert gas atom.

Ferrocene has the classical aromatic properties of resistance toward reaction with acids and bases, even concentrated sulfuric acid. It is sensitive toward oxidizing acids because the iron is oxidized to the ferric state and the resulting metallocene cation is much less stable (the iron no longer has the krypton electronic structure). Ferrocene does not undergo addition reactions typical of cyclopentadiene, but readily undergoes electrophilic substitutions, such as Friedel-Crafts acylation.

Depending on the catalyst and the conditions, either acetylferrocene **(24)** or 1,1'-diacetylferrocene **(25)** may be produced as the major product of acetylation. The struc-

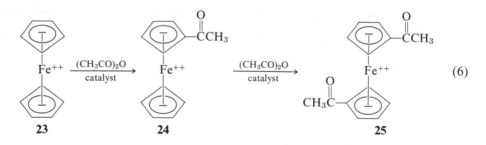

(6)

23   24   25

ture of the disubstituted product has been established by degradation studies, using reactions which removed the iron from the organic part of the molecule and which gave no products indicative of the presence of two acetyl groups on a cyclopentadienyl ring. This confirmed the expectation that the second acetyl group would not substitute on the same ring as the first, since the acetyl group would be expected to deactivate the ring toward a second electrophilic attack, by analogy to reactions of benzene derivatives. The fact that only one such diacetylferrocene has been found indicates that the cyclopentadienyl rings are able to rotate about the axis of the bonds to the metal, although the staggered conformation indicated in formula **23** is probably preferred.

The acetylation of ferrocene offers an excellent opportunity for a demonstration of the application of thin-layer chromatography to monitor the course of a reaction, and of column chromatography to effect separations of products from one another and from starting material on a preparative scale. The utility of these methods will be demonstrated in the experiments described below.

EXPERIMENTAL PROCEDURE

A. ACETYLFERROCENE AND 1,1'-DIACETYLFERROCENE

1. Preparation from ferrocene. Place 0.0054 mole of ferrocene and 10 ml of acetic anhydride in a 50-ml Erlenmeyer flask and add 2 ml of 85% phosphoric acid dropwise, with stirring. Attach a calcium chloride drying tube to the flask and heat it gently for about 15–20 min on a steam bath. Pour the reaction mixture onto about 20 g of chipped ice (ignoring any tarry material that may cling to the bottom of the flask) and, when the ice has melted, neutralize the solution by adding solid sodium bicarbonate. Add the bicarbonate with swirling until bubbles of carbon dioxide are no longer produced when small portions are added.[4] About 20–25 g of bicarbonate will be required. Approximately 20 g can be added as rapidly as foaming will allow; additional bicarbonate should be added more slowly as an excess of base is to be avoided. Cool the reaction mixture for about 30 min in an ice-water bath; then add a little solid sodium bicarbonate to the mixture. If reaction occurs, add more bicarbonate until no further bubbling occurs. (See Exercise 1 below.) Now collect the orange-brown solid that has precipitated by suction filtration, and wash it with water until the washings are pale orange. Dry the product in air.★

2. Purification by "wet-column" chromatography. Prepare a chromatographic column using a 50-ml buret filled about halfway with acid-washed alumina.[5] Dissolve the dry crude product from part 1 in a minimum amount of benzene and introduce it onto the alumina column with a capillary pipet. Elute with petroleum ether (bp 60–80°) first, which should bring a band of yellow-orange material down the column much more rapidly than a red-orange material. After the yellow-orange band has been removed from the column with petroleum ether, the red-orange material may be removed more rapidly with petroleum ether–ethyl ether (1:1 by volume).

Evaporate the solvents from the fractions of eluate containing the yellow-orange and red-orange materials and determine the melting point of the crystalline residues. The reported melting point of ferrocene is 173°, of acetylferrocene, 85°. Recrystallize the acetylferrocene from petroleum ether (bp 60–80°), and determine the yield.

Carry out a test with sodium hypoiodite on a small sample of the pure product to furnish evidence of its structure; see Chapter 16.2 for details of the iodoform test.

---

[4]The use of pH paper is a more exact way to determine when neutralization is complete. However, if the reaction mixture has darkened so much that the use of pH paper is impractical, the absence of bubbling upon addition of bicarbonate is a satisfactory criterion of neutralization.

[5]See Chapter 3.5 for details of preparing a chromatographic column. The column should be used the same day it is prepared, so it may be convenient to allow the crude acetylferrocene to dry in the desk until the next laboratory period, when the column will be prepared and used.

3. Purification by "dry-column" chromatography. Dissolve the brown solid product from part 1 in a small quantity of acetone (*ca.* 5 ml) and add this solution to 2 g of activity III alumina.[6] Evaporate the solvent by stirring the resulting slurry under a gentle stream of air. When completely dry, the remaining solid should be granular and free-flowing.

Prepare a chromatography column according to the directions given in Chapter 3.8, add the alumina on which the reaction mixture is adsorbed to the top of the column, and tap the column gently to provide a level top. Complete the preparation of the column by adding enough fresh alumina to provide a level layer 6 mm in depth on top of the sample.

Develop the column in the manner described in Chapter 3.8 with the exception that methylene chloride rather than a mixture of benzene and petroleum ether is used as the solvent. Upon completion of development of the column, immediately measure the R_f values for ferrocene, acetylferrocene, and, if you can detect it, diacetylferrocene.[7]

Isolate the purified products by scraping the individual bands of products from the column with the aid of a spatula and extracting the alumina containing each band three times with 20- to 30-ml portions of technical ether. After each extraction, decant the supernatant liquid into an appropriately sized flask; following the final extraction, evaporate the solvent to obtain the solid product. (*Caution:* No open flames should be used during the evaporation.) Determine the yield and melting point of the isolated acetylferrocene. The average yield of isolated acetylferrocene is about 30% of the theoretical.

B. Acetylation of ferrocene: thin-layer chromatography

1. Phosphoric acid catalyst. Dissolve 0.0054 mole of ferrocene in 20 ml of acetic anhydride in a 125-ml Erlenmeyer flask, fitted with a calcium chloride drying tube, by heating on a steam bath and swirling. Cool the solution to room temperature, and then add dropwise and with swirling 3 ml of 85% phosphoric acid. Reattach the drying tube, note the time, and allow the mixture to stand at room temperature. After two hours,[8] measure out a 5-ml aliquot of the reaction mixture and pour it onto about 20 g of chipped ice. After the ice has melted, neutralize the mixture with solid sodium bicarbonate until effervescence ceases. Extract the resulting mixture with ether, dry the ether extract with calcium chloride, and evaporate the ether.★ Dissolve the solid residue in a small amount of benzene and use the benzene solution to spot a thin-layer chromatographic (tlc) plate.[9] Dissolve some ferrocene in benzene and place a spot alongside the spot from the reaction mixture. Use a mixture of benzene and absolute ethanol (30 : 1 by volume) to develop the tlc plate, and determine the R_f values of ferrocene and acetylferrocene.

---

[6] See footnote 15 of Chapter 3.8 for a preparation of this grade of alumina.

[7] The characteristic colors of the individual components are: ferrocene—yellow; acetylferrocene—reddish-orange; diacetylferrocene—tan.

[8] This period may be used for the preparation of tlc plates (see reference B.6 in Chapter 3), or if these are furnished ready for use, part 2 of this experiment (using aluminum chloride catalyst) may be initiated.

[9] Commercial silica-gel plates (without fluorescent indicator) may be used conveniently. For details of tlc technique, see Chapter 3.7.

Allow the remainder of the reaction mixture to stand with the drying tube attached and, after periods of one or two days and/or a week, measure out 5-ml aliquots and pour them onto 20 g of chipped ice and repeat the work-up and tlc procedure described for the two-hour reaction mixture. Compare the thin-layer chromatograms obtained from the aliquots of the reaction mixture after longer reaction times. Determine the R_f for 1,1′-diacetylferrocene, which should be found as a product from the longer reaction periods.

2. Aluminum chloride catalyst.   Dissolve 1.0 g of ferrocene in *ca.* 15 ml of dichloromethane. Calculate amounts of acetyl chloride and powdered *anhydrous* aluminum chloride ($AlCl_3$) which will provide 1.5 moles of each of these compounds per mole of ferrocene. Weigh out the correct amounts of these reagents within ± 0.02 g and dissolve both immediately in 25 ml of dichloromethane. *Caution:* Neither of these reagents should be exposed to moist air any longer than necessary. Do not allow either of them to come in contact with the skin. The vapor of acetyl chloride is irritating and harmful to breathe.

Add the solution of ferrocene to the solution of acetyl chloride and aluminum chloride, swirl to mix, note the time, and allow the mixture to stand in a flask protected by a calcium chloride drying tube. After intervals of 1, 15, 30, and 60 min, measure out a 5-ml aliquot of the reaction mixture and pour it onto about 15 g of chipped ice. Neutralize the aliquot with solid sodium bicarbonate until effervescence ceases, add about 10 ml of dichloromethane, and separate the layers. Wash the organic solution with distilled water and dry it over calcium chloride.★ Evaporate the solvent from the extract and dissolve the solid residue in a small amount of benzene.★ Carry out tlc analysis of the benzene solution as described in part 1 above. Determine R_f values for ferrocene, acetylferrocene, and 1,1′-diacetylferrocene.

As an interesting extension of this experiment, some of the class members may use different molar ratios of ferrocene, acetyl chloride, and aluminum chloride, and the effect on the relative rates of production of acetylferrocene and 1,1′-diacetylferrocene may be determined. Other suggested molar ratios are ferrocene:acetyl chloride:aluminum chloride = 1:1:1 and 1:2:2.

EXERCISES

1. Why does more sodium bicarbonate react after the reaction mixture from ferrocene, acetic anhydride, and phosphoric acid has been neutralized once and then allowed to stand 30 min? (*Hint:* It is not necessary to wait and neutralize a second time when acetyl chloride is used as the acetylating agent.)

2. Judging from the procedure for the column chromatography of crude acetylferrocene, (a) which is more strongly adsorbed on alumina, ferrocene or acetylferrocene? (b) In which solvent, petroleum ether or ethyl ether, is acetylferrocene more soluble? Why?

3. Ferrocene cannot be nitrated successfully by the usual mixed nitric acid-sulfuric acid medium. Why not?

4. If ferrocene were locked in the conformation shown in **23**, how many isomers of 1,1′-diacetylferrocene would there be? It may be helpful to draw the structure as it would appear from a position above the molecule as shown in **23**, *i.e.*, in line with the axis of the bonds between the rings and the iron.

SPECTRA OF STARTING MATERIAL AND PRODUCT

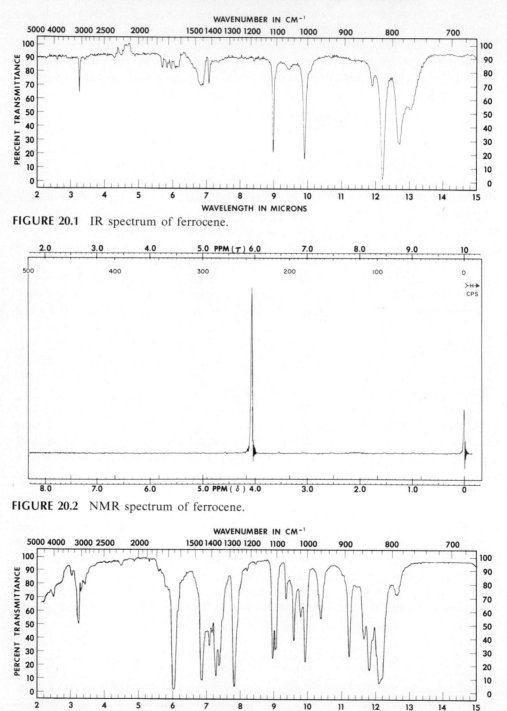

FIGURE 20.1 IR spectrum of ferrocene.

FIGURE 20.2 NMR spectrum of ferrocene.

FIGURE 20.3 IR spectrum of acetylferrocene.

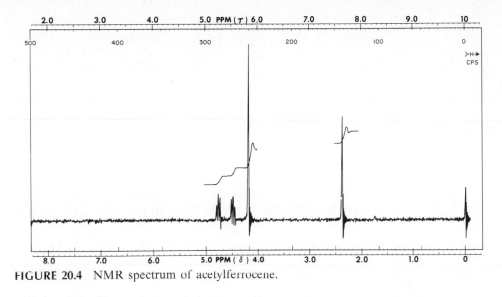

FIGURE 20.4 NMR spectrum of acetylferrocene.

20.3 Tropylium Iodide

In contrast to the accidental discovery of ferrocene, the successful synthesis of salts containing the cycloheptatrienyl cation moiety in 1954 was the result of intense competitive research efforts. As mentioned in Section 20.1, Hückel predicted in 1931 that the cycloheptatrienyl, or tropylium, cation should exhibit unusual stability indicative of aromaticity. Many chemists set out on syntheses designed to produce the cation with the intention of proving Hückel's theory to be either correct or incorrect. Ironically, the isolation of a salt containing the cation was delayed by initial failure to anticipate its water-solubility, which might have been expected if Hückel's prediction were correct! However, since no other carbonium ions had previously been found to be stable in water solution,[10] it is not surprising that this possibility was at first overlooked.

In this experiment, a sequence of reactions is used which demonstrates the fact that the tropylium cation is much more stable than the triphenylmethyl cation (refer to Section 20.1 for the comparison in terms of $pK_R\oplus$). Triphenylmethyl fluoborate (**27**) is first prepared from triphenylmethanol [**26**, equation (7)], and then allowed to react with 1,3,5-cycloheptatriene [**28**, equation (8)]. Hydride ion (H:$\ominus$) is trans-

$$(C_6H_5)_3COH + HBF_4 \xrightarrow{(CH_3CO)_2O} (C_6H_5)_3C\oplus BF_4\ominus + H_2O \qquad (7)$$
$$\textbf{26}\textbf{27}$$

$$(C_6H_5)_3C\oplus BF_4\ominus + \begin{array}{c}\text{H}\\\text{H}\end{array} \longrightarrow (C_6H_5)_3C\text{—H} + \text{—H } BF_4\ominus \qquad (8)$$

$$\textbf{27}\textbf{28}\textbf{29}\textbf{30}$$

---

[10] Although the tripropylcyclopropenyl cation is more stable than the tropylium cation (see Section 20.1), it was not prepared until 1962.

$$\text{30} \quad \text{BF}_4^{\ominus} + \text{Na}^{\oplus}\text{I}^{\ominus} \xrightarrow{\text{H}_2\text{O}} \text{31} \quad \text{I}^{\ominus} + \text{Na}^{\oplus}\text{BF}_4^{\ominus} \tag{9}$$

ferred from the triene to the triphenylmethyl cation to produce the more stable tropylium cation, which precipitates from ether solution in the form of the fluoborate, **30**. Tropylium iodide **(31)** is much less soluble in water than the bromide (which was the first salt isolated), and can be precipitated by metathetical reaction of tropylium fluoborate with sodium iodide [equation (9)].

EXPERIMENTAL PROCEDURE

In a 150-ml beaker, place 0.10 mole of acetic anhydride and add 3.0 g of 48% fluoboric acid dropwise with stirring and cooling. *Caution: Fluoboric acid is a highly corrosive liquid; do not allow it to come in contact with the skin.* Now add 0.015 mole of triphenylmethanol to this mixture in small portions with stirring; a yellow precipitate should be present after the addition is complete.[11] Pour the mixture into 200 ml of anhydrous ether in a 400-ml beaker, stir for a few minutes, and then permit the precipitate of triphenylmethyl fluoborate to settle. Pour a few milliliters of ether carefully down the side of the beaker; if any turbidity appears, the precipitation was incomplete and more ether should be added.

The next step is to collect the triphenylmethyl fluoborate on a filter. Since it is sensitive to moisture in the air, be prepared to dissolve it *immediately* in 10 ml of *anhydrous* nitromethane by rapidly carrying out the operations described below. Use a 4-cm Büchner funnel and a well-fitting paper to collect the triphenylmethyl fluoborate by suction filtration. As soon as the filtrate has run through the funnel, release the vacuum and insert the funnel stem into a clean, dry 25-ml Erlenmeyer flask. Pour about 7 or 8 ml of anhydrous nitromethane onto the triphenylmethyl fluoborate on the filter paper and stir with a glass rod until solution is complete. Lift the paper slightly with the stirring rod so that the solution runs through the funnel and into the flask below. Add another 2 or 3 ml of anhydrous nitromethane to wash the filter paper and funnel, and allow the wash liquid to drain into the flask with the main part of the solution.

To the nitromethane solution of triphenylmethyl fluoborate, add 0.010 mole of 1,3,5-cycloheptatriene. After stirring for two or three minutes, pour the solution, with stirring, into 200 ml of anhydrous ether. Allow the white precipitate of tropylium fluoborate to settle and test for completeness of precipitation by adding a little ether as before.

Collect the precipitate by suction filtration, allowing time for removal of all of the ether. Transfer the precipitate to a 50-ml Erlenmeyer flask and then dissolve

---

[11] It may appear that the triphenylmethanol does not dissolve or react; by careful observation it may be noticed that the white crystals of triphenylmethanol do dissolve, but they are replaced immediately by the yellow crystals of triphenylmethyl fluoborate.

it in a *minimum amount* of hot water. To the warm solution, add dropwise about 5 ml of a saturated solution of sodium or potassium iodide; an immediate precipitation of red crystals of tropylium iodide should occur. If crystallization does not occur immediately, stopper the flask and cool it in an ice-water bath. Collect the crystals on a filter and wash them with a little *ice-cold* water. Air dry these crystals and determine the yield, based on 1,3,5-cycloheptatriene.

EXERCISES

1. What is the function of the acetic anhydride that is added to the 48% fluoboric acid? Why does the mixture require cooling?
2. What would happen to triphenylmethyl fluoborate if it were allowed to stand in moist air for several hours? What *weight* of water is required to react with 0.015 mole of triphenylmethyl fluoborate?
3. Apparently $C_{10}H_{10}$ ([10]annulene) and $C_{14}H_{14}$ ([14]annulene) are much less stable than [18]annulene, even though these compounds fit Hückel's Rule for $n = 2$ and $n = 3$. Suggest an explanation, considering possible geometries of [10]- and [14]annulene.

SPECTRA OF STARTING MATERIALS

For the nmr spectrum of triphenylmethanol, see Figure 14.6.

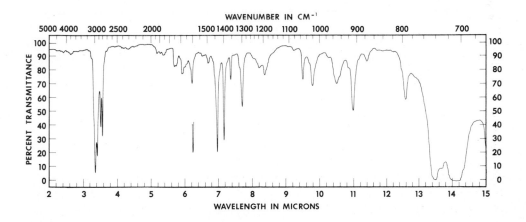

FIGURE 20.5 IR spectrum of 1,3,5-cycloheptatriene.

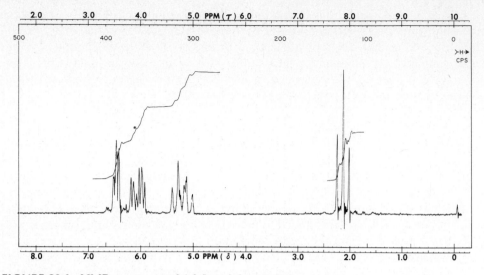

FIGURE 20.6 NMR spectrum of 1,3,5-cycloheptatriene.

optical activity

Resolution of Racemic α-Phenylethylamine

21.1 Polarimetry

As we noted in Chapter 7, dissymmetric molecules are optically active; that is, they have a rotational effect on the plane of polarized light[1] as it passes through the substance. The phenomenon may be explained as follows. Consider that a dissymmetric molecule has a dissymmetric distribution of electron density. As the polarized light passes through the molecule, its electric component will necessarily be affected in a dissymmetric manner by the electrons of the molecule. The result is that the plane of the polarized light is twisted about the axis of propagation. Any substance having this effect on polarized light is termed optically active. A symmetric molecule (one that is superimposable with its mirror image) will be optically inactive, since the light, as it passes through the substance, will on the average[2] encounter a symmetric distribution of electrons and will thus be unaffected.

In a symmetric environment, enantiomers are identical with regard to all physical constants and chemical properties, with one exception: although each member of the enantiomeric pair rotates the plane of polarized light by angles of the same magnitude, they do so in opposite directions. By convention, when looking through the sample *into* the beam of light, a rotation which is observed to be clockwise, or to the right, is termed a *positive* rotation, and a counterclockwise rotation is termed a *negative* rotation. Should the sample under consideration be a mixture of exactly equal amounts of the

---

[1]Electromagnetic radiation (including light) is composed of propagated electric and magnetic fields. From a wave-mechanical point of view, a ray of light is composed of two plane waves, one electric and one magnetic, oriented at right angles to one another. Considering for simplicity only the electric waves, ordinary light is completely unordered, with these waves oriented at all possible angles in the plane perpendicular to the direction of propagation. In plane-polarized light, each ray has the plane of its electric wave oriented parallel to the corresponding plane of all other rays.

[2]It must be realized that such an experiment is macroscopic rather than microscopic, for even with a very small sample of the substance under consideration, a very large number of molecules are present. Any measurement must necessarily be an average observation of the effect on all rays within the polarized light by all molecules through which they pass.

two enantiomers, the rotation of one of the components will be exactly offset by the equal but opposite rotation of the other. The net rotation will be observed as zero. Such a mixture is called a *racemic modification*. In order that there be any net observed rotation in a mixture of enantiomers, one member of the pair must be present in amounts greater than the other.

The observed angle of rotation depends on a number of factors: the nature of the compound, its concentration (in solution) or density (neat liquid), the length of the sample through which the light must pass (the path length), temperature, solvent, and wavelength of light. The concentration, or density, and path length of the sample are important considerations because they determine the average number of optically active molecules through which a beam of light will pass. The magnitude of rotation is the result of a cumulative effect, and if the light, on the average, passes through a greater number of active molecules, the observed magnitude of rotation will be greater. The dependence is essentially linear with respect to both concentration and path length.

It is certainly desirable to present the optical rotation as a physical constant of the active compound. Therefore, it is common practice to report the *specific rotation* rather than the observed angle of rotation. The specific rotation is calculated by correcting the observed rotation to *unit* concentration and path length with the following equation:

$$[\alpha] = \frac{\alpha}{l \times c} \text{ or } \frac{\alpha}{l \times d} \tag{1}$$

$[\alpha]$ = specific rotation (degrees)
α = observed rotation (degrees)
l = path length (decimeters)
c = concentration (g/ml of solvent, solution)
d = density (g/ml, neat)

In order to specify the other variables upon which the rotation depends, the temperature and wavelength (nm) employed are presented as superscript and subscript, respectively, on the symbol for specific rotation, and the solvent used is denoted in parentheses following the numerical value and sign of the specific rotation, *e.g.*, $[\alpha]_{490}^{25°} + 23.4°$ (CH_3OH). Usually, a sodium lamp emitting light at 589 nm, the sodium D line, is used as a light source, and in this case it is common to use D in place of the numerical value of the wavelength in the expression, *e.g.*, $[\alpha]_D^{25°} - 15.2°$ (H_2O).

Various types of commercial polarimetric apparatus are available. Some, such as the Rudolph polarimeter, are manually operated and require direct readings by the operator, whereas other, newer varieties are automatic and utilize photoelectric cells for measurements of greater accuracy and precision. The fundamental attributes of these instruments are similar however, and may be amply understood within the context of the following discussion.

A very simply constructed student polarimeter has been described by Shaw.[3,4] The

[3] William H. R. Shaw, "A Mailing-Tube Polarimeter," *Journal of Chemical Education,* **32,** 10 (1955).
[4] The Edmund Scientific Co., Barrington, New Jersey, is marketing a student polarimeter modified from Shaw's original design according to our specifications. Its sample tube has a path length of approximately 2 dm and a capacity of about 35 ml.

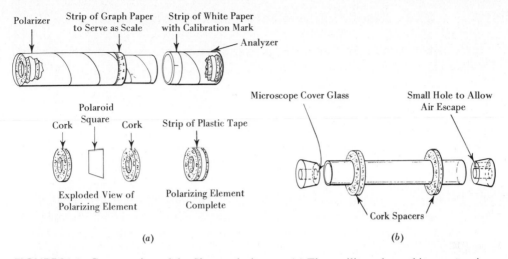

FIGURE 21.1 Construction of the Shaw polarimeter. (a) The mailing tube and its accessories; (b) the sample tube.

polarimeter is constructed from a mailing tube in each end of which a piece of Polaroid film has been permanently affixed [Figure 21.1(a)]. The tube is such that the two ends may be rotated relative to one another. The sample is held within a glass tube placed within the mailing tube [Figure 21.1(b)]. Light passing into the polarimeter at one end will be polarized by the Polaroid film at that end (the *polarizer*); as the polarized light passes on through the sample tube containing a solution of an optically active substance, the plane of the polarized light will be rotated through an angle. The angle of rotation is determined by rotating the section of the tube through which the observer is viewing until the light intensity observed is the same as that observed when the sample tube contains only pure solvent. The number of degrees through which this *analyzer* section of the tube must be turned is then equivalent to the optical rotation, α, of the sample (see Figure 21.2). This can be converted to the specific rotation by applying equation (1).

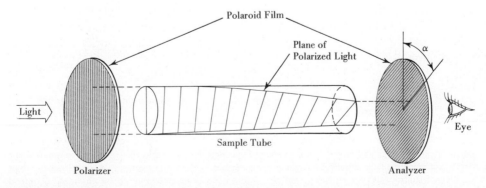

FIGURE 21.2 Schematic illustration of optical rotation ($\alpha = ca.\ -45°$).

21.2   Resolution of Racemic α-Phenylethylamine

α-Phenylethylamine (3) may be prepared from acetophenone (1) and ammonium formate (2).[5]

$$C_6H_5-\overset{\overset{\text{O}}{\|}}{C}-CH_3 + H-\overset{\overset{\text{O}}{\|}}{C}-O^{\ominus}NH_4^{\oplus} \longrightarrow C_6H_5-\underset{\underset{\text{H}}{|}}{\overset{\overset{\text{NH}_2}{|}}{C}^*}-CH_3 + H_2O + CO_2$$

$$\quad\quad\quad 1 \quad\quad\quad\quad\quad\quad\quad 2 \quad\quad\quad\quad\quad\quad\quad\quad 3$$

The product as obtained is not optically active, despite the fact that it contains an asymmetric carbon atom (marked in formula **3** with an asterisk), because the enantiomers are produced in exactly equal amounts; *i.e.*, a racemic modification is formed.

Enantiomers cannot be separated from one another by the standard methods of crystallization, distillation, or chromatography. This is because they have identical physical properties, with the exception of their rotation of polarized light in opposite directions, as noted in the preceding section. However, *diastereomeric isomers (diastereomers)* do differ in physical properties such as solubility, boiling point, and chromatographic adsorption characteristics. The most generally useful procedure for *resolving a racemic modification* of enantiomers involves converting the enantiomers to diastereomers, separating the diastereomers by one of the standard experimental procedures, and then regenerating the enantiomers from the separated diastereomers.

This procedure is illustrated for resolution of racemic α-phenylethylamine in Chart 21.1. Using optically active (+)-tartaric acid (one pure enantiomer), the racemic amine is converted to a mixture of diastereomeric salts, which have different solubilities in methanol, and hence can be separated by fractional crystallization. To obtain *both* enantiomeric isomers in a state of high optical purity may require a large number of careful and tedious crystallization steps, but it is usually possible to obtain the enantiomer which gives the less-soluble salt in reasonable optical purity by only one or two crystallization steps.[6]

By measuring the optical rotation of the α-phenylethylamine recovered from the less-soluble salt in this experiment it will be possible to determine not only whether it is the (+) or (−) enantiomer, but also the extent of optical purity achieved in the resolution.

[5] A. W. Ingersoll in *Organic Syntheses, Collective Vol. II*, A. H. Blatt, Editor, John Wiley & Sons, Inc., New York, 1943, p. 503.

[6] In some cases it is also possible to obtain the other enantiomer without resorting to many repeated crystallizations of the more-soluble diastereomer. For example, in the research on which this experiment is based [W. Theilacker and H.-G. Winkler, *Chemische Berichte*, **87**, 690 (1954)], it was found that the *hydrogen sulfate salt* of the enantiomer which gave the more-soluble tartrate salt was *less soluble* than the hydrogen sulfate salt of the racemic amine, so that both enantiomers could be obtained in optical purity quite conveniently.

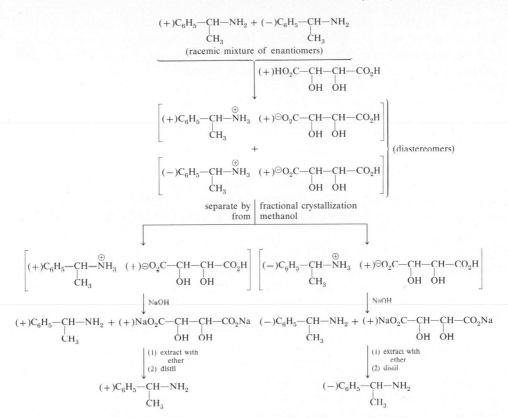

CHART 21.1   Resolution of (±)-α-phenylethylamine by means of (+)-tartaric acid.

EXPERIMENTAL PROCEDURE

Dissolve an accurately weighed sample of approximately 5 g of racemic α-phenyl-ethylamine in 35 ml of methanol and determine its specific rotation, [α], using a polarimeter and equation (1). Detailed instructions for the use of a polarimeter are given in Chapter 22.2.

In a 1-liter Erlenmeyer flask, place 0.208 mole of (+)-tartaric acid and 415 ml of methanol and heat to boiling. To the hot solution, add cautiously (to avoid foaming) with stirring the 35 ml of solution recovered from the polarimeter and enough additional racemic α-phenylethylamine to make a total of 0.206 mole. Allow the solution to cool slowly to room temperature and to stand undisturbed for 24 hr, or until the next laboratory period.★ The amine hydrogen tartrate should separate in the form of white prismatic crystals. (If the salt separates in the form of needle-like crystals, the mixture should be reheated until *all* of the crystals have dissolved and then allowed to cool slowly. If any prismatic crystals of the salt are available, they should be used to seed the solution.)

Collect the crystals of the amine hydrogen tartrate (17–19 g) on a filter and wash them with a small volume of methanol.[7] Dissolve the crystals in about four times their weight of water and add 15 ml of 14 N sodium hydroxide solution. Extract the resulting mixture with four 75-ml portions of ethyl ether,[8] wash the combined ether extracts with 50 ml of saturated sodium chloride solution, and dry the ether solution over anhydrous magnesium sulfate.★

Distil the ether solution from a steam bath using an unpacked Hempel fractionating column and then arrange to distil the residue (α-phenylethylamine) under aspirator pressure, using no fractionating column. It will be necessary to use a burner or an electric heater because the amine has a bp of 94–95° (28 mm). The yield of optically active α-phenylethylamine should be 5–6 g.

Weigh the distilled product accurately, dissolve it in *ca.* 35 ml of methanol (measure the volume accurately), and determine its specific rotation, $[\alpha]$, using a polarimeter and equation (1). The reported specific rotation of optically pure α-phenylethylamine, $[\alpha]_D^{25}$, is 40.1° (neat).

21.3   Optional Projects

A. PREPARATION OF A CRYSTALLINE DERIVATIVE
OF RACEMIC α-PHENYLETHYLAMINE
AND OF OPTICALLY ACTIVE α-PHENYLETHYLAMINE

In order to demonstrate that resolution of a racemic modification does not change the chemical properties of the compound (in a symmetric environment) or the physical properties, other than the effect on polarized light, the crystalline benzoyl derivative, the benzamide **4**, may be prepared from a sample of racemic α-phenylethylamine and from a sample of amine which has been resolved. The reaction is

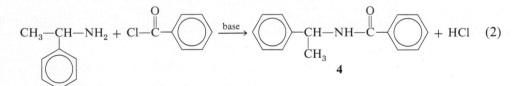

The experimental procedure for the preparation of benzamides from amines is described in Chapter 25.2. The melting point of this benzamide has been variously reported by different workers to be between 120° and 125°. It may be recrystallized from aqueous ethanol or from mixtures of benzene and heptane. Compare the melting points of derivatives prepared from racemic and optically active samples of α-phenylethylamine.

In order to determine if the formation of the benzamide takes place with retention of optical activity, the derivative produced from the optically active sample of α-phenylethylamine should be examined for optical rotation. This may be done by

[7] Do not discard the filtrate before you ask your instructor if you are to turn it in for recovery of the other enantiomer of α-phenylethylamine.

[8] Do not discard the aqueous solution; pour it into a bottle provided for recovery of tartaric acid.

dissolving an accurately weighed sample of approximately 0.9 g of the benzamide in 38 ml of benzene and measuring the rotation in the polarimeter, calculating $[\alpha]$ by using equation (1). The $[\alpha]_D^{27}$ has been reported to be $+39.2°$ (benzene).

Determine not only the extent of optical purity of the benzamide, but also whether the sign of rotation is the same as that of the resolved α-phenylethylamine.

B. PREPARATION OF DI-α-PHENYLETHYLAMINE (α,α'-DIMETHYLDIBENZYLAMINE)

The reaction of optically active α-phenylethylamine with acetophenone to produce the imine **5** [equation (3)] followed by catalytic hydrogenation might be expected

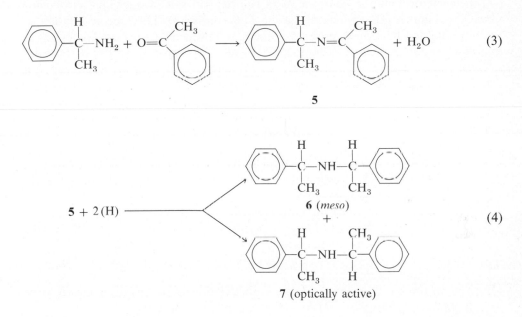

to produce the diastereomers of di-α-phenylethylamine, **6** and **7**. These should be capable of separation by ordinary means, since they should have significantly different physical properties. However, when these reactions were carried out and the di-α-phenylethylamine was converted to its hydrochloric acid salt, an attempted separation of the diastereomeric isomers by fractional crystallization failed, and the high specific rotation of the salt indicated that the hydrogenation took place stereoselectively, so that the optically active isomer, **7,** was almost exclusively the product formed.[9] When the reduction was carried out with lithium aluminum hydride instead of by catalytic hydrogenation, a lower specific rotation was found, but the optically active isomer was the major product.

The reactions of equations (3) and (4) are analogous to those of equations (4) to (6) in Chapter 16. However, the reaction of acetophenone, a ketone, is slower

[9] C. G. Overberger, N. P. Marullo, and R. G. Hiskey, *Journal of the American Chemical Society*, **83,** 1374 (1961).

than the reactions of the aldehydes in Chapter 16, and it may not be possible to complete the reaction in one laboratory period. The reduction of **5** with lithium aluminum hydride was also carried out at a higher temperature and for a longer time than the reactions of sodium borohydride. *This reagent* (LiAlH$_4$) *is pyrophoric and must be handled with care.*

It is quite possible that NaBH$_4$ may be used in place of LiAlH$_4$, and that the conditions reported for the formation of the imine (**5**) are more severe than necessary. A trial with the procedures of Chapter 16 would be worth a try. In any case, with proper supervision, the conversion of optically active α-phenylethylamine to di-α-phenylethylamine and the determination of the specific rotation of the secondary amine makes an interesting project.

The specific rotation of the pure hydrochloric acid salt of optically active di-α-phenylethylamine, $[\alpha]_D^{20}$, is reported to be 71.8° (ethanol). The specific rotation of the salt obtained by reduction of the imine **5**, followed by a work-up analogous to that described in the article of footnote 9, should be compared with 71.8° in order to calculate the degree of optical purity of **7** obtained, and thus the proportion of **6** and **7** produced in the reduction of **5**.

EXERCISES

1. Report the specific rotation of your resolved product as ($+$) or ($-$). Suppose the observed rotation was found to be 180°. How could you determine whether the rotation was ($+$) or ($-$)?
2. How could you increase the optical purity of your product?
3. Describe clearly the point in the experimental procedure at which the major part of the other enantiomer was removed from your product.
4. How could the ($+$)-tartaric acid be recovered so as to be used over again to resolve more racemic amine?
5. The absolute configuration of ($+$)-α-phenylethylamine has been shown to be R. Make a perspective drawing of this configuration.
6. Suppose you had prepared a racemic organic acid and wished to obtain an optically active form of it. How could you do this?
7. Suggest a possible procedure for resolving a racemic alcohol.

REFERENCE

A. Ault, *Journal of Chemical Education,* **42,** 269 (1965); *Organic Syntheses,* **49,** 93 (1969).

carbohydrates

22.1   Introduction

Carbohydrates, also referred to as "sugars" or saccharides (Sanskrit, sárkarā, grit, gravel, sugar), are an extremely important class of naturally occurring polyhydroxy aldehydes (aldoses) and ketones (ketoses), or substances that yield aldoses or ketoses upon hydrolysis. Many of the simplest carbohydrates, *monosaccharides,* display the general formula $C_nH_{2n}O_n$ [or $C_n(H_2O)_n$, a historical and misleading representation of the general formula, because it is hardly true that these substances are hydrates of carbon]. However, other monosaccharidic sugars, *e.g.,* the deoxy-sugars such as deoxyribose, and compounds containing heteroatoms such as nitrogen, sulfur, or phosphorus do not adhere to this formula.

More complex carbohydrates yield monosaccharides upon hydrolysis. Thus, the disaccharide sucrose (table sugar) hydrolyzes to provide one molecule of D-glucose and one of D-fructose.[1] Polysaccharides, depending on their constitution, may be degraded to a mixture of monosaccharides, or to only a single product; *e.g.,* starch is a mixture of the polymers amylose and amylopectin, each constituted only of D-glucose units. With particular reference to the structural and storage polysaccharides (*vide infra*), it should be noted that polysaccharides are by far the most ubiquitous bioorganic compounds on earth. Although the experiments in this chapter of necessity utilize the very simplest of carbohydrates, the impression should be avoided that these may be the most abundant, or even the most significant of the carbohydrates.

Carbohydrates provide the ultimate energy source in the food chain. D-glucose is synthesized in green leaves from carbon dioxide and water by the process of photosynthesis and the action of chlorophyll. This thermodynamically unfavorable process is made possible by the energy of sunlight. D-glucose is combined in the

[1] The D is a symbol used to designate the configuration of these sugars relative to that of D-glyceraldehyde, the standard of configuration for carbohydrates. See any modern organic textbook for further explanation.

plant to provide starch and cellulose. Starch, and in some animals cellulose, is broken down again into D-glucose after ingestion. One function of the liver is to recombine D-glucose from the blood stream to form glycogen, which serves to store energy within the body. Glycogen as needed is reconverted into D-glucose, which is metabolized ultimately to carbon dioxide and water, providing the animal with the energy originally stored during photosynthesis. Those other monosaccharides that are found, usually in combined form, in living systems are thought to be produced from D-glucose by the actions of various enzymes.

Carbohydrates are utilized within a living organism in many ways in addition to those of storage and transference of energy. In plants, for example, the polysaccharide cellulose is an important structural component providing rigidity and form. In animals, polysaccharides in combination with protein are important constituents of connective and other tissues. For example, the chondroitin sulfates are found in mammalian cartilages, tendons, heart valves, cornea, etc. A segment of the structure of chondroitin A, one of the three chondroitin sulfates that have been isolated, is shown below. Protein linkages are found at certain points along the chain of the polysaccharide.

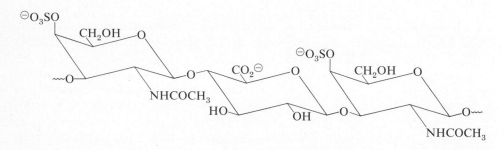

Additionally, carbohydrates serve as precursors in the biochemical formation of several other important bioorganic compounds. For example, they are involved in the biosynthesis of certain α-amino acids, a possible pathway for which is shown in broad generality in equation (1), where [N] represents some nitrogen containing

$$\text{Carbohydrate} \longrightarrow \underset{O}{R-\overset{\parallel}{C}-CO_2H} \xrightarrow{[N]} \underset{NH}{R-\overset{\parallel}{C}-CO_2H} \xrightarrow{[H]} \underset{NH_2}{R-\overset{\mid}{CH}-CO_2H} \qquad (1)$$

compound and [H] represents a reducing agent. Although this equation is vastly oversimplified in that a large number of steps intervene, particularly in the formation of the α-keto acid, it does serve to exemplify the precursoral significance of carbohydrates to a living organism. Many other examples of the use of carbohydrates as biochemical "building blocks" may be cited; these include the occurrence of ribose and deoxyribose as structural constituents of nucleosides and deoxynucleosides necessary in the formation of RNA and DNA and the incorporation of ribitol in the biosynthesis of riboflavin, one of the B vitamins.

The discussion in the preceding paragraphs, although it has been cursory, hopefully provides an appreciation for the multiple functionality of carbohydrates in biochemistry and suggests that the proper and balanced "operation" of an organism as a chemical system is dependent on this class of compounds in many ways. In the sections that follow, some of the chemical and physical properties of carbohydrates will be investigated, and certain of the techniques used in the isolation and proof of structure of this interesting group of substances will be described.

22.2   Mutarotation of Glucose

Figure 22.1 shows the open-chain structures, in Fischer projection form, of several of the more common monosaccharides.[2,3] Only a few of the monosaccharides shown are known to be naturally occurring; the remainder have been synthesized, however.

ALDOTRIOSE

```
      CHO
  H —— OH
     CH₂OH
```
D-(+)-glyceraldehyde

ALDOTETROSES

```
      CHO              CHO
  H —— OH         HO —— H
  H —— OH          H —— OH
     CH₂OH            CH₂OH
```
D-(−)-erythrose   D-(−)-threose

ALDOPENTOSES

```
      CHO            CHO            CHO            CHO
  H —— OH       HO —— H        H —— OH       HO —— H
  H —— OH        H —— OH       HO —— H       HO —— H
  H —— OH        H —— OH        H —— OH       H —— OH
     CH₂OH          CH₂OH          CH₂OH          CH₂OH
```
D-(−)-ribose   D-(−)-arabinose   D-(+)-xylose   D-(−)-lyxose

ALDOHEXOSES

```
    CHO       CHO       CHO       CHO       CHO       CHO       CHO       CHO
 H—OH    HO—H    H—OH    HO—H    H—OH    HO—H    H—OH    HO—H
 H—OH     H—OH   HO—H    HO—H    H—OH     H—OH   HO—H    HO—H
 H—OH     H—OH    H—OH    H—OH   HO—H    HO—H    HO—H    HO—H
 H—OH     H—OH    H—OH    H—OH    H—OH    H—OH    H—OH    H—OH
 CH₂OH   CH₂OH   CH₂OH   CH₂OH   CH₂OH   CH₂OH   CH₂OH   CH₂OH
```
D-(+)-allose   D-(+)-altrose   D-(+)-glucose   D-(+)-mannose   D-(+)-gulose   D-(−)-idose   D-(+)-galactose   D-(+)-talose

KETOHEXOSES

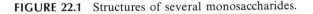

```
     CH₂OH           CH₂OH
       |               |
      C=O             C=O
  HO —— H         H —— OH
  H —— OH        HO —— H
  H —— OH        H —— OH
     CH₂OH           CH₂OH
```
D-(−)-fructose   D-(+)-sorbose

FIGURE 22.1 Structures of several monosaccharides.

[2] Only the D isomers are shown. It may be noted that the sign of optical rotation, shown as (+) or (−), is *independent* of the relative configuration (D or L).

[3] Although the monosaccharides are shown in their *open-chain* form, those containing a chain of at least four carbon atoms exist predominately in *cyclic* hemiacetal or hemiketal forms (*vide infra*).

By examination of the structures, it can be seen that monosaccharides generally contain more than one asymmetric carbon atom and are therefore subject to extensive stereoisomerism. An aldohexose, for example, has four asymmetric atoms and consequently may exist in sixteen different stereoisomeric forms ($2^4 = 16$; it may be valuable to review that part of Chapter 7 concerning stereoisomerism). Within this group of sixteen isomers there are eight enantiomeric pairs.

As is shown in the figure, D-glucose is one specific stereoisomer of the sixteen aldohexoses. Consideration of the functional groups present in D-glucose, and recollection that aldehydes react with alcohols to produce hemiacetals and acetals, suggest that the open-chain form of this sugar might undergo an intramolecular reaction to produce a hemiacetal. Note that the conversion of an open-chain structure with its four asymmetric carbon atoms to a cyclic hemiacetal by, for example, reaction between the aldehyde group at C-1 and the C-5 hydroxyl group creates an additional asymmetric center at C-1. Because there are two possible configurations about this new center, there must be possible two isomeric hemiacetals: **1**, the α-form, and **2**, the β-form (see Figure 22.2). Saccharides such as **1** and **2** that differ only in

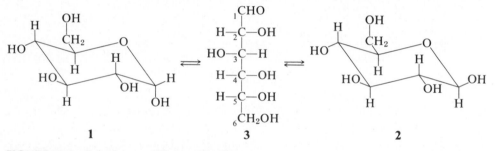

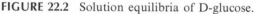

FIGURE 22.2 Solution equilibria of D-glucose.

configuration at the carbon atom involved in the cyclization (C-1 for aldoses and C-2 for ketoses) are called *anomers*.

D-glucose does, in fact, exist in the two *diastereomeric* cyclic forms, **(1)** and **(2)**, the two isomers being in equilibrium with one another in aqueous solution by way of the intermediacy of the open-chain structure **3** (Figure 22.2). The cyclic forms are the major components of the equilibrium mixture of **1**, **2**, and **3**. Because compounds **1** and **2** are diastereomers rather than enantiomers, they have different physical properties and can be separated from one another by rather specific techniques of crystallization, as is done in the experiment described below.

The α-form has a specific rotation of $[\alpha]_D^{25°} + 112°$. When it is placed in water solution the specific rotation gradually changes until it reaches a constant $[\alpha]_D^{25°} + 52.7°$. The β-form, in water solution, undergoes a change of specific rotation from $[\alpha]_D^{25°} + 19°$ to $[\alpha]_D^{25°} + 52.7°$. Quite obviously, $+52.7°$ represents the specific rotation of an equilibrium mixture of **1** and **2**. The equilibration of diastereomeric isomers which gives rise to an equilibrium rotation is called *mutarotation*.

By following the rate of change of optical rotation of α- or β-D-glucose, information concerning the *rate* of mutarotation can be obtained. For a reaction of the type shown in equation (2),

$$A \underset{k_2}{\overset{k_1}{\rightleftarrows}} B \tag{2}$$

the following expression can be written:[3a]

$$2.303 \log \frac{A_e - A_o}{A_e - A_t} = (k_1 + k_2)t \tag{3}$$

where A_e is the concentration of A at equilibrium, A_o is the concentration of A at time $t = 0$, and A_t is the concentration at time t during the equilibration process. Note that the term $(A_e - A_o)$ represents the total extent of reaction from the beginning to the final establishment of equilibrium, that is, the total change in concentration of A, whereas the term $(A_e - A_t)$ represents the extent of reaction remaining at time t. We may substitute for these terms any other terms that also represent the ratio of the total extent of reaction and the extent remaining at time t. Any terms utilized must vary linearly during the course of reaction, as does concentration. Discussion in Chapter 21 indicated that optical rotation is linearly related to concentration; thus, we may substitute the optical rotation of the solution measured at the beginning, end, and at time t for the concentrations in equation (3) to give equation (4):

$$2.303 \log \frac{\alpha_e - \alpha_o}{\alpha_e - \alpha_t} = (k_1 + k_2)t \tag{4}$$

Thus we see that measurement of the optical rotation of the initial solution of α-D-glucose, of the rotation of the same solution at equilibrium, and of the rotations at a series of times in between gives the data needed for calculation of $(k_1 + k_2)$. This is accomplished by plotting the left-hand term against t; the slope of the straight line obtained is $(k_1 + k_2)$. It is not possible to obtain k_1 and k_2 individually unless the value of the equilibrium constant is also known (see Experimental Procedure).

EXPERIMENTAL PROCEDURE

A. PREPARATION OF α-D-GLUCOSE
α-D-glucose is the form obtained when D-glucose is crystallized slowly at room temperature from a mixture of acetic acid and water. Dissolve 50 g of anhydrous D-glucose (dextrose) in 25 ml of distilled water contained in a 500-ml Erlenmeyer flask. Heat the mixture on a steam bath to effect complete solution. The solution will be quite viscous and should be continuously stirred during this step. When solution is complete (*no* crystals remaining), remove the syrup from the steam bath, add 100 ml of *glacial* acetic acid which has been previously cooled in an ice bath to 17°,[4] and swirl the flask until the mixture becomes homogeneous. Stopper the

[3a] The interested student can find the derivation of this expression in A. A. Frost and R. G. Pearson, *Kinetics and Mechanism,* 2nd Ed., John Wiley & Sons, Inc., New York, 1961, p. 186.

[4] Cooling below 17° may result in crystallization of acetic acid, whose freezing point is 16°.

flask with a cork and allow it to remain undisturbed in the desk until the next laboratory period.

Collect the α-D-glucose by suction filtration, wash the crystals thoroughly with 50-60 ml of 95% ethanol, followed by 50–60 ml of absolute ethanol. Either air-dry the crystals or dry them in an oven at approximately 80° for 1–2 hr.

B. RATE OF MUTAROTATION

Before coming to class, prepare in the laboratory record book a chart for collecting the following information during the performance of this experiment: temperature, concentration, path length of sample tube, time of mixing of solution, calibration factor for Shaw polarimeter (if used),[5] and a series of measurements of time and optical rotation.

Obtain from the instructor a polarimeter and sample tube. Note the scale attached to the polarizer section of the mailing tube. Count the number of divisions on that scale about the circumference of the tube and divide into 360° in order to determine the calibration factor.

Carefully fill the sample tube with distilled water and put the end-piece in place in such a way that no air bubbles remain trapped within. Place the sample tube in the polarimeter and, viewing through the analyzer section, point the assembly at the day-time sky. Turn the analyzer end of the polarimeter until the observed field turns deep purple and place a light pencil mark on the polarizer scale in alignment with the zero pointer on the analyzer section. This mark will serve as a zero point for later determinations of rotation. Empty and carefully dry the sample tube and end-pieces.

Weigh 10–15 g of α-D-glucose accurately to the nearest 0.05 g and transfer the sample *completely* to a dry 100-ml volumetric flask. Fill the flask to within a few milliliters of the volumetric mark with distilled water; tightly stopper the flask and shake the contents to effect complete solution. The fine solid should dissolve completely within about 30 sec. When the solid is approximately one-half dissolved, note the time to the nearest minute and record it. As soon as the solid is dissolved, carefully fill the flask to the mark with distilled water using a dropper. Again stopper and shake the flask until the solution is homogeneous (approximately 5–10 sec). Working carefully but rapidly, fill the polarimeter sample tube with this solution and insert the end-piece leaving no air bubbles. Place a thermometer in the volumetric flask and, as soon as there is time, read and record the temperature. Insert the sample tube in the polarimeter, put the analyzer in place, setting the pointer at the zero point. Point the polarimeter at the daytime sky and turn the analyzer until the field turns deep purple.[6] Since α-D-glucose has a positive rotation, the analyzer will need to be turned to the right (clockwise). Note the time to the nearest minute and the number of

---

[5] The directions given here are suitable for use with a polarimeter of the type described by Shaw which has a sample tube with a path length of 2–4 dm and a capacity of 35–70 ml. If a polarimeter is used which has a sample tube appreciably larger or smaller than this, the volume and concentration of the solution should be adjusted accordingly. See footnotes 3 and 4, p. 370.

[6] Due to reflections within the tube, the field will appear as a "bulls-eye target." Pay particular attention to the color within the "eye." Because further readings will be taken and it is desired that they be consistent, note carefully the color shade at which the reading has been taken, and for all subsequent readings adjust the polarimeter to the same shade.

divisions[7] the pointer is displaced from the zero point; record these data. During the next 6 or 7 min take a similar reading once a minute, recording the data obtained. Then take additional readings every 3 or 4 min for the next 40–50 min. Remove the sample tube from the polarimeter and measure and record the path length of the solution through which the polarized light passes.

Approximately 4.5–5 hr will be required for the rotation to drop to within experimental error of the equilibrium value. Empty the solution from the sample tube back into the volumetric flask, tightly stopper the flask, and place it in the desk until the next meeting. The instructor should have placed identifying numbers on each of the polarimeters available. Note which polarimeter was used in this experiment and record the number in the notebook.

At the next laboratory period and using *the same* polarimeter as before, refill the sample tube with the equilibrated solution and take a final reading, which will be α_e.

Treatment of data

1. Define the first recorded time at which a measurement was taken as time $t = 0$. The observed rotation at that time is then α_o. For each reading following the defined zero point, determine and record the elapsed time Δt (from $t = 0$) in minutes.

2. Perform the calculations needed for the graphical plot in the next operation.

3. On a piece of graph paper, plot $2.303 \log (\alpha_e - \alpha_o)/(\alpha_e - \alpha_t)$ (vertically) *vs.* elapsed time Δt (horizontally). Determine the slope of the best straight line that can be drawn through these points;[8] the value of the slope (in $\min^{-1}$) is equivalent to the value $(k_1 + k_2)$.

4. From the following equation,[9] calculate the value of $(k_1 + k_2)$ to have been expected at the temperature at which the kinetic determination was made. Compare this with the value obtained.[10]

$$\log (k_1 + k_2) = 11.0198 - 3873/T(^\circ K) \tag{5}$$

5. On a separate piece of graph paper, plot α_t (vertically) *vs.* Δt (horizontally). A curve should be obtained. Draw the best curved line through these points and extrapolate the curve from time $t = 0$ back to the time of initial mixing. Using the extrapolated value of α obtained at that time, calculate the initial specific rotation, using equation (1) of Chapter 21, of the α-D-glucose used in the experiment. Remember, pure α-D-glucose has a specific rotation of $[\alpha]_D^{25^\circ} + 112^\circ$.

6. Using the value of the equilibrium constant, K_e, for the equilibrium between α- and β-D-glucose obtained by other workers, and the value of $(k_1 + k_2)$ obtained in this experiment, calculate the individual values of k_1 and k_2.

$$K_e = 1.762 = \frac{k_1}{k_2} \tag{6}$$

---

[7] If the scale has been constructed from millimeter-lined graph paper, estimation of the displacement to 0.2 division should be possible.

[8] See footnote 12 in Chapter 12 (p. 213).

[9] C. S. Hudson and J. K. Dale, *Journal of the American Chemical Society,* **39**, 320 (1917).

[10] From the authors' experience with this experiment, the experimental and calculated values of $(k_1 + k_2)$ should compare within a factor of two.

C. SPECIFIC ROTATION OF SACCHARIDES

Prepare aqueous solutions of accurately known concentration (*ca.* 0.1 g/ml) of one or more pure sugars, as assigned by the instructor. Following the directions for the measurement of optical rotation given in part B and using the Shaw polarimeter, determine the specific rotations of each of these sugars. Compare with the values recorded in Table 25.14.

EXERCISES

1. Would any different experimental results have been expected if one had started with β-D-glucose rather than with α-D-glucose? Explain.
2. From the specific rotation values of $+112°$ for α-D-glucose, $+19°$ for β-D-glucose, and $+52.7°$ for the equilibrium mixture, show by calculation that the value for the equilibrium constant given in equation (6) is correct.
3. Write a mechanism for the isomerization of α- to β-D-glucose.
4. An experimentally observed rotation appearing as 60° could just as correctly be interpreted as a rotation of 240°, 420°, 600°, $-120°$, $-300°$, etc. How could it be determined which rotation is the correct one?

22.3 The Hydrolysis of Sucrose

Sucrose, a familiar foodstuff, is a carbohydrate having the structure **4**. As shown in the structure, the sucrose molecule is formed by a linkage between the hemiacetal form of α-D-glucose and one of the hemiketal forms of D-fructose (**5**). The observation that

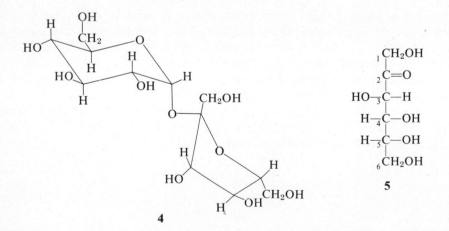

sucrose does not undergo mutarotation is evidence that the linkage is through the glycosidic hydroxyl groups of each sugar.

Sucrose, $[\alpha]_D^{25°}$ $+66.5°$, undergoes acid-catalyzed hydrolysis to give a mixture of D-glucose and D-fructose [equation (7)]. Under the conditions of the hydrolysis, the

$$\text{Sucrose} \xrightarrow{\text{H}_3\text{O}^\oplus} \text{D-Glucose} + \text{D-Fructose} \qquad (7)$$

glucose undergoes rapid mutarotation to give the equilibrium mixture of α- and β-forms, $[\alpha]_D^{25°} +52.7°$, whereas the fructose is formed as an equilibrium mixture of the isomers shown in Figure 22.3 and has $[\alpha]_D^{25°} -92°$. Isomers **6** and **7**, five-membered

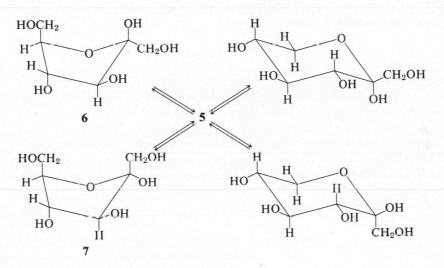

FIGURE 22.3 Solution equilibria of D-fructose.

ring hemiketals, predominate in the equilibrium mixture. The observation that the sign of optical rotation changes upon hydrolysis of sucrose has led to the name *invert sugar* for the product mixture. The enzyme *invertase* accomplishes the same chemical result as does the acid-catalyzed hydrolysis of sucrose.

EXPERIMENTAL PROCEDURE

Place 10 g of pure sucrose, whose weight has been accurately determined, in a 250-ml round-bottomed flask. Add *ca.* 100 ml of water, swirl the contents of the flask to effect solution, and then add about 0.5 ml of concentrated hydrochloric acid. Heat the solution at reflux for about 2 hr.

At the end of the period of reflux, cool the reaction mixture to room temperature and carefully transfer *all* of the solution to a 250-ml volumetric flask. Use small amounts of water to rinse the round-bottomed flask and add the rinses to the rest of the solution. Dilute the mixture to a volume of 250 ml. With the Shaw polarimeter, whose use is described in Section 22.2, part B, determine the specific rotation of the product mixture from the hydrolysis of sucrose. Save at least 5 ml of this solution for later use (see part A, Section 22.4).

EXERCISES

1. Offer an explanation for the change in sign of the optical rotation of sucrose following hydrolysis. Calculate the specific rotation of *invert* sugar from the known equilibrium rotations of D-glucose and D-fructose. How does this number compare with that determined experimentally?

2. In what way is the specific rotation of invert sugar analogous to the specific rotation of a racemic mixture?

22.4   Carbohydrates: Their Characterization and Identification

The determination of the complete structure of an unknown monosaccharide was a formidable problem to the early organic chemists owing to the stereochemical questions that had to be answered; *e.g.,* it was insufficient to know that, in its open-chain form, (+)-glucose was a 2,3,4,5,6-pentahydroxyhexanal; the configuration at each of the four asymmetric carbon atoms had to be defined. Furthermore, once it was realized that intramolecular hemiacetal or hemiketal formation occurred, it became necessary to determine the size of the ring that was formed; *i.e.,* did the sugar exist in the form of a furanose (five-membered) or pyranose (six-membered) ring? Nevertheless, the complete structures of most monosaccharides containing as many as seven carbon atoms have now been determined.

If the question of structure elucidation is extended to polysaccharides, the complexity of the problem is greatly increased. Such questions as how many and which monosaccharides constitute the unknown, which carbon atoms are bonded to the oxygen atom that serves as the intermolecular linkage between the monosaccharide units, the stereochemistry of linkages at the anomeric carbon atoms, and the size of the ring formed by each monosaccharide contained in the polysaccharide (the ring size adopted by a particular monosaccharide unit in a polysaccharide may not be the same as the size found for the monosaccharide itself!) require answers.

It is not surprising that organic chemists have developed a variety of techniques for eliciting information about unknown carbohydrates that will aid in their systematic characterization. The following paragraphs discuss various possible tests which may be used for the classification of carbohydrates according to structural types. Note that the judicious interpretation of the results of these tests will provide much, if not all, of the information required to prove the structure of an unknown carbohydrate. It should be mentioned that in performing the tests for classification of monosaccharides there is nothing which precludes the test's being applied to a mixture of monosaccharides, such as might be obtained from hydrolysis of a sample of one or more disaccharides. Thus, a Seliwanoff test for ketoses applied to the disaccharide sucrose would be negative, whereas the same test applied to invert sugar, the mixture of monosaccharides resulting from hydrolysis of sucrose, would be positive, owing to the presence of fructose.

A. REDUCING SUGARS

All monosaccharides and *many* disaccharides are reducing sugars. This property is specifically based on the presence of an aldehyde group which may participate in an oxidation-reduction reaction with various oxidizing agents. Although all aldoses of four or more carbon atoms exist predominantly in cyclic hemiacetal form, their reversible equilibrium with the aldehydo form allows the reaction to proceed, with a shift in the equilibrium (see Figure 22.2). Disaccharides in which one of the rings is a hemiacetal are reducing sugars for the same reason:

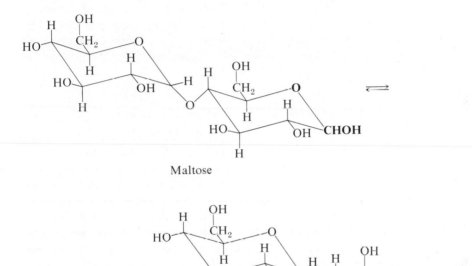

Maltose (8)

Disaccharides in which both rings are in acetal (or ketal) form are *not* reducing sugars because they cannot be in equilibrium with aldehydo forms. Note, for example, that in sucrose **(4)** both potential aldehyde functions are tied up in the glycosidic (intermonosaccharidic) linkage.

Ketoses yield positive tests for reducing sugars because of the base-catalyzed Lobry de Bruyn tautomerization. Thus, for example, the base-catalyzed enolization of fructose provides an enediol which may reketonize to provide a mixture of glucose and mannose, both of which are reducing sugars. Equation (9) shows (with partial structures) the processes that occur.

$$
\begin{array}{c}
\text{CH}_2\text{OH} \\
| \\
\text{C}{=}\text{O} \\
| \\
\text{CHOH}
\end{array}
\xrightarrow{\text{HO}^{\ominus}}
\begin{array}{c}
\text{CHOH} \\
\| \\
\text{C} \\
| \quad \text{OH}\\
\text{CHOH}
\end{array}
\xrightarrow{\text{HO}^{\ominus}}
\begin{array}{c}
\text{CHO} \\
| \\
\text{H}{-}\text{C}{-}\text{OH} \\
| \\
\text{CHOH}
\end{array}
+
\begin{array}{c}
\text{CHO} \\
| \\
\text{HO}{-}\text{C}{-}\text{H} \\
| \\
\text{CHOH}
\end{array}
\qquad (9)
$$

Note that each of the tests for reducing sugars is carried out under basic conditions under which the equilibria shown in equation (9) are established.

Tollens' test. This procedure is included within the classification tests for aldehydes on p. 440.

Benedict's test. The reagent that is used in this test is prepared as a solution of cupric sulfate, sodium citrate, and sodium carbonate. The formation of a cupricitrate complex ion prevents $Cu(OH)_2$ from precipitating in the basic solution. Cupric ion serves as an oxidizing agent of aliphatic aldehydes, including α-hydroxyaldehydes, but not of aromatic aldehydes.[11] A positive test for these types of aldehydes is evidenced by the formation of a yellow to red precipitate [equation (10)]. The red precipitate is cuprous oxide, Cu_2O. The yellow precipitate occasionally observed has apparently not been characterized; its formation seems to depend on the amount of oxidizing agent present.

$$RCHO + 2\,Cu^{\oplus\oplus} + 5\,HO^{\ominus} \xrightarrow{\text{citrate}} RCO_2^{\ominus} + \underset{\substack{\text{brick-}\\\text{red}}}{Cu_2O} + 3\,H_2O \qquad (10)$$

Barfoed's test for monosaccharides. As is true of the tests just described, this test likewise depends on the reducing properties of the saccharides being tested. The conditions are such, however, as to allow a significant selectivity between monosaccharides and disaccharides. The test reagent consists of an aqueous solution of cupric acetate and acetic acid. Thus, in contrast to the previous tests, the reaction is carried out under *acidic* conditions. A positive test for monosaccharides is constituted by the formation of the brick-red precipitate of Cu_2O within *two or three* minutes. Disaccharides require a longer time, providing the precipitate only after about ten minutes or more. Nonreducing sugars, *e.g.,* sucrose, apparently undergo slow hydrolysis under the test conditions, as they also give a precipitate after *extended* time. It is not fully clear why reducing disaccharides oxidize more slowly than monosaccharides; however, it is this fact on which the test is based.

B. Dehydration of monosaccharides: group color tests

Under strongly acidic conditions, particularly with heating, monosaccharides undergo dehydration to provide furfural (8) from pentoses, and 5-hydroxymethylfurfural (9) from hexoses. Furfural and its derivatives may react with various phenolic compounds under these acidic conditions to give highly colored products. In many cases, these colors are diagnostic of the structural type of the original sugar. Disaccharides may undergo hydrolysis during the reaction and give colors corresponding to the monosaccharides present.

There are several possible mechanisms for the dehydration. It is, however, generally accepted that ketuloses (2-ketoses) undergo fairly rapid reaction in a manner at least similar to that shown in equation (11).

[11] J. D. Morrison, *Journal of Chemical Education,* **42,** 554 (1965).

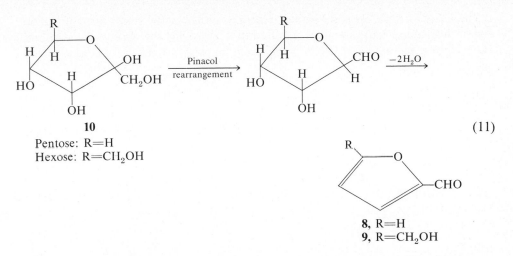

Aldoses, on the other hand, are not structurally capable of providing ready dehydration to furfural or its derivatives. There have been several mechanisms proposed for the dehydration of aldoses; of these, only the one shown in equation (12) for

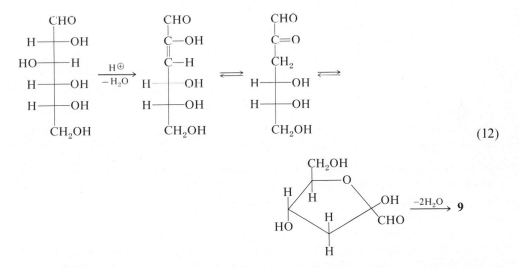

glucose has received any experimental support. Aldopentoses dehydrate more rapidly than aldohexoses, and in fact form **8** at about the same rate as ketohexuloses form **9**. Mechanistic investigations of the dehydration of aldopentoses are lacking; however, the observation that they react faster than aldohexoses indicates that they may dehydrate by a mechanism other than that shown in equation (12).

The Molisch test: An apparent universal test for carbohydrates.   Although there has not been a systematic study of this test, the Molisch test has generally been regarded as a universal test for saccharides. There is no doubt that it does give positive results for all saccharides that are likely to be encountered in a course of this type.

The basis for this test is the known reaction, under acidic conditions, of furfural and its derivatives with two moles of α-naphthol to provide the violet-colored quinoid **11,** as shown in Figure 22.4. All of the reactions with α-naphthol are examples of

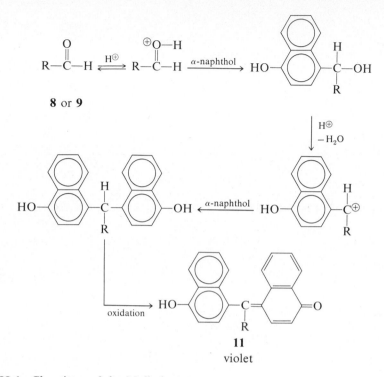

FIGURE 22.4 Chemistry of the Molisch test.

electrophilic aromatic substitutions which occur at the *para* position of the ring activated by the phenolic hydroxy group. Polysaccharides undergo at least partial hydrolysis under the conditions of the test, and the resulting monosaccharides, as well as any other monosaccharides present, are dehydrated to provide furfural or 5-hydroxymethylfurfural according to the mechanisms previously described. Concentrated sulfuric acid is used as the acid catalyst and dehydrating agent.

The formation of the violet color is a positive test for most, and perhaps all, saccharides. Take care to note, however, that if this test is being used to classify an unknown that *may or may not be a carbohydrate,* and if that unknown happens to be furfural or one of its derivatives, a *positive* test will be observed. Other testing procedures, such as, for example, the solubility tests of Chapter 25, should prevent possible confusion, if they are systematically applied.

Seliwanoff's test: A selective test for ketoses. Ketoses, particularly pentuloses (2-ketopentoses) and hexuloses (2-ketohexoses), which are the ketoses most frequently encountered, often exist in the form of five-membered rings (furanose form). Those ketoses found predominantly in the form of six-membered rings (pyranose form) may, under the acidic conditions of the test, equilibrate rapidly with furanose forms, through ring-opening and reclosure. Thus, ketoses generally either exist in or may readily attain the 2-ketofuranose structure [**10**, equation (11)], which favors formation of furfural and its homologs by dehydration.

On the other hand, aldoses require more extensive structural changes in order to provide furfural [equation (12)]. These changes apparently constitute a kinetic "bottle-neck," with the result that 2-ketoses dehydrate to furfural more rapidly and under milder conditions than aldoses.[12]

The Seliwanoff test takes advantage of the difference in rates of dehydration of aldoses and ketoses to provide a qualitative test to distinguish the two types of monosaccharides. The dehydrating medium is 6 N hydrochloric acid, and the reaction is carried out at elevated temperature. These are milder dehydration conditions than those used in the Molisch test and allow a greater kinetic selectivity. In the Molisch test, even though ketoses are more easily dehydrated than aldoses, concentrated sulfuric acid is such a strong dehydrating agent that *all* monosaccharides dehydrate rapidly, an example of a kinetic leveling effect.

Resorcinol is used as the complexing agent to provide the color test. The exact nature of the colored products that form is not known; however, because resorcinol (1,3-dihydroxybenzene) is strongly activated toward electrophilic aromatic substitution, these products probably arise from reactions at least approximately similar to those in Figure 22.4 for the Molisch test:

$$\text{furfural} + \text{resorcinol} \xrightarrow{H^{\oplus}} \text{colored products} \tag{13}$$

Ketuloses provide positive tests, as evidenced by the development of color within 2 min. Aldoses and polysaccharides require somewhat longer time. The hexuloses fructose and sorbose yield cherry-red solutions. Pentuloses give colors ranging from blue to green.

Bial's test: A test for pentoses.   Bial's test has been used to distinguish pentoses from hexoses, and is based entirely on the nature of the color that develops when the monosaccharide is dehydrated in the presence of orcinol (5-methylresorcinol) and ferric chloride. Furfural, from dehydration of pentoses, condenses with orcinol to provide a color variously described as blue or green; the exact color probably varies with the concentration of ferric chloride. 5-Hydroxymethylfurfural, from hexoses, reacts with orcinol to give a muddy-brown to gray solution which is easily distinguishable from the green color from pentoses. Some difficulty may be encountered with mixtures of pentoses and hexoses, for the green color may be masked. Polysaccharides containing pentose or ketohexose units yield these colors owing to hydrolysis of the saccharide linkage. Pentoses and ketoses provide a positive test in shorter time than aldohexoses because of their more rapid dehydration.

---

[12] This would suggest that ketoses with the carbonyl at positions other than the 2-position would dehydrate at rates roughly equivalent to aldoses; however, there is apparently no information available on this point. Ketoses other than ketuloses are encountered only rarely.

Although it is unlikely that a student will encounter a heptose or, with the exception of glyceraldehyde, a triose in this course, a point of caution should be noted: Heptoses and trioses yield solutions virtually indistinguishable from those produced by pentoses. Because trioses cannot form furfural, some other species is involved in the reaction with orcinol in these cases.

As in the Seliwanoff test, the structures of the colored products with orcinol are not known. However, because orcinol is activated for electrophilic attack (at the 2-position, between the hydroxyl groups), some type of aromatic substitution product is probably involved:

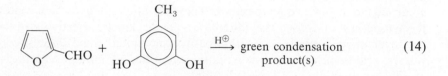

$$\text{(14)}$$

C. The mucic acid test: A specific test
   for galactose

Hot 8 N nitric acid behaves somewhat differently with a saccharide than hot hydrochloric acid and concentrated sulfuric acid. Because nitric acid is an oxidizing agent, the sugar is oxidized rather than dehydrated. Aldoses are oxidized at both ends of the ring-opened form to provide saccharic acids. Ketoses oxidize to give a mixture of dicarboxylic acids resulting from chain fragmentation. Galactaric acid 12, equation (15)] from oxidation of galactose tends to be much less soluble in the

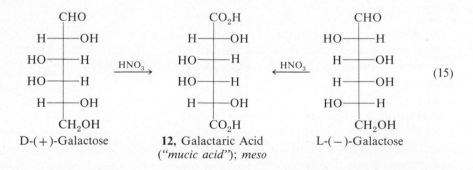

$$\text{(15)}$$

oxidizing medium (and water) than the saccharic acids obtained from other aldoses. This may be attributed partly to the high molecular symmetry of galactaric acid. Thus, the formation of a fine crystalline precipitate is considered a positive test for galactose. The majority of other saccharic acids, particularly those encountered in this course, are soluble in the oxidizing medium used.

Additional useful information may be obtained about an unknown aldose by determining whether or not those saccharic acids which are water soluble are also optically active. Stereoisomeric aldoses may provide different results. For example, glucose will yield an optically active glucaric acid, whereas galactose, the C-4 epimer of glucose, will yield the optically inactive (meso) galactaric acid.

D. FORMATION OF OSAZONES

Most carbohydrates, particularly when impure, crystallize from solution only with difficulty, tending instead to form syrups; this frequently makes their direct characterization difficult. Many carbohydrates react with phenylhydrazine to form bright yellow, crystalline derivatives called *osazones*. These derivatives may usually be identified readily both by their temperatures of decomposition (or melting points) and their crystalline forms.

Not all sugars form osazones, for only those that have an aldehydo or keto carbonyl group, either free or in equilibrium with a hemiacetal or hemiketal, will react with phenylhydrazine. For example, α- and β-D-glucose (**1** and **2**) in water solution are in dynamic equilibrium with the ring-opened form **3**, which, because of its aldehydo group, will react with phenylhydrazine (see Chapter 16.1).

Equation (16) shows the formation of glucosazone (**13**), the osazone derivative of glucose. Note that three equivalents of phenylhydrazine are required and that two of these reagent molecules are incorporated into the osazone structure. Additional products include aniline and ammonia.

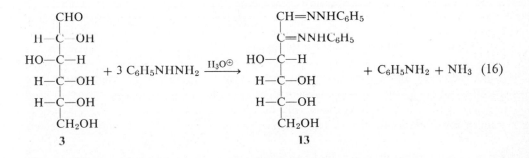

The accepted mechanism for the reaction is presented in Figure 22.5. Following the formation of a phenylhydrazone, **14**, an internal oxidation-reduction reaction is accomplished by tautomeric migration of two hydrogens from C-2 to the hydrazone moiety to give **15**. The newly formed carbonyl group condenses with a second equivalent of phenylhydrazine to give **16** which undergoes subsequent tautomerization to **17**. Following a 1,4-elimination of aniline, which produces either **18** or **19**, a third equivalent of phenylhydrazine condenses with the imine group to give the osazone, **20**, and ammonia. Although it may appear that **20** should suffer further reaction by intramolecular oxidation-reduction between the secondary alcohol group at C-3 and the hydrazone group at C-2, the reaction stops at this point so that only two phenylhydrazine units are introduced. The formation of the intramolecular hydrogen bond shown in **20** has been established as the cause of limitation of the reaction to the first two carbons of the chain.[13]

---

[13] When 1-methylphenylhydrazine $[H_2NN(CH_3)C_6H_5]$ is used in place of phenylhydrazine, the reaction proceeds down the chain readily at least as far as C-5. Note that in this reagent, the N-H proton involved in the hydrogen bonding in **20** has been replaced with a methyl group. O. L. Chapman, W. J. Welstead, Jr., T. J. Murphy, and R. W. King, *Journal of the American Chemical Society,* **86,** 732, 4968 (1964).

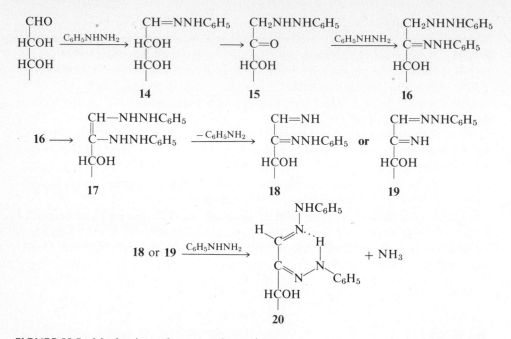

FIGURE 22.5 Mechanism of osazone formation.

As indicated above, osazones may frequently be identified by their crystalline forms. Osazones from different saccharides tend to form highly distinctive crystalline clusters which may most conveniently be observed under a microscope. Strongly supporting evidence for the identification of an unknown saccharide may be collected by comparing the osazone of an unknown with those of several known saccharides. Further, under closely controlled conditions, the time required for precipitation of many osazones may be significant in tentative identification of the saccharide.

E. Other derivatives

Other useful derivatives of saccharides, in addition to osazones, include polyacetate and polybenzoate esters. The preparation of the acetate derivatives is described in the section on carbohydrates in Chapter 25.2.

EXPERIMENTAL PROCEDURE

A. Formation of osazones

Using the procedure described below, attempt preparation of osazone derivatives of D-glucose, D-fructose, and sucrose. It will be most efficient to carry out these reactions simultaneously.

Heat a large beaker of water to boiling. In each of three test tubes, dissolve 0.2 g of one of the above sugars in 4 ml of water and 0.5 ml of saturated sodium bisulfite solution,[14] add to this solution *either* 0.6 ml of glacial acetic acid, 0.6 g of sodium

---

[14] The addition of sodium bisulfite prevents oxidation of the phenylhydrazine during the reaction and the contamination of the osazone by the resulting tarry products.

acetate, and 0.4 g of phenylhydrazine *or* 0.6 g of sodium acetate and 0.6 g of phenyl-hydrazine hydrochloride. (*Caution:* Phenylhydrazine is toxic and care should be exercised in its handling.) Stir the solutions thoroughly, place the test tubes in the beaker of boiling water, and discontinue heating. Allow the test tubes to remain in the cooling water bath for 30 min.

After *ca.* 30 min, remove the test tubes, cool the contents to room temperature, and collect any precipitates by suction filtration. Recrystallize each of the osazones obtained from ethanol-water and determine their melting points. Determine the melting point of a *mixture* of the osazone derivatives of glucose and fructose. Insofar as the melting points of osazones may depend on the rate of heating, it is advisable to carry out these melting point determinations simultaneously.

Prepare an osazone derivative of the product of hydrolysis of sucrose. Substitute 4 ml of the solution which was diluted for rotation measurements (Section 22.3) for the 0.2 g of sugar and 4 ml of water used in the above procedure. Determine the melting points of this osazone *and* of a mixture of this osazone with the osazone of fructose.

Table 25.14 includes the melting points of the osazones of many saccharides.

B. Microscopic examination of osazones and their time of formation

In addition to the melting points of osazones, their crystalline structure and times of precipitation may be quite useful in the tentative identification of saccharides. In order to examine these properties of an osazone, follow the procedure given in part A with the following exceptions: (1) *omit* the 0.5 ml of saturated sodium bisulfite solution; and (2) maintain the water bath at boiling temperature throughout the reaction. Monosaccharides will form a precipitate within 15–20 min after the time the test tubes are placed in the beaker of boiling water. The elapsed time to precipitation is fairly reproducible for several saccharides: arabinose, 10 min; galactose, 15–19 min; glucose, 4–5 min; fructose, 2 min; mannose, 0.5 min (as hydrazone, not osazone); xylose, 7 min; and sucrose, 30 min (requires hydrolysis prior to the formation of the osazone). Osazones of disaccharides are soluble in hot water and do not precipitate until the solution is cooled.

To examine the osazone crystal habit, allow the solution to cool *slowly* to room temperature (rapid cooling may cause distortion of the crystal structures of the osazones). Using a wide-tip dropper, transfer a drop of the solution containing the suspended osazone to a microscope slide and examine the crystals under a low power microscope. Either compare the unknown osazone with those osazones of known sugars in this manner, or consult sources of published photomicrographs.[15]

It is possible to view individually the osazones formed from a mixture of a monosaccharide and a disaccharide by viewing a drop of the *hot* reaction solution under the microscope, and then filtering the hot solution to remove the precipitated osazone of the monosaccharide. Slow cooling of the solution will allow the osazone of the disaccharide to precipitate and subsequently to be examined. In the first

---

[15] Photomicrographs of the osazones of most of the common sugars may be found in W. Z. Hassid and R. M. McCready, *Industrial and Engineering Chemistry, Analytical Edition,* **14,** 683 (1942). (This edition has, in recent years, been retitled *Analytical Chemistry.*)

instance, one must work relatively rapidly to avoid unwanted precipitation of the more soluble osazone on the microscope slide. Preheating of the slide is helpful.

C. IDENTIFICATION OF AN UNKNOWN SACCHARIDE

Obtain a sample of an unknown saccharide. For use in the tests below, prepare a 1% solution of the saccharide in distilled water by dissolving 0.5 g in 50 ml of water. Apply the classification tests discussed previously to identify the structural type of the unknown saccharide. Assuming the tests are carried out properly, the scheme shown in Figure 22.6 should accomplish this goal. Note that the diamonds

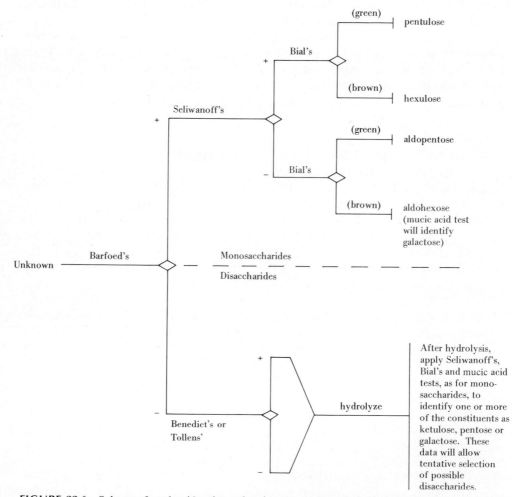

FIGURE 22.6 Scheme for classification of unknown mono- and disaccharides.

at the several junctures in the scheme represent decisions that need to be made, based on the test result. (*Note:* It is advisable to run all tests simultaneously on the unknown and on at least two known saccharides, one of which is known to give a positive test and the other a negative test. Label all test tubes, to avoid confusion.)

Through these tests, having established the general structural nature of the unknown saccharide, final identification may be attained through measurement of physical properties and preparation of derivatives. Virtually all saccharides decompose upon melting; however, if the saccharide is pure, the temperature at which the decomposition occurs is useful in identification. The specific rotation of the saccharide is an important and highly useful physical constant. Experimental details are given in part C of Section 22.2.

The crystal structure, melting point, and time of formation of osazones provide useful information for the *tentative* identification of a saccharide (part B). This information in conjunction with the physical data discussed in the last paragraph will normally allow final identification. If the compound does not form an osazone or provides a mixture of osazones, *e.g.*, a nonreducing disaccharide containing two different monosaccharide units, or if a second derivative is desired in addition to the osazone, the acetate derivative may be prepared; directions are given in Chapter 25.2, p. 464.

Table 25.14 includes lists of decomposition points, specific rotations, and melting points of both osazone and acetate derivatives.

1. Molisch test.   This procedure is described in Chapter 25.2, p. 463.

2. Barfoed's test.   Place 3 ml of Barfoed's reagent[16] and 1 ml of a 1% sugar solution in a test tube; place the test tube in a beaker of boiling water for *5 min* (*no longer*). Remove the test tube and cool it under running water. A red precipitate of cuprous oxide is a positive test. It may be necessary to view the tube against a dark background in good light. For comparison, run tests simultaneously on the unknown, glucose, and lactose. Record the results.

3. Seliwanoff's test.   Place 0.5 ml of 1% sugar solution, 1.5 ml of distilled water, and 9 ml of Seliwanoff's reagent[17] in a test tube; place the test tube in a beaker of boiling water. *Heat for no more than 2.5 min,* and remove the tube from the hot water. Hexuloses provide cherry-red solutions, and pentuloses give blue-to-green solutions. Aldoses and disaccharides produce no color. Although the color may give tentative indication of the number of carbon atoms in the monosaccharide, Bial's test should be run for confirmation. For comparison, run the test simultaneously on glucose and fructose, as well as on the unknown.

4. Bial's test.   Place 3 ml of Bial's reagent[18] and 2 drops of a 1% sugar solution in a test tube. Place the tube in a beaker of boiling water and heat until a color develops. A pentose will provide a green (or blue) solution within about 2 min; the muddy-brown or gray color from a ketohexose or disaccharide containing ketohexose,

---

[16] Barfoed's reagent is prepared by dissolving 66 g of cupric acetate and 10 ml of glacial acetic acid in water and diluting this solution to 1 liter.

[17] Seliwanoff's reagent is prepared by dissolving 0.25 g of resorcinol in 500 ml of 6 N hydrochloric acid (1:1, concentrated hydrochloric acid: water).

[18] Bial's reagent is prepared by dissolving 0.6 g of orcinol in 200 ml of concentrated hydrochloric acid, and adding 8–10 drops of 10% aqueous ferric chloride solution.

e.g., sucrose, also forms in about 2 min. An aldohexose gives a brown color only after heating for several more minutes. As an aid in examining the solution, remove the test tube from the hot water after any color develops and add 4 ml of water and 2 ml of 1-butanol. Shake the tube and record the results. For comparison, run the test simultaneously on the unknown, xylose, and glucose.

5. Mucic acid test. Place about 50 mg of the sugar in a test tube, and add 1 ml of water and 1 ml of concentrated nitric acid. Heat the tube in a boiling water bath for 1 hr; then allow the bath and tube to cool slowly. If caramelization occurs during the heating period, invalidating the test, repeat the test keeping the water bath at a lower temperature. Scratch the tubes with a clean stirring rod to hasten crystallization. If a precipitate forms, confirm its insolubility by adding 2 ml of water and stirring. An insoluble precipitate of mucic acid is a positive test for the presence of galactose, whether pure or as a component in a mixture of other sugars. Disaccharides hydrolyze under these conditions and provide positive tests if galactose was part of the structure. Run the test simultaneously on the unknown, galactose, and glucose, for purposes of comparison.

6. Tollens' test. This procedure is included within the classification tests for aldehydes on p. 440.

7. Benedict's test. Place about 0.2 g of the compound in a test tube and add *ca.* 5 ml of water. Stir to dissolve and then add 5 ml of Benedict's reagent.[19] Heat the solution to boiling. The formation of a yellow (green when viewed in the blue solution of the reagent) to red precipitate is a positive test for aliphatic aldehydes and α-hydroxyaldehydes.

8. Hydrolysis of a disaccharide. Dissolve 0.15 g of the disaccharide in 15 ml of water and add 2 ml of 3 *N* hydrochloric acid. Heat the solution for 5–7 min in a beaker of water held at about 70°. Cool the solution. This solution may be used directly in tests to be run on the hydrolysate as described above.

22.5   Isolation of α-α-Trehalose

Trehalose is the name given to the D-glucosyl-D-glucosides. It derives from the isolation of α-α-trehalose, **21,** from the trehala manna, an oval shell built by certain insects, which has been shown to consist of 25–30% α-α-trehalose. The anomeric forms α-β- and β-β-trehalose have not been found in nature.

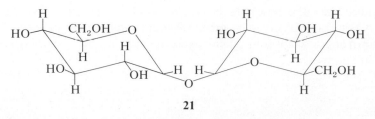

21

<hr>

[19] Benedict's reagent is prepared by dissolving 86.5 g of hydrated sodium citrate and 50 g of anhydrous sodium carbonate in 400 ml of water. To this solution is added with stirring a solution of 8.65 g of cupric sulfate in 50 ml of water. This solution is diluted to 500 ml and filtered if necessary.

This disaccharide was probably first isolated from the ergot of rye, a fungus, in 1832. It has since been shown to occur in other fungi, bacteria, the blood of insects, certain algae and lichens, some of the higher plants, such as the resurrection plant, and yeast, as well as the trehala manna. α-α-Trehalose also occurs in combined form in human tubercle bacilli. The lipids from these bacilli may be separated into free fatty acids and natural fats, the fats containing no glycerol and being esters of fatty acids with α-α-trehalose. These types of fats are termed *microsides:* esters of sugars with fatty acids.

This experiment involves the isolation of α-α-trehalose from dried baker's yeast, in which its content may reach 10–15%. α-α-Trehalose is formed biosynthetically by yeast enzymes from D-glucose and stored within the cell, as is glycogen also. These cells also contain an enzyme, trehalase, capable of enzymatically hydrolyzing the glycosidic linkage. Proliferating and active yeast cells utilize available D-glucose virtually exclusively, producing carbon dioxide and water, and metabolize stored α-α-trehalose only after all D-glucose has been consumed. Older yeast cells, on the other hand, ferment α-α-trehalose at least as fast, and perhaps faster, than D-glucose. These observations probably offer at least a partial explanation for the observation that the α-α-trehalose content of baker's yeast decreases on storage.

Interestingly, baker's yeast has been shown to ferment α-α-trehalose which has been added to the yeast, while leaving the α-α-trehalose stored within the yeast unaffected. Apparently there is a spatial separation in the cell between trehelase and its stored trehalose. There is some evidence that the enzyme may be at the cell surface.

The procedure for the isolation of trehalose from yeast offers interesting insight into the requirements for separating cellular components. Most of the materials extracted from the yeast are systematically removed before the trehalose is finally precipitated. Insoluble materials, such as polysaccharides, fibrous protein, etc., and various aromatic compounds, such as aromatic amino acids and heterocyclic compounds, are separated from the extract by filtration and treatment with activated charcoal. Globular proteins, of which the enzyme trehalase is one, are removed by heating to cause coagulation (denaturization) followed by precipitation as their insoluble zinc salts. Phosphorylated sugars are separated from the extract through the addition of barium hydroxide and filtration of their insoluble barium salts. Thus, at the final precipitation, α-α-trehalose is essentially the only ethanol-insoluble component in the extract.

EXPERIMENTAL PROCEDURE

A. ISOLATION OF α-α-TREHALOSE

Prepare a paste from 32 g of dried baker's yeast with 68 ml of water. Add 250 ml of 95% ethanol and, with occasional stirring, allow the mixture to stand for *ca.* 30 min. Filter the mixture by suction filtration and wash the filter cake with three 30-ml portions of 70% ethanol. Combine the washings with the main solution. To the filtrate, add 20 ml of 20% aqueous zinc sulfate, 1 ml of 1% phenolphthalein solution, and a sufficient quantity of saturated barium hydroxide solution to make the solution basic (about 50 ml will be required). Add 2 g of activated charcoal and heat the mixture to 70° on a steam bath. Filter the hot solution through a Büchner funnel previously layered with filter-aid. (To form the layer of filter-aid, filter a slurry of filter-aid in ethanol through the funnel, and then discard the ethanol.) Adjust the filtrate to *ca.* pH 7 with 0.1 N hydrochloric acid, and concentrate the solution to approximately 10 ml by *gentle* heating under vacuum. Slowly stir 80 ml of 95% ethanol into the resulting syrup, stopper the flask, and leave it undisturbed. Crystals will normally form within a couple of days; however, a week or more may occasionally be required. The crystals are frequently quite large owing to their slow growth. If desired, crystallization may be hastened by the addition of a little more ethanol or by cooling in an ice-water bath.

Apply the Molisch test, Chapter 25.2, to verify that the isolated substance is a saccharide. Determine that it is α-α-trehalose through measurement of its decomposition point (203°; dihydrate, 97°) and its specific rotation (Section 22.2; the $[\alpha]_D^{20}$ of trehalose is $+178.3°$), and perhaps also by formation of its octaacetate derivative (Chapter 25.2), whose reported melting point is 100–102°.

Verify that α-α-trehalose is a nonreducing polysaccharide by application of Barfoed's and Benedict's tests, Section 22.4. Hydrolyze a portion of the trehalose by preparing a 0.5% solution in 1 N hydrochloric acid and boiling for 20 min. Carry out any necessary tests to demonstrate that the hydrolysate consists solely of glucose. That D-glucose is the sole monosaccharidic constituent of trehalose may be demonstrated by preparing an osazone of the hydrolysate (if sufficient trehalose was isolated), or by thin-layer chromatographic analysis.

B. THIN-LAYER CHROMATOGRAPHY
OF MONOSACCHARIDES[20]

Obtain a 12-cm strip of cellulose chromatogram sheet (without fluorescent indicator). Spot the strip about 1 cm from the bottom with the dilute solution of an unknown sugar or sugar mixture and also with solutions of any desired known sugars for comparison purposes (use glucose for the trehalose hydrolysate). Spots should be separated by about 1 cm. Develop the plate using as developing solvent a mixture of pyridine-ethyl acetate-acetic acid-water in the respective ratios 5:5:1:3. Development may require nearly 1.5 hr. The spots may be visualized by either spraying with *p*-anisidine phthalate reagent[21] or by leaving the plate in contact with iodine vapor. With the spray reagent, hexoses yield green spots and pentoses give red-violet spots after heating the plate at 100° for 10 min. Record the results by drawing a picture of the developed plate in the notebook.

C. OPTIONAL EXPERIMENTS

The tests and experiments performed above have allowed partial verification of the structure of α-α-trehalose. Structural points not investigated include: the size of the saccharide (di, tri, etc.); the ring size(s); and the anomeric configuration (α-α, α-β,

---

[20] For an excellent compilation of techniques and procedures for the thin-layer chromatographic analysis of saccharides and their derivatives, consult B. A. Lewis and F. Smith, in *Thin-Layer Chromatography,* 2nd ed., E. Stahl, editor, Springer-Verlag, New York, 1969, Chapter 10.

[21] This reagent spray, for reducing sugars, is prepared as a solution of 1.23 g of *p*-anisidine and 1.66 g of phthalic acid in 100 ml of 95% ethanol.

or β-β). The instructor may wish to assign to interested students, as a project, the complete structural determination of this saccharide. Information is available in various literature sources as to approaches to investigations of this type. A suggested chemical solution of at least the first two points mentioned includes preparation of the fully methylated derivative of trehalose and its subsequent hydrolysis to provide 2,3,4,6-tetra-O-methyl-D-glucose. Necessary procedures and information concerning this derivative of glucose may be located by initial consultation of Beilstein (see Chapter 26). Alternatively, spectroscopic techniques (nmr and ir) provide excellent approaches to the solution of this type of structural problem. Discussions of the correlations between spectral data and structure of carbohydrates are available in several sources of literature Classes B and E (Chapter 26).

amino acids and peptides

23.1   Introduction

The amino acids constitute a highly important class of naturally occurring organic compounds. They are the monomeric units which are joined through amide linkages, called peptide bonds, to produce the important biopolymers upon which every living system depends: the proteins **(1)**.

$$\overset{\oplus}{H_3N}-CH-\overset{\overset{O}{\|}}{C}\left(\!-NH-CH-\overset{\overset{O}{\|}}{C}\!-\right)_n\!\!NH-CH-\overset{\overset{O}{\|}}{C}-O^{\ominus}$$

$$\underset{R}{}\qquad\qquad\underset{R}{}\qquad\qquad\underset{R}{}$$

1

Proteins are polyamides in the molecular weight range above 5000; those polyamides of molecular weight below 5000 are more usually referred to as polypeptides. These types of compounds serve a variety of biological functions. Some, the *fibrous proteins,* are structural in nature, composing such tissues as hair, skin, and muscle fiber. They possess quite appreciable mechanical strength, are insoluble in water, and chemically are relatively inert. Fibrous proteins generally possess very high, somewhat indefinite molecular weights and are sometimes polymer-like. Others, the *globular proteins* and smaller natural *peptides,* have much smaller molecular weights and exist as discrete chemical entities, often obtainable in crystalline form. They are water soluble and have characteristic reactivity. The globular proteins serve a variety of roles ranging from catalytic functions (enzymes) and overall regulatory functions (hormones), through immunological defense functions (antibodies). To a limited extent, differences in function are reflected in differences in molecular weight; *e.g.,* compare fibrous and globular proteins. However, particularly among the globular proteins, differences in biological properties are more completely determined by the exact sequence of different amino acids in the peptide chain: the *primary structure.*

The number of such possible arrangements is vast. For example, in a pentapeptide **(1,** $n = 3$**)** composed of five different amino acids, the number of different sequential arrangements of amino acids is 120. Each of these in principle would possess different biochemical reactivities. However, the biological properties of a peptide or protein is made more complex as a result of its three-dimensional structure, as described below.

Although each individual amide linkage is planar, owing to conjugation as shown in (**2**), conformational differences in structure may arise through rotation about the remaining single bonds (those to C_α in **2**). These rotations allow the chain to coil

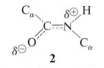

2

and to achieve stabilization through hydrogen bonding between amido hydrogens and carbonyls on separated peptide units. Such coiling constitutes the *secondary structure* of peptides.[1] Additionally, due to convolutions and gross foldings of the coiled chain, amino acid residues (peptide units) in widely separated positions of the chain may be brought into close proximity, acting together in concert and providing the peptide with its characteristic reactivity. This folding, which may be the result of a variety of structural influences, constitutes the *tertiary structure* of the peptide. Lastly, the spatial relationship of one polypeptide chain to another results in the *quarternary structure* of the overall protein structure.[2]

A complete understanding of the biochemical behavior of a peptide, from a molecular and mechanistic point of view, must depend on the determination of its total structure. Of the various levels of structural complexity, determination of the primary structure (sequencing) must be deemed the most important. This is a logical consequence of the realization that the higher degrees of structural complexity are, in the first instance, dependent on the primary structure.

Determination of the primary structure requires initial knowledge of the numbers of each kind of amino acid involved in the chain. This may be accomplished by the total hydrolysis of the peptide to provide a mixture of the amino acids constituting the structure.

$$\overset{\oplus}{N}H_3-\underset{\underset{R_1}{|}}{C}H-\overset{\overset{O}{\|}}{C}-NH-\underset{\underset{R_2}{|}}{C}H-\overset{\overset{O}{\|}}{C}-NH-\underset{\underset{R_3}{|}}{C}H-\overset{\overset{O}{\|}}{C}-O^\ominus \xrightarrow{H_3O^\oplus}$$

$$\overset{\oplus}{N}H_3-\underset{\underset{R_1}{|}}{C}H-CO_2^\ominus + \overset{\oplus}{N}H_3-\underset{\underset{R_2}{|}}{C}H-CO_2^\ominus + \overset{\oplus}{N}H_3-\underset{\underset{R_3}{|}}{C}H-CO_2^\ominus \quad (1)$$

Qualitative and quantitative analysis of this mixture then determines the identities and relative numbers of each amino acid present. If the molecular weight is known,

[1] See any modern organic textbook for further information about the three-dimensional structural properties of peptides.

[2] For example, the tobacco mosaic virus, with an overall molecular weight of 41,000,000, is composed of many identical polypeptide subunits, each with a molecular weight of 17,500, held together by noncovalent interactions.

then the exact numbers of each amino acid in the chain may be determined. For example, if the hydrolysis of a peptide of unknown structure provided a mixture of amino acids analyzed to contain only alanine (ala) and glycine (gly, see Table 23.1) in a ratio of 2:1, respectively, the peptide could be a tripeptide (ala$_2$, gly),[3] a hexapeptide (ala$_4$, gly$_2$), etc. If the unknown peptide was found to have a molecular weight of approximately 200, a general structure of unknown sequence would be established. Note that the molecular weight of (ala$_2$, gly) is 203 (2 ala + gly − 2H$_2$O), while the molecular weight of (ala$_4$, gly$_2$) is 388 (4 ala + 2 gly − 5H$_2$O). Experimental approaches to the determination of sequence will be discussed in Section 23.3.

Peptide bonds may be hydrolyzed under either acid- or base-catalyzed conditions. Although both procedures suffer from some disadvantages, the acid-catalyzed hydrolysis is preferable, primarily because alkaline conditions result in extensive racemization of the chiral center at the α-position as well as in degradation of some amino acid residues, *e.g.,* arginine (arg) and threonine (thr, see Table 23.1). Although certain of the amino acid residues are sensitive to acid and undergo partial destruction during acid-catalyzed hydrolysis (serine and tryptophan, for example), these effects are well understood and quantitative corrections may be applied during careful work in the research laboratory. The mechanism of acid hydrolysis of the peptide linkage, which is simply that of an amide, is qualitatively similar to that for the hydrolysis of an ester [see equation (5), Chapter 17.2].

Total hydrolysis of a peptide is normally accomplished by treatment with 6 N HCl at 100–110° over a period of 16–20 hr. To avoid drying of the mixture through evaporation, and consequential charring, the hydrolysis is effected in a sealed tube. An experiment involving peptide hydrolysis in the determination of structure of an unknown dipeptide is included in the experimental part of Section 23.3.

23.2 Analysis of Amino Acids

Although the number of conceivable structures containing both amino and carboxylic acid functional groups on the same carbon atom is vast indeed, fortunately only twenty or so are actually found in polypeptides from living sources. Nearly all of these are α-amino acids of the type shown in **3.** The occasional exception contains an α-amino function as part of a ring, *e.g.,* proline. All have an α-hydrogen, so that the α-carbon atom is asymmetric except in the case of glycine, the simplest α-amino acid, in which there are two α-hydrogen atoms. With the exception of glycine and of a few D-amino acids derived from microorganisms, all of the important amino acids found in polypeptides from living sources are of the L-configuration **(4).** Table 23.1 includes many of the common, naturally occurring amino acids.

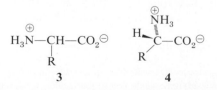

3   **4**

---

[3] The formula indicates that the peptide is composed of two alanine units and one glycine unit. The comma, by convention, indicates that the *sequence* of these units in the peptide is *unknown.*

TABLE 23.1 The Common Amino Acids

Name	Abbreviation	Formula	Isoelectric Point	Color from Pyridine-Isatin Reagent	Numerical Key to Figures 23.3 and 23.4
Alanine	Ala	$CH_3CH(NH_2)CO_2H$	6.0	Pink-red	10
Arginine	Arg	$\begin{array}{c}H_2N\\HN\end{array}C-NH(CH_2)_3CH(NH_2)CO_2H$	11.2	Deep pink	4
Asparagine	Asn	$NH_2COCH_2CH(NH_2)CO_2H$	5.4	...	...
Aspartic acid	Asp	$HO_2CCH_2CH(NH_2)CO_2H$	2.8	Bright red	6
Cysteine	Cys	$HSCH_2CH(NH_2)CO_2H$	5.1	Yellow-brown	1
Glutamic acid	Glu	$HO_2C(CH_2)_2CH(NH_2)CO_2H$	3.2	Bright red	9
Glutamine	Gln	$NH_2COCH(CH_2)_2CH(NH_2)CO_2H$	5.7	...	...
Glycine	Gly	$H_2NCH_2CO_2H$	6.0	Orange-red	7
Histidine	His	$\begin{array}{c}N\\NH\end{array}CH_2CH(NH_2)CO_2H$	7.5	Orange-red	3
Isoleucine	Ile	$CH_3CH_2CH(CH_3)CH(NH_2)CO_2H$	6.0	Bright red	16
Leucine	Leu	$(CH_3)_2CHCH_2CH(NH_2)CO_2H$	6.0	Orange-red	18
Lysine	Lys	$NH_2(CH_2)_4CH(NH_2)CO_2H$	9.6	Red	2
Methionine	Met	$CH_3S(CH_2)_2CH(NH_2)CO_2H$	5.7	Pink	14
Phenylalanine	Phe	$C_6H_5CH_2CH(NH_2)CO_2H$	5.5	Red-brown	17
Proline	Pro	$\begin{array}{c}N\\H\end{array}CO_2H$	6.3	Intense blue	11
Serine	Ser	$HOCH_2CH(NH_2)CO_2H$	5.7	Pink	5
Threonine	Thr	$CH_3CH(OH)CH(NH_2)CO_2H$	5.6	Pink	8
Tryptophan	Try	$\begin{array}{c}N\\H\end{array}CH_2CH(NH_2)CO_2H$	5.9	Red-brown	15
Tyrosine	Tyr	$p\text{-}HOC_6H_4CH_2CH(NH_2)CO_2H$	5.7	Light brown	12
Valine	Val	$(CH_3)_2CHCH(NH_2)CO_2H$	6.0	Red	13

405

Amino acids as acids and bases.   Note that the amino acids have both basic and acidic functional groups. As a result, in their crystalline forms they exist as zwitterions (internal salts, *e.g.,* **3**). Consequently, they are high melting solids that are generally insoluble in organic solvents but soluble in water.

Because of the acidic and basic character of an amino acid, there are established in aqueous solution, pH-dependent equilibria among the forms shown in equation (2).

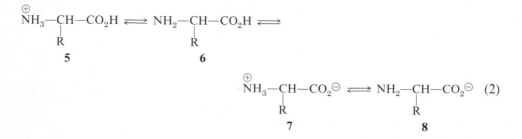

The equilibria are displaced toward **8** in more alkaline solutions; in more acidic solutions, **5** becomes more predominant. The equilibrium between **6** and **7** results in no change in hydrogen ion concentration; thus, the ratio of **7** to **6** in solution is pH independent. The pH-dependent component of the equilibria in equation (2) is the relative concentrations of the species **5** and **8**. If electrodes are placed in a solution of an amino acid, there will be a net migration of the solute toward either the cathode or the anode, depending on whether **5** and **8**, respectively, is predominant.

At a certain pH, specific for each amino acid, the concentrations of **5** and **8** will be equal, and there will be no net migration of the solute toward the electrodes. This pH value is called the *isoelectric point.* Table 23.1 lists the isoelectric points for the common amino acids. Note that the values fall into three ranges: 2–3 for those amino acids containing additional acid groups as part of the side chain R (the acidic amino acids); 5.5–6.5 for those amino acids containing neutral side chains; and 9–11 for those amino acids containing an additional basic site in the side chain (the basic amino acids). It should be noted that it is at the isoelectric point that amino acids have their *minimum* solubility in water.

Chromatographic techniques are utilized nearly universally for the analysis of mixtures of amino acids. In order of importance, as gauged from work carried out in research laboratories, these procedures involve ion exchange chromatography, paper chromatography, and thin-layer chromatography. Because amino acids are colorless, each of these techniques necessarily requires methods of detecting the separated amino acids. The most important detecting agent currently in use is ninhydrin **(9).**

The ninhydrin color-forming reaction of amino acids.   Ninhydrin reacts with amino acids of type **3** to produce characteristic blue-violet colors. The sensitivity and reliability of the test is such that 0.1 μmole of amino acid gives a color intensity that is reproducible to a few percent, so long as a reducing agent such as stannous chloride is present to prevent oxidation of the colored salt by dissolved oxygen. Although not all amino acids give the same color (for example, proline gives a pale yellow color), most do, indicating that the colored product formed is the same in most cases, irrespective of the structure of the original amino acid. The sequence of steps involved in the color-forming reaction is shown in Figure 23.1.

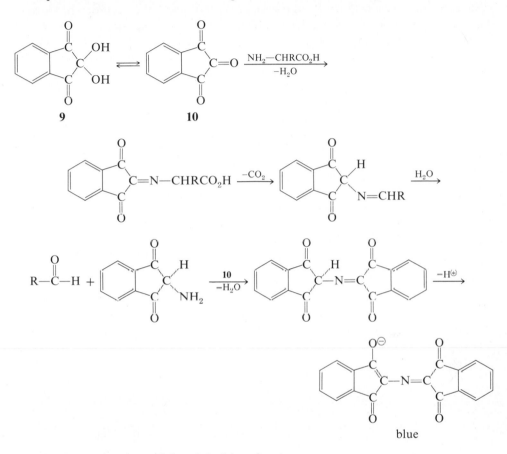

FIGURE 23.1   Chemistry of the ninhydrin color test.

The automatic amino acid analyzer.   In the past, quantitative amino acid analyses were highly time-consuming, extremely tedious, and required considerable amounts of peptide (about 25 g of protein were required for a full amino acid analysis). More recently, however, the use of ion exchange chromatography in conjunction with the *automatic amino acid analyzer* has revolutionized the practice of protein analysis. A complete analysis on automated equipment may be completed in 4–5 hr on no more than 0.1 μmole of protein! The amino acids are separated by elution ion exchange chromatography (see below). The eluent is passed through the column at a constant rate. Because the amino acids elute at different rates, the flow of eluent at the base of the column contains different amino acids at different elapsed time intervals. As it leaves the column the effluent is admixed with a solution of ninhydrin and then passed through a Teflon tube immersed in a boiling water bath to speed up the color-forming reaction (Figure 23.1). The eluent stream then continues through a photoelectric colorimeter which continuously measures the color intensity; an electronic recorder is used to record the color intensity as a function of time. Because the color intensity in the effluent stream as a function of time is directly related to the elution times for the various amino acids, the recorder produces a chromatogram which is qualitatively similar to a gas chromatogram (Chapter 3.4). When correction factors are applied which relate the sensitivity of each amino acid to the ninhydrin color-forming reaction, the areas under the peaks are proportional to the molar ratios of the amino acids in the original mixture. Thus, this procedure allows both qualitative and quantitative analysis of mixtures of amino acids.

The ion exchange resin employed consists of a sulphonated cross-linked polymer produced by copolymerization of styrene and *para*-divinylbenzene (Figure 23.2). The

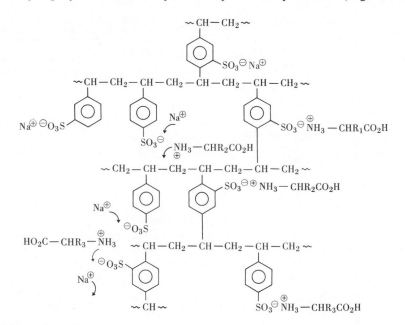

FIGURE 23.2   Cationic exchange chromatography of amino acids. The arrows represent displacement of one ion by another as the eluent passes down the column.

sample is applied as a solution buffered at pH 2 to a column containing this resin. At this pH, below the isoelectric point of all of the amino acids, the amino acids are present in the conjugate acid form **5**. Acting as cations, they displace sodium ions and are held at the head of the column by the sulfonate groups of the resin. An empirical pattern of elution involving aqueous buffered solutions of sodium ion of increasing pH has been worked out that allows separation of all of the common amino acids as a result of their individual abilities to be displaced from the resin by sodium ion. Although the factors which control resolution of the amino acids are complex, to a first approximation the differential elution pattern of two amino acids is controlled by the relative extent to which, at a given pH, they are present in the cationic form **5**. The more basic amino acids would then be expected to be eluted at higher pH values than the neutral or acidic amino acids.

Paper chromatography of amino acids.   Paper chromatography is an especially valuable tool for the relatively rapid qualitative analysis of mixtures of amino acids. As noted in Chapter 3, paper chromatography is a type of partition chromatography in which the substrate is partitioned between the water which is tightly bound within the cellulose fibers of paper (the stationary phase) and an organic solvent which is allowed to migrate upward across the paper by capillary action (the mobile phase). Those acids having the highest solubility in the organic solvent relative to their solubility in water will have the greatest mobility and will migrate upward on the paper most rapidly. Because different amino acids will migrate at different rates, they will separate at different vertical displacements on the paper. Although the R_f values (see Figure 3.11) which may be used to identify the amino acid components present in the mixture are fairly reproducible, there is sufficient variation in the quality and type of paper used, in the temperature from run to run, and in the purity and exact composition of solvent as to make sole reliance on the values published by other workers somewhat risky. Consequently, it is standard practice to run samples of known amino acids simultaneously with the unknown mixture in order to compare R_f values under identical conditions.

A very large variety of organic solvent systems and types of paper have been investigated for the purpose of separating mixtures of amino acids. The solvent and paper used are generally defined by the nature of the particular determination to be performed; that is, the experimental conditions are dependent on which amino acids are present in the mixture.

Because the separability of different amino acids depends greatly on the solvent system used, it is frequently found that some amino acids will separate into individual spots while others will remain unresolved in overlapping spots [see Figure 23.3(a)]. The use of two-dimensional paper chromatography is standard practice to overcome the problem of overlapping. In this procedure, the mixture is spotted on the paper in the lower left-hand corner, and the paper is developed by irrigation with one solvent. The paper is then removed from the developing chamber and allowed to dry. It is then turned 90° in orientation to the original direction of solvent flow and developed with a different solvent, chosen for its ability to separate those amino acids that did not separate with the first solvent. Thus, the spots are displaced in two directions from the original spot rather than in one.

Figure 23.3 shows idealized reproductions of actual paper chromatograms. In Figure 23.3(a), a mixture of 18 amino acids has been developed in one-dimension using a mixture of 1-butanol-acetic acid-water as solvent in a ratio of 4:1:5 (by volume). Note that although there is significant separation of amino acids, several of the spots remain unresolved and overlapped [serine (5), aspartic acid (6), and glycine (7), for example]. In Figure 23.3(b), the same chromatogram is turned 90° and irrigated with m-cresol-phenol in a 1:1 (by weight) ratio buffered with a pH 9.3 borate solution. Observe that all of the spots have now been resolved by the two-dimensional chromatography. The numbers which identify the amino acid responsible for each spot are keyed in Table 23.1.

Although two-dimensional paper chromatography provides greater resolution of the amino acids in a mixture, it suffers the disadvantage of allowing only one sample to be run at a time. Because it is desirable to run samples of known amino acids

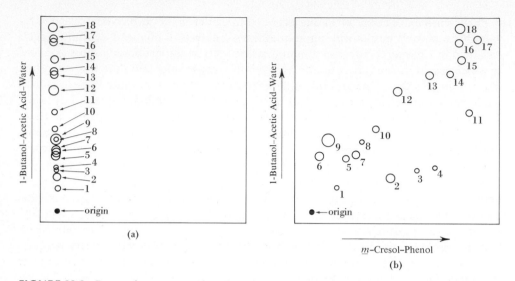

FIGURE 23.3   Paper chromatography of a mixture of 18 amino acids. (a) In one dimension; (b) in two dimensions using different solvents.

simultaneously with the unknown to facilitate identification, the experiment that follows will involve only one-dimensional chromatography. The solvent system 1-butanol-acetic acid-water (4:1:5) is quite good for this purpose and will be used.

EXPERIMENTAL PROCEDURE[4]

Pour approximately 20 ml of solvent consisting of a 4:1:5 (by volume) mixture of 1-butanol, acetic acid, and water, respectively, into a 800-ml or 1-liter beaker. The depth should be about 6 mm. Avoid splashing the liquid onto the sides of the beaker. Cover the beaker with a watch glass. If a piece of cotton is available, use it to close the opening at the lip of the beaker. Avoid unnecessary exposure to the vapors; work in the hood if convenient. Allow *at least 10 min* after preparing the chamber for the atmosphere within to become reasonably saturated with the solvent vapor.

Standard aqueous amino acid solutions (*ca.* 0.07 *M*) will be provided in the laboratory. Among those amino acids which may be included are: DL-aspartic acid, L-isoleucine, L-arginine, glycine, DL-serine, L-proline, DL-phenylalanine, DL-methionine, DL-threonine, L-lysine, L-cysteine, L-histidine, DL-alanine, L-leucine, DL-tryptophan, L-tyrosine, L-glutamic acid, and DL-valine. Each student will be provided with two unknown amino acid mixtures, each of which will contain two or more amino acids chosen from those listed above. The instructor will indicate whether any additional amino acids, other than those listed, have been included.

---

[4]The experimental directions that are provided are adapted in part from those given by L. Slaten, M. Willar, and Sister A. A. Green, *Journal of Chemical Education,* **33,** 140 (1956), which may be consulted for additional experimental details.

Obtain a piece of chromatographic paper measuring 13×15 cm. The size necessary should be determined by the number of samples to be run. Place a series of light pencil dots about 1–1.5 cm apart along a line parallel to a long edge of the paper and about 2 cm from this edge. The outermost dots should be about 1 cm from the short edges of the paper. Label each of the spots so that the solution spotted at each may readily be identified later from information recorded in the notebook. Avoid the common student error of losing track of the positions at which given solutions are spotted. Such errors eventually lead to confusion and probable mis-identification of the components of the mixture. (*Caution:* Fingerprints contain significant amounts of amino acids, often enough to be easily detected by the methods used in this experiment. This is particularly true after the solutions of amino acids have been handled. Avoid touching the surface of the chromatographic paper. Handle it as little as possible, and then only along a thin strip on the edge opposite to that along which the samples are to be spotted.)

Spots are applied to the paper with a capillary pipet. If these are not available, the instructor will demonstrate how to draw them from capillary melting point tubes. In spotting, a small drop of the solution is placed at the mark indicated and allowed to dry before development. If the drop is too small, the amount of sample may be insufficient for eventual detection of some of the components; if it is too large, it may lead to loss of resolution during development. The ideal size is 2–3 mm in diameter. Spots will increase in size by a factor of three or four times in the direction of mobility during development; they should not spread laterally very much. Practice spotting samples on scrap filter paper, and when you feel comfortable with your technique, spot the chromatographic paper with unknowns and any samples of knowns to be run. Be sure to use a *new* capillary pipet for each spot to avoid contamination of the samples being spotted.

After spotting, allow the spots to dry and then coil the paper into a cylinder, fasten it with staples or paper clips, and insert it, sample edge down, into the developing chamber. The paper should not touch the sides of the beaker. When the solvent front has migrated to within about 2 cm from the upper edge of the paper, remove the paper and lightly mark the position of the solvent front. Stand the cylinder on a watch glass for a few minutes to dry. The drying should then be completed by spreading the paper out under a heat lamp, or by placing it in an oven at about $105°$, or perhaps by hanging it in a gentle stream of air in the hood.

When the paper is dry, spray it lightly and evenly with ninhydrin solution[5] and redry it either under a heat lamp or in the oven (heat is necessary to the color-forming reaction). The colors should be readily visible after 20–30 min. It may be desirable to run additional chromatograms in order to obtain R_f values for additional standard samples, or to develop the chromatogram using different detecting agents. An excellent detecting agent that may be used in addition to ninhydrin is isatin.[6] This reagent produces colors by a reaction similar to that shown for ninhydrin in Figure 23.1. Isatin may be useful in order to distinguish between amino acids of similar

[5] The ninhydrin spray reagent is prepared as a 0.1% solution of ninhydrin in ethyl alcohol.

[6] The isatin spray reagent is prepared as a solution of 1 g of isatin and 1.5 g of zinc acetate in 100 ml of isopropyl alcohol and 1 ml of pyridine. The solution is effected by warming on a water bath at $80°$, after which the solution must be kept cool.

R_f values because the different amino acids produce a greater variety of colors. The colors produced by isatin spray for the common amino acids are provided in Table 23.1. These colors develop after heating the chromatogram to 80–85° for 30 min.

Calculate R_f values for each of the standard samples and for each of the spots resolved in the unknowns (consult, if necessary, the legend to Figure 3.11 for the procedure of calculation). Systematically tabulate in the notebook the colors observed for all spots and the calculated R_f values. Identify the constituents of the unknowns, giving the justification for your conclusions. Fasten the chromatograms in your laboratory notebook as a permanent record.

EXERCISES

1. Rationalize the large observed difference in the isoelectric points of lysine (9.6) and glutamic acid (3.2).
2. Proline reacts with ninhydrin to produce a yellow color; the other amino acids in Table 23.1 produce a blue color. What structural feature of proline do you expect is responsible for this distinction in behavior? (*Hint:* Consult the mechanism of the color-forming reaction with ninhydrin as given in Figure 23.1.)
3. The empirical formula of the protein ribonuclease as obtained from cattle is: $C_{566}H_{890}O_{168}N_{192}S_{13}$. It is obviously meaningless to use such a formula in the classical way as a basis for structural determinations. Suggest a more practical way of measuring the composition of a protein.
4. Why is the solubility of an α-amino acid at a minimum in a solution having a pH corresponding to the isoelectric point of the acid?

23.3 Determination of Primary Structure

The most difficult aspect of the determination of primary structure of a polypeptide is the establishment of the sequence of amino acid residues in the chain. The total hydrolysis and amino acid analyses discussed in the last section allow determination of the number and types of amino acids present; however, all information regarding sequence is lost at the hydrolysis step. The standardized approach to establishing sequence is discussed in the following paragraphs.

Terminal residue analyses. Note that the amino acid residues at the termini of the polypeptide chain differ from the remainder of the residues: One, the *N-terminal residue,* is the only residue which contains a free *alpha* amino group; the other, the *C-terminal residue,* is the only residue which contains a free carboxyl group *alpha* to a peptide linkage (see **1**, for example). The special significance of these residues is that it is relatively simple to determine their identity. The importance of identifying these residues may be illustrated in the following example: There are 720 possible sequential arrangements for a hexapeptide containing six *different* amino acid residues. If *either* of the terminal residues is determined, the remaining number of possible sequences is only 120, and if both are known, this number is reduced to 24. Thus, in this example, terminal residue analyses would result in a reduction of the number of structures to be considered by a factor of 30! The results of terminal residue analysis, together with information gained by partial hydrolysis (below), will usually allow the sequence of polypeptides to be determined.

A very successful method of identifying the N-terminal residue utilizes 2,4-dinitrofluorobenzene (DNFB). DNFB reacts by nucleophilic aromatic substitution in weakly alkaline aqueous solutions with free amino (N-terminal and lysyl), phenol (tyrosyl), and imidazole (histidyl) groups to provide dinitrophenyl (DNP) derivatized peptides **(11)** [equation (3)]. Excess DNFB may be removed from the alkaline reaction mixture by extraction with ether. The DNP-peptide remains water

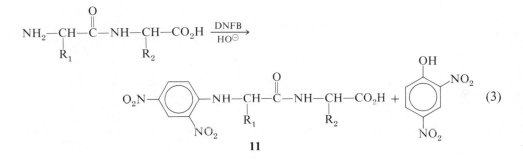

11

soluble at alkaline pH, as does 2,4-dinitrophenol, which is formed by hydrolysis of DNFB during the reaction. The DNP-peptide (and, unfortunately, 2,4-dinitrophenol) is separated by adjustment of the reaction mixture to pH 1 and extraction with ether. The DNP-peptide no longer contains basic amino groups and consequently is not soluble in the acidic aqueous medium. Following removal of the ether, the DNP-peptide is totally hydrolyzed with 6 N HCl at 100° for 24 hr to provide a mixture of both derivatized and underivatized amino acids. Table 23.2 portrays the chemical results at this stage, with especial attention drawn to lysine, histidine, and tyrosine, those amino acids containing side-chain functional groups reactive to DNFB.

The N-terminal residue may now readily be distinguished from the other amino acid residues of the original peptide because it provides the only amino acid derivatized at the alpha position; the other acids bear free alpha amino groups. The N-terminal DNP-amino acid (with the exception of DNP-arginine) may be separated from the acidic hydrolysis mixture by extraction with ether and identified by standard chromatographic procedures.

C-terminal residue analysis is normally accomplished in either one of two ways. Reaction of the peptide with anhydrous hydrazine results in hydrazinolysis of each of the peptide linkages, providing the C-terminal amino acid residue as the only free alpha amino acid in the product mixture [equation (4)].

$$NH_2-CH-C(=O)-NH-CH-CO_2H \xrightarrow{H_2NNH_2}$$

$$\underset{R_1}{} \quad \underset{R_2}{}$$

$$NH_2-CH-C(=O)-NHNH_2 + \overset{\oplus}{N}H_3-CH-CO_2^{\ominus} \quad (4)$$

$$\underset{R_1}{} \qquad\qquad \underset{R_2}{}$$

TABLE 23.2   Chemical Results of 2,4-Dinitrophenylation and Hydrolysis of the Resulting DNP-Peptide

Amino Acid Residue	As N-Terminal Residue	Other
Lysyl	DNP—NH—CH—CO$_2$H $\vert$ (CH$_2$)$_4$ $\vert$ DNP—NH	$\overset{\oplus}{\text{NH}}_3$—CH—CO$_2$H $\vert$ (CH$_2$)$_4$ $\vert$ DNP—NH
Histidyl	DNP—NH—CH—CO$_2$H $\vert$ CH$_2$ (imidazole ring) N—DNP	$\overset{\oplus}{\text{NH}}_3$—CH—CO$_2$H $\vert$ CH$_2$ (imidazole ring) N—DNP
Tyrosyl	DNP—NH—CH—CO$_2$H $\vert$ CH$_2$ (phenyl ring) DNP—O	$\overset{\oplus}{\text{NH}}_3$—CH—CO$_2$H $\vert$ CH$_2$ (phenyl ring) DNP—O
All other amino acids	DNP—NH—CH—CO$_2$H $\vert$ R	$\overset{\oplus}{\text{NH}}_3$—CH—CO$_2$H $\vert$ R

As the only water soluble fragment of the original peptide, it may be isolated and identified.

A second method of C-terminal residue determination makes use of the enzyme carboxypeptidase, a pancreatic enzyme whose characteristic reactivity is the hydrolysis of peptide bonds adjacent to free alpha-carboxyl groups. Because the peptide bond to the C-terminal residue is the only such bond in a peptide, carboxypeptidase selectively removes this residue as a free alpha amino acid, producing a shortened peptide chain

$$\overset{\oplus}{\text{NH}}_3\text{—CH—}\overset{\overset{\text{O}}{\|}}{\text{C}}\text{—NH—CH—}\overset{\overset{\text{O}}{\|}}{\text{C}}\text{—NH—CH—CO}_2^{\ominus} \xrightarrow{\text{carboxypeptidase}}$$
$$\underset{\text{R}_1}{} \qquad \underset{\text{R}_2}{} \qquad \underset{\text{R}_3}{}$$

$$\overset{\oplus}{\text{NH}}_3\text{—CH—}\overset{\overset{\text{O}}{\|}}{\text{C}}\text{—NH—CH—CO}_2^{\ominus} + \overset{\oplus}{\text{NH}}_3\text{—CH—CO}_2^{\ominus} \quad (5)$$
$$\underset{\text{R}_1}{} \qquad \underset{\text{R}_2}{} \qquad \underset{\text{R}_3}{}$$

Carboxypeptidase will then remove the *new* C-terminal residue, and so on. Analysis by paper chromatography may be used to identify the free amino acids present in the mixture. The amino acid corresponding to the original C-terminal residue will develop maximum concentration, as judged from spot color intensity on the chromatogram, at a shorter elapsed reaction time than acids from positions at successive positions in from that end of the peptide chain, because it is the first residue removed.

Partial hydrolysis and sequence determination. In principle, the sequence of residues in a polypeptide might be determined by devising a procedure for selectively removing a terminal residue, identifying it, then removing and identifying the next, and so forth until all have been identified (a terminal sequence determination). A variety of chemical procedures are available which may be used in just this way. These procedures, however, are in general feasible only for relatively short peptides; *e.g.*, the primary structures of peptides containing up to 60 amino acid residues have been determined in this fashion with the aid of automated apparatus. Consequently, the sequence of larger polypeptides and proteins is determined by effecting only their *partial* hydrolysis to produce a mixture of smaller peptides (dipeptides, tripeptides, etc.) which are separated and whose sequences are determined by terminal sequence analysis. When the sequences of a sufficient number of fragments are known, the primary structure of the original polypeptide may be logically deduced.[7]

The experiment that follows involves the determination of structure of an unknown dipeptide. It is apparent that the extent of accumulated information necessary to deduce the structure of a dipeptide is somewhat less than would be required for larger peptides, yet the procedures in the experiment demonstrate many of the techniques commonly used by protein chemists. The structural determination will be completed with milligram quantities of the unknown peptide; thus, experience will be gained in the microtechniques of handling materials and solutions.

EXPERIMENTAL PROCEDURE

A. HYDROLYSIS OF AN UNKNOWN DIPEPTIDE

Obtain a 10-cm length of soft glass tubing having an internal diameter of 1.0–1.5 mm. If this is not available, it may be cut from a drawn-out piece of larger diameter tubing. Seal one end by drawing it out using a microburner. (*Caution:* The seal must be complete.) Place about 1 mg of an unknown dipeptide either on a procelain spot plate or in a small test tube. Using either a syringe or a 0.1-ml pipet, add to the sample 30 μl (0.03 ml) of 6 N hydrochloric acid. Mix well and, using a disposable type pipet, transfer the solution to the previously prepared hydrolysis tube. Seal the tube by drawing it out, affix an identification label, and heat the tube in an oven set at 110° for 10–12 hr.

After allowing the hydrolysis tube to cool, carefully open it and, with a disposable pipet, transfer the solution to either a small watch glass or a spot plate. Evaporate the sample to dryness with a heat lamp and, to remove the last traces of hydrogen

---

[7] The interested student may consult any modern organic or biochemistry textbook for additional information, details, and examples of these and other procedures and their applications.

chloride, add 20 μl of water and reevaporate. (*Caution:* In each of these evaporative steps, be careful not to char the sample.) Add 50 μl of water and use 5–10 μl of this solution per spot in analyzing for the component amino acids of the dipeptide by paper chromatography using the procedure provided in Section 23.2.

B. N-TERMINAL RESIDUE ANALYSIS

To a 12-ml conical centrifuge tube add 2 mg of the unknown dipeptide, 0.2 ml of water, 0.05 ml of 4.2% aqueous sodium bicarbonate solution, and 0.4 ml of stock 2,4-dinitrofluorobenzene solution.[8] (*Caution:* 2,4-dinitrofluorobenzene is a vesicant; avoid contact of it with the skin. *Do not pipet DNFB solution by mouth.*) Stopper the tube and shake the mixture frequently over 1 hr. The pH should be maintained at 8–9 by adding, as necessary, additional 4.2% sodium bicarbonate solution. Large amounts of precipitates indicate that the pH is too low.

After the 1-hr reaction period, add 1 ml of water and 0.05 ml of sodium bicarbonate solution. Extract this solution three times with equal volumes of peroxide free ether to remove unchanged DNFB. If necessary the solution may be centrifuged to hasten the separation of layers. During each extraction, stir the heterogeneous mixture thoroughly in order to obtain efficient mixing of the phases. The ether layer may be removed each time with a disposable pipet.

Using pH paper as a guide, adjust the pH of the aqueous layer to about pH 1 by adding approximately 0.1 ml of 6 N hydrochloric acid, and extract three times with 2-ml portions of ether. Combine the ether extracts in a test tube and evaporate the ether. The evaporation may conveniently be accomplished by placing the test tube in a beaker of warm water and blowing a gentle stream of air into the test tube.

All traces of ether must be removed from the DNP-peptide before its hydrolysis. To accomplish the removal of traces of ether simultaneously with the transfer of the DNP-peptide to a hydrolysis tube (which may be prepared by sealing one end of a 10-cm length of 5 mm glass tubing), add 0.2 ml of acetone to the dried DNP-peptide and transfer the resulting solution to the tube. Evaporate the solution in the hydrolysis tube to dryness as before, using a disposable pipet to channel a gentle air stream into the tube. Add 0.5 ml of 6 N hydrochloric acid, seal and label the tube, and heat it at 100° in an oven for 10–12 hr.

Open the hydrolysis tube and transfer the solution to a small test tube. After adding 2 ml of water, extract the hydrolysis solution three times with 2-ml portions of ether. Combine the ether extracts and evaporate to dryness as before. Dissolve the DNP-amino acid in 0.5 ml of acetone and use this solution for chromatographic analysis (below).

If it is desired to confirm the identity of the C-terminal residue, the aqueous phase of the hydrolysate may be evaporated to dryness using a heat lamp. After adding 0.1 ml of water and redrying, the C-terminal amino acid is dissolved in 50 μl of water and identified by paper chromatography following the procedure of Section 23.2. Note that if the C-terminal residue is lysine, histidine, or tyrosine (see Table 23.2), the procedure of Section 23.2 is not applicable.

---

[8] The 2,4-dinitrofluorobenzene solution is prepared by dissolving 0.25 g of DNFB in 4.75 ml of absolute ethanol.

C. Identification of N-terminal DNP-amino acids
by thin-layer chromatography[9]

Obtain a 4 × 10-cm strip of polyamide chromatogram sheet for qualitative analysis of the N-terminal DNP-amino acid. On a line about 1.5 cm from a short side, and with 1-cm spacings from each other and from the long sides, spot 5 μl each of the unknown DNP-amino acid-acetone solution, and the two appropriate standard DNP-amino acid solutions provided in the laboratory. (Note that the appropriate solutions may be identified from the results of the amino acid analysis performed above.) One of two solvents may be used to develop the chromatogram, according to the requirements of the analysis. Consult Figure 23.4 for information to aid in

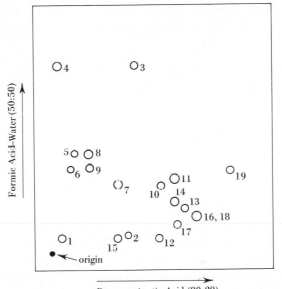

FIGURE 23.4 Two-dimensional chromatogram of 18 N-terminal DNP-amino acids (see second column of Table 23.2) and 2,4-dinitrophenol (spot number 19). The numbers corresponding to each of the derivatized amino acids are keyed in Table 23.1.

this decision. Use either benzene-glacial acetic acid (80:20, by volume) or 90% formic acid-water (50:50, by volume). The first solvent will require about 1.5 hr for development, and the second about 1 hr. The DNP-amino acids produce yellow spots as does 2,4-dinitrophenol. The phenol, in contrast to the DNP-amino acids, is colorless below pH 4. If you are uncertain which spot from the unknown is 2,4-dinitrophenol, add a drop of dilute hydrochloric acid to each spot. If, because of low concentration, the DNP spots are hard to find, examination under ultraviolet light may be helpful. [*Caution:* DNP-amino acids are light sensitive; the chromatographic development should be performed in the dark by placing the chamber

[9]For additional information, consult K. Wang and I. S. Y. Wang, "Chromatographic Identification of Dinitrophenylamino Acids on Polyester Film Supported Polyamide Layers," *Journal of Chromatography and Data,* **27,** 318 (1967).

in the desk drawer.] Because the spot colors will fade with time, they should be circled with a pencil shortly after development. A drawing of the plate should be recorded in the notebook.

Using the information collected in the above procedures, assign a structure for the unknown dipeptide.

EXERCISES

1. The rate of hydrolysis of gelatin, a protein, increases linearly with acid concentration over the range of 3.0–10.4 M hydrochloric acid. Explain, with reference to the mechanism of amide hydrolysis, why this should be the case.

2. Account for the observation that glycylvaline is hydrolyzed under acidic conditions much more rapidly than valylglycine.

3. When asparagine is present in peptides, there sometimes is observed an acid-catalyzed rearrangement during hydrolysis to a mixture of alpha- **(12)** and beta-aspartyl **(13)** peptides. Account mechanistically for these transformations.

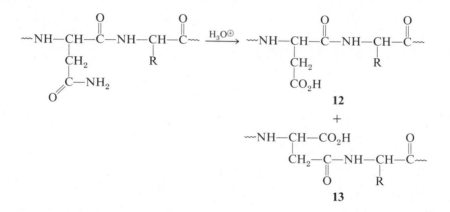

4. A pentapeptide (Glu$_2$, Gly, Val, Ile), obtained by the partial hydrolysis of insulin, gave, upon further hydrolysis, five peptides of the following compositions: (Glu · Gly · Ile · Val); (Glu · Ile · Val); (Glu · Val); (Glu$_2$); and (Ile · Val). Deduce the two possible primary structures for the pentapeptide which would be in accord with this evidence and consider how they might be distinguished experimentally.

5. Provide a mechanism for the reaction of 2,4-dinitrofluorobenzene with a peptide to produce an N-terminal DNP-peptide [**11**, equation (3)]. Would you expect 2,4-dinitrochlorobenzene to require more or less vigorous conditions if used in place of DNFB?

6. Provide a flow diagram which outlines the various separations which are effected on the basis of solubilities during the preparation of the N-terminal DNP-amino acid from a dipeptide.

7. What is the precipitate which forms at low pH during the 2,4-dinitrophenylation of a peptide?

REFERENCES

1. R. E. Dickerson and I. Geis, *The Structure and Actions of Proteins,* Harper & Row, Publishers, New York, 1969.
2. S. Blackburn, *Protein Sequence Determination: Methods and Techniques,* Marcel Dekker, New York, 1970.
3. J. L. Bailey, *Techniques in Protein Chemistry,* Elsevier Publishing Company, New York, 1962.

chapter twenty-four
natural products
Classification, Isolation, and Characterization

24.1   Introduction

Naturally occurring organic compounds, that is, those substances found in and produced by living organisms, have been a source of fascination to man for centuries. The interest in these substances exists for a variety of reasons ranging from the practical applications of such compounds in daily life to the scientific challenges presented by them. Thus, man has used natural products to alleviate pain and to cure diseases, to provide colorful dyes for his body and his clothing, to flavor his foods, and to cause death, both of the animals on which he preys and of his enemies. From the standpoint of science, it was the natural products, because of their ready availability, that provided chemists with one of their first experimental challenges during the period when chemistry was developing from alchemy into a more exact science. It is noteworthy that natural products still present some of the greatest challenges to modern organic chemists.

As the scientific field now labeled "natural products chemistry" evolved, it was found convenient and profitable to place a given compound in one of four general categories defined on the basis of characteristic structural features found in most natural products. These categories are the carbohydrates (sugars), the acetogenins, the terpenes and steroids, and the alkaloids. Because the carbohydrates are discussed in some detail in Chapter 22, they will not be further discussed here.

The *acetogenins* are a group of compounds that share the distinction that their *biosynthesis*[1] involves the head-to-tail polymerization of the two-carbon acetate unit to generate a *linear* polyacetyl chain, *e.g.,* **1**; as a class, then, the acetogenins are characterized by the absence of extensive branching in their carbon skeletons. Further transformations of the basic structure **1** can produce, among other substances, stearic acid **(2)**, a fatty acid, and rhamnetin **(3)**, a yellow pigment of the flavone type. The polymerization of acetate units to generate the general structure

---

[1] Biosynthesis is the synthesis of a chemical compound by a living organism.

1 is related mechanistically to the Claisen ester condensation, the active form of acetate in biological systems being a thioester, acetyl coenzyme A (CH$_3$COSEn, where En represents the coenzyme).

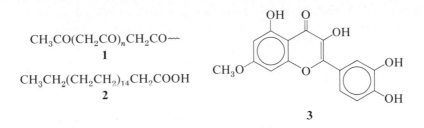

$$CH_3CO(CH_2CO)_nCH_2CO\sim$$
1

$$CH_3CH_2(CH_2CH_2)_{14}CH_2COOH$$
2

3

The substances that fall into the category designated *terpenes* and *steroids* can be viewed as having carbon skeletons constructed by 1,4-polymerization of the five-carbon, branched-chain isopentyl unit, **4**. The active form of this unit in biosynthesis is isopentenyl pyrophosphate **(5)**, which itself is produced in nature by condensation of three acetyl coenzyme A units, with one carbon atom being lost in the form of carbon dioxide. Thus, in contrast to the generally linear character of the acetogenins, the terpenes and steroids possess carbon skeletons that are branched at regular intervals, as illustrated by the structure of vitamin A **(6)**. It should be noted that substances in the terpene class contain two or more of the five-carbon units, two examples being **6** and camphor **(7)**, whereas steroids such as cholesterol **(8)** consist of at least four isopentyl units.

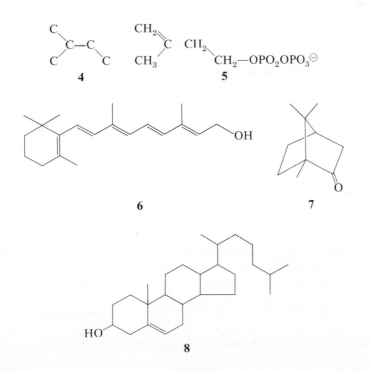

The final category of natural products to be considered here, the *alkaloids*, probably contains the most members and is the most diverse in terms of structural types. This is a consequence of the fact that all natural products which are basic (alkaline) are classed as alkaloids. The basicity of the substances in this category results from the presence in them of one or more nitrogen atoms; the nitrogen arises from incorporation of an α-amino acid unit, **9**, as one of the basic structural units during the biosynthesis of the alkaloid. Examples of alkaloids include the relatively simple nicotine **(10)** and the considerably more complex strychnine **(11)**.

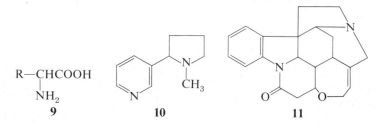

$$R{-}\underset{\underset{\textstyle 9}{\overset{\textstyle |}{NH_2}}}{\overset{\textstyle |}{CH}}COOH$$

Historically, most natural products have been extracted from plants rather than from animals, owing to the generally greater chemical simplicity and availability of the former; microorganisms, which are "simple" in the same sense as are plants and also can be propagated rapidly and in large quantities, are assuming increasing importance as sources of natural products.

The isolation of a natural product in pure form normally presents a considerable challenge to the experimentalist, mainly because even the simplest plants and microorganisms represent mixtures of many organic compounds. The general approach that is taken in the isolation of a natural product can be summarized in the following way. The plant or microorganism is ground or homogenized into fine particles and the resulting material is then usually extracted with a solvent or mixture of solvents in which the desired natural product is expected to be soluble. If the natural product is volatile (as would be the case, for example, if an attempt were being made to characterize the substance(s) responsible for a particular odor), the volatile compounds in the extract can be detected and possibly isolated by application of glpc techniques. More often, however, the natural product is a relatively nonvolatile substance, so that removal of the solvent used in the extraction leaves an oil or gum that requires further manipulation to achieve the resolution of the mixture into its various components. It should be noted that in rare instances some natural products that when pure are crystalline compounds will precipitate from solution during the removal of the solvent, and it was often just this type of fortuitous purification that facilitated the first investigations of natural products.

The more common situation with natural products isolation is that the residual oil or gum might be treated with acids or bases in order to separate basic and acidic components, respectively, from neutral substances; slightly volatile compounds might be separated from nonvolatile ones by subjecting the residue to steam distillation.

One of the most powerful techniques developed as a result of attempts to purify natural products is that of chromatography of various types. Paper and column chromatography have been of dramatic importance to this branch of chemistry; more recently thin- and thick-layer, liquid-liquid, and gas-liquid chromatographic techniques have increasingly been used to aid in the resolution of a crude mixture of natural products into its various components.

The next stage facing the chemist working in the field of natural products is determination of the structure of the isolated product. Here again traditional procedures, such as qualitative tests for various functional groups, and chemical degradations to known substances were and are of great importance; more recently spectroscopic techniques such as mass, infrared, and nuclear magnetic resonance spectroscopy have greatly facilitated the determination of the structure.

In many instances, the final stage or goal for chemists working in this field is to develop a synthetic pathway that permits synthesis of the natural product. The synthesis of some natural products represents mainly an intellectual challenge and/or an opportunity to demonstrate the utility of new synthetic techniques. In some cases, particularly those in which the natural product has medicinal uses, the development of an efficient synthesis may be of importance because of the severely limited supply of the material from the natural source.

Some of the various techniques required for isolation of pure natural products are described in the experiments below. Hopefully they will provide an introduction to this fascinating branch of organic chemistry.

24.2   Citral from Lemon Grass Oil

Terpenes are responsible for the characteristic flavors, odors, and colors of many substances encountered in nature. As an example, citral (**12**) is an aldehydic terpene that possesses a pleasant lemon-like odor and taste. It is amusing to note that whereas citral evokes pleasant odor and taste responses in humans, it apparently is less attractive to other organisms, because certain insects such as ants are known to employ citral as one component of a secretion used to ward off potential predators. As might be expected from the nature of its odor, citral is of commercial importance as a constituent of perfumes in which a lemon-like essence is desired; in addition it is employed as an intermediate for the synthesis of vitamin A (**6**).

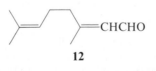

12

Owing to the commercial importance of citral, an extensive search for its presence in natural products has been made. Not too surprisingly, one source turns out to be oil from the skins of lemons and oranges although it is only a minor component of the oil. Citral is the major component, however, of the oil that results from pressing of lemon grass; in fact, 75–85% of this oil is the desired natural product.

Citral contains carbon-carbon double bonds, one of which is conjugated with a carbonyl (aldehyde) function. The presence of the double bond conjugated with the carbonyl group makes citral subject to polymerization, and the aldehyde function contained in it is rather easily oxidized to a carboxylic acid group (RCHO $\longrightarrow$ RCOOH), a reaction that is common with aldehydes. Thus, citral is an extremely labile substance that reacts under conditions such as heat or light or the presence of reagents such as acids, bases, and oxygen that induce its polymerization and/or oxidation. The isolation of citral, therefore, potentially presents a significant challenge to the experimentalist. The task is greatly simplified, however, by the fact that citral is relatively volatile (bp 229°/760 mm) and has a low solubility in water. These two properties make it a suitable candidate for steam distillation, a technique that allows distillation of citral from crude lemon grass oil at a temperature less than 100°, far below citral's normal boiling point, and in a neutral medium. It is worth noting that steam distillation is often the method of choice when reactive, volatile substances are to be separated from nonvolatile (or water-soluble) contaminants.

The citral isolated in this experiment is actually a mixture of the geometric isomers **12a,** geranial, and **12b,** neral. The separation of these two isomers is extremely difficult to achieve using standard techniques and is not attempted in this experiment.

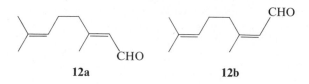

12a   **12b**

EXPERIMENTAL PROCEDURE

Add 10 ml of lemon grass oil and 100 ml of water to a 250-ml round-bottomed flask and attach the flask to an apparatus set for steam distillation using an external source of steam (Figures 2.11 and 2.12). Steam distil the mixture as rapidly as possible, continuing the distillation until droplets of oil no longer appear in the distillate; approximately 250 ml of distillate is required. Drain water from the steam trap (Figure 2.13) whenever necessary.★ After allowing the distillate to cool to room temperature or below, add a portion of it to a separatory funnel containing 50 ml of technical ether, shake the funnel, and separate the aqueous and organic phases. Add another portion of the distillate to the organic phase in the separatory funnel, shake, and separate the layers as before. Repeat these steps until all of the distillate has been shaken in portions with the ether, so that the citral has been removed from the aqueous phase. After the extraction is complete, the aqueous phases may be discarded.

Dry the organic phase over anhydrous calcium chloride,★ decant the dried organic solution into a 250-ml round-bottomed flask, and evaporate the solvent under aspirator vacuum. It may be advantageous to place the flask in a pan of water *at room temperature* during the evaporation of the ether. The residue is citral, bp 229°. Determine the percentage recovery of citral from the sample of lemon grass oil.

The isolated product can be characterized by obtaining IR and NMR spectra and comparing them with those in Figures 24.1 and 24.2. Alternatively, a glpc analysis can be performed, using an authentic sample of citral for comparison, to assess the nature and purity of the product that has been isolated. A typical glpc trace of the product is provided in Figure 24.3.

Chemical characterization of the product can be achieved by testing for unsaturation according to the procedures described in Chapter 6.1 (see p. 124) and for the presence of an aldehyde function by the chromic acid method outlined in Chapter 15.2. Solid derivatives of **12a** such as its 2,4-dinitrophenylhydrazone (mp 134–135°)

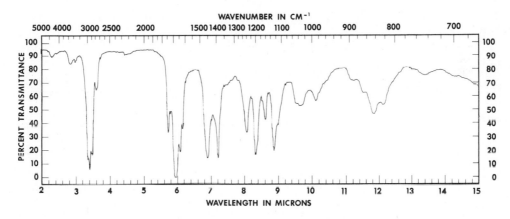

FIGURE 24.1   IR spectrum of citral.

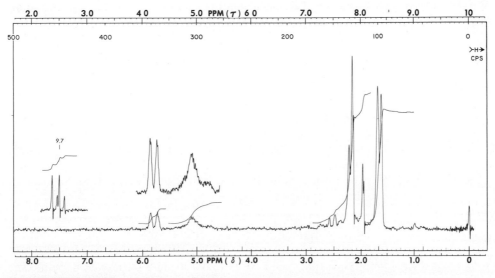

FIGURE 24.2   NMR spectrum of citral.

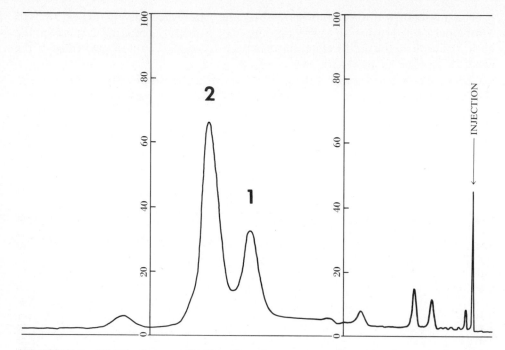

FIGURE 24.3 GLPC of steam distillate from lemon grass oil; peak 1 is neral and peak 2 is geranial.

and its semicarbazone (mp 164–165°) can be prepared by the procedures presented in Chapter 16.1.

Recrystallization normally removes any of the corresponding derivatives resulting from the minor amount of **12b** contained in the isolated citral. Note that the 2,4-dinitrophenylhydrazone of **12b** has a melting point of 171–172°, and the semicarbazone of this isomer melts at 125–126°.

EXERCISES

1. Calculate the relative amounts of geranial and neral present in the sample of citral as indicated by the glpc trace of Figure 24.3. By evaluating the relative areas of the aldehydic protons in the NMR spectrum (Figure 24.2) of citral, perform a similar calculation of the ratio of the two isomers present.
2. Why is the ether not removed from citral by distillation at atmospheric pressure, the more usual procedure?
3. Geranial **(12a)** is thermodynamically more stable than neral **(12b)**. Suggest an explanation for this.

24.3 Piperine from Black Pepper

A weakly basic substance that could be extracted from a variety of peppers was isolated and characterized in 1882 and given the name piperine, from the Latin name for pepper (*piper*). Piperine **(13)** is a 1,4-disubstituted butadiene having a specific geometry about the double bonds. This substance along with minor amounts of chavicine, a geometric isomer of piperine, constitutes about 10% of the weight of black pepper. Among other components of black pepper are starches (20–40%), volatile oils (1–3%), and water (8–13%).

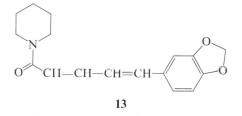

13

Because both piperine and chavicine are relatively nonvolatile substances, they are not responsible for the aroma of black pepper. The taste of black pepper, however, is at least partially attributable to these substances; for example, although piperine is tasteless at first, it does ultimately produce a burning sensation and sharp aftertaste. The initial tastelessness of this substance may be a consequence of its extremely low solubility in water, so that it cannot penetrate the layer of saliva on the tongue and reach the taste buds; certainly, piperine that has been wetted with ethanol produces an immediate sharp taste when placed on the tongue. Some investigators have postulated that chavicine is responsible for the characteristic taste of pepper, but there still appears to be no general agreement regarding this point.

The isolation of piperine can be accomplished in an uncomplicated manner by extraction of ground pepper with 95% ethanol. Ideally the extraction would be performed using a Soxhlet apparatus like that shown in Figure 3.3 so that only a relatively small volume of solvent would be required. Such apparatus is often not available in the undergraduate laboratory so that a round-bottomed flask fitted with a reflux condenser will be used instead; as a consequence, much more solvent will be required than would be with a Soxhlet apparatus.

The crude extract obtained by heating black pepper in ethanol contains, in addition to piperine and chavicine, some acidic, resinous materials that must not be allowed to precipitate with the piperine and thereby contaminate it. In order to prevent co-precipitation of piperine and the resin acids, dilute ethanolic potassium hydroxide is added to the concentrated extract to keep acidic materials in solution as their potassium salts.

Hydrolysis of piperine. When an attempt to determine the structure of an unknown natural product is being made, it is often found useful to cleave the unknown substance into smaller fragments by a chemical reaction; these fragments can generally be more readily identified than the original molecule. Once they have been identified, these smaller molecules represent pieces of a "jigsaw puzzle" which the chemist must put together in a rational way, the goal being, of course, to fit the pieces back together so that the structure of the original substance is duplicated.

In the case of piperine, the gross structure is that shown in **13**; as noted above, however, the stereochemistry about the double bonds has not been specified. In the original proof of structure of this substance, it was recognized that the single nitrogen atom present in the molecule was part of an amide linkage (see exercise 2) so that hydrolysis of this functional group should produce an acid and an amine, both of which might be of known structure. In fact, base-catalyzed hydrolysis followed by appropriate work-up allowed isolation of piperidine **(14),** a known cyclic amine, and

14

piperic acid, a substance that could be synthesized from cinnamaldehyde [equation (1)]. Consequently, the gross structure **13** could be proposed for piperine.

$$C_6H_5CH{=}CH{-}CHO + (CH_3\overset{\overset{\displaystyle O}{\|}}{C})_2O \xrightarrow[\Delta]{NaOAc} C_6H_5CH{=}CH{-}CH{=}CHCOOH \qquad (1)$$
Piperic Acid

Assignment of the full structure to piperine of course requires that the geometry about the double bonds be specified. Given that piperic acid is one of the four geometric isomers shown below and that the melting point of each isomer is that given, isolation of piperic acid and determination of its melting point should allow assignment of the required stereochemistry (see exercise 1).

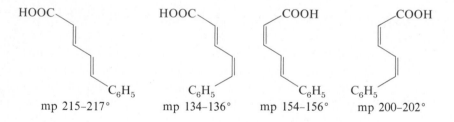

<table>
<tr><td>mp 215–217°</td><td>mp 134–136°</td><td>mp 154–156°</td><td>mp 200–202°</td></tr>
</table>

EXPERIMENTAL PROCEDURE

A. Isolation of Piperine

Place 30 g of finely ground black pepper in a suitably sized round-bottomed flask, add 300–350 ml of 95% ethanol, and gently heat the mixture under reflux for *ca.* 3 hr.[2]★ Because there is solid present in the boiling mixture, bumping may occur, particularly if heating is too vigorous. Filter the mixture by suction and concentrate the filtrate to a volume of 20–30 ml by distillation. Add 30 ml of warm 2 *N* ethanolic potassium hydroxide solution to the residue from the distillation, stir the warm mixture well, and decant or filter the solution to remove any insoluble materials.★ While keeping the solution warm on a steam bath, add 15–20 ml of water; the solution should become turbid and yellow needles may be discerned. Allow the resulting solution to stand until the next laboratory period and then isolate the yellow

---

[2]The period of reflux need not be done during a single laboratory session.

precipitate of piperine that has formed. Recrystallize the crude piperine from acetone. The resulting crystals should be formed as fine yellow needles. Determine the yield and melting point of the product. The reported melting point of piperine is 129–131°.

B. HYDROLYSIS OF PIPERINE

(*Note:* Adjust quantities in the following procedure according to the amount of piperine that is available.) Heat a mixture of 1 g of piperine and 10 ml of 2 *N* ethanolic potassium hydroxide at reflux for 1.5 hr.★ Evaporate the ethanolic solution to dryness by performing a vacuum distillation with the aid of a water aspirator and a steam bath.★ Cool the receiver in an ice-salt bath during the distillation. Suspend the solid potassium piperate that remains in the stillpot in about 20 ml of hot water and carefully acidify this suspension with 6 *N* hydrochloric acid. Collect the precipitate that results, wash it with cold water, and recrystallize the crude piperic acid from absolute ethanol. Determine the yield and melting point of the isolated product.

Qualitative detection of an amine (piperidine) in the distillate can be accomplished by dissolving a few drops of the distillate in a few milliliters of water and determining the pH of the solution. The distillate should also have an odor characteristic of an amine.

The piperidine can be isolated as its hydrogen chloride salt by saturating the distillate with *gaseous* hydrogen chloride (in the hood!), removing the ethanol by performing a vacuum distillation, and recrystallizing the residue from absolute ethanol. Because salts of amines are generally very hygroscopic, the isolated salt should not be exposed to atmospheric moisture any more than is necessary. The reported melting point of piperidine hydrochloride is 242–244° (do not attempt to determine this melting point if the melting point apparatus being used contains a heating fluid such as mineral oil which may ignite above 200°).

EXERCISES

1. On the basis of the melting point observed for piperic acid, assign a stereo-chemically complete structure to piperine; given that chavicine has stereo-chemistry about both double bonds that is the opposite of that in piperine, write out the structure of this geometric isomer.
2. What approaches, both chemical and spectroscopic, might be taken to demonstrate the presence of an amide function in an unknown compound?
3. Is the piperine that is isolated in this experiment expected to be optically active? Why or why not?
4. Amides can be hydrolyzed with aqueous acid as well as with aqueous base. After considering the nature of the other functional groups present in piperine, suggest a reason why base catalyzed hydrolysis is the method of choice for hydrolysis of this particular substance.
5. What portion of the piperine molecule is responsible for its color? Would piperic acid be expected to be colored? Why or why not?
6. In the removal of ethanol that follows the hydrolysis of piperine, why is it important that the distillation receiver be cooled in an ice-salt bath?
7. In what class of natural products should piperine be placed? Why?

SPECTRA OF PRODUCTS

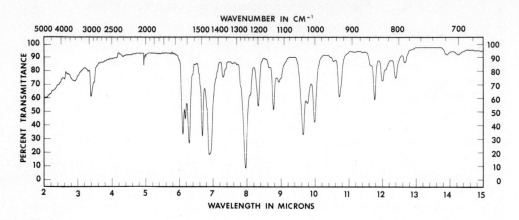

FIGURE 24.4   IR spectrum of piperine.

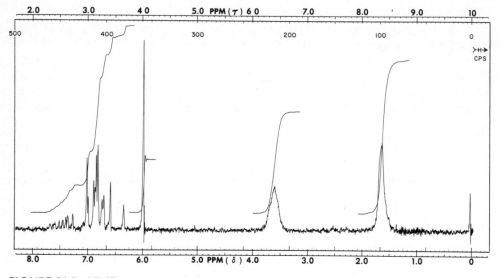

FIGURE 24.5   NMR spectrum of piperine.

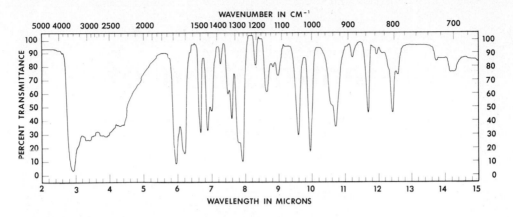

FIGURE 24.6   IR spectrum of piperic acid.

REFERENCES

1. J. B. Hendrickson, D. J. Cram, and G. S. Hammond, *Organic Chemistry,* 3rd ed., McGraw-Hill Book Company, Inc., New York, 1970, Chapter 27.
2. J. B. Hendrickson, *The Molecules of Nature,* W. A. Benjamin, Inc., New York, 1965.
3. P. Yates, *Structure Determination,* W. A. Benjamin, Inc., New York, 1966.
4. R. Ikan, *Natural Products: A Laboratory Guide,* Academic Press, Inc., New York, 1969, p. 185.

chapter twenty-five
identification of
organic compounds

Systematic procedures for the identification of organic compounds were developed much later than those for inorganic compounds and elements. The first successful scheme of organic qualitative analysis was developed by Professor Oliver Kamm and culminated in his textbook published in 1922. This scheme is the one on which most textbooks are still based,[1] and we shall refer to it, including the modifications and modernizations that have been made, as *classical qualitative organic analysis*.

As we have mentioned earlier (Chapters 3 and 4), in recent years the development of instrumental methods of separation and analysis (particularly chromatographic and spectroscopic techniques) has revolutionized the laboratory practice of organic chemistry. However, the interest in classical qualitative organic analysis remains high because it is recognized by teachers and students alike as the most effective as well as the most interesting means of teaching fundamental organic chemistry. For this reason, in this book we have retained an adequate outline of the classical scheme and have also provided an introduction to some of the more *modern spectroscopic methods of identification and structure determination*.

25.1 Separation of Mixtures of Organic Compounds

In this chapter we shall be concerned with the identification of a pure organic compound. However, it must be recognized that when a chemist is faced with the problem of identifying an organic compound, it is seldom pure, but is usually mixed with by-products and starting materials. Modern methods of separation, particularly chromatographic procedures, make the isolation of a pure compound easier than it used to be, but one must not lose sight of the importance of classical techniques of separation.[2] These have been utilized in many of the experimental procedures in the preceding chapters of this book.

---

[1]See references 1 and 2 at the end of Section 25.3.

[2] Methods of separation and purification of organic compounds are treated in detail in Chapters 2 and 3.

The common basis of the procedures most often used to separate mixtures of organic compounds is the difference in *polarity* which exists or may be induced in the components of the mixture. This difference in polarity is exploited in nearly all of the separation techniques, including distillation, recrystallization, extraction, and chromatography. The greatest differences in polarity, which make for the simplest separations, are those which exist between salts and nonpolar organic compounds. Whenever one or more of the components of a mixture can be converted to a salt, it can be separated easily and efficiently from the nonpolar components by extraction or distillation.

We shall not attempt in this chapter to give additional general or specific directions for separating mixtures. If you are given an unknown mixture to separate into its components before identifying them individually, review Chapters 2 and 3, and study the experimental procedures of the preparative experiments of the book. Examine each procedure carefully to determine what impurities are removed from the desired product and what separation techniques are utilized to accomplish this in each step of the "work-up." These same techniques may be employed in separating the components of an unknown mixture.

25.2 Classical Qualitative Organic Analysis

The system consists of six fundamental steps, which are usually best carried out in the sequence listed below.
1. **Preliminary examination** of physical and chemical characteristics
2. Determination of **physical constants**
3. **Elementary analysis**
4. **Solubility tests,** including acid-base reactions
5. **Classification tests;** functional reactivity other than acid-base reactions
6. **Preparation of derivatives**

It is a tribute to the effectiveness of the system that, although the number of organic compounds is many thousands of times the number of common inorganic ions, it is usually possible to identify a known organic compound with more certainty than an inorganic compound. However, with the exception of a few guidelines, there is no rigid regimen of "cook-book" directions to be followed; a student must rely on his own judgment and initiative in choosing a course of attack on his unknown.

DISCUSSIONS OF THEORY
AND DIRECTIONS FOR EXPERIMENTAL PROCEDURES
1. The preliminary examination. The preliminary examination may provide more information with less effort than any other part of the identification procedure, if it is carried out intelligently! The simple observation that the unknown is a *crystalline* solid, for example, eliminates from consideration a major fraction of all organic compounds, since most of them are liquids at room temperature. The *color* is also informative; most pure organic compounds are white (or colorless). A brown color is most often characteristic of small amounts of impurities; *e.g.*, aromatic amines and phenols quickly become discolored by the formation of trace amounts of highly colored air-

oxidation products. Color in a pure organic compound is usually attributable to conjugated double bonds.

The *odor* of many organic compounds is highly distinctive, particularly among those of lower molecular-weight. A conscious effort should be made to learn and recognize the odors which are characteristic of several classes of compounds, such as the alcohols, esters, ketones, aliphatic and aromatic hydrocarbons. The odors of certain compounds demand respect, if not admiration, even when they are encountered in small amounts and at considerable distance; for example, the unpleasant odors of thiols (mercaptans), isonitriles, and higher carboxylic acids and diamines cannot be described definitively, but they are recognizable, once encountered. The more pleasant odors may also be recognized more easily than described. *Be cautious* in smelling unknowns since some compounds are not only disagreeable but also irritating to the mucous membranes. Large amounts of organic vapors should never be inhaled since many compounds are toxic.

The *ignition test* is a highly informative procedure. Heat a small amount (one drop of a liquid or *ca.* 50 mg of a solid) gently on a small spatula or crucible cover, at first above or to the side of a microburner flame. Make a note as to whether a solid melts at low temperature or only upon heating more strongly. Observe the flammability and the nature of any flame. A yellow, sooty flame is indicative of an aromatic or highly unsaturated aliphatic compound; a yellow but nonsooty flame is characteristic of aliphatic hydrocarbons. Oxygen content in a substance makes its flame more colorless (or blue); extensive oxygen content lowers or prevents flammability, as does halogen content. The unmistakable odor of sulfur dioxide indicates the presence of sulfur in the compound.

If a white, nonvolatile residue is left after ignition, add a drop of water and test the solution with litmus or pH paper; a sodium (or other metal) salt is indicated by an alkaline test.

2. Physical constants. If the unknown is a solid, determine its melting point by the capillary tube method (Chapter 1.4). If the melting range is more than 2°, recrystallize the sample. (Chapter 2.7).

If the unknown is a liquid, determine its *boiling point* by the micro boiling-point procedure (Chapter 1.5). If the boiling point is indefinite or non-reproducible, or if the unknown sample is discolored or inhomogeneous, distil it (Chapter 2) and determine the boiling point. Other physical constants which are useful for liquids, particularly in the case of hydrocarbons, ethers, and other less reactive compounds, are the *refractive index* and the *density* (Chapter 1.6 and 1.7, respectively). Consult your instructor about the advisability of making these measurements; full directions are given in references 1 and 2 at the end of Section 25.3.

3. Elementary analysis. A knowledge of the elements other than carbon and hydrogen present in an unknown organic compound is an enormous advantage in identification. The most commonly occurring elements are oxygen, nitrogen, sulfur and the halogens. There are no simple tests for oxygen, and the other elements are most commonly held by covalent bonds, so that they do not respond directly to the usual ionic tests. However, if the organic unknown is fused with molten sodium, most compounds react so that N, S, and X are converted to the ions CN^-, $S^=$, CNS^-, and X^-. After the excess sodium is carefully decomposed, the aqueous solution containing these anions is analyzed by conventional methods of inorganic analysis.

Sodium fusion. Support a small, soft-glass test tube in a vertical position in a clamp with an asbestos liner (or none—*no rubber*). Place a clean cube of sodium (about 3 to 4 mm on an edge) in the tube and heat it gently with a micro burner flame until the sodium melts, and the vapors rise about 1 cm. Then remove the flame and drop a small amount of the organic compound (2 to 3 drops of a liquid or the equivalent amount of a solid) *directly* onto the molten sodium. Avoid allowing the sample to drop onto the sides of the test tube. (*Caution: Safety glasses must be worn and the face should be kept away from the mouth of the tube, as some compounds react violently or explosively with the molten sodium.*) A brief flash of fire is normally observed. Heat the bottom of the tube again, then remove the flame and add a second equivalently sized portion of the organic compound. Now heat the bottom of the tube until it is a dull red color. Next remove the flame and allow the tube to cool. Add about 1 ml of ethanol to decompose the excess sodium. After a few minutes, if you are not certain all of the sodium has been destroyed, add another 1 ml of ethanol. After the reaction has subsided, apply gentle heat to boil the ethanol and place an inverted funnel connected to a vacuum line (or aspirator) over the mouth of the test tube. When the ethanol has been removed, heat the bottom of the test tube again and then drop the hot tube into a small beaker containing about 15 ml of distilled water (*caution*).[3] Break up the tube further with a stirring rod, heat the solution to boiling, and filter. The filtrate is used for the following tests.

a. Sulfur. Acidify a 1- or 2-ml sample of the filtrate with acetic acid and add a few drops of 5% lead acetate solution. A black precipitate of lead sulfide indicates the presence of sulfur in the original organic compound.

b. Nitrogen. Adjust the pH of 1 ml of the filtrate to 13 with dilute sulfuric acid (an indicator paper such as Hydrion E may be used). Add two drops each of a saturated solution of ferrous ammonium sulfate and of 30% potassium fluoride solution. Boil the mixture gently for about 30 sec and then carefully acidify it by the dropwise addition of 30% sulfuric acid until the precipitate of iron hydroxide just dissolves. An excess of acid should be avoided. Appearance of the deep blue precipitate of potassium ferric ferrocyanide (Prussian blue) indicates the presence of nitrogen. If the presence of a blue precipitate is uncertain and the result is only what appears to be a green or blue solution, filter it; a blue color remaining on the filter paper is a weak but positive test for nitrogen. Nitrogen present in an organic molecule is converted by sodium fusion into cyanide ion.

c. Halogens. Acidify about 2 ml of the filtrate by dropwise addition of nitric acid (test with litmus) and boil the solution gently for several minutes to expel any hydrogen sulfide or cyanide that may be present. Cool the solution and add several drops of 5% silver nitrate. A heavy precipitate of silver halide indicates the presence of chlorine, bromine, or iodine in the original organic compound. (A faint turbidity may be due to

---

[3] Alternatively, a Pyrex test tube may be used for the fusion. After the tube has cooled slightly, water may be added to the tube and the mixture stirred and then poured out into a small beaker. Rinse the tube with more water, combining the rinse with the main solution. The total amount of water should be about 15 ml. Boil the mixture, filter, and use the filtrate as described.

the presence of traces of halogen impurities in the original compound, the reagents, or even the glass of the sodium-fusion test tube.) Tentative identification of the particular halogen may be made on the basis of color: silver chloride is white, silver bromide is pale yellow, and silver iodide is yellow, but as all three halides darken rapidly in light, positive identification must be made by standard inorganic qualitative procedures,[1,4] or by means of thin-layer chromatography (Chapter 3.7), the procedure for which follows.

Obtain from the instructor a 2.5×7.5-cm strip of fluorescent silica gel chromatogram sheet. About 1-cm from one end, place four equivalently spaced spots as follows: At the left, using a capillary to provide the sample, spot the original test solution. Because this solution is likely to be relatively dilute in halide ion, it may need to be respotted several times; allow the spot to dry following each application. This may be hastened by blowing on the plate. Take care, however, to keep the spot as small as possible. Next, in order, spot samples of 1 M potassium chloride, 1 M potassium bromide, and 1 M potassium iodide. Develop the plate in a solvent mixture of acetone, n-butyl alcohol, concentrated ammonium hydroxide, and water in the volume ratio of $13:4:2:1$ (see Chapter 3.7 for details). Following development, allow the plate to air dry and, in a hood, spray the plate lightly with an indicator spray prepared by dissolving 1 g of silver nitrate in 2 ml of water and adding this solution to 100 ml of methanol containing 0.1 g of fluorescein and 1 ml of concentrated ammonium hydroxide. Allow the yellow strip to dry and then irradiate it for several minutes with a long wavelength ultraviolet lamp (3660 Å). Compare the spots formed from the test solution with those formed from the solutions of known halides. (*Note:* Iodide gives two spots.)

4. Solubility tests.   Solubility tests, including acid-base reactions, were a fundamental part of the original Kamm system and still may be used to give indicative information; however, it should be recognized that definite assignment of an unknown to a formal solubility class is rather arbitrary because of the large number of compounds that exhibit borderline behavior, no matter where the dividing lines are drawn. The classification scheme is outlined in Figure 25.1.

Make the tests with the solvents or reagents in left-to-right sequence; *i.e.,* determine the solubility in water first. In considering water-solubility, for our purposes a compound is said to be soluble if it dissolves to the extent of 3 g in 100 ml of water or, more practically, 100 mg in 3 ml of water. If an unknown dissolves to this extent in water, do not test it in sodium hydroxide solution or any of the other reagents, but test its aqueous solution with litmus paper and determine its subclassification as shown at the top right of Figure 25.1.

If an unknown is insoluble in water, test it in 1.5 N NaOH solution. In this, and in the other reagent solutions, the definition of solubility is not 3%, but any *greater solubility in the acid or base than in water*[5], which reflects the presence of an acidic

---

[4] See reference 3 at the end of Section 25.3.

[5] If the unknown does not completely dissolve in the reagent solution, do the following: After shaking the unknown with dilute acid or base, decant the liquid or filter it from undissolved sample and carefully *neutralize* the filtrate; the formation of a precipitate or turbidity is indicative of greater solubility in the reagent than in water. It is important only to neutralize the filtrate, because an unknown may show enhanced solubility in both acid and base solution if it contains *both* basic and acidic functional groups. To test for increased solubility in concentrated sulfuric acid relative to water, cautiously dilute the solution.

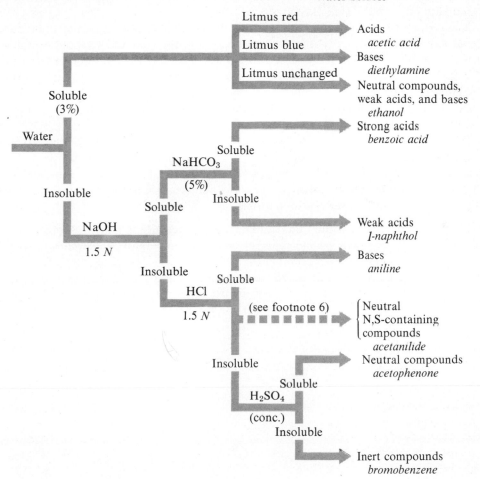

FIGURE 25.1 Classification by means of solubility and acid-base reactions, with typical examples of each class.

or basic functional group in the water-insoluble unknown compound. Unknowns that are more soluble in dilute base than in water are tested for increased solubility (relative to water) in 5% $NaHCO_3$ solution to distinguish between acids weaker or stronger than carbonic acid. Unknowns that are insoluble in both water and dilute base are next tested for solubility in 1.5 N hydrochloric acid; no attempt is made to distinguish between the weak and strong bases that dissolve in dilute acid solution.

Many compounds that are too weakly basic to dissolve in dilute aqueous acid will dissolve or react with concentrated H_2SO_4. They are termed "neutral" on the basis of their behavior in aqueous solutions. However, there is no point in testing compounds that are soluble in dilute HCl, or neutral water-insoluble compounds containing N or S, in concentrated H_2SO_4, since they will invariably dissolve or react with it.[6] Any de-

[6]This is the reason for the dashed arrow in Figure 25.1.

tectable reaction such as evolution of a gas or formation of a precipitate is considered "solubility" in concentrated H_2SO_4.

The borderline for water-solubility of monofunctional organic compounds, by our definition, is most commonly at or near the member of the homologous series containing five carbon atoms. Thus, butanoic acid is soluble, pentanoic acid is borderline, and hexanoic acid is insoluble; 1-butanol is soluble, 1-pentanol is borderline, and 1-hexanol insoluble. The relationship between molecular surface area and solubility is pointed up, however, by the observation that 2-methyl-2-butanol is clearly soluble (12.5 g per 100 ml of water).

We strongly recommend that at this stage of the investigation each student present a *preliminary report* to his instructor of the results of his determination of physical constants, elementary analysis and solubility tests. The instructor should advise the student of any errors (in the case of melting and boiling points a satisfactory limit of accuracy is $\pm 5°$) so as to prevent serious loss of time on "wild goose chases."

5. Classification tests.   Under this heading are grouped all the reactions, other than acid-base reactions, which are indicative of other functional groups and allow assignment of the unknown to a structural class, such as alkene, alkyne, alcohol, aldehyde, etc. Several reactions of this sort have been encountered previously in this book, and the remainder of these classification tests are presented below. The structural classes (and reference pages) that will be considered are:

A. NEUTRAL COMPOUNDS
    Carbonyl Compounds (aldehydes and
        ketones), p. 439
    Unsaturated Compounds (alkenes), p. 441
    Alkyl Halides, p. 441
    Aryl Halides, p. 443
    Alcohols, p. 445
    Esters, p. 458
    Amides, p. 462
    Carbohydrates, p. 463

    Nitriles, p. 461
    Nitro Compounds, p. 457
    Aromatic Hydrocar-
        bons, p. 443
B. ACIDIC COMPOUNDS
    Carboxylic Acids, p. 450
    Phenols, p. 448
C. BASIC COMPOUNDS
    Amines, p. 451

6. Preparation of derivatives.   In the classical approach, it is often useful to convert the suspected compound into another compound (a derivative) which is a solid. The melting points of two such solid derivatives, along with the melting point or boiling point of the unknown compound, often serve to identify the unknown completely. Some methods for preparing derivatives have been given previously, and in addition, other representative and dependable procedures for preparing solid derivatives of the other types of compounds will be presented here. Short tables of some selected compounds, along with their melting or boiling points and melting points of several solid derivatives, are also presented.

Each of the classes of compounds tabulated above will be considered in the following manner. *Classification tests* for the functional group characteristic of the type of compound will be given (or in those cases where these tests have already been presented, reference will be made to suitable places in the text), and these will be followed by brief descriptions of methods of *preparation of several representative solid derivatives.* Reference will be made to the tables of compounds and derivatives which follow.

As mentioned in connection with the outline of the six fundamental steps, in general it is wise to carry these out in the order listed. On the other hand, there is nothing necessarily sacred about the order of the first four steps. It usually makes little difference whether one carries out an elementary analysis before doing solubility tests, or *vice versa*. For example, if one found a water-insoluble compound to be soluble in dilute hydrochloric acid, an amine is a possibility, and an immediate test for nitrogen is indicated. Conversely, if a positive nitrogen test had been found first, a solubility test for an amine would be a logical next step.

There is one warning that should be emphasized strongly, however. Do not go directly to the preparation of a derivative on the basis of a "hunch" as to the class of compound to which your unknown belongs, or even as to its specific identity, but rather make certain of the type of functional group present by obtaining one or more positive classification tests before attempting the preparation of any derivative. "Jumping the gun"—going directly from a melting point or odor to a derivative is the most common error of beginners in qualitative organic analysis. For example, it is very frustrating to try to make a ketone derivative when the unknown is actually an ester.

In deciding which of the classification tests to try first, one should use the information gained in the first four steps. To continue with the example used above, if the unknown has been found to contain nitrogen and to be soluble in dilute acidic solution, a classification test for an amine should be applied first. In the absence of clear indication of the type of compound, the most reliable classification tests should be applied first. Among the most reliable, certainly, are those for aldehydes and ketones.

ALDEHYDES AND KETONES
(Tables 25.1 and 25.2)

A. CLASSIFICATION TESTS

1. 2,4-Dinitrophenylhydrazine. This reagent will give a positive test for either an aldehyde or a ketone. Experimental details for this test are given in Chapter 16.1.

2. Chromic acid. This reagent serves as a clear-cut method for distinguishing between aldehydes and ketones. The reactions are discussed in detail in Chapter 15.2, and the experimental procedures are given there.

3. Iodoform test using sodium hypoiodite (NaOI). This test is used to detect the presence of methyl ketones (compounds having a terminal $-COCH_3$ group), as compared to a ketone which does not contain this specific functional group. Hydroxymethyl compounds ($-CHOHCH_3$) also give iodoform with NaOI, so that it is always necessary to consider this alternative before making a final decision; the problem is solved by determining that a keto group is present or absent. This reaction is discussed in Chapter 16.2, and experimental procedures accompany that discussion.

4. Tollens' test. Another method for distinguishing between aldehydes and ketones (an alternative to the chromic anhydride test) is the Tollens' test. A positive test determines the presence of an aldehyde group, whereas no reaction occurs with ketones.

Tollens' reagent consists of silver ammonia complex, $Ag(NH_3)_2^{\oplus}$, in an ammonia solution. This reagent, on reaction with an aliphatic or aromatic aldehyde, is reduced, whereas the aldehyde is oxidized to the corresponding carboxylic acid; the silver is reduced from the +1 state to elemental silver, and the silver often is deposited as a silver mirror on the side of the test tube. Thus, the formation of the silver mirror is considered a positive test. The equation for the reaction which occurs is given below.

$$RCHO + 2\ Ag(NH_3)_2^{\oplus} + 2\ HO^{\ominus} \longrightarrow 2\ \underline{Ag} + RCO_2^{\ominus}NH_4^{\oplus} + H_2O + 3\ NH_3 \quad (1)$$

Similar tests for aldehydes make use of Fehling's and Benedict's reagents which contain complex salts (tartrate and citrate, respectively) of cupric ion as the oxidizing agents. With these reagents, a positive test is the formation of a brick red precipitate of cuprous oxide (Cu_2O) which forms when Cu^{++} is reduced to Cu^+ by the aldehyde. These two tests are more useful in distinguishing between aliphatic and aromatic aldehydes, since the aliphatic compounds give a fast reaction. Benedict's test is discussed in more detail and the procedure given in Section 22.4.

PROCEDURE

Prepare Tollens' reagent by mixing solution A (2.5 g of silver nitrate in 43 ml of distilled water) and solution B (3 g of potassium hydroxide in 42 ml of distilled water) according to the following directions: Obtain 3 ml of both solutions A and B. To 3 ml of solution A, add concentrated ammonium hydroxide solution dropwise until the initial brown precipitate begins to clear. The solution should be grayish and almost clear. Add 3 ml of solution B, and again by dropwise addition add concentrated ammonium hydroxide until the solution is almost clear.

To carry out the test, add 0.5 ml of the reagent to 3 drops or 50–100 mg of the unknown compound; the formation of a silver mirror or black precipitate constitutes a positive test.[7] If no reaction takes place at room temperature, warm the solution slightly in a beaker of warm water.

At the end of the laboratory period, discard any unused Tollens' reagent; *do not store the solution because it decomposes on standing and yields an explosive precipitate.*

B. DERIVATIVES

Two very suitable solid derivatives of aldehydes and ketones are the *2,4-dinitrophenylhydrazones* and the *semicarbazones*. Oximes are also sometimes useful. Methods for preparing and purifying each of these are given in Chapter 16.1; see Tables 25.1 and 25.2 for lists of possible compounds and their derivatives.

---

[7] If the test tube is not scrupulously clean, the silver will not deposit on the wall as a mirror, but will appear as a black precipitate.

UNSATURATED COMPOUNDS

CLASSIFICATION TESTS

Two common types of unsaturated compounds are alkenes and alkynes, characterized by the carbon-carbon double and triple bond, respectively, as the functional group. There are no simple direct ways to prepare solid derivatives of unsaturated aliphatic compounds having no other functional groups, but it is often useful to detect the presence of these two functional groups. The two common qualitative tests for unsaturation are the reaction of the compounds with *bromine in carbon tetrachloride* and with *potassium permanganate*. In both cases, a positive test is denoted by decoloration of the reagent. Discussion and experimental procedures for these two tests may be found in Chapter 6.1.

ALKYL HALIDES

CLASSIFICATION TESTS

Qualitative tests for alkyl halides are useful in deciding whether the compound in question is a primary, secondary, or tertiary halide. In general, it is quite difficult to prepare solid derivatives of alkyl halides, so that we shall limit this discussion to the two qualitative tests: (1) the reaction with *alcoholic silver nitrate* solution; and (2) the reaction with *sodium iodide in acetone*.

1. Alcoholic silver nitrate. If a compound is known to contain a halogen (bromine, chlorine or iodine), information concerning its environment may be obtained from observation of its reaction with alcoholic silver nitrate. The overall reaction is shown in equation (2). Such a reaction will be of the S_N1 type. As has been asserted earlier

$$RX + AgNO_3 \xrightarrow{\text{ethanol}} \underline{AgX} + RONO_2 \qquad (2)$$

(Chapter 12.2), tertiary halides are more reactive than secondary halides, which are in turn more reactive than primary halides, in an S_N1 reaction. Differing rates of silver halide precipitation would be expected from halogen in each of these environments, *viz.*, primary $<$ secondary $<$ tertiary. These differences are best determined by testing authentic samples of primary, secondary, and tertiary halides with silver nitrate in separate test tubes and observing the results.

Alkyl bromides and iodides react more rapidly than chlorides, and the latter may require warming to produce a reaction in a reasonable period of time.

Aryl halides are unreactive toward the test reagent as are, in general, any vinyl or alkynyl halides. Allylic halides, even when primary, show reactivities as great as or greater than tertiary halides due to resonance stabilization of the resulting allyl cation.

PROCEDURE

Add 1 drop of the alkyl halide to 2 ml of a 2% solution of silver nitrate in ethanol. If no reaction is observed within 5 min at room temperature, warm the mixture in a beaker of boiling water and observe any change. Note the color of any precipitates; silver chloride is white, silver bromide is pale yellow, and silver iodide is yellow. If there is any precipitate, add several drops of 1 N nitric acid solution to it and note any changes; the silver halides are insoluble in acid. To determine expected reactivities, test known primary, secondary, and tertiary halides in this manner. If possible, use alkyl iodides, bromides, and chlorides so that differences in halogen reactivity can also be observed.

2. Sodium iodide in acetone.   Another method for distinguishing between primary, secondary, and tertiary halides makes use of sodium iodide dissolved in acetone. This test complements the alcoholic silver nitrate test, and when these two tests are used together, it is possible to determine fairly accurately the gross structure of the attached alkyl group.

The test depends on the fact that both sodium chloride and sodium bromide are not very soluble in acetone, whereas sodium iodide is. The reactions which occur are:

$$RCl + NaI \xrightarrow{\text{acetone}} RI + \underline{NaCl} \qquad (3)$$

$$RBr + NaI \xrightarrow{\text{acetone}} RI + \underline{NaBr} \qquad (4)$$

This reaction is an S_N2 substitution in which iodide ion is the nucleophile; the order of reactivity is primary > secondary > tertiary.

With this test reagent, primary bromides give a precipitate of sodium bromide in about 3 min at room temperature, whereas the primary and secondary chlorides must be heated to about 50° before reaction occurs. Secondary and tertiary bromides react at 50°, but the tertiary chlorides fail to react in a reasonable time. It should be noted that this test is necessarily limited to bromides and chlorides.

PROCEDURE

Place 1 ml of the sodium iodide–acetone test solution in a test tube and add 2 drops of the chloro or bromo compound. If the compound is a solid, dissolve about 50 mg of it in a minimum volume of acetone, and add this solution to the reagent. Shake the test tube and allow it to stand for 3 min at room temperature. Note whether a precipitate forms; if no change occurs after 3 min, warm the mixture in a beaker of water at 50°. After 6 min of heating, cool to room temperature and note whether any precipitate forms. Occasionally a precipitate forms immediately after combination of the reagents; this represents a positive test only if the precipitate remains after shaking the mixture and allowing it to stand for 3 min.

Carry out this reaction with a series of primary, secondary, and tertiary halides, both chlorides and bromides. Note in all cases the differences in reactivity as evidenced by the rate of formation of sodium bromide or chloride.

ARYL HALIDES AND AROMATIC HYDROCARBONS
(Tables 25.3 and 25.4)

A. CLASSIFICATION TEST

This test for the presence of an aromatic ring should be performed only on compounds which have been shown to be insoluble in concentrated sulfuric acid (see solubility tests). The test involves the reaction between an aromatic compound and chloroform in the presence of anhydrous aluminum chloride catalyst. The colors produced in this type of reaction are often quite characteristic for certain aromatic compounds, whereas aliphatic compounds give little or no color with this test. Some typical examples are tabulated below.

Type of Compound	Color
Benzene and homologs	Orange to Red
Aryl Halides	Orange to Red
Naphthalene	Blue
Biphenyl	Purple

Often these colors change with time and ultimately yield brown-colored solutions. Carbon tetrachloride may be used in place of chloroform; it yields similar colors.

The test is based upon a series of Friedel-Crafts alkylation reactions; for benzene, the ultimate product is triphenylmethane [equation (5)].

$$3\ C_6H_6 + CHCl_3 \xrightarrow{\text{AlCl}_3} (C_6H_5)_3CH + 3\ HCl \tag{5}$$

The colors arise due to formation of species such as triphenylmethyl cations $(C_6H_5)_3C^+$, which remain in the solution as $AlCl_4^-$ salts; ions of this sort are highly colored due to the extensive delocalization of charge which is possible throughout the three aromatic rings.

The test is significant if positive, but a negative test does not rule out an aromatic structure; some compounds are so unreactive that they do not readily undergo Friedel-Crafts reactions (Chapter 8.1).

PROCEDURE

Heat about 100 mg of *anhydrous* aluminum chloride in a Pyrex test tube held almost horizontally until the material has sublimed to 3 or 4 cm above the bottom of the tube. Allow the tube to cool until it is almost comfortable to touch and then add down the side of the tube about 20 mg of a solid or 1 drop of a liquid unknown, followed by 2 or 3 drops of chloroform. The appearance of a bright color, ranging from red to blue, where the sample and chloroform come in contact with the aluminum chloride, is a positive indication of an aromatic ring.

Positive tests for aryl halides are difficult to obtain directly, and some of the best evidence for their presence involves indirect methods. Elemental analysis will indicate the presence of halogen. If *both* the silver nitrate and sodium iodide–acetone tests are negative, then the compound is most likely a vinyl or aromatic halide, both of which are very unreactive towards silver nitrate and sodium iodide. Distinction between a vinyl and an aromatic halide can be made by means of the aluminum chloride–chloroform test.

B. DERIVATIVES

Two types of derivatives can be used to characterize aromatic hydrocarbons and aryl halides. These are prepared by (1) nitration and (2) side-chain oxidation. The second method involves oxidation of a side chain to a carboxylic acid group. Since carboxylic acids are often solids, they themselves serve as suitable derivatives. The mechanism and equations for this oxidation reaction have been discussed (Chapter 17.1). The two common methods of oxidation use either chromic acid or potassium permanganate; although the experimental details of a permanganate oxidation are given in Chapter 17, procedures for doing both types of oxidations will be given here since much smaller amounts of compound normally must be used.

PROCEDURE

1. Nitration.   Some of the best solid derivatives of aryl halides are mono- and di-nitration products. Two general procedures can be used for nitration. Whenever nitrating a compound, whether it is known or unknown, use care since many of these compounds react vigorously under typical nitration conditions.

Method A: This method yields *m*-dinitrobenzene from benzene or nitrobenzene and the *p*-nitro derivative from chloro- or bromobenzene, benzyl chloride, or toluene. Dinitro derivatives are obtained from phenol, acetanilide, naphthalene, and biphenyl. Add about 1 g of the compound to 4 ml of concentrated sulfuric acid. Add 4 ml of concentrated nitric acid dropwise to this mixture; shake after each addition. Carry out the reaction in a large test tube or small Erlenmeyer flask and heat at 45° for *ca.* 5 min, using a beaker of water to supply the heat. Then pour the reaction mixture onto 25 g of ice and collect the precipitate on a filter. The solid may be recrystallized from aqueous ethanol if needed.

Method B: This method is the best one to use for nitrating halogenated benzenes, since dinitration occurs to give compounds which have higher melting points and are easier to purify than are mononitration products obtained from Method A. The xylenes, mesitylene, and pseudocumene yield trinitro compounds. Follow the procedure used for Method A, except that 4 ml of *fuming* nitric acid should be used in place of the *concentrated* nitric acid. Warm the mixture using a steam cone for 10 min and work up as above. If little or no nitration occurs, substitute *fuming* sulfuric acid for the *concentrated* sulfuric acid. *Carry out this reaction in a hood.*

2. Side-chain oxidation. a. Permanganate method.   Add 1 g of the compound to a solution prepared from 80 ml of water and 4 g of potassium permanganate. Add 1 ml of 3 *N* sodium hydroxide solution and heat the mixture at reflux until the purple color characteristic of the permanganate has disappeared; this will normally take from 30 min to 3 hr. At the end of the reflux period, cool the mixture and carefully acidify it with dilute sulfuric acid. Now heat the mixture for an additional 30 min and cool it again; remove excess brown manganese dioxide (if any) by addition of sodium bisulfite solution. The bisulfite serves to reduce the manganese dioxide to manganous ion, which is water soluble. Collect the solid acid which remains by suction filtration. Recrystallize the acid from benzene or aqueous ethanol. If little or no solid acid is formed, this may be due to the fact that the acid is somewhat water-soluble. In this case, extract the aqueous layer with chloroform, ether, or dichloromethane, dry the organic extracts and remove the organic solvent by means of a steam bath in the hood. Recrystallize the acid which remains. In this particular method, the presence of base during the oxidation often means that some silicic acid will form on acidification; thus, purification before determining the melting point is necessary.

b. Chromic acid method. Dissolve 7 g of sodium dichromate in 15 ml of water, and add 2 to 3 g of the compound to be oxidized. Add 10 ml of concentrated sulfuric acid to the mixture with mixing and cooling. Attach a reflux condenser to the flask and heat gently until a reaction ensues; as soon as the reaction begins, remove the flame and cool the mixture if necessary. After spontaneous boiling subsides, heat the mixture at reflux for 2 hr. Pour the reaction mixture into 25 ml of water and collect the precipitate by filtration. Transfer the solid to a flask and add 20 ml of 2 N sulfuric acid, and then warm the flask on a steam cone with stirring. Cool the mixture, collect the precipitate, and wash it with about 20 ml of cold water. Dissolve the residue in 20 ml of 1.5 N sodium hydroxide solution and filter the solution. Add the filtrate, with stirring, to 25 ml of 4 N sulfuric acid. Collect the new precipitate, wash with water, and recrystallize from either toluene or aqueous ethanol.

Lists of possible aromatic hydrocarbons and halides appear in Tables 25.3 and 25.4, along with derivatives.

ALCOHOLS (Table 25.5)

A. Classification tests

The tests for the presence of a hydroxy group not only detect the presence of the group but may also give an indication as to whether it occupies a primary, secondary, or tertiary position.

1. Chromic acid in acetone can be used to detect the presence of a hydroxy group, provided that it has been shown previously that the molecule does not contain an aldehyde function. The reactions and experimental procedures for this test are given in Chapter 15.2. It has been pointed out that chromic acid does not distinguish between primary and secondary alcohols, for primary and secondary alcohols *both* give a positive test whereas tertiary alcohols do not.

2. The Lucas test can be used to distinguish between primary, secondary, and tertiary alcohols. The reagent used is a mixture of concentrated hydrochloric acid and zinc chloride, which on reaction with alcohols converts them to the corresponding alkyl chlorides. With this reagent, primary alcohols give no appreciable reaction, secondary alcohols react more rapidly, and tertiary alcohols react very rapidly. A positive test depends on the fact that the alcohol is soluble in the reagent, whereas the alkyl chloride is not; thus, the formation of a second layer or an emulsion comprises a positive test.

1 line long

The solubility of the alcohol in the reagent places limitations on the utility of the test, and in general, only monofunctional alcohols with six or less carbon atoms, as well as

PRIMARY: $$RCH_2OH + HCl \xrightarrow{ZnCl_2} \text{No reaction} \tag{6}$$

SECONDARY: $$R_2CHOH + HCl \xrightarrow{ZnCl_2} R_2CHCl + H_2O \tag{7}$$

TERTIARY: $$R_3COH + HCl \xrightarrow{ZnCl_2} R_3CCl + H_2O \tag{8}$$

polyfunctional alcohols, can be used. Note the similarity of this reaction with the nucleophilic displacement reactions between alcohols and hydrohalic acids which have been discussed in Chapter 12.2. In the Lucas test the presence of zinc chloride, which is a Lewis acid, greatly increases the reactivity of alcohols toward hydrochloric acid.

PROCEDURE

Add 10 ml of the hydrochloric acid–zinc chloride reagent (Lucas reagent) to about 1 ml of the compound in a test tube. Stopper the tube and shake; allow the mixture to stand at room temperature. Try this test with a primary, a secondary, and a tertiary alcohol and note the time which is required for the formation of the alkyl chloride, which will appear either as a second layer or as an emulsion.

3. The ceric nitrate reagent can also be used as a qualitative test for alcohols. Although this reagent has been used primarily for phenols, it does give a positive test with alcohols. Discussion about and experimental procedures for this test are given under Phenols, p. 448.

B. Derivatives

Two common derivatives of alcohols are the urethans and the benzoate esters; the former are best for primary and secondary alcohols, whereas the latter are useful for all types of alcohols.

1. Urethans. When an alcohol is allowed to react with an aryl substituted isocyanate, $ArN{=}C{=}O$, addition of the alcohol occurs to give a urethan [equation (9)].

$$ArN{=}C{=}O + R'OH \longrightarrow ArNH{-}\overset{\displaystyle O}{\overset{\displaystyle \|}{C}}{-}OR' \tag{9}$$

Some commonly used isocyanates are α-naphthyl, p-nitrophenyl, and phenyl isocyanate. A major side reaction which occurs is that of water with the isocyanate; water hydrolyzes the isocyanate to an amine, and the amine reacts with more isocyanate to give a disubstituted urea [equations (10) and (11)]. Since the ureas are high melting

$$ArN{=}C{=}O + H_2O \rightarrow (ArNH{-}COOH) \rightarrow ArNH_2 + CO_2 \tag{10}$$
$$\text{A Carbamic Acid}$$
$$\text{(unstable)}$$

$$ArN{=}C{=}O + ArNH_2 \longrightarrow ArNH{-}\overset{\overset{\displaystyle O}{\|}}{C}{-}NHAr \qquad (11)$$

A Disubstituted Urea

amides, owing to their symmetry, their presence makes purification of the desired urethan quite difficult. In using this procedure, take precautions to insure that the alcohol is anhydrous. This procedure works best for alcohols which are water-insoluble and can therefore be obtained easily in anhydrous form.

This type of derivative can also be useful for phenols; the procedure given here has been generalized so that it can be used for alcohols and phenols. Other derivatives of phenols will be given below.

PROCEDURE

Place 1 g of the *anhydrous* alcohol or phenol in a test tube and add 0.5 ml of phenyl isocyanate or α-naphthyl isocyanate. (If a phenol is used, add 2 or 3 drops of pyridine as a catalyst.) If the reaction is not spontaneous, affix a drying tube to the test tube, and warm the reaction mixture over a steam bath for 5 min. Cool the mixture in a beaker of ice and scratch the sides of the test tube to induce crystallization. Purify the crude solid so obtained by recrystallizing it from carbon tetrachloride or petroleum ether. (*Note:* 1,3-Di-(α-naphthyl)-urea has mp 293° and 1,3-diphenylurea (carbanilide) has mp 237°; if your product shows one of these melting points, repeat the preparation taking greater care to maintain anhydrous conditions.)

2. 3,5-Dinitrobenzoates.   The reaction between 3,5-dinitrobenzoyl chloride and an alcohol gives the corresponding ester. This method is useful for primary, secondary, and tertiary alcohols, especially those which are water-soluble and which are likely to contain traces of water.

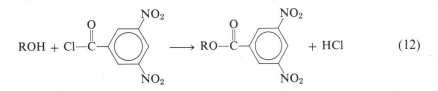

$$ROH + Cl{-}\overset{\overset{\displaystyle O}{\|}}{C}{-}\underset{NO_2}{\overset{NO_2}{\bigcirc}} \longrightarrow RO{-}\overset{\overset{\displaystyle O}{\|}}{C}{-}\underset{NO_2}{\overset{NO_2}{\bigcirc}} + HCl \qquad (12)$$

PROCEDURE

Method A.   (*Note:* 3,5-Dinitrobenzoyl chloride is highly reactive toward water; it should be used immediately after weighing, taking care to minimize its exposure to air and to keep the bottle tightly closed.) Mix 2 ml of the alcohol with about 0.5 g of 3,5-dinitrobenzoyl chloride and 0.5 ml of pyridine in a small flask bearing a reflux condenser. Boil the mixture gently for 30 min (15 min is sufficient for a primary alcohol). Cool the solution and add *ca.* 10 ml of 5% aqueous sodium bicarbonate solution. Cool this solution in an ice bath and collect the crude crystalline product. Recrystallize the product from aqueous ethanol. A minimum volume of solvent should be used; adjust the composition of the solvent by adding just enough water to the alcohol so that the product dissolves in the hot solution but yields crystals when cooled.

Method B.   Carry out this reaction in the hood, if possible, or use a gas trap such as that shown in Figure 5.1. To a 50-ml flask, fitted with a reflux condenser (bearing a gas trap if a hood is not available) is added 1 g of 3,5-dinitrobenzoic acid, 3 ml of thionyl chloride, and 1 drop of pyridine. The mixture is heated at reflux until the acid has dissolved and then for an additional 10 min. The total reflux time should be about 30 min. Equip the flask for simple distillation as in Figure 2.2. Cool the receiving flask with an ice-salt bath and attach the vacuum adapter to an aspirator by means of a safety trap such as that shown in Figure 2.16. Evacuate the system and distil the excess thionyl chloride by heating with a steam bath. When the thionyl chloride has been removed, cautiously release the vacuum. Discard the excess thionyl chloride by pouring slowly down a drain *in the hood* or by putting it in a container for recovered thionyl chloride. To the pot residue, which is 3,5-dinitrobenzoyl chloride, add in the same flask 2 ml of the alcohol and 0.5 ml of pyridine. Fit the flask with a reflux condenser bearing a drying tube. Proceed with the period of reflux following the procedure of Method A.

A list of alcohols and their derivatives is given in Table 25.5.

PHENOLS (Table 25.6)

A. Classification tests

There are several tests which can be used to detect the presence of a phenolic hydroxy group: (1) bromine water, (2) ceric nitrate reagent, and (3) ferric chloride solution. In addition to these, solubility tests give a preliminary indication of a phenol since phenols are soluble in 1.5 N sodium hydroxide solution but generally insoluble in 5% sodium bicarbonate solution. However, care must be exercised here because phenols containing highly electronegative groups are stronger acids and may be soluble in 5% sodium bicarbonate. Examples are 2,4,6-tribromophenol and 2,4-dinitrophenol.

1. Bromine water.   Phenols, in general, are highly reactive toward electrophilic substitution, and consequently are brominated readily by bromine water. The rate of bromination is much greater in water than in carbon tetrachloride solution. The water, being more polar than carbon tetrachloride, increases the ionization of bromine and thus enhances the ionic bromination mechanism. Although hydrogen bromide is liberated, it is not observed when using water as the solvent. Phenols are so reactive

$$\text{(13)}$$

that all unsubstituted positions *ortho* and *para* to the hydroxy group are brominated. The brominated compounds so formed are often solids and can also be used as derivatives (see below). Aniline and substituted anilines are also very reactive toward bromine and react analogously; however, solubility tests can be used to distinguish between anilines and phenols.

PROCEDURE

Prepare a 1% aqueous solution of the unknown. If necessary, dilute sodium hydroxide solution may be added dropwise to bring the phenol into solution. Add a saturated solution of bromine in water dropwise to this solution; continue addition until the bromine color remains. Note how much bromine water was used and try this experiment on phenol and aniline for purposes of comparison.

2. Ceric nitrate reagent. Alcohols and phenols are capable of replacing nitrate ions in complex cerate anions, resulting in a change from a yellow to a red solution:

$$(NH_4^{\oplus})_2Ce(NO_0)_6^{\ominus} + ROH \rightarrow (NH_4^{\oplus})_2Ce(OR)(NO_3)_5^{\ominus} + HNO_3 \qquad (14)$$

yellow   red

Alcohols and phenols having no more than ten carbon atoms give a positive test. Alcohols give a red solution. Phenols give a brown to greenish-brown precipitate in aqueous solution; in dioxane a red to brown solution is produced. Aromatic amines may be oxidized by the reagent and give a color indicating a positive test.

PROCEDURE

Dissolve about 20 mg of a solid or 1 drop of a liquid unknown in 1–2 ml of water and add 0.5 ml of the ceric ammonium nitrate reagent; shake and note the color. If the unknown in insoluble in water, dissolve it in 1 ml of dioxane before adding 0.5 ml of the reagent.

3. Ferric chloride test. Most phenols and enols react with ferric chloride to give colored complexes. The colors vary, depending not only on the nature of the phenol or enol, but also on the solvent, concentration, and time of observation. Some phenols which do not give coloration in aqueous or alcoholic solution do so in chloroform solution, especially after addition of a drop of pyridine The nature of the colored complexes is still uncertain; they may be ferric phenoxide salts that absorb visible light to give an excited state in which electrons are delocalized over both the iron atoms and the conjugated organic system. The production of a color is typical of phenols and enols; however, many of them do *not* give colors, so that a negative ferric chloride test must not be taken as significant without supporting information (*e.g.*, the ceric nitrate and bromine water tests).

PROCEDURE

Dissolve 30 to 50 mg of the unknown compound in 1–2 ml of water (or a mixture of water and ethanol if the compound is not water-soluble) and add several drops of a 2.5% aqueous solution of ferric chloride. Most phenols produce red, blue, purple, or green colorations; enols give red, violet, or tan colorations.

B. DERIVATIVES

Two useful solid derivatives of phenols are α-naphthyl urethan and bromo derivatives. The preparation of urethans has been discussed previously in this chapter (Alcohols, page 446) and whereas any substituted urethan could be prepared, the majority of the derivatives reported are the α-naphthylurethans. They are suggested as the urethans of choice.

PROCEDURE

The bromination of phenols may be carried out as follows. Prepare a brominating solution by dissolving 10 g of potassium hydroxide in 60 ml of water and adding 6 g of bromine. Dissolve 1 g of the phenol in water, ethanol, acetone, or dioxane, and add the brominating solution to it dropwise. Add just enough of the bromine solution to impart a yellow bromine color to the solution. Add about 50 ml of water to the mixture. Shake the mixture vigorously and remove the solid by filtration. Wash the solid with a dilute solution of sodium bisulfite to remove excess bromine and recrystallize it from ethanol or aqueous ethanol.

A listing of phenols and their derivatives is given in Table 25.6.

CARBOXYLIC ACIDS (Table 25.7)

A. CLASSIFICATION TEST

One of the best qualitative tests for the carboxylic acid group is solubility in basic solutions. Carboxylic acids are soluble in both 1.5 N sodium hydroxide solution and in 5% sodium bicarbonate solution, from which they can be regenerated by addition of acid. Solubility properties have been discussed previously in this chapter.

Information of a quantitative nature can be obtained about carboxylic acids, this being the *equivalent weight* or *neutralization equivalent*. This value can be obtained by titrating a known weight of the acid with a known volume of standardized base solution. Further discussion of the equivalent weight is given in Chapter 17.1.

PROCEDURE

Dissolve an accurately weighed sample (about 0.2 grams) of the acid in 50 to 100 ml of water or ethanol or a mixture of the two. It may be necessary to warm the mixture to dissolve the compound completely. Titrate the solution with a previously standardized sodium hydroxide solution having a concentration of about 0.1 N. Use phenolphthalein as the indicator, and from these data calculate the equivalent weight as described in Chapter 17.1.

B. Derivatives

Three good solid derivatives of carboxylic acids are: (1) amides, (2) anilides, and (3) p-toluidides. Each of these three derivatives is prepared from the corresponding acid chloride by treatment of the latter with ammonia, aniline, or p-toluidine. The amides are generally less satisfactory than the other two, for they tend to be more soluble in water and as a result, harder to isolate. The acid chlorides are most conveniently prepared from the acid, or its salt, and thionyl chloride.

$$RCOOH \text{ (or } RCOO^{\ominus} Na^{\oplus}) + SOCl_2 \rightarrow RCOCl + SO_2 + HCl \text{ (or NaCl)} \quad (15)$$

AMIDES: $RCOCl + 2 NH_3 \xrightarrow{\text{cold}} RCONH_2 + NH_4^{\oplus} Cl^{\ominus}$ (16)

ANILIDES: $RCOCl + 2 C_6H_5NH_2 \rightarrow RCONHC_6H_5 + C_6H_5NH_3^{\oplus} Cl^{\ominus}$ (17)

TOLUIDIDES: $RCOCl + 2\ p\text{-}CH_3C_6H_4NH_2 \rightarrow RCONHC_6H_4CH_3$

$$+ p\text{-}CH_3C_6H_4NH_3^{\oplus} Cl^{\ominus} \quad (18)$$

PROCEDURE

The acid chloride may be prepared by heating, under reflux, 5 ml of thionyl chloride and 1 g of the acid or its sodium salt. Heat for 15–30 min, and use care to protect the apparatus from moisture; keep the vapors from filling the laboratory [see Chapter 17 (Experimental Procedure) for additional comments about using thionyl chloride].

1. Amides. In the hood, pour the mixture containing the acid chloride and unchanged thionyl chloride into 15 ml of *ice-cold*, concentrated ammonia; be very careful, for a vigorous reaction occurs. Collect the precipitated amide on a filter and recrystallize from water or aqueous ethanol. Determine the melting point of the product.

2. Anilides and p-toluidides. Prepare the acid chloride as before, using 1 g of acid or its salt and 2 ml of thionyl chloride. Heat the mixture at gentle reflux for 30 min, and then cool it. To the cold mixture, add a solution of 1 to 2 g of the appropriate amine (aniline or p-toluidine) dissolved in 30 ml of benzene. Heat the mixture on a steam bath for *ca.* 2 min, and then decant the benzene solution into a separatory funnel. Wash the organic layer with 2 ml of water, 5 ml of 1.5 N hydrochloric acid, 5 ml of 1.5 N sodium hydroxide solution, and finally with 2 ml of water. Evaporate the benzene and recrystallize the amide from either water or aqueous ethanol.

A list of carboxylic acids and their derivatives is given in Table 25.7.

AMINES (Tables 25.8 and 25.9)

A. Classification tests

Two common qualitative tests for amines are the Hinsberg test and the nitrous acid reaction. In addition to the initial solubility test, which may indicate the presence of an amine, these tests, if properly performed and intelligently interpreted, will usually allow the classification of an amine as either primary, secondary, or tertiary.

1. The Hinsberg test.   The reaction between primary or secondary amines and benzenesulfonyl chloride [equations (19) and (21), respectively] yields the corresponding substituted benzenesulfonamide. The reaction is carried out in excess base; if the amine is primary, the sulfonamide, which has an acidic amido hydrogen, is converted by base [equation (20)] to the normally soluble potassium salt. Thus, with few exceptions (see below), primary amines react with benzenesulfonyl chloride to provide homogeneous reaction mixtures. Acidification of this solution regenerates the insoluble primary benzenesulfonamide. On the other hand, the benzenesulfonamides of secondary amines bear no acidic (amido) hydrogens: They typically are insoluble in both acid and base. Therefore, secondary amines react to yield heterogeneous reaction mixtures, with production of either an oily organic layer or a solid precipitate.

PRIMARY:

$$\mathrm{RNH_2} + \underset{\text{Insoluble in water}}{\langle \bigcirc \rangle - \mathrm{SO_2Cl}} \xrightarrow{\text{KOH}} \langle \bigcirc \rangle - \mathrm{SO_2NHR} + \mathrm{KCl} + \mathrm{H_2O} \qquad (19)$$

$$\overset{\text{excess}}{\underset{\text{HCl}}{}} \updownarrow \overset{\text{excess}}{\underset{\text{KOH}}{}}$$

$$\underset{\text{Soluble in water}}{\langle \bigcirc \rangle - \mathrm{SO_2\overset{\ominus}{N}R} \ \mathrm{K^{\oplus}} + \mathrm{H_2O}} \qquad (20)$$

SECONDARY:

$$\mathrm{R_2NH} + \underset{\text{Insoluble in water}}{\langle \bigcirc \rangle - \mathrm{SO_2Cl}} \xrightarrow{\text{KOH}} \langle \bigcirc \rangle - \mathrm{SO_2NR_2} + \mathrm{KCl} + \mathrm{H_2O} \qquad (21)$$

$$\downarrow \overset{\text{excess}}{\underset{\text{KOH}}{}}$$

No reaction

The distinction between primary and secondary amines then depends on the different solubility properties of their benzenesulfonamide derivatives. However, the potassium salts of certain primary sulfonamides are not completely soluble in basic solution. Examples are generally found among those primary amines of higher molecular weight and those having cyclic alkyl groups.[8] To avoid confusion and possible misassignment of a primary amine as secondary, the basic solution is separated from the oil or solid and acidified. The formation of an oil or precipitate indicates that the derivative is partially soluble and that the amine is primary. It is important not to over-acidify the solution because this may precipitate certain side products which may form, resulting in an ambiguous test. The original oil or solid should be tested for solubility in water and acid to substantiate the test for a primary or secondary amine.

[8] P. E. Fanta and C. S. Wang, *Journal of Chemical Education*, **41**, 280 (1964).

Tertiary amines behave somewhat differently.[9] Typically, under the conditions of the Hinsberg test, the processes shown in equation (22) provide for the conversion of benzenesulfonyl chloride to potassium benzenesulfonate with recovery of the tertiary amine. Because tertiary amines are nearly always insoluble in the aqueous potassium hydroxide solution, the test mixture remains heterogeneous. It is worthwhile to note relative densities of the oil layer and of the test solution. Benzenesulfonamides are generally more dense than the solution, whereas the amines are less dense. The oil is separated and tested for solubility in aqueous acid; solubility usually indicates a tertiary amine.

TERTIARY:

$$R_3N + \text{C}_6\text{H}_5-SO_2Cl \longrightarrow \text{C}_6\text{H}_5-SO_2-NR_3^{\oplus}Cl^{\ominus} \xrightarrow{KOH}$$

$$\text{C}_6\text{H}_5-SO_3^{\ominus}K^{\oplus} + NR_3 + H_2O \quad (22)$$

$$\xrightarrow{NR_3} \text{C}_6\text{H}_5-SO_2NR_2 + NR_4^{\oplus}Cl^{\ominus} \quad (23)$$

$$\text{C}_6\text{H}_5-SO_2Cl \xrightarrow{KOH} \text{C}_6\text{H}_5-SO_3^{\ominus}K^{\oplus} + KCl + H_2O \quad (24)$$

$$\xrightarrow[\text{including:}]{ArNR_2} \begin{array}{l}\text{Complex}\\ \text{mixture,}\end{array} \quad \text{C}_6\text{H}_5-SO_2-NRAr \quad (25)$$

The test procedure should be followed as closely as possible. The procedure is designed to minimize complications which may arise because of side reactions of tertiary amines with benzenesulfonyl chloride. As shown in equation (23), the initial adduct is subject to further reaction with another molecule of amine to produce the benzenesulfonamide of a secondary amine. The relative competing rates of reaction of the adduct with hydroxide ion and with amine do not favor equation (23), particularly when excess amine is avoided, yet the formation of *small* amounts of an insoluble product may, through confusion, cause an amine to be incorrectly designated as secondary. Adduct formation such as shown in equation (22) is generally less of a problem with tertiary arylamines, because they are normally much less soluble in the test solution and are less nucleophilic than trialkylamines. The competing hydrolysis of benzenesulfonyl chloride by hydroxide ion [equation (24)] allows the recovery of most of the amine. In addition, however, tertiary arylamines are subject to other side reactions producing a complex mixture of mainly insoluble products [equation (25)]. Because benzenesulfonyl chloride reacts more slowly with

[9]Historically and nearly invariably, sources of information have asserted that tertiary amines do not react with benzensulfonyl chloride. For an interesting and timely refutation of this widely accepted myth, see C. R. Gambill, T. D. Roberts, and H. Shechter, *Journal of Chemical Education,* **49**, 287 (1972).

tertiary arylamines than with hydroxide ion, it is also possible to minimize the attendant ambiguity caused by these reactions of the tertiary amine by keeping the reaction time short and the temperature low.

In summary of the discussion in the preceding paragraph: Tertiary amines may produce small amounts of insoluble products if the concentration of the amine in the test solution is too high and if the reaction time is too long. By following the directions of the procedure, and *taking care not to interpret small amounts of insoluble product as a positive test for secondary amines,*[10] the Hinsberg test may be used with confidence to designate an amine as primary, secondary, or tertiary.

PROCEDURE

Mix 10 ml of 2 N aqueous potassium hydroxide, 0.2 ml or 0.2 g of the amine, and 0.7 ml of benzenesulfonyl chloride (*Caution:* It is a lachrymator) in a test tube. Stopper the tube and shake the mixture *vigorously,* with cooling if necessary, until the odor of benzenesulfonyl chloride is gone. This, in even the slowest case, should take no more than about 5 min. Test the solution to see that it is still basic; if it is not, add sufficient 2 N potassium hydroxide solution until it is.

If the mixture has formed two layers or a precipitate, note the relative densities, and separate the oil or solid by decantation or filtration. Test an oil for solubility in dilute hydrochloric acid. The sulfonamide of a secondary amine will be insoluble, whereas an amine will dissolve, at least partially. If this solubility test indicates an amine, it may be either a tertiary amine or one of certain secondary amines that react with benzenesulfonyl chloride only very slowly due to steric bulk. Test a solid for solubility in water and in dilute acid. The potassium salt of a sulfonamide which is insoluble in base solution will usually be soluble in water; the sulfonamide which forms from the potassium salt when placed in acid will be insoluble in that medium. A solid sulfonamide of a secondary amine will be insoluble in both water and acid. Acidify the solution from the original reaction mixture to pH 4 using pH indicator paper or a few drops of Congo red indicator solution; the formation of a precipitate or oil indicates a primary amine.

If the original mixture has not formed two layers, the test is indicative of a primary amine. Acidify the solution to pH 4; a sulfonamide of a primary amine will precipitate.

2. Nitrous acid test. This test should always be used to confirm and extend the benzenesulfonyl chloride test.

PROCEDURE

Dissolve about 50 mg of a solid or 2 drops of a liquid amine in 2 ml of 2 N hydrochloric acid and cool the solution to 0–5° in an ice bath. Add about 5 drops of a cold 20% aqueous solution of sodium nitrite. Maintain the solution at 0–5°.

---

[10] Tertiary amines often contain quantities of secondary amines as impurities. If it was not possible to obtain a reliable boiling point, and the amine was not carefully distilled, small quantities of precipitate may form for this reason also, obscuring the test results.

(a) Immediate evolution of a *colorless* gas (nitrogen) is indicative of a primary aliphatic amine.

$$RNH_2 \xrightarrow[\text{HCl}]{\text{NaNO}_2} [RN_2^{\oplus}Cl^{\ominus}] \xrightarrow{\text{H}_2\text{O}} N_2 + ROH + RCl + \text{alkenes} \qquad (26)$$

(b) If no nitrogen was evolved, but an insoluble yellow or orange liquid separated, the unknown amine is indicated to be a secondary amine, either aliphatic or aromatic. The colored product is an N-nitroso compound

$$R_2NH + HONO \longrightarrow R_2N-N=O + HOH \qquad (27)$$

Tertiary amines also occasionally yield a small amount of yellow oil, although this is not common under the conditions of this test.

$$2\,R_2NCH_2R' + 4\,HONO \longrightarrow 2\,R_2N-N=O + 2\,R'CHO + N_2O + 3\,H_2O \quad (28)$$

Trace formation of a yellow to orange liquid should not be regarded as a conclusive test for a secondary amine. It is good practice to perform the nitrous acid test on known amines for purposes of comparison.

(c) Because primary aromatic amines give diazonium salts which are stable at low temperatures, the absence of gas evolution, oil formation, and yellow coloration of the solution (see below) may indicate this type of amine. In this case, add a few drops of the cold reaction mixture to a cold solution of 50 mg of β-naphthol in 2 ml of 2 N sodium hydroxide. The formation of an orange to red azo dye is indicative of a primary aromatic amine.

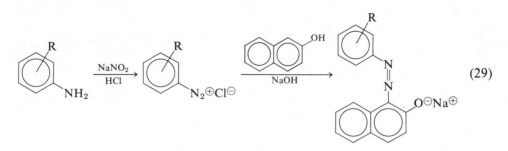

$$(29)$$

(d) If the cold reaction mixture has developed a yellow coloration, this may be indicative of an N,N-dialkylaniline with an unsubstituted *para* position; the *p*-nitroso derivative is green, but in acidic solution, its salt is yellow. Neutralization should provide a green solution or precipitate.

$$R_2N-\bigcirc + HONO + HCl \longrightarrow R_2\overset{\oplus}{N}H-\bigcirc-N=O + H_2O \qquad (30)$$
$$Cl^{\ominus}$$

(e) If no appreciable reaction was observed, the unknown is indicated to be a tertiary aliphatic amine or a tertiary aromatic amine with a substituent at the *para* position.

B. DERIVATIVES

Suitable derivatives of primary and secondary amines are the benzamides and benzenesulfonamides.

$$RNH_2 \text{ (or } R_2NH) + C_6H_5COCl \xrightarrow{\text{Pyridine}} C_6H_5\overset{\overset{\displaystyle O}{\|}}{C}-NHR \text{ (or } C_6H_5\overset{\overset{\displaystyle O}{\|}}{C}-NR_2) \qquad (31)$$

$$RNH_2 \text{ (or } R_2NH) + C_6H_5SO_2Cl \xrightarrow{\text{NaOH}} C_6H_5SO_2-NHR \text{ (or } C_6H_5SO_2NR_2) \qquad (32)$$

PROCEDURE

1. Benzenesulfonamides. The method of preparing the benzenesulfonamides has been discussed under the Hinsberg test. The derivatives can be prepared using that method, but sufficient amounts of material should be used so that the final product can be purified by recrystallization from 95% ethanol. If the derivative is obtained as an oil, it *may* crystallize with scratching in the presence of the mother liquor with a stirring rod. If the oil cannot be made to crystallize, separate it and dissolve in a minimum quantity of hot ethanol and allow to cool. Note that some amines do not give *solid* benzenesulfonamide derivatives.

2. Benzamides. The benzamides can be prepared as follows. Dissolve 0.5 g of the amine in 5 ml of dry pyridine and 10 ml of dry benzene. Add, dropwise, 0.5 ml of benzoyl chloride. Heat the resulting mixture in a water bath for 30 min at about 60–70°, and then pour the solution into 100 ml of water. Separate the benzene layer and wash the aqueous layer once with 10 ml of benzene. Combine the organic extracts and wash them with water followed by 5% sodium carbonate solution. After drying over magnesium sulfate and removing the drying agent by filtration, distil the mixture to remove most of the benzene. When the volume is 3 to 4 ml, add 20 ml of hexane to the mixture. Remove the solid which is formed by filtration and wash it with a small volume of hexane. Recrystallize the product from one of the following solvents: cyclohexane–hexane mixtures, cyclohexane–ethyl acetate mixtures, ethanol, or aqueous ethanol.

Although one or both of the above methods are satisfactory with most primary and secondary amines, tertiary amines do not undergo the same reactions. In general, one must take advantage of the fact that tertiary amines do form salts. Two useful crystalline salts are the ones from methyl iodide and picric acid:

$$R_3N\colon + CH_3I \longrightarrow [R_3\overset{\oplus}{N}-CH_3]\ I^{\ominus} \qquad (33)$$

$$R_3N\colon + HO\!\!-\!\!\langle\text{ring, NO}_2\rangle\!\!-\!\!NO_2 \longrightarrow R_3\overset{\oplus}{N}H\ ^{\ominus}O\!\!-\!\!\langle\text{ring, NO}_2\rangle\!\!-\!\!NO_2 \qquad (34)$$

3. Methiodides. The methyl iodide derivative may be prepared as follows: Mix 0.5 g of the amine with 0.5 ml of methyl iodide and warm the test tube with a water bath for several minutes. Cool the test tube in an ice-water bath; the tube may be scratched with a rod to help induce crystallization. Purify the product by recrystallization from absolute ethanol or methanol or from ethyl acetate.

4. Picrates. The picric acid derivative (often called the picrate) may be prepared by mixing 0.3 to 0.5 g of the compound with 10 ml of 95% ethanol. If solution is not complete, remove the excess solid by filtration. Add 10 ml of a saturated solution of picric acid in 95% ethanol to the mixture and heat the mixture to boiling. Cool the solution slowly and remove the yellow crystals of the picrate salt by filtration. Recrystallize the salt from ethanol.

A list of primary and secondary amines and their derivatives can be found in Table 25.8, whereas tertiary amines and derivatives appear in Table 25.9.

NITRO COMPOUNDS (Table 25.10)

A. CLASSIFICATION TEST

Ferrous hydroxide test. Organic compounds that are oxidizing agents will oxidize ferrous hydroxide (blue) to ferric hydroxide (brown). The most common organic compounds that function in this way are the *nitro compounds*, both aliphatic and aromatic, which are in turn reduced to amines in the reaction. Other less common types of com-

$$RNO_2 + 6 \underline{Fe(OH)_2} + 4 H_2O \longrightarrow RNH_2 + 6 \underline{Fe(OH)_3} \qquad (35)$$
$$ \text{blue} \text{brown}$$

pounds which give the same test are nitroso compounds, hydroxylamines, alkyl nitrates, alkyl nitrites, and quinones.

PROCEDURE

In a 10 × 75-mm or smaller test tube,[11] mix about 20 mg of a solid or 1 drop of a liquid unknown with 1.5 ml of freshly prepared 5% ferrous ammonium sulfate solution. Add 1 drop of 6 N sulfuric acid and 1 ml of 2 N potassium hydroxide in methanol. Stopper the tube immediately and shake it. A positive test is indicated by the blue precipitate turning rust-brown within one minute. (A slight darkening or greenish coloration of the blue precipitate should not be considered a positive test.)

B. DERIVATIVES

Two different types of derivatives of nitro compounds can be prepared. Aromatic nitro compounds can be di- and trinitrated with nitric acid and sulfuric acid. Discussion of and procedures for nitration have been given under "Aryl Halides," and the student is referred there for additional information.

---

[11] A small test tube is used in order to lessen the exposure of the reagent to air, which may bring about some oxidation.

The other method for preparation of a derivative can be utilized for both aliphatic and aromatic nitro compounds. This involves the reduction of the nitro compound to the corresponding primary amine, followed by conversion of the amine to a benzamide or benzenesulfonamide, as described under "Amines." The reduction is most often carried out with tin and hydrochloric acid.

$$RNO_2 \text{ or } ArNO_2 \xrightarrow[\text{HCl}]{\text{Sn}} RNH_2 \text{ or } ArNH_2 \tag{36}$$

PROCEDURE

Carry out the reduction of the nitro compound by combining 1 g of the compound and 2 g of granulated tin in a small flask. Attach a reflux condenser and add, in small portions, 20 ml of 3 N hydrochloric acid. Shake after each addition. After addition is complete, warm the mixture using a steam bath for 10 min. If the nitro compound is insoluble, add 5 ml of 95% ethanol to increase its solubility. Decant the warm, homogeneous solution into 10 ml of water, and add enough 12 N sodium hydroxide solution so that the tin hydroxide completely dissolves. Extract the basic solution with several 10-ml portions of ether. Dry the ether solution over potassium hydroxide pellets and remove the ether by distillation.

The residue contains the primary amine. Convert it to one of the derivatives described under "Amines."

A list of some nitro compounds and their derivatives appears in Table 25.10.

ESTERS (Table 25.11)

A. Classification test

A test for the presence of the ester group involves the use of hydroxylamine and ferric chloride. The former converts the ester to a hydroxamic acid, which then complexes with Fe(III) to give a colored species.

$$\underset{\substack{\| \\ O}}{R-C-OR'} + H_2NOH \longrightarrow \underset{\substack{\| \\ O}}{R-C-NHOH} + R'OH \tag{37}$$

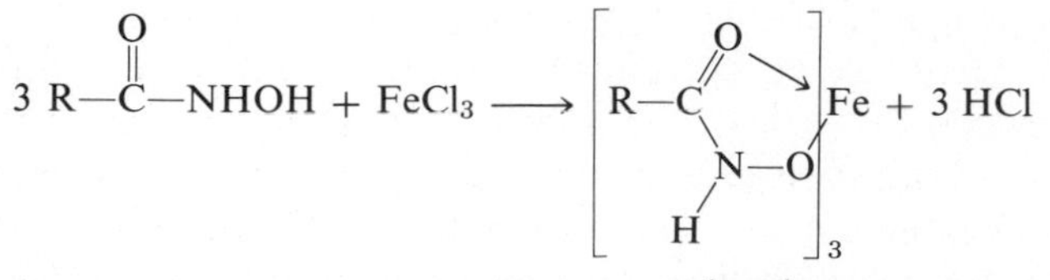

Hydroxamic
Acid

$$3\ \underset{\substack{\| \\ O}}{R-C-NHOH} + FeCl_3 \longrightarrow \left[R-C \underset{\substack{N-O \\ | \\ H}}{\overset{O}{\diagup}} Fe \right]_3 + 3\ HCl \tag{38}$$

colored

All carboxylic acid esters (including polyesters and lactones) give magenta colors which vary in intensity depending on structural features in the molecule. Acid chlorides and anhydrides also give positive tests. Formic acid produces a red color, but other free acids give negative tests. Primary or secondary aliphatic nitro compounds give a positive test because ferric chloride reacts with the *aci* form which is present in basic solution. Most imides give positive tests. Some amides, but not all, give light magenta coloration, whereas most nitriles give a negative test. (A modification of the procedure below will be given later which will yield a positive test for amides and nitriles.)

PROCEDURE

Before performing the final test, it is necessary to run a preliminary (or "blank") test since some compounds will give a positive test even though they do not contain an ester linkage.

1. Preliminary test. Mix 1 ml of 95% ethanol and 50–100 mg of the compound to be tested, and add 1 ml of 1 *N* hydrochloric acid. Note the color which is produced when 1 drop of 5% aqueous ferric chloride solution is added. If the color is orange, red, blue, or violet, the following test for the ester group does not apply and cannot be used.

2. Final test. Mix 40–50 mg of the unknown, 1 ml of 0.5 *M* hydroxylamine hydrochloride in 95% ethanol, and 0.2 ml of 6 *N* sodium hydroxide. Heat the mixture to boiling, and after cooling it slightly, add 2 ml of 1 *N* hydrochloric acid. If the solution is cloudy, add more (about 2 ml) 95% ethanol. Add 1 drop of 5% ferric chloride and observe the color. Add more ferric chloride solution if the color does not persist, and continue to add it until it does. Compare the color which you obtain here with that from the preliminary test. If the color is burgundy or magenta, as compared to the yellow color in the preliminary experiment, the presence of an ester group is indicated.

B. QUANTITATIVE TEST

Once an ester group has been detected, it is possible to obtain information about its equivalent weight. This value, termed the *saponification equivalent*, has already been discussed in Chapter 17.2; the principles and equations involved in this technique are presented there.

PROCEDURE

Dissolve approximately 3 g of potassium hydroxide in 60 ml of 95% ethanol. Allow the small amount of insoluble material to settle to the bottom, and fill a 50-ml buret with the clear solution by decantation. Measure exactly 25.0 ml of the alcoholic solution into each of two 125-ml Erlenmeyer flasks. Weigh *accurately* into one of the flasks a 0.3–0.4-g sample of pure, dry ester; the other basic solution will be used as a "blank." Fit each of the flasks with a reflux condenser by means of a rubber stopper which has been washed with dilute sodium hydroxide solution and rinsed with distilled water.

Heat the solutions in both flasks at a gentle reflux for 1 hr. When the flasks have cooled, loosen the stoppers in the flasks, and rinse each condenser and stopper with about 10 ml of distilled water, catching the rinse water in the flask. Add phenolphthalein and titrate each of the solutions with *standardized* hydrochloric acid which is approximately 0.5 *N*.

The difference in the volumes of hydrochloric acid required to neutralize the base in the flask containing the sample and in the "blank" flask represents the amount of potassium hydroxide which reacted with the ester. The volume difference (in milliliters) multiplied by the normality of the hydrochloric acid equals the number of *milli*equivalents of potassium hydroxide consumed. Using the titration data, calculate the saponification equivalent of the unknown ester.

Should the ester not completely saponify in the allotted time, heat under reflux for longer periods (2–4 hr). In some cases, higher temperatures may be required; if so, diethylene glycol must be used as a solvent. Consult the instructor for details about this method.

C. Derivatives

In order to characterize an ester completely, it is necessary to prepare solid derivatives of both the acid and the alcohol components. The problem here is to isolate both of these components in pure form so that suitable derivatives can be prepared. One such way is to carry out the ester hydrolysis in a high-boiling solvent. If the alcohol is low-boiling, it can be distilled from the reaction mixture and characterized. The acid which remains in the mixture can be isolated also. Derivatives of acids and alcohols have already been discussed.

$$\text{RCOOR}' + \text{HO}^{\ominus} \rightarrow \text{RCOO}^{\ominus} + \text{R}'\text{OH} \tag{39}$$

PROCEDURE

In a small reaction vessel mix 3 ml of diethylene glycol, 0.6 g (2 pellets) of potassium hydroxide, and 10 drops of water. Heat until the solution is homogeneous, and cool to room temperature again. Add 1 ml of the ester and equip the apparatus with a condenser. Heat to boiling again with swirling and, after the ester layer dissolves (3–5 min), recool the solution. Equip for a simple distillation and heat the flask strongly so that the alcohol distils. The distillate, which is fairly pure and dry, can be used for the preparation of a solid derivative. This method is generally quite useful, and all but high-boiling alcohols can be removed by direct distillation.

The residue which remains after distillation contains the salt of the carboxylic acid. Add 10 ml of water to the residue and mix thoroughly. Acidify the solution with 6 *N* sulfuric acid. If a solid acid is expected, allow the mixture to stand until crystals of the acid form, and collect them by filtration. If crystals do not form, extract the aqueous acidic solution with ether or dichloromethane, dry the organic solution, and evaporate the solvent. Use the residual acid to prepare a derivative.

A list of esters and their boiling or melting points is given in Table 25.11. Alcohols and carboxylic acids and their derivatives are given in Tables 25.5 and 25.7, respectively.

NITRILES (Table 25.12)

A. CLASSIFICATION TEST

A qualitative test which may be used for nitriles is similar to that for esters. Common nitriles (as well as amides—see below) give a colored solution on treatment with hydroxylamine and ferric chloride:

$$
\begin{array}{c}
\overset{\displaystyle \overset{NH}{\|}}{R-C-N} + H_2NOH \longrightarrow R-\overset{\displaystyle \overset{NH}{\|}}{C}-NHOH \xrightarrow{FeCl_3} \left[R-C \begin{array}{c} \nearrow NH \\ \searrow \\ N-O \\ | \\ H \end{array} Fe \right]_3 + 3\,HCl \quad (40)
\end{array}
$$

PROCEDURE

Prepare a mixture consisting of 2 ml of 1 M hydroxylamine hydrochloride in propylene glycol, 30–50 mg of the compound which has been dissolved in a minimum amount of propylene glycol, and 1 ml of 1 N potassium hydroxide. Heat the mixture to boiling for 2 min and cool to room temperature; add *ca.* 0.5–1.0 ml of a 5% alcoholic ferric chloride solution. A red to violet color is a positive test. Yellow colors are negative, and brown colors and precipitations are neither positive nor negative.

B. DERIVATIVE

On hydrolysis, in either acidic or basic solution, nitriles are ultimately converted to the corresponding carboxylic acids. Using methods given previously, it is then possible to prepare a derivative of the acid.

$$\text{Basic Hydrolysis: } RCN + NaOH \xrightarrow{H_2O} RCOO^{\ominus}Na^{\oplus} + NH_3 \qquad (41)$$

$$\text{Acidic Hydrolysis: } RCN \xrightarrow[H_2SO_4]{H_2O} RCONH_2 \xrightarrow[H_2SO_4]{H_2O} RCOOH + NH_4^{\oplus} \qquad (42)$$

PROCEDURE

1. Basic hydrolysis. Mix 10 ml of 3 N sodium hydroxide solution and 1 g of the nitrile. Heat the mixture to boiling and note the odor of ammonia or hold a piece of moist red litmus paper over the container and note the color change. After the mixture is homogeneous, cool it and make it acidic to litmus. If the acid is a solid, collect the crystals by filtration. If it is a liquid, extract the acidic solution with ether; after drying the ether solution, remove the ether by distillation. The residue that remains is the acid. Prepare a suitable derivative of the acid using procedures given previously.

2. Acidic hydrolysis. Treat 1 g of the nitrile with 10 ml of concentrated sulfuric acid or concentrated hydrochloric acid, and warm the mixture to 50° for *ca.* 30 min. Dilute the mixture with water (*Caution:* Add the mixture slowly to water if sulfuric acid has been used) and heat the mixture at gentle reflux for 30 min to 2 hr. The organic layer will be the acid. Cool the mixture and either collect the crystals by filtration or extract the liquid with ether. Prepare derivatives of the acid.

A list of some nitriles is given in Table 25.12; carboxylic acids and their derivatives are given in Table 25.7.

AMIDES (Table 25.13)

A. CLASSIFICATION TEST

A qualitative test for an amide group is the same as that given for a nitrile (see above).

$$R-\overset{O}{\underset{\|}{C}}-NH_2 + H_2NOH \longrightarrow R-\overset{O}{\underset{\|}{C}}-NHOH \xrightarrow{FeCl_3} \left[R-\overset{O}{\underset{\underset{H}{N-O}}{C}} Fe \right]_3 + 3\ HCl \qquad (43)$$

PROCEDURE

Follow exactly the procedure given for nitriles. The colors which are observed with amides are the same as those with nitriles.

B. DERIVATIVES

Like nitriles, amides must be hydrolyzed (acidic or basic) to give an amine and a carboxylic acid. In the case of unsubstituted amides, ammonia is liberated, but with substituted amides, a substituted amine is obtained. In those cases, it is necessary to classify the amine as being primary or secondary, and to prepare derivatives of both the acid and the amine.

$$RCONR'_2 \xrightarrow[H_2O]{H^{\oplus}\ or\ HO^{\ominus}} RCOOH + HNR'_2 \qquad (44)$$
$$(R' = alkyl,\ aryl,\ or\ H)$$

PROCEDURE

1. Basic hydrolysis. Carry out the procedure described for the hydrolysis of nitriles. Distil the ammonia or volatile amine from the alkaline solution into a container of dilute hydrochloric acid. Neutralize this acidic solution, carry out the Hinsberg test, and prepare a derivative of the amine. If the amine is not volatile, it may be extracted from the aqueous layer with ether, the solution dried over potassium hydroxide pellets, and the ether removed to give the amine. After the amine has been obtained, either by distillation or extraction from the hydrolysis mixture, make the alkaline solution acidic and isolate the acid (either by filtration if a solid or by extraction if a liquid). Characterize the acid by preparing a suitable solid derivative.

2. Acidic hydrolysis.   Carry out the hydrolysis using the method described for nitriles. In this case, the free acid is liberated and can be removed by filtration or extraction with ether. Prepare a derivative of the acid. Make the acidic hydrolysis mixture alkaline to liberate the amine. Collect the amine by distillation or extraction, characterize it by the Hinsberg test and make a derivative.

A list of amides is given in Table 25.13. Lists of amines and acids and their derivatives are given in Tables 25.7 and 25.8.

CARBOHYDRATES (Table 25.14)

A. CLASSIFICATION TESTS

Carbohydrates will, in general, provide positive tests characteristic of aldehydes or ketones and alcohols. Several possible observations should, however, obviate confusion in the characterization of a carbohydrate. Nearly all saccharides are soluble in water, with the exception of the polysaccharides such as starch, cellulose, and glycogen. This is because of the polar and polyfunctional nature of the compounds. The same factors also render saccharides insoluble in ethyl ether, a relatively nonpolar solvent, in which polar monofunctional compounds are generally soluble. These solubility characteristics provide initial hints that the unknown may be a carbohydrate. Further, saccharides undergo extensive charring when treated with concentrated sulfuric acid. Treatment of a small sample of the solid unknown with a drop or two of concentrated sulfuric acid and observation of charring is tentative evidence for a carbohydrate. Apply the Molisch test for confirmation.

A number of other tests, useful in characterizing a carbohydrate according to its structural type, are discussed with procedures provided in Chapter 22.4.

The Molisch test.   This test was discussed in Chapter 22.4 in connection with other similar tests. The procedure is given below.

PROCEDURE

Place 0.5 ml of a 1% solution (10 mg/ml) of the unknown in a small test tube and add 2 drops of a 5% solution of α-naphthol in ethanol.[12] Incline the tube and, with a dropper, carefully add 1 ml of concentrated sulfuric acid so that it flows down the side of the tube and forms a layer beneath the aqueous solution. Note any color formed at the interface of the layers after a short time (*ca.* 2 or 3 min). A green ring may form initially, owing to nitrate or nitrite impurities in the sulfuric acid; a positive test for a carbohydrate is constituted by the slightly slower formation of a red-violet color at this interface. Gentle shaking, but not enough to mix the layers, will cause the lower acid layer to assume a violet color. For comparison, run the test also with a known mono- or disaccharide and with butyraldehyde.

[12] This solution, the Molisch reagent, is prepared by dissolving 8 g of α-naphthol in 202 ml of 95% ethanol.

B. Derivatives

Two useful derivatives of carbohydrates are the osazones and the acetate esters. Benzoate esters are also occasionally used. The procedure for the preparation of osazones is given in Chapter 22.4. Reaction of saccharides with acetic anhydride, an acetylating agent, leads to the complete acetylation of the sugar. The reaction may be effected in the presence of either sodium acetate or pyridine, with one or the other of these procedures sometimes being more satisfactory for a given sugar. Frequently, in the case of reducing sugars, these procedures may lead to anomeric acetates. For example, acetylation of D-glucose with acetic anhydride and sodium acetate leads to β-D-glucopyranose pentaacetate [equation (45)], whereas the use of acetic anhydride in pyridine provides α-D-glucopyranose pentaacetate [equation (46)].

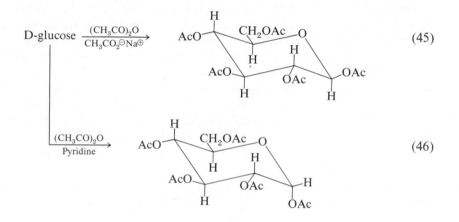

$$(45)$$

$$(46)$$

PROCEDURE

Method A.   Mix 3 g of the *dry* saccharide with 1.5 g of *anhydrous* powdered sodium acetate and 15 ml of acetic anhydride. Heat the mixture on a steam bath with occasional stirring for 2 hr. Pour the hot solution, with vigorous stirring, into 100 ml of ice water. Allow the mixture to stand, with occasional stirring, until the excess acetic anhydride has hydrolyzed and separate the acetate by filtration. Thoroughly wash the crystals with water. Purification may be effected by recrystallization from ethanol.

Method B.   Add 2 g of the *dry* saccharide to 20 ml of *anhydrous* pyridine and then, with shaking, add 7.5 ml of acetic anhydride. After any initial reaction has subsided, boil the solution using a reflux condenser for *ca.* 5 min. Cool the solution and pour it into about 70 ml of ice water. Filter the crystals and wash them successively with cold 1 *N* hydrochloric acid and with water. Purification may be effected by recrystallization from ethanol.

A list of carbohydrates and their derivatives is given in Table 25.14.

25.3 Modern Spectroscopic Methods of Analysis

One major limitation to the classical system of qualitative organic analysis which has not been mentioned but which is inherent to the system is that only *known compounds* can be identified. The research chemist is constantly faced with the task of identifying new compounds. Although much information about the *type* of a compound may be derived from the classical system, until recently the complete identification of an unknown organic structure required a combination of degradation and synthesis which was usually a lengthy and laborious task.

The advent of modern spectroscopy has changed this picture dramatically, because not only known compounds but also new and unknown compounds may be identified quickly and certainly by a combination of spectroscopic methods, such as those described in Chapter 4. A number of examples might be cited of the structural identification in a matter of weeks or months of molecules of a complexity greater than those of other compounds which defied the lifework of several of the great 19th and early 20th century organic chemists. One of the significant early examples of this was the application of infrared spectroscopy and X-ray diffraction to the determination of the structure of the penicillin-G molecule during World War II.

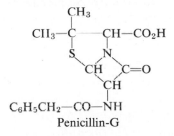

Penicillin-G

Modern procedure for identification of organic compounds usually involves a combination of classical and spectroscopic methods. Ideally a student should be introduced to applications of spectroscopy in organic chemistry through use of the instruments which produce the spectra. However, some of these instruments are very expensive, so that this is not feasible in most instances. The next best alternative is for students to have access to the spectra of typical known compounds for study, and then to be provided with spectra of "unknowns" to identify.

In the preceding chapters of this book over 90 ir spectra and over 70 nmr spectra of starting materials and products have been presented. A careful study of these spectra, aided by the material of Chapter 4 and Appendices III and IV, and by additional discussion on the part of the instructor, should enable a student to make meaningful use of ir and nmr spectra in the identification of unknown organic compounds. The spectral data may serve to complement or supplement the "wet" classification tests, or may in many cases substitute for these tests. For example, a strong ir absorption in the 1690–1760 cm^{-1} region is certainly as indicative of a carbonyl as formation of a 2,4-dinitrophenylhydrazone, and nmr absorptions in the 6–8.5 δ region are a more reliable indication of any aromatic compound than a color test with $CHCl_3$ and $AlCl_3$.

However, it must be emphasized that in spectroscopic analysis, just as in classical qualitative organic analysis, *careful interpretation of the data* and a certain amount of *chemical intuition* or common sense must be exercised. The student should be cautioned not to "go overboard" in his enthusiasm with modern spectroscopy. Although there are some problems that can be resolved quickly and uniquely by spectroscopy, there remain some problems that can be resolved just as simply, and much more economically, by classical qualitative organic analytical procedures, and still others that require the intelligent application of both the modern and the classical methods. Examples are provided below which should provide insight into the complementary application of classical testing procedures and of spectral analysis in the determination of structure of an unknown compound.

A student was given a liquid of unknown structure which had a boiling range of 143–145°, and provided negative tests for halogen, nitrogen, and sulfur when subjected to elemental analysis by sodium fusion. The compound dissolved in water to give a neutral solution. Quite certainly, owing to its water solubility, the compound contains oxygen-bearing polar functional groups. As an aid to the functional group analysis, the student obtained ir and nmr spectra of the substance. These are shown in Figure 25.2. Immediately evident from the ir spectrum is the presence of an ester group because of the strong absorption at *ca.* 1750 cm^{-1}. Although this band could, for example, be indicative of a five-membered ring ketone, this and other possibilities are apparently negated by the presence of the peak at 2.0 δ, one of two peaks in the nmr spectrum integrating for three protons and apparently corresponding to methyl groups with no adjacent hydrogens. The higher field peak (2.0 δ) is characteristic of a methyl ketone or acetate, the latter being consistent with the 1750 and 1230 cm^{-1} bands in the ir spectrum. The second methyl peak at 3.3 δ is shifted to lower field as would be expected if the methyl group were bonded to oxygen. This peak is apparently at too high a field for a methyl ester, since methyl esters more normally show the methyl absorption at about 3.7–4.1 δ. It is, however, within the range (3.3–4.0 δ) frequently observed for the aliphatic alpha-hydrogens of an alcohol or ether. Because neither of the spectra show any evidence for an —OH group, the compound probably contains a CH$_3$O— group. The presence of an aliphatic ether is consistent with the C—O absorption observed at 1050 cm^{-1} in the ir spectrum. The multiplets centered at 4.1 and 3.5δ each integrate for two hydrogens and show mutual spin-spin coupling. This pattern is almost certainly diagnostic of differently shielded, adjacent methylenes, *e.g.,* X—CH$_2$CH$_2$—Y. The low field absorption for each of the CH$_2$ groups indicates that each is bonded to oxygen. Only a single structure is seemingly compatible with all of these spectral observations: 2-methoxyethyl acetate (**1**, bp 145°).

$$CH_3O-CH_2CH_2-O\overset{\displaystyle O}{\overset{\|}{C}}CH_3$$

1

The student (wisely) confirmed the structural assignment through the hydrolysis of the ester and formation of the 3,5-dinitrobenzoate derivative of the resulting

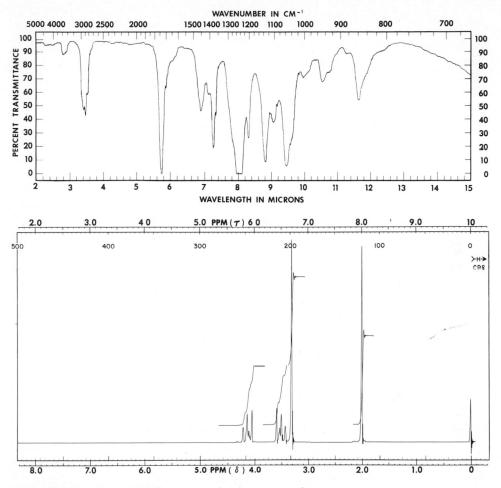

FIGURE 25.2   IR and NMR spectra of unknown number one.

alcohol; the derivative showed the correct melting point for 2'-methoxyethyl 3,5-dinitrobenzoate.

A second liquid unknown was obtained which had a boiling range of 227–230°. The compound showed solubility in only concentrated sulfuric acid, according to the scheme of Figure 25.1. Sodium fusion analyses were negative for nitrogen, sulfur, and halogen. The compound provided the nmr and ir spectra shown in Figure 25.3. In an initial attempt to identify potential functional groups, these spectra were analyzed as follows.

The nmr spectrum shows four regions of absorption at 6.75, 4.78, 2.1–3.0, and 1.75 δ, with relative areas of 14:28:75:88, respectively. This integrational pattern is consistent with a relative hydrogen abundance of 1:2:5:6. The large singlet at 1.75 δ, integrating for six protons, is apparently caused by two methyl groups in closely similar environments. They are probably attached to carbon-carbon double bonds because of their chemical shift and the presence in the ir spectrum of ab-

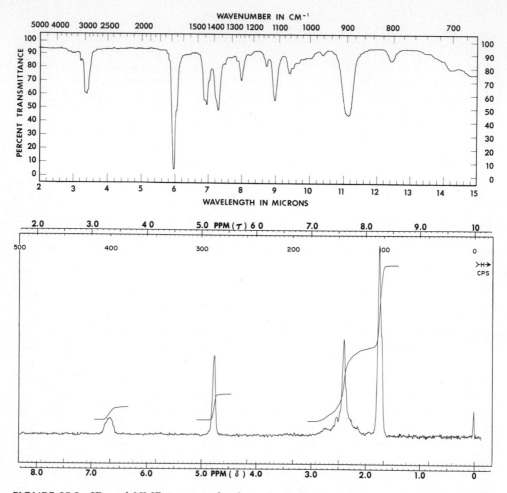

FIGURE 25.3   IR and NMR spectra of unknown number two.

sorptions at 3000–3100, 1645, and 895 cm^{-1}, all characteristic olefinic absorptions. The band at 895 cm^{-1} as well as the two-proton nmr absorption at 4.78 δ are at the positions expected for a terminal double bond (C=CH$_2$). The strong ir band at 1680 cm^{-1} indicates that the compound is probably either an α,β-unsaturated ketone or aldehyde, the latter being effectively eliminated by the absence of any additional evidence for —CHO in either of the spectra.

Although the one-proton nmr peak at 6.75 δ is at an unusually low field, its position is quite in keeping with the possible presence of a conjugated ketone. Olefinic hydrogens beta to the carbonyl in these functions are strongly deshielded because of charge separation in contributing resonance structures:

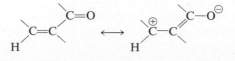

The four-proton multiplet from 2.1–3 δ represents aliphatic hydrogens (probably methylene and/or methine on the basis of the shift and general appearance of the multiplet) deshielded by adjacent groups such as vinyl or carbonyl.

The compound was chemically shown to contain *at least* one double bond through the observation that it decolorized both bromine and potassium permanganate solutions. That the compound was a ketone was verified through the formation of a 2,4-dinitrophenylhydrazone derivative (mp 190–191°). Consultation of Table 25.2 led to the identification of the unknown as carvone **(2)**, an essential oil isolated from caraway seed. (The reader should find it highly instructive to examine this structure carefully while reviewing the spectral analysis discussed above.)

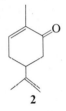

2

The preceding example should foster appreciation of the complementary aspects of spectral analysis and the "wet" classification scheme. Although the spectra were not in themselves specifically definitive in the assignment of structure to the unknown, they implicated the possible presence of certain functional groups and suggested appropriate classification tests. This allowed our imaginary student to avoid the effort and time which might conceivably have been spent in running through a series of negative tests.

The instructor will indicate how he wishes any spectroscopic data provided to be utilized. In some instances he may wish that it be used in place of, or to supplement, classification tests on an actual unknown sample issued, although a derivative may still be required. In other instances, he may allow identification of the unknown on the basis of physical constants that have been determined on a sample, plus the spectroscopic data provided, but no derivative will be required. Or he may issue some "paper unknowns" which consist of ir and nmr spectra, with some additional information such as elements present, molecular weight, etc., but no actual sample will be issued.

REFERENCES

1. R. L. Shriner, R. C. Fuson, and D. Y. Curtin, *The Systematic Identification of Organic Compounds,* 5th ed., John Wiley & Sons, Inc., New York, 1964.
2. N. D. Cheronis, J. B. Entrikin, and E. M. Hodnett, *Semimicro Qualitative Organic Analysis,* 3rd ed., Interscience Publishers, New York, 1965.
3. T. R. Hogness, W. C. Johnson, and A. R. Armstrong, *Qualitative Analysis and Chemical Equilibrium,* 5th ed., Holt, Rinehart and Winston, Inc., New York, 1966.
4. D. J. Pasto and C. R. Johnson, *Organic Structure Determination,* Prentice-Hall, Inc., Englewood Cliffs, N.J., 1969.

25.4   Tables of Derivatives

The following abbreviated tables of common organic compounds and their usual derivatives are arranged according to classes of compounds. The melting and boiling points listed represent the highest points in the range actually observed.

TABLE 25.1   Aldehydes

			Derivatives	
Name of Compound	*BP*	*MP*	*Semi-carbazone*	*2,4-Dinitrophenyl-hydrazone*
Propionaldehyde	49°		154°	156°
Glyoxal	50		270	328
Acrolein	52		171	165
Isobutyraldehyde	64		126	182
n-Butyraldehyde	75		106	123
Trimethylacetaldehyde	75		190	209
2-Methylbutanal	93		103	120
Chloral	98		90*d*	131
Pentanal	103			107
Crotonaldehyde	104		199	190
Hexanal	131		106	104
Heptanal	156		109	108
Furfural	161		202	230(214)
3-Cyclohexenecarboxaldehyde	165		155	
Benzaldehyde	179		222	237
Phenylacetaldehyde	194		156	121
Salicylaldehyde	197		231	248
o-Tolualdehyde	200		212	195
p-Tolualdehyde	204		215	234
Citronellal	207		92	78
o-Chlorobenzaldehyde	208		225	214
p-Chlorobenzaldehyde	214	47°	230	270
o-Anisaldehyde	246	39	215	254
p-Anisaldehyde	248		210	254*d*
Cinnamaldehyde	252		215	255*d*
p-Bromobenzaldehyde		57	228	
Vanillin		81	230	271*d*
p-Nitrobenzaldehyde		106	221	320*d*

TABLE 25.2 Ketones

| Name of Compound | BP | MP | Derivatives | |
			Semi-carbazone	2,4-Dinitrophenyl-hydrazone
Acetone	56°		190°	126°
Butanone	80		136	117
3-Methyl-2-butanone	94		113	120
3-Pentanone	102		139	156
2-Pentanone	102		112	144
Pinacolone	106		158	125
4-Methyl-2-pentanone	117		132	95
Chloroacetone	119		164d	125
2, 4-Dimethyl-3-pentanone	125		160	95
2-Hexanone	129		122	110
4-Methyl-3-penten-2-one	130		164	203
Cyclopentanone	131		203	146
2,4-Pentanedione	139		122(mono); 209(di)	209
4-Heptanone	145		133	75
5-Methyl-2-hexanone	145		147	95
3-Heptanone	148		103(152)	
2-Heptanone	151		127	89
Cyclohexanone	156		167	162
6-Methyl-3-heptanone	160		132	
5-Methyl-3-heptanone	160		102	
4-Hydroxy-2-pentanone	166			203
2,6-Dimethyl-4-heptanone	169		122	92
2-Octanone	173		123	58
Acetophenone	203	20°	199	240
l-Menthone	209		189	146
3,5,5-Trimethylcyclohexen-1-one	215		200	
Benzyl methyl ketone	216	27	198	156
Propiophenone	218	20	174	191
p-Methylacetophenone	226	28	205	258
2-Undecanone	228		122	63
Carvone	230		163(143)	191
Dibenzyl ketone	330	34	146	100
p-Methoxyacetophenone	258	38	198	220
Benzophenone		48	167	239
p-Bromoacetophenone		51	208	235
Desoxybenzoin	320	60	148	204
Benzalacetophenone	348	58(62)	168	245
Benzoin		133	206d	245
p-Hydroxybenzophenone		135	194	242
d,l-Camphor		178	248	164

TABLE 25.3 Aromatic Hydrocarbons

| | | | Derivatives | | |
| | | | Nitration Product | | |
Name of Compound	BP	MP	Position	MP	Oxidation Product
Benzene	80°		1,3	89°	
Toluene	111		2,4	70	Benzoic Acid
Ethylbenzene	136		2,4,6	37	Benzoic Acid
p-Xylene	138		2,3,5	13	Terephthalic Acid
m-Xylene	139		2,4	83	Isophthalic Acid
o-Xylene	142		4,5	71	Phthalic Acid
Isopropylbenzene	153		2,4,6	109	Benzoic Acid
t-Butylbenzene	169		2,4	62	Benzoic Acid
m-Diethylbenzene	181		2,4,6	62	Isophthalic Acid
Tetralin	206		5,7	95	Phthalic Acid
Cyclohexylbenzene	237		4	58	Benzoic Acid
Diphenylmethane		25°	2,4,2',4'	172	
Dibenzyl		53	4,4'	180	
Biphenyl		70	4,4'	237	
Naphthalene		80	1	61	
Triphenylmethane		92	4,4',4''	206	
Acenaphthene		96	5	101	
Fluorene		115	2,7	199	

TABLE 25.4 Aryl Halides

			Derivatives			
			Nitration Product		Sidechain Oxidation	
Name of Compound	BP	MP	Position	MP	Product	MP
Chlorobenzene	132°		2,4	52°		
Bromobenzene	156		2,4	75		
o-Chlorotoluene	159		3,5	63	o-Chlorobenzoic acid	140°
m-Chlorotoluene	162		4,6	91	m-Chlorobenzoic acid	158
p-Chlorotoluene	162		2	38	p-Chlorobenzoic acid	242
o-Dichlorobenzene	179		4,5	110		
o-Bromotoluene	182		3,5	82	o-Bromobenzoic acid	147
2,4-Dichlorotoluene	200		3,5	104	2,4-Dichloro-benzoic acid	160
α-Chloronaphthalene	259		4,5	180		
p-Bromotoluene	184	28°	2	47	p-Bromobenzoic acid	251
p-Dichlorobenzene		53	2	54		
β-Chloronaphthalene		56	1,8	175		
p-Dibromobenzene		89	2,5	84		

TABLE 25.5 Alcohols

Name of Compound	BP	MP	Derivatives 3,5-Dinitro-benzoate	Phenylurethan
Methanol	65°		108°	47°
Ethanol	78		93	52
Isopropyl Alcohol	82		123	88
t-Butyl Alcohol	83		142	136
n-Propyl Alcohol	97		74	51
sec-Butyl Alcohol	99		76	65
t-Pentyl Alcohol	102		117	42
Isobutyl Alcohol	108		87	86
3-Pentanol	116		97	48
n-Butyl Alcohol	118		64	57
1-Methoxy-2-propanol	119		85	
2-Pentanol	119		61	
3-Methyl-1-butanol	132		61	57
1-Pentanol	138		46	46
Cyclopentanol	141		115	132
2-Methyl-1-pentanol	148		51	
1-Hexanol	156		58	42
Cyclohexanol	161		113	82
Furfuryl Alcohol	172		81	45
2,6-Dimethyl-4-heptanol	173			62
1-Heptanol	177		47	68
2-Octanol	179		32	114
Ethylene Glycol	198		169	157
Methylphenylcarbinol	203		95	94
Benzyl Alcohol	206		113	78
β-Phenethyl Alcohol	220		108	80
Cinnamyl Alcohol		33°	121	90
1-Tetradecanol		39	67	74
(−)-Menthol		42		111
1-Hexadecanol		50	66	73
Neopentyl Alcohol		53		144
1-Octadecanol		60	66	80
Benzoin		133		165
(−)-Cholesterol		148		168

TABLE 25.6 Phenols

			Derivatives		
Name of Compound	BP	MP	α-Naphthyl-urethan	Bromo Derivatives	
o-Chlorophenol	175°	7°	120°		
Phenol	180	42	133	Tribromo	95°
o-Cresol	190	31	142	Dibromo	56
o-Bromophenol	195		129	Tribromo	95
p-Cresol	202	36	146	Dibromo	49
m-Cresol	202	3	128	Tribromo	84
2-Methoxyphenol	205	32	118	4,5,6-Tribromo	116
p-Chlorophenol	217	43	166		
o-Nitrophenol		45	113	4,6-Dibromo	117
p-Ethylphenol		47	128		
p-Bromophenol		63	169	Tribromo	95
3,5-Dichlorophenol		68		Tribromo	189
3,5-Dimethylphenol		68		Tribromo	166
2,4,6-Trimethylphenol		69		Dibromo	158
2,5-Dimethylphenol		74	173	Tribromo	178
α-Naphthol		94	152	2,4-Dibromo	105
m-Nitrophenol		97	167	Dibromo	91
p-t-Butylphenol		100	110	Bromo	50
				Dibromo	67
o-Hydroxyphenol		105	175	Tetrabromo	193
m-Hydroxyphenol		110		Tribromo	112
p-Nitrophenol		114	151	2,6-Dibromo	142
p-Hydroxybenzaldehyde		115		3,5-Dibromo	181
β-Naphthol		122	157	Bromo	84
2,3-Dihydroxyphenol		133	230 (tri)	Dibromo	158

TABLE 25.7 Carboxylic Acids

| Name of Compound | BP | MP | Derivatives | | |
			Anilide	p-Toluidide	Amide
Acetic	118°		114°	147°	
Propionic	140		106	126	
Pivalic	164	35°	129	120	154°
Chloroacetic	185	63	134	120	118
α-Chloropropionic	186		92	124	
Dichloroacetic	189		118	153	
Cyclohexylcarboxylic		30	146		186
β-Phenylpropionic (Hydrocinnamic)		48	92	135	82
Trichloroacetic		57	94	113	141
Phenylacetic		76	118	136	154
Glycolic		80	97	143	120
o-Methoxybenzoic		100	131		128
o-Toluic		104	125	144	142
m-Toluic		111	126	118	97
Benzoic		122	163	158	128
Cinnamic		133	153	168	147
meso-Tartaric		140			190
o-Chlorobenzoic		142	118	131	139
m-Nitrobenzoic		146	155		142
o-Bromobenzoic		150	141		155
Benzilic		150	175		154
Salicylic		158	136		139
m-Chlorobenzoic		158	122		134
p-Toluic		178	145	160	158
Anisic		184	169	186	162
Phthalic		208d	169	201(di)	149
p-Hydroxybenzoic		215	202	204	162
p-Nitrobenzoic		241	217	204	201
p-Chlorobenzoic		242	194		179

TABLE 25.8 Primary and Secondary Amines

Name of Compound	BP	MP	Derivatives	
			Benzamide	Benzene-sulfonamide
Isopropylamine	33°		100°	26°
n-Propylamine	49		84	36
sec-Butylamine	63		76	70
n-Butylamine	77		42	
Piperidine	106		48	93
Di-n-propylamine	110			51
n-Hexylamine	129		40	96
Morpholine	130		75	118
Cyclohexylamine	134		149	89
N-Methylcyclohexylamine	147		86	
Aniline	183		163	112
Benzylamine	184		105	88
N-Methylaniline	196		63	79
o-Toluidine	199		146	124
m-Toluidine	203		125	95
N-Ethylaniline	205		60	
o-Ethylaniline	211		147	
p-Ethylaniline	216		151	
o-Anisidine	225		60	89
m-Chloroaniline	230		120	121
Tetrahydroisoquinoline	250	20	75	67
Dibenzylamine	300		112	68
p-Toluidine	200	45	158	120
α-Naphthylamine		50	160	167
2,5-Dichloroaniline		50	120	
Diphenylamine	302	54	180	124
p-Anisidine	240	58	154	95
2-Aminopyridine		60		165(di)
N-Phenyl-α-naphthylamine		62	152	
p-Bromoaniline		66	204	134
p-Chloroaniline	232	70	192	122
o-Nitroaniline		71	110	104
N-Phenyl-β-naphthylamine		108	148	
m-Nitroaniline	284	114	155	136
Anthranilic Acid		147	182	214
p-Nitroaniline		147	199	139
p-Nitro-N-methylaniline		152	112	121
Carbazole	351	246	98	

TABLE 25.9 Tertiary Amines

Name of Compound	BP	MP	Derivatives	
			Picric Acid	Methyl Iodide
Triethylamine	89°		173°	
Pyridine	116		167	117°
2-Picoline	129		169	230
3-Picoline	143		150	92
4-Picoline	143		167	152
Tri-n-propylamine	156		116	208
2,4-Lutidine	159		183	113
2,4,6-Collidine	172		156	
2-Methyl-4-ethylpyridine	173		123	
N,N-Dimethylaniline	193		163	228d
Tri-n-butylamine	211		106	186
N,N-Diethylaniline	218		142	102
Quinoline	239		203	72(133)
Isoquinoline	243		222	159

TABLE 25.10 Aromatic Nitro Compounds

Name of Compound	BP	MP	Acyl Derivatives of Reduced Amine		Other Derivatives
			Benzene-sulfonamide	Benzamide	
Nitrobenzene	210°		112°	160°	m-Dinitro-benzene 90°
o-Nitrotoluene	224		124	147	2,4-Dinitro-toluene 70
m-Nitrotoluene	231	16°	95	125	m-Nitro-benzoic Acid 140
o-Nitroanisole	265		89		2,4,6-Tri-nitroanisole 68
o-Chloronitrobenzene	246	32	129	99	2,4-Dinitro-chloroben-zene 52(50)
o-Bromonitrobenzene	261	43		116	2,4-Dinitro-bromobenzene 72
2,4-Dinitrochlorobenzene		52		178	2,4-Dinitro-phenol 114
p-Nitroanisole	258	54	95	154	2,4-Dinitro-anisole 89
β-Nitronaphthalene		78	136	162	
p-Chloronitrobenzene	242	83	121	192	p-Nitro-phenol 114
Picryl chloride		83	211		Picric Acid 122
m-Dinitrobenzene	302	90	194	240	m-Nitro-aniline 114
4-Nitrobiphenyl		114		230	Acetamide 171
p-Bromonitrobenzene	259	126	136	204	p-Nitro-phenol 114

TABLE 25.11 Esters

Name of Compound	BP	MP	Name of Compound	BP	MP
Methyl acetate	57°		n-Propyl benzoate	230°	
Ethyl acetate	77		Methyl m-chlorobenzoate	231	
Isopropyl acetate	91		Ethyl salicylate	234	
tert-Butyl acetate	98		n-Butyl benzoate	249	
n-Butyl acetate	126		Methyl p-toluate		33°
n-Amyl acetate	149		Methyl cinnamate		36
Cyclohexyl acetate	175		Benzyl cinnamate		39
n-Heptyl acetate	192		Phenyl salicylate		42
Phenyl acetate	197		Methyl p-chlorobenzoate		44
o-Cresyl acetate	208		Ethyl m-nitrobenzoate		47
m-Cresyl acetate	212		α-Naphthyl acetate		49
p-Cresyl acetate	213		Ethyl p-nitrobenzoate		56
Ethyl benzoate	213		Phenyl benzoate		69
Methyl o-toluate	215		β-Naphthyl acetate		71
Benzyl acetate	217		p-Cresyl benzoate		71
Isopropyl benzoate	218		Methyl m-nitrobenzoate		78
Ethyl phenylacetate	229				

TABLE 25.12 Nitriles

Name of Compound	BP	MP	Name of Compound	BP	MP
Acetonitrile	81°		m-Chlorobenzonitrile		41°
Propionitrile	97		o-Chlorobenzonitrile	232°	47
Chloroacetonitrile	127		p-Methoxybenzonitrile		62
Benzonitrile	191		p-Chlorobenzonitrile		92
o-Tolunitrile	205		o-Nitrobenzonitrile		110
m-Tolunitrile	212		m-Nitrobenzonitrile		118
p-Tolunitrile	217	27°	p-Nitrobenzonitrile		147
Phenylacetonitrile	234				

TABLE 25.13 Amides

Name of Compound	MP	Name of Compound	MP
N-n-Propylacetanilide	50°	o-Bromobenzanilide	141°
m-Acetotoluide	66	Salicylamide	142
Propionamide	81	o-Toluamide	142
Acetamide	82	m-Nitrobenzamide	143
o-Nitroacetanilide	92	p-Toluanilide	145
n-Butyranilide	95	Cinnamanilide	151
Trichloroacetanilide	97	m-Nitrobenzanilide	154
N-Methylacetanilide	102		
Acetanilide	114	o-Bromobenzamide	155
Dichloroacetanilide	118	m-Nitroacetanilide	155
o-Toluanilide	125	p-Toluamide	160
m-Toluanilide	126	Benzanilide	163
p-Methoxyacetanilide	127	p-Bromoacetanilide	167
Benzamide	130	p-Chloroacetanilide	179
m-Chlorobenzamide	134	p-Chlorobenzanilide	194
m-Bromobenzanilide	136	p-Nitrobenzamide	201

TABLE 25.14   Carbohydrates

Name of Compound	Decomposition Point	Specific Rotation, $[\alpha]_D^{20}(H_2O)$	Derivatives	
			Osazone	Acetate
Raffinose (hyd)	79°	+104.5°		99°
β-Melibiose (hyd)	85	+129.5	178	
Glucose (hyd)	90	+47.7	205	132 (β)
				112 (α)
L-Rhamnose (hyd)	94	+8.0	182	
D-Ribose	94	−21.5	166	
Maltose (hyd)	100	+129.0	206	158
D-Fructose	104	−92.0	205	109 (β)
				70 (α)
L-Rhamnose (anh)	105	+9.4	182	
D-Lyxose	105	+13.7	163	
Raffinose (anh)	119	+123		99
D-Mannose	132	+14.2	205	
D,L-Glyceraldehyde	142ᵃ	0	132	154
D-Xylose	145	+18.7	163	141
D-Glucose (anh)	146	+52.8	205	132 (β)
				112 (α)
L-Arabinose	160	+104.5	166	86 (β)
				95 (α)
L-Sorbose	160	−43.4	156(168)	97
D,L-Arabinose	164	0	169	
Arbutin (hyd)	165	−60.3		136
Maltose	165	+130.4	206	158
D-Galactose	169	+80.2	201	142 (β)
				95 (α)
Sucrose	185	+66.5		72
Lactose	203	+52.5	200d	100
α,α-Trehalose	203	+178.3		102
Cellobiose	225	+34.6	198	202 (β)
				229 (α)

ᵃMelts without decomposition.

chapter twenty-six
the literature of organic chemistry

The purpose of this chapter is to provide assistance to the interested student and teacher in obtaining additional information on experimental organic chemistry—information which will be useful in amplifying, modifying, and extending the introductory organic laboratory course for which this book was designed. This chapter is not intended to be a comprehensive guide to the literature of organic chemistry; an excellent series of articles which serves this purpose has been written by Professor J. E. H. Hancock, and reference to these articles is given at the end of this chapter.

Hancock divided all of the literature of organic chemistry into 18 classes. Seven of these which should be of most value to the organic laboratory student (and also, probably, to the practicing organic chemist) will be listed here, and then selected examples of each will be given, with brief explanatory notes.

Class A. Primary Research Journals
Class B. Review Journals
Class C. Encyclopedias and Dictionaries
Class D. Abstract Journals
Class E. Advanced Textbooks
Class F. Reference Works on Synthetic Procedures and Techniques
Class G. Catalogs of Physical Data

Class A. Primary Research Journals

These journals publish original research, with theoretical discussion and experimental details.

1. *Journal of the American Chemical Society.* In recent years the articles on organic chemistry have been limited to those which are especially timely ("Communications to the Editor") or of wide interest to all chemists.

2. *Journal of Organic Chemistry.*   Articles and notes on organic chemistry only.

3. *Tetrahedron* and *Tetrahedron Letters.*   Articles and brief communications on organic chemistry, some in German and French, as well as in English.[1]

Class B. Review Journals

Some review journals publish reviews in all areas of chemistry, but many cover only specific areas; all give references to primary journal articles.

1. *Chemical Reviews.*   Published bimonthly by the American Chemical Society since 1924. General in scope.

2. *Annual Reports on the Progress of Chemistry.*   Published by The Chemical Society, London. Annual reviews of all areas of chemistry, classified well so that organic work can easily be identified. It is being superseded by a series of "Specialist Subject Reports," covering many specific areas, such as alkaloids, nuclear magnetic resonance, organometallic compounds.

3. *Journal of Chemical Education.*   Often contains reviews written by experts at a level such that students and others unfamiliar with the subjects may understand them. New, tested experiments and modifications of old experiments suitable for organic laboratory courses are frequently published.

4. *Accounts of Chemical Research.*   Published monthly by the American Chemical Society. Concise reviews of active research areas.

Other reviews may be found in publications entitled "Advances in Carbohydrate Chemistry," "Annual Reviews of Biochemistry," and "Progress in Infrared Spectroscopy," etc.

Class C. Encyclopedias and Dictionaries

1. *Beilstein's Handbuch der Organischen Chemie* is perhaps the most complete reference work in any branch of science.[2] First published in 1883, the subsequent fourth edition contains data on all of the organic compounds known in 1909, about 140,000. This edition is referred to as the "Hauptwerk," or main work. Instead of printing further editions, the German Chemical Society has issued supplements ("Ergänzungwerke") covering certain periods of time. The first supplement (E I) extended the coverage through 1919 and the second (E II) through 1929. The third supplement will cover the period 1930–1949, but it is still in the process of publication and completion is not expected before 1980.

There is good reason for the delay in bringing this encyclopedic coverage of organic chemistry up to date. The total number of organic compounds known in 1968 was estimated to be 2.5 million, increasing by about 800 new ones every day. Despite

---

[1] Unless specified otherwise, all publications mentioned are in English.

[2] Although *Beilstein* is written in German, because so many organic chemical terms are the same in German and English, only a rudimentary knowledge of German will suffice for practical use of this work.

this staggering statistic, because the organic chemical literature is so well organized, it is possible for a practiced student to determine in no more than an hour or so spent in a good library whether a particular organic compound has ever been prepared. If it has, he may also learn how it was synthesized, and the physical properties of the pure compound.

The main work and the first and second supplements to *Beilstein* are covered by the formula index ("Formelregister") in Vol. 29. This means that by referring to this one index, a person may determine whether or not an organic compound was known up to and including the year 1929. A name index ("Sachregister") is also provided for *Beilstein,* but its use is not as simple or reliable as that of the formula index. For this reason, the beginner is advised to use the formula index when it is feasible. However, after a little familiarity with the system has been acquired, one may actually extrapolate the use of the *Formelregister* and *Sachregister* into the third and fourth supplements, thus possibly finding material published as late as 1949 and 1959. For more information about the organization of *Beilstein* and the use of its indexes one should refer to the third of the articles by Hancock.

Searches for organic compounds that have appeared in the literature since 1929 are best made using the abstract journals discussed in the next section (Class D).

2. *Other compilations of organic compounds.*

(a) *Heilbron's Dictonary of Organic Compounds,* 4th ed., edited by Pollock and Stevens, Oxford University Press, 1965; supplements have appeared annually since 1965. The 4th edition contains an alphabetical listing of about 25,000 compounds with selected derivatives, reactions, and references.

(b) *Handbook of Chemistry and Physics,* annual or biennial editions, Chemical Rubber Publishing Co., Cleveland, Ohio. Gives physical properties on about 14,000 organic compounds.

(c) *Lange's Handbook of Chemistry,* various editions, Handbook Publishers, Sandusky, Ohio.

(d) *Handbook of Tables for Identification of Organic Compounds,* 3rd ed., Z. Rappoport, editor, Chemical Rubber Publishing Co., Cleveland, Ohio, 1967. Gives physical properties and derivatives for over 4000 compounds; organized according to functional groups.

(e) *CRC Atlas of Spectral Data and Physical Constants for Organic Compounds,* J. G. Grasseli, editor, Chemical Rubber Publishing Co., Cleveland, Ohio, 1973.

(f) *Merck Index of Chemicals and Drugs,* Merck and Co., Inc., Rahway, N. J., 8th ed., 1968. Gives concise summaries of physical and biological properties of about 10,000 compounds, with some literature references. Organization is alphabetical by name; synonyms and trademarked names are provided.

Class D. Abstract Journals

Abstract journals provide concise summaries of articles of Class A, listings of reviews (Class B), and announcements of new books (Classes E and F), with references to the original articles or books.

1. *Chemical Abstracts.* This journal began publication in 1907. It now abstracts articles from about 12,000 scientific publications worldwide. Author, subject, and formula indexes presently appear semiannually. A collective formula index covers the years 1920–1946; the next one covers the 10-year period 1947–1956; subsequent collective formula indexes have appeared at 5-year intervals. The most recent one covers the years 1967–1971. In the years between the issuances of the 5-year collective indexes, semiannual indexes must be consulted.

To use the subject indexes of *C. A.* effectively, one must acquire a working knowledge of the system used for naming and indexing of organic compounds. This is described fully in the subject index of Vol. **56** (1962). In particular, it should be noted that the subject indexes are compiled under chief headings, followed by subheadings; for example, 1-benzyl-3,4-dihydroisoquinoline is listed under *Isoquinoline* (which appears at the top of the page).

_____, 1-benzyl-3,4-dihydro, **43**; 3742e, 5026a. In this reference, the first number **(43)** refers to the volume of *C. A.* containing the abstract; the second number (3742) refers to the column number (there have been two columns per page since 1934) where the abstract is found; and the terminal letter refers to an approximate distance down the column on a scale from a to i. This use of *letters* began in 1947; previously a terminal superscript *number* was used to indicate column position.

Beginning in 1967 a new system was introduced. Abstracts are now coded by separate numbers assigned consecutively through each volume. Letters are still appended to the abstract number, however, not to denote position in the column, but rather as a computer check character. For example, two consecutive abstracts are numbered 6865w and 6866d; the letters w and d were calculated by the computer on the basis of the sequence of digits in the abstract number. The letter serves as a check-character to prevent errors such as digit transposition or miscopying in the computer handling of abstract numbers. The letter has no bearing on the information in an abstract or on the arrangement of abstracts within sections.

C. A. Subject Indexes have been substantially reorganized recently and now include the following sections: "Index Guide," whose introductions should be consulted first; "General Subject Index" (headings which do not refer to specific chemical substances, *e.g.,* amines, chromatography, ferrite substances); "Chemical Substances Index" (headings for completely defined chemical substances, *e.g.,* elements, chemical compounds, specific minerals and alloys); "Formula Index" (listed by molecular formula); and "Index of Ring Systems." A companion to these indexes is the *C. A. S. Registry Index* (a computer-generated numbering system for unique unambiguous chemical substances, a kind of Social Security number for each compound).

2. *Chemisches Zentralblatt* is the other major abstract journal, published in German as indicated by the name. It predates *C. A.* by over 50 years, and prior to 1940 it was more complete and reliable than *C. A.* For this reason in an exhaustive search for an organic compound, in the years between 1929 (the last year covered by *Beilstein's* formula index) and 1940, it would be well to consult the collective formula indexes of *C. Z.,* one for the period 1929–1934 and one for the period 1935–1939. *Chemisches Zentralblatt* ceased publication in 1970.

Class E. Advanced Textbooks

These will be subdivided according to subject and function.

1. GENERAL

(a) *Chemistry of Carbon Compounds,* E. H. Rodd, editor, Vols. 1–5, 1951–1962, Elsevier Publishing Company, New York; 2nd ed., S. Coffey, editor, still in the process of publication; Vols. I-IIIA had appeared by 1972. A comprehensive survey of all classes of organic compounds, giving properties and syntheses for many individual compounds.

(b) *Organic Chemistry,* by J. B. Hendrickson, D. J. Cram, and G. S. Hammond, McGraw-Hill Book Company, Inc., New York, 1970. The organization is mainly by reaction type. There are ten advanced chapters on special topics.

(c) *Organic Chemistry,* by N. L. Allinger, M. P. Cava, D. C. DeJongh, C. R. Johnson, N. A. LeBel, and C. A. Stevens, Worth Publishers, Inc., New York, 1971.

(d) *Advanced Organic Chemistry: Reactions, Mechanisms, and Structures,* by J. March, McGraw-Hill Book Company, Inc., New York, 1968. Many references to the original work are given.

(e) *Structure and Mechanism in Organic Chemistry,* by C. K. Ingold, 2nd ed., Cornell University Press, Ithaca, N. Y., 1969. A classic treatise on the title subject.

2. IDENTIFICATION AND ANALYSIS
OF ORGANIC COMPOUNDS
See references at end of Chapter 25.3.

3. INSTRUMENTAL TECHNIQUES OF ANALYSIS
See references at end of Chapter 4.

Class F. Reference Works on Synthetic Procedures and Techniques

1. *Survey of Organic Syntheses,* by C. A. Buehler and D. E. Pearson, Wiley-Interscience, New York, 1970. This extensive one-volume book covers the principal methods of synthesizing the main types of organic compounds. The limitations of the reactions, the preferred reagents, the newer solvents, and experimental conditions are considered.

2. *Organic Syntheses,* Editor-in-chief, Henry Gilman, John Wiley & Sons, Inc., New York, 1932–present, 51 volumes through 1971, collected and indexed every ten volumes into "Collective Volumes." Detailed directions for the synthesis of over 1000 compounds. Procedures have all been thoroughly checked by independent investigators before publication. Many of the general methods may be applied to synthesis of related compounds other than those described. Collective volumes have indexes of formulas, names, types of reaction, types of compounds, purification of solvents and reagents, and illustrations of special apparatus.

3. *Organic Reactions,* by various contributors, John Wiley & Sons, Inc., New York, 1942–present, 18 volumes through 1970. Each volume contains from five to twelve chapters, each of which deals with an organic reaction of wide applicability. Typical experimental procedures are given in detail and there are extensive tables of examples with references. Each volume contains a cumulative author and chapter-title index.

4. *Reagents for Organic Synthesis,* by Mary and Louis Fieser, Wiley-Interscience, New York, Vol. 1, 1967; Vol. 2, 1969; Vol. 3, 1971. In the three volumes, 415 reagents and solvents are described in terms of methods of preparation or source, purification, and utilization in typical reactions. Ample references to primary literature are given.

5. *Technique of Organic Chemistry,* edited by A. Weissberger, 3rd ed., Interscience Publishers, New York, 1959. Revised volumes have appeared at regular intervals; in 1970 the general title was widened to *Techniques of Chemistry.* Examples of titles of individual volumes are: Vol. III, *Separation and Purification;* Vol. IV, *Distillation;* Vol. VIII, *Investigation of Rates and Mechanisms of Reaction;* Vol. XII, *Thin-layer Chromatography;* Vol. XIII, *Gas Chromatography;* Vol. IV, 2nd ed., part 1, *Elucidation of Organic Structures by Physical and Chemical Methods.*

Class G. Catalogs of Physical Data

1. NMR SPECTRA

(a) *High Resolution NMR Spectra Catalog,* compiled by the staff of Varian Associates, Palo Alto, Calif., Vol. I, 1962; Vol. II, 1963. Hydrogen nmr spectra of 587 representative organic molecules are depicted and the peaks are assigned to the hydrogen nuclei responsible for the absorptions.

(b) *Nuclear Magnetic Resonance Spectra,* published by Sadtler Research Laboratories, Inc., Philadelphia. Hydrogen nmr spectra of over 14,000 compounds had been published by 1972 and about 1200 were being added annually. Assignment of peaks are made as in the Varian spectra and integration of the signals is shown on many of the spectra.

2. INFRARED SPECTRA

Sadtler Standard Spectra, Midget Edition, published by Sadtler Research Laboratories, Inc., Philadelphia. In 1971 about 41,000 spectra had been published and about 2000 were being added annually.

The Aldrich Library of Infrared Spectra, C. J. Pouchert, Aldrich Chemical Co., Inc., Milwaukee, Wisc., 1970.

Some of the texts of Class E, such as those listed at the end of Chapter 4, contain numerous nmr and ir spectra with molecular assignments. Several "problem books" of spectroscopic analysis have been published more recently; these give various combinations of ir, nmr, uv, and mass spectra of "unknown" organic compounds, with answers provided. Among these are the following: (a) *Spectral Exercises in Structural Determination of Organic Compounds,* by R. H. Shapiro, Holt, Rinehart and Winston, Inc., New York, 1969; (b) *Organic Spectral Problems,* by J. R. Dyer,

Prentice-Hall, Inc., Englewood Cliffs, N. J., 1972; (c) *More Spectroscopic Problems in Organic Chemistry,* by A. J. Baker, T. Cairns, G. Eglinton, and F. J. Preston, Heyden and Son, London, 1967.

Notes on Use of the Literature of Organic Chemistry in an Introductory Laboratory Course

The literature outline that has been given in this chapter may be used in a variety of ways, according to the aims and needs of different courses and the library facilities available, ranging from no use at all to extensive application. Even in those cases where the pressure of time and/or lack of facilities preclude the use of literature beyond the pages of this textbook itself, we feel that the presence of this chapter may be valuable to the serious students who may decide to go farther in the study of organic chemistry.

In many organic laboratory courses, instructors are interested in making part of the experimentation "open-ended"—encouraging the students to plan and carry out experiments with some independence. Although this is highly desirable, it has an element of danger unless the plans are checked and the work is monitored carefully. In several chapters of this text additional or alternative experiments are provided or suggested. The inclusion of the literature outline of this chapter now provides a wide source of information for additional experiments.

The most likely class of literature to yield appropriate **synthetic experiments** is Class F. The experiments from *Organic Syntheses* are particularly suitable; although they are usually on a large scale, they can easily be scaled down. Also deserving special mention are the experiments that appear from time to time in the *Journal of Chemical Education* (Class B). Useful improvements or modifications of experiments are often found first, however, in the primary literature (Class A); for example, the modification of the experiment on the preparation of triptycene (Chapter 13) which avoids the dangerous handling of the dry diazonium salt intermediate is based on a report in the *Journal of the American Chemical Society* which appeared after the first edition of this text was published. For more experience in **identification** of unknown organic compounds, the books of Class C.2 will be most useful. The catalogs of spectra listed in Class G represent a tremendous reservoir from which to draw for **paper unknowns** and problems. They should be used with discretion, however, because many molecules give ir and nmr spectra that are not easily interpreted by beginners. If an instructor wishes his students to learn how to make a **comprehensive search** for a specific compound in the literature, to learn its properties or a preferred method of synthesis, he can refer them to the second and third articles by Hancock for a fuller introduction to the use of the Class C and D literature.

The following example is given as an illustration of how one might proceed to solve problems such as those proposed in Exercise 11: "Mustard gas" is one of the names that has been applied to the compound $ClCH_2CH_2—S—CH_2CH_2Cl$. Find the answers to the following questions: 1. When was this compound first synthesized or isolated? 2. By whom? 3. When? 4. Where can the most recent information on this compound be found?

First, write the molecular formula as $C_4H_8Cl_2S$, and look in *Beilstein's General Formelregister, Zweites Ergänzungswerk*. On p. 65 will be found the entry β,β'-Dichlor-diäthylsulfid, Senfgas **1**, 349, I 175, II 348, 940," and just below it, the entry "α,α'-Dichlor-diäthylsulfid **1** II 685." The first entry will be recognized as that pertaining to the subject compound. The references are to p. 349 in Vol. 1 of the main work, p. 175 in Vol. 1 of the first supplement, and to pp. 348 and 940 in Vol. 1 of the second supplement. Although the third supplement was published after the general index, by noting the "System No." of the subject compound, 23, one may locate it on p. 1382 of the third supplement.

Referring to p. 349 in Vol. 1 of the main work, one finds "β,β"-Dichlor-diäthylsulfid $C_4H_8Cl_2S = (CH_2Cl \cdot CH_2)_2S$. B. Aus Thiodiglykol S $(CH_2 \cdot CH_2 \cdot OH)_2$ und PCl_3 (V. Meyer, *B*. **19**, 3260).-." This translates: "B. = Bildung, Formation. From thiodiglycol and PCl_3 (V. Meyer, *Berichte*, **19**, 3260)." Looking up the reference in the *Berichte der Deutschen Chemischen Gesellschaft*, Vol. 19, p. 3260 (published in 1886) one finds that Victor Meyer first prepared this compound in two steps as follows:

$$2\ Cl{-}CH_2CH_2{-}OH \xrightarrow{K_2S} HO{-}CH_2CH_2{-}S{-}CH_2CH_2{-}OH$$

$$HO{-}CH_2CH_2{-}S{-}CH_2CH_2{-}OH \xrightarrow{PCl_3} Cl{-}CH_2CH_2{-}S{-}CH_2CH_2{-}Cl$$

The entry in the main work of *Beilstein* (Vol. 1, p. 349) gives the boiling point and solubility properties of the subject compound, and one chemical reaction in a total of six lines. The final two words are "Sehr giftig," very poisonous—a terse commentary on a material that was much feared as a lethal military weapon in World War II, but was never used. (It is interesting to note the statement by Meyer in his *Berichte* article that although his laboratory assistant developed skin eruptions and eye inflammation after preparing this compound, he himself suffered no ill effects even though he took no precautions in handling it!)

In the first supplement (p. 175) there are 15 lines devoted to "β,β'-diäthylsulfid" and in the second supplement (p. 348), five and one-half pages, indicating the increased interest in this compound during the 1920–1929 decade. Turning next to the *Chemical Abstracts Collective Formula Index* for 1920–1946, one finds under $C_4H_8Cl_2S$ the entry "(See also Sulfide, bis(chloroethyl).) Sulfide, 1-chloroethyl 2-chloroethyl, **25**: 2114.[8]" Since "Sulfide, 1-chloroethyl 2-chloroethyl" is not the compound of interest, we look in the *C. A. Decennial Subject Index*, 1917–1926, for the entry "Sulfide, bis (β-chloroethyl)," which is followed by the names "*mustard gas; yperite*" and by six general references and two columns of more specific references beginning with "absorption by skin, mechanism of, **14**: 300[4]" and ending with "toxicity and skin-irritant effects of, **15**: 1943[6]."

One could presumably use the *C. A. Decennial Subject Indexes* for more recent decades, but one must be on guard for changes in nomenclature. For this reason, it is usually advantageous to use formula indexes first. For example, when we go to the *C. A.* Jan.–June 1972 *Formula Index*, under $C_4H_8Cl_2S$ we find "Ethane, 1,1'-thiobis[2-chloro-" as the name for our compound, with seven references to abstracts. Under this name in the *Chemical Substance Index* for the same period, we find the same seven references, but with specific subjects headings; e.g., "DNA cross-linking induced by, 95344w."

REFERENCE

J. E. H. Hancock, "An Introduction to the Literature of Organic Chemistry," *Journal of Chemical Education,* **45,** 193–199, 260–266, 336–339 (1968).

EXERCISES

1. Find the melting points of the following crystalline derivatives (none of these are listed in the tables of Chapter 25): (a) 2,4-dinitrophenylhydrazone of isovaleraldehyde; (b) semicarbazone of methyl vinyl ketone; (c) 3,5-dinitro-benzoate of 2-methyl-2-pentanol; (d) *p*-toluidide of isobutyric acid; (e) benzamide of ethylamine.

2. Locate an article or chapter on each of the following types of organic reactions: (a) the aldol condensation; (b) the Wittig reaction; (c) reactions of diazoacetic esters with unsaturated compounds; (d) hydration of alkenes and alkynes via hydroboration; (e) metalation with organolithium compounds.

3. Give a reference for a practical synthetic procedure for each of the following compounds and state the yield that may be expected: (a) 1,2-dibromocyclo-hexane; (b) α-tetralone; (c) 3-chlorocyclopentene; (d) 2-carbethoxycyclo-pentanone; (e) norcarane; (f) cycloheptatriene; (g) 1-methyl-2-tetralone; (h) adamantane.

4. Locate descriptions of procedures for the preparation or purification of the following reagents and solvents used in organic syntheses: (a) Raney nickel catalysts; (b) sodium borohydride; (c) dimethyl sulfoxide; (d) sodium amide; (e) diazomethane.

5. Find ir spectra for the following compounds: (a) N-cyclohexylbenzamide; (b) 4,5-dihydroxy-2-nitrobenzaldehyde; (c) benzyl acetate; (d) isopropyl ether; (e) 3,6-diphenyl-2-cyclohexen-1-one; (f) 4-amino-1-butanol.

6. Find nmr spectra of the following compounds: (a) benzyl acetate; (b) isopropyl ether; (c) 4-amino-1-butanol; (d) *n*-propyl alcohol; (e) indane.

7. N-mesityl-N'-phenylformamidine

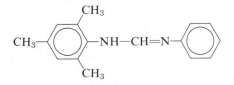

was first synthesized between 1950 and 1960. Find the primary research article in which this compound is described and write an equation for the reaction used to prepare it.

8. N-phenyl-N'-*p*-tolylformamidine[3]

is reported in *Beilstein* to have a melting point of 86°. If you check the first reference given in *Beilstein,* however, you will find the surprising fact that the same chemist who reported this pure compound to have a melting point of 86° had described it as melting at 103.5–104.5° two years previously. The discrepancy between these reports was not explained until the ambiguity was reexamined in the period 1947–1956.

Find the article which solved this mystery.

9. The benzoyl derivative of α-phenylethylamine (formula **4** in Chapter 21) was first described in the form of the optically active (−)-isomer in 1905. Using the formula index of *Beilstein* and given the German name "benzosäure-1-α-phenyläthylamid," find the first reference to this compound in a primary research journal. If you can read the German, give (a) the method of preparation by writing the equation, (b) the melting point of the pure compound and the recrystallization solvent, and (c) the $[\alpha]_D$. For a description in English, find an article published between 1910 and 1920.

10. The name used for the compound described in Exercise 9 in the formula indexes of *Chemical Abstracts* is "benzamide, N-methylbenzyl" or "benzamide, N-α-methylbenzyl." (a) Find a second reference to the (−)-isomer which was published between 1930 and 1940, and compare the physical constants given there with the earlier data. (b) Find a reference to a paper published in Czechoslovakia between 1950 and 1960 giving data on the racemic form of the compound. (c) Find a reference to data on the (+)-isomer of the compound in a paper published between 1960 and 1970.

11. Determine whether or not each of the following compounds has ever been synthesized and, if it has, give the reference to its first appearance in the literature.

(a) Vitamin A   (b) Strychnine

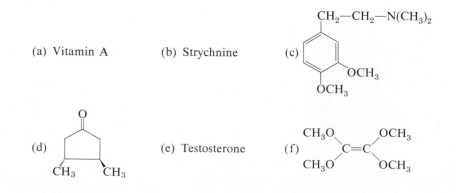

(d)   (e) Testosterone

---

[3] The German name for this compound is the same as in English except that the final "e" is omitted.

(g) Penicillin-G

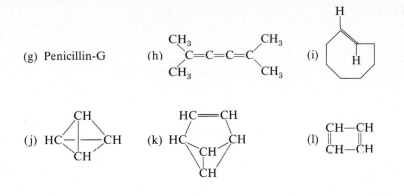

(h) $\begin{array}{c}CH_3\\ \\CH_3\end{array}\!\!C\!\!=\!\!C\!\!=\!\!C\!\!=\!\!C\begin{array}{c}CH_3\\ \\CH_3\end{array}$

(i)

(j) $HC\!\!-\!\!\!\!\begin{array}{c}CH\\ | \\CH\end{array}\!\!\!\!-\!\!CH$

(k) $HC\begin{array}{c}HC\!\!=\!\!CH\\ \\CH\\ | \\CH\end{array}CH$

(l) $\begin{array}{c}CH\!\!-\!\!CH\\ \| \quad\quad \| \\CH\!\!-\!\!CH\end{array}$

appendix one
drying agents
Desiccants

The important procedure of drying either a reagent, solvent, or product will be encountered at some stage of nearly every reaction performed in the organic chemistry laboratory. The techniques of drying solids and liquids and some of the drying agents which are commonly used will be described in this appendix.

Solids. It is important to dry solid organic compounds because water and organic solvents not only may affect melting points and quantitative elemental analyses, but also may cause difficulties if the wet solid is used in a reaction whose success depends on the absence of small amounts of water or other liquids. If the solid has been recrystallized from a volatile organic solvent, it can usually be dried to an extent satisfactory for most purposes by air-drying at room temperature. The process can be accelerated by spreading the solid out on a piece of filter paper or on a clay plate, either of which will serve to absorb traces of water or any excess solvent present. Another useful method to enhance the rate of drying is to collect the solid on a Büchner funnel held in a filter flask, to press the solid as dry as possible with a clean cork, and then to pull air through the filter cake by means of the aspirator (water pump).

If the solid compound is hygroscopic or has been recrystallized from water or a high-boiling solvent, it is usually necessary to dry the substance at atmospheric pressure or under vacuum in an oven operating at a temperature below the melting or decomposition point of the solid. For air-sensitive solids, drying must be done either in an inert atmosphere, such as that provided by nitrogen, or under vacuum. For samples which are to be submitted for quantitative elemental analysis, the solid is normally dried to constant weight by heating under vacuum.

It is often convenient to store dried organic solids in desiccators containing desiccants such as silica gel, phosphorus pentoxide, calcium chloride, etc., although tightly stoppered bottles make good storage vessels. Of course, it is also possible to use a desiccator to dry an organic solid wet with water. A solid that is contaminated with a hydrocarbon solvent can be dried by placing it in a desiccator containing a block of paraffin, which serves to absorb the hydrocarbon.

Liquids.   The drying of organic liquids is particularly important because their most common method of purification is by distillation. Any water present in an organic compound being distilled may react with the compound at or below the temperature of the distillation, or it may co- or steam distil with the liquid and thereby contaminate the distillate.[1]

Two of the more important general requirements for a drying agent are that neither it nor its hydrolysis product will react chemically with the organic liquid which is to be dried and that it can be easily and *completely* separated from the dried liquid. Another important consideration for a drying agent is that it be efficient in its action so that most or all of the water will be removed by the desiccant.

The more commonly used drying agents and some of their properties are listed in the accompanying table. Of the agents listed, calcium chloride, sodium sulfate, and magnesium sulfate will generally suffice for the needs of the basic course in organic chemistry laboratory.

It should be noted that the desiccants given in the table function in one of two ways: Either the drying agent interacts reversibly with water, by the process of adsorption or absorption [equation (1)], or it reacts irreversibly, with water serving as an acid or a base. In the case of reversible hydration, a certain amount of water will remain in equilibrium with the hydrated drying agent; the lower the amount of water left at equilibrium, the greater is the *efficiency* of the desiccant. Thus, magnesium sulfate is found to be more efficient in its drying power than is sodium sulfate but less efficient than calcium sulfate. Desiccants which remove water by irreversible chemical reaction clearly have very high efficiencies, but are generally more expensive than the other types of drying agents.

$$\text{Drying Agent} + H_2O \rightleftharpoons (\text{Drying Agent}) \cdot x\ H_2O \tag{1}$$

With those drying agents that operate by formation of hydrates, it is imperative that the drying agent be *completely* removed by filtration or decantation *before* distillation of the dried liquid, since most of the hydrates decompose with loss of water at temperatures above 30–40°. For those desiccants, such as calcium hydride, sodium, and phosphorus pentoxide, which react vigorously with water and for those like calcium sulfate, which have a low capacity for removing water, a preliminary drying using a less reactive and efficient desiccant with a high capacity is normally required. When a drying agent which reacts with water to evolve hydrogen is used, appropriate precautions must be taken to vent the hydrogen and thereby prevent buildup of this highly flammable gas.

It is poor technique to use an unnecessarily large quantity of drying agent when drying a liquid since the desiccant may adsorb or absorb the organic liquid along with water. Also, mechanical losses upon filtration or decantation of the dried solution may become significant. Of course, the amount of drying agent required depends upon the quantity of water present and upon the capacity of the desiccant. In general, a portion

[1] The drying of some organic solvents makes profitable use of the codistillation of water with the solvent in a process termed azeotropic distillation. For example, benzene can be dried in this manner. Solids can also sometimes be dried by dissolving them in a suitable solvent, removing any water by azeotropic distillation, and then recovering the solid by removal of the solvent.

of drying agent that covers the bottom of the vessel in which the liquid is contained should suffice. If additional desiccant is required, more can be added. Swirling the container of liquid and desiccant enhances the rate of drying when desiccants such as calcium chloride and magnesium sulfate are being used, since it hastens the establishment of the equilibrium for hydration.

Drying Agent	Acid-Base Properties	Product(s) with water	Comments [2]
$CaCl_2$	Neutral	$CaCl_2 \cdot H_2O$ $CaCl_2 \cdot 2H_2O$ $CaCl_2 \cdot 6H_2O$	High capacity and fast action; not highly efficient; good preliminary drying agent; readily separated from dried solution because $CaCl_2$ is available as large granules; cannot be used to dry either alcohols and amines (because of compound formation) or phenols, esters, and acids [because drying agent contains some $Ca(OH)_2$]; hexahydrate decomposes (loses water) above 30°.
Na_2SO_4	Neutral	$Na_2SO_4 \cdot 7H_2O$ $Na_2SO_4 \cdot 10H_2O$	Inexpensive; high capacity; slow action and low efficiency; good general preliminary drying agent; physical form is that of a powder so filtration required for removal of drying agent from dried solution; decahydrate decomposes above 33°.
$MgSO_4$	Neutral	$MgSO_4 \cdot H_2O$ $MgSO_4 \cdot 7H_2O$	Faster and more efficient than Na_2SO_4; good general drying agent; requires filtration for removal of drying agent from dried solution; heptahydrate decomposes above 48°.
$CaSO_4$	Neutral	$CaSO_4 \cdot \frac{1}{2}H_2O$	Low capacity but extremely fast action and high efficiency; preliminary drying of solution with drying agent of higher capacity recommended; hemihydrate can be dehydrated by heating at 235° for two to three hours.
$CuSO_4$	Neutral	$CuSO_4 \cdot H_2O$ $CuSO_4 \cdot 3H_2O$ $CuSO_4 \cdot 5H_2O$	More efficient than $MgSO_4$ or Na_2SO_4 but more expensive than either.
K_2CO_3	Basic	$K_2CO_3 \cdot 1\frac{1}{2}H_2O$ $K_2CO_3 \cdot 2H_2O$	Fair efficiency and capacity; good for esters, nitriles and ketones; cannot be used with acidic organic compounds.

[2] *Capacity*, as used in this table, refers to the amount of water which can be removed by a given weight of drying agent; *efficiency* refers to the amount of water, if any, in equilibrium with the hydrated desiccant.

Drying Agent	Acid-Base Properties	Product(s) with water	Comments [2]
H_2SO_4	Acidic	$H_3O^{\oplus}HSO_4^{\ominus}$	Good for alkyl halides and aliphatic hydrocarbons; cannot be used with even such weak bases as alkenes and ethers; high efficiency.
P_2O_5	Acidic	HPO_3 $H_4P_2O_7$ H_3PO_4	See comments under H_2SO_4; also good for ethers, aryl halides and aromatic hydrocarbons; extremely high efficiency; preliminary drying of solution recommended; dried solution can be distilled from drying agent.
CaH	Basic	$H_2 + Ca(OH)_2$	High efficiency but somewhat slow action; good for basic, neutral or weakly acidic compounds; cannot be used for base-sensitive substances; preliminary drying of solution recommended; dried solution can be distilled from drying agent.
Na	Basic	$H_2 + NaOH$	Good efficiency but slow action; cannot be used on compounds sensitive to alkali metals or to base; care must be exercised in destroying excess drying agent; preliminary drying required; dried solution can be distilled from drying agent.
BaO or CaO	Basic	$Ba(OH)_2$ or $Ca(OH)_2$	Slow action but high efficiency; good for alcohols and amines; cannot be used with compounds sensitive to base; dried solution can be distilled from drying agent.
KOH or NaOH	Basic	Solution	Rapid and efficient but use limited almost exclusively to drying of amines.
Molecular Sieve #3A or #4A [3]	Neutral	Water strongly adsorbed	Rapid and highly efficient; preliminary drying recommended; dried solution can be distilled from drying agent if desired. Molecular Sieve is the trade name for aluminosilicates whose crystal structure contains a network of pores of uniform diameter. The pore size of sieves #3A and 4A is such that only water and other small molecules such as ammonia can pass into the sieve; water is strongly adsorbed as water of hydration; hydrated sieves can be reactivated by heating at 300–320° under vacuum or at atmospheric pressure.

[3] The numbers refer to the nominal pore size, in Ångstrom units, of the sieve.

the laboratory notebook

A notebook used in the laboratory is to be a *complete* record of the experimental work. An 8 × 10-inch bound book is normally satisfactory for this purpose. The criterion used in judging what should be in the notebook is that the record should be so thorough and so well organized that anyone who reads the experiment can understand it and can see exactly what has been done, and thus could repeat it in precisely the same way the original work was done, if necessary.

All data are to be recorded in the notebook *at the time that they are obtained.* There is no reason for recording anything on odd pieces of paper to be transcribed into the notebook later. Neatness is desirable, but it is less important than having a notebook that is complete. Copying over experimental data costs time, cannot possibly improve them and, at times, may even worsen the data because errors may be made in copying. The notes made at the time of performance necessarily constitute the primary record.

In setting up the notebook at the beginning of the laboratory work, the following general structure should be used:

1. Leave room at the beginning of the notebook for a Table of Contents and keep it up-to-date.
2. Number the pages, if they are not already numbered.
3. Start every new experiment on a fresh page. Make all entries *in ink.*
4. It is not necessary to copy out the details of experimental procedure when you make no variation whatsoever from it. However, every experiment should have some *reference* (so that if the procedure is a standard one, it can be checked). In essence, your notebook should be a log of your laboratory operations; dates, times, and other pertinent conditions should be entered regularly.

It is particularly important that any *variations* from standard procedure and the reasons for them be noted; if this is not done, it is not possible to reconstruct later just what was done and why it was done. Nothing should ever be deleted from the notebook; merely draw a line through something *considered* wrong and amend it appropriately. The reasons which seem adequate at a particular time may later be found to be inadequate, and a result rejected at an early stage because it seemed impossible may eventually turn out to be the only worthwhile outcome of a particular experiment.

The notebook should not only contain a complete record of any observations but should also reflect precisely what was *concluded* from these observations and the *reasons* for these conclusions. Often one finds in laboratory work that data have been interpreted incorrectly, and only a careful and complete record of the experimental observations will allow one to discover the error in interpretation.

In organic chemistry, two distinctly different "types" of experiments are normally carried out in the laboratory. The first of these is the "preparative-type" of experiment, in which one compound is converted into another. The second of these is the "investigative-type" of experiment, in which one studies physical properties such as boiling point or melting point, or chemical properties, such as qualitative tests for various functional groups in a molecule. These two types of experiments require slightly different types of notebook write-up. As an aid in preparing an acceptable notebook, suggested formats for both preparative- and investigative-type experiments are presented below. However, the amount of advance preparation and the notebook form that is to be used will probably be discussed by the instructor. The formats given here are illustrative only.

A. Notebook format
for preparative-type experiments

The important information that should be included in describing preparative-type experiments is the following:

I. Introduction.   Prepare a brief statement regarding the work to be done.

II. Main reaction(s) and mechanisms.   Give the main reaction(s) leading to the preparation of the desired compound. Where possible, include the mechanisms of the reactions.

III. Table of reagents and products.   List, in tabular form, the molecular weights of each reagent and product, and calculate and enter the number of moles of each reagent to be used. Use the chemical equation(s) for the main reactions to derive the theoretical molar ratio of reagents and products, and enter these values in the list.

Determine which reagent, if any, is used in less than the theoretical molar ratio required; if more than one such reagent is found, find which of these deviates most from the molar ratio needed. This will be the *limiting reagent* in the reaction, *i.e.*, the reagent which will determine the maximum amount of product that can be formed. After the limiting reagent has been consumed, the desired reaction will cease no matter how much of the other reagents remain.

IV. Yield data.   Using the table of reagents and products, calculate the maximum theoretical yield of product that one can expect to obtain from the starting materials. The theoretical yield of product (in moles) can be calculated by determining the moles of product formed per mole of limiting reagent used:

Theoretical yield (in moles) = (theoretical molar ratio of product to limiting reagent) $\times$ (moles of limiting reagent actually used)

Multiplying this theoretical yield by the molecular weight of the product gives the expected yield (in grams).

Determine the actual yield of product, in grams, that is obtained in the experiment. Use the actual yield and the theoretical yield to determine the percent yield, which is:

$$\% \text{ yield} = \frac{\text{actual yield (grams)}}{\text{theoretical yield (grams)}} \times 100$$

V. *Observed properties of the product.* Enter the physical properties of the product that is obtained from the preparation. Important data include melting point or boiling point, color, crystalline form, etc. It is important to record these properties accurately.

VI. *Side reactions.* List all possible side reactions that may occur in the reaction. To help determine what these might be, it is often necessary to consult additional sources, such as lecture notes or the lecture textbook.

VII. *Other methods of preparation.* List alternate methods of preparation that could be used to prepare this same compound. Consult a textbook, and comment briefly on the advantages of the other methods as opposed to the one used. Suggest a reason why the present method of preparation is preferable, if it is.

VIII. *Method of purification.* A very important phase of synthetic organic chemistry is the purification of the desired product. Because of the many and varied reactions that organic compounds undergo, frequently the most difficult step in the preparation of a compound is not its actual formation but its isolation in pure form from side products and unchanged starting materials. In order to accomplish this, list all the compounds that could possibly be present on the basis of a consideration of the main and side reactions. Devise a purification scheme, using a flow sheet, to show how various experimental procedures eliminate the undesired substances and yield the pure product. By examining a purification scheme in this manner, the purpose of each step in the procedure becomes clear, and one may be able to predict what impurities, if any, may contaminate the product.

IX. *Answers to exercises.*

This type of format is shown in Figure 1 (pages 502 and 503), using an experiment involving the chlorination of cyclohexane (Chapter 5.1, page 104) as an illustration. The figure represents an open notebook; it should be noted that a new experiment is normally started on the left-hand page.

B. NOTEBOOK FORMAT
FOR INVESTIGATIVE-TYPE EXPERIMENTS

A format that can be used for investigative-type experiments is similar to that shown in Figure 1. The description of a new experiment should normally start on the left-hand page, and each new experiment should contain a title and reference. The important information which should be included in the notebook is:

   I. *Introduction.* Give a brief introduction to the experiment and state the purpose(s) of it. This should take up no more than half of the page.

  II. *Experiments and results.* Give a brief statement (one or two lines) of each experiment that is to be performed, and leave sufficient room to enter the results as they are obtained. Do *not* recopy the experimental details from the text—just identify each experiment that is to be done. Continue this section as far as needed.

 III. *Conclusions.* After completing the assigned experiments, state briefly the conclusions which have been reached on the basis of the results. If the experiment was the identification of an unknown, summarize the results of the findings here.

  IV. *Answers to exercises.*

FIGURE 1.   Sample for preparative-type experiment.

Notebook Page No. 3.

CHLORINATION OF CYCLOHEXANE

Reference: Chapter 5.1 in Roberts, Gilbert, Rodewald and Wingrove, page 104

I. INTRODUCTION.

Chlorocyclohexane is to be prepared from cyclohexane using sulfuryl chloride as reagent.

II. MAIN REACTION(S) AND MECHANISM(S).

$$C_6H_{12} + SO_2Cl_2 \rightarrow C_6H_{11}Cl + HCl + SO_2$$

(Mechanism intentionally omitted)

Notebook Page No. 4.

III. TABLE OF REAGENTS AND PRODUCTS

Compound	M.W.	Wt. Used, g.	Moles Used	Ratio of Moles: Theory	Used	Other Data
Cyclohexane	84	33.6	0.4	1	2	Density = 0.779 bp 81°
Sulfuryl chloride	135	27.0	0.2	1	1	Density = 1.667 bp 69°
Chlorocyclohexane	118.5			1		Density = 1.016 colorless bp 142.5°
Benzoyl peroxide	—	0.1	—	(Initiator)		

Limiting reagent: *Sulfuryl chloride*

IV. YIELD DATA.

(To determine the *theoretical yield*, it is necessary first to determine the *limiting reagent*. The equation for the main reaction shows that 1 mole of cyclohexane reacts with 1 mole of sulfuryl chloride. Observe, however, that in this experiment 0.4 mole of cyclohexane is allowed to react with 0.2 mole of sulfuryl chloride, so that sulfuryl chloride is the *limiting reagent*. Since the theoretical molar ratio of chlorocyclohexane to sulfuryl chloride is 1:1, the number of moles of chlorocyclohexane produced will equal the number of moles of sulfuryl chloride used.)

Theoretical yield of chlorocyclohexane = (moles $C_6H_{11}Cl$) (M.W. $C_6H_{11}Cl$) = (0.2 mole) (118.5 g/mole) = 23.7 g.

If, in an actual experiment, 15.0 g of pure chlorocyclohexane was obtained, the percent yield would be:

$$\% \text{ Yield} = \frac{15.0 \text{ g}}{23.7 \text{ g}} \times 100 = 63\%$$

FIGURE 1. Sample for preparative-type experiment (continued).

Notebook Page No. 5.

V. OBSERVED PROPERTIES OF THE PRODUCT.
bp 138–140°; colorless liquid; insoluble in water.

VI. SIDE REACTIONS.
$C_6H_{11}Cl + SO_2Cl_2 \rightarrow C_6H_{10}Cl_2 + SO_2 + HCl$
$\qquad\qquad\qquad\qquad$ + other polychlorinated cyclohexyl compounds.

VII. OTHER METHODS OF PREPARATION.
$C_6H_{10} + HCl \rightarrow C_6H_{11}\text{—}Cl$
$C_6H_{11}\text{—}OH + HCl \rightarrow C_6H_{11}\text{—}Cl + H_2O$
$C_6H_{11}\text{—}OH + SOCl_2 \rightarrow C_6H_{11}\text{—}Cl + HCl + SO_2$

Notebook Page No. 6.

VIII. METHOD OF PURIFICATION.

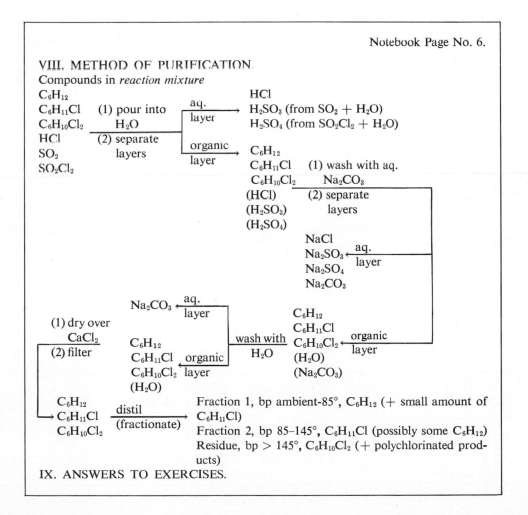

IX. ANSWERS TO EXERCISES.

compilation
of nmr absorptions

Listed below are the chemical shifts observed for the protons of a number of different types of organic compounds. The shifts are classified according to whether they are (a) methyl, (b) methylene, or (c) methine, type of hydrogen atoms. The italicized hydrogen are those responsible for the absorptions listed.

Compound	Chemical Shift (ppm)		Compound	Chemical Shift (ppm)	
(a) Methyl Hydrogen Atoms	τ	δ		τ	δ
CH_3NO_2	5.7	4.3	CH_3CHO	7.8	2.2
CH_3F	5.7	4.3	CH_3I	7.8	2.2
$(CH_3)_2SO_4$	6.1	3.9	$(CH_3)_3N$	7.9	2.1
$C_6H_5COOCH_3$	6.1	3.9	$CH_3CON(CH_3)_2$	7.9	2.1
$C_6H_5-O-CH_3$	6.3	3.7	$(CH_3)_2S$	7.9	2.1
CH_3COOCH_3	6.4	3.6	$CH_2=C(CN)CH_3$	8.0	2.0
CH_3OH	6.6	3.4	CH_3COOCH_3	8.0	2.0
$(CH_3)_2O$	6.8	3.2	CH_3CN	8.0	2.0
CH_3Cl	7.0	3.0	CH_3CH_2I	8.1	1.9
$C_6H_5N(CH_3)_2$	7.1	2.9	$CH_2=CH-C(CH_3)=CH_2$	8.2	1.8
$(CH_3)_2NCHO$	7.2	2.8	$(CH_3)_2C=CH_2$	8.3	1.7
CH_3Br	7.3	2.7	CH_3CH_2Br	8.4	1.6
CH_3COCl	7.3	2.7	$C_6H_5C(CH_3)_3$	8.7	1.3
CH_3SCN	7.4	2.6	$C_6H_5CH(CH_3)_2$	8.8	1.2
$C_6H_5COCH_3$	7.4	2.6	$(CH_3)_3COH$	8.8	1.2
$(CH_3)_2SO$	7.5	2.5	$C_6H_5CH_2CH_3$	8.8	1.2
$C_6H_5CH-CHCOCH_3$	7.7	2.3	CH_3CH_2OH	8.8	1.2
$C_6H_5CH_3$	7.7	2.3	$(CH_3CH_2)_2O$	8.8	1.2
$(CH_3CO)_2O$	7.8	2.2	$CH_3(CH_2)_3Cl, Br, I$	9.0	1.0
$C_6H_5OCOCH_3$	7.8	2.2	$CH_3(CH_2)_4CH_3$	9.1	0.9
$C_6H_5CH_2N(CH_3)_2$	7.8	2.2	$(CH_3)_3CH$	9.1	0.9

Compound	Chemical Shift (ppm)		Compound	Chemical Shift (ppm)	
(b) Methylene Hydrogen					
Atoms	τ	δ		τ	δ
$EtOCOC(CH_3)\!-\!CH_2$	4.5	5.5	$EtCH_2Cl$	6.6	3.4
CH_2Cl_2	4.7	5.3	$(CH_3CH_2)_4N^+I^-$	6.6	3.4
CH_2Br_2	5.1	4.9	CH_3CH_2Br	6.6	3.4
$(CH_3)_2C\!=\!CH_2$	5.4	4.6	$C_6H_5CH_2N(CH_3)_2$	6.7	3.3
$CH_3COO(CH_3)C\!=\!CH_2$	5.4	4.6	$CH_3CH_2SO_2F$	6.7	3.3
$C_6H_5CH_2Cl$	5.5	4.5	CH_3CH_2I	6.9	3.1
$(CH_3O)_2CH_2$	5.5	4.5	$C_6H_5CH_2CH_3$	7.4	2.6
$C_6H_5CH_2OH$	5.6	4.4	CH_3CH_2SH	7.6	2.4
$CF_3COCH_2C_3H_7$	5.7	4.3	$(CH_3CH_2)_3N$	7.6	2.4
$Et_2C(COOCH_2CH_3)_2$	5.9	4.1	$(CH_3CH_2)_2CO$	7.6	2.4
$HC\!\equiv\!C\!-\!CH_2Cl$	5.9	4.1	$BrCH_2CH_2CH_2Br$	7.6	2.4
$CH_3COOCH_2CH_3$	6.0	4.0	Cyclopentanone $\alpha\!-\!CH_2$	8.0	2.0
$CH_2\!=\!CHCH_2Br$	6.2	3.8	Cyclohexene $\alpha\!-\!CH_2$	8.0	2.0
$HC\!\equiv\!CCH_2Br$	6.2	3.8	Cycloheptane	8.5	1.5
$BrCH_2COOCH_3$	6.3	3.7	Cyclopentane	8.5	1.5
CH_3CH_2NCS	6.4	3.6	Cyclohexane	8.6	1.4
CH_3CH_2OH	6.4	3.6	$CH_3(CH_2)_4CH_3$	8.6	1.4
			Cyclopropane	9.8	0.2

Compound	Chemical Shift (ppm)		Compound	Chemical Shift (ppm)	
(c) Methine Hydrogen					
Atoms	τ	δ		τ	δ
C_6H_5CHO	0.0	10.0	C_6H_5Cl	2.8	7.2
$p\text{-}ClC_6H_4CHO$	0.1	9.9	$CHCl_3$	2.8	7.2
$p\text{-}CH_3OC_6H_4CHO$	0.2	9.8	$CHBr_3$	3.2	6.8
CH_3CHO	0.3	9.7	p-Benzoquinone	3.2	6.8
Pyridine (α)	1.5	8.5	$C_6H_5NH_2$	3.4	6.6
$p\text{-}C_6H_4(NO_2)_2$	1.6	8.4	Furan (β)	3.7	6.3
$C_6H_5CH\!=\!CHCOCH_3$	2.1	7.9	$CH_3CH\!=\!CHCOCH_3$	4.2	5.8
C_6H_5CHO	2.4	7.6	Cyclohexene	4.4	5.6
Furan (α)	2.6	7.4	$(CH_3)_2C\!=\!CHCH_3$	4.8	5.2
Naphthalene (β)	2.6	7.4	$(CH_3)_2CHNO_2$	5.6	4.4
$p\text{-}C_6H_4I_2$	2.6	7.4	Cyclopentyl bromide	5.6	4.4
$p\text{-}C_6H_4Br_2$	2.7	7.3	$(CH_3)_2CHBr$	5.8	4.2
$p\text{-}C_6H_4Cl_2$	2.8	7.2	$(CH_3)_2CHCl$	5.9	4.1
C_6H_6	2.7	7.3	$C_6H_5C\!\equiv\!C\!-\!H$	7.1	2.9
C_6H_5Br	2.7	7.3	$(CH_3)_3C\!-\!H$	8.4	1.6

table of characteristic infrared frequencies

A. *The hydrogen stretch region (3600–2500 cm⁻¹).* Absorption in this region is associated with the stretching vibration of hydrogen atoms bonded to carbon, oxygen, and nitrogen. Care should be exercised in the interpretation of very weak bands because these may be overtones of strong bands occurring at frequencies one-half the value of the weak absorption, *i.e.*, 1800–1250 cm⁻¹. Overtones of bands near 1650 cm⁻¹ are particularly common.

$\bar{v}(cm^{-1})$	Functional Group	Comments
(1) 3600–3400	O—H stretching Intensity: variable	3600 cm⁻¹ (sharp) unassociated O—H, 3400 cm⁻¹ (broad) associated O—H; both bands frequently present in alcohol spectra; with strongly associated O—H (CO_2H or enolized β-dicarbonyl compound) band is very broad (*ca.* 500 cm⁻¹ with its center at 2900–3000 cm⁻¹).
(2) 3400–3200	N—H stretching Intensity: medium	3400 cm⁻¹ (sharp) unassociated N—H, 3200 cm⁻¹ (broad) associated N—H; an NH_2 group usually appears as a doublet (separation *ca.* 50 cm⁻¹); the N—H of a secondary amine is often very weak.
(3) 3300	C—H stretching of an alkyne Intensity: strong	The *complete* absence of absorption in this region 3300–3000 cm⁻¹ indicates the absence of hydrogen atoms bonded to C≡C or C=C and *usually* indicates the lack of unsaturation in the molecule. Because this absorption may be very weak in large molecules, some care should be exercised in this interpretation.
(4) 3080–3010	C—H stretching of an alkene Intensity: strong to medium	
(5) 3050	C—H stretching of an aromatic compound Intensity: variable; usually medium to weak	
(6) 3000–2600	OH strongly hydrogen-bonded Intensity: medium	A very broad band in this region superimposed on the C—H stretching frequencies is characteristic of carboxylic acids [see (1) above].

$\bar{v}$(cm^{-1})	Functional Group	Comments
(7) 2980–2900	C—H stretching of an aliphatic compound Intensity: strong	Just as in the previous C—H entries [(3)–(5) above], *complete* absence of absorption in this region indicates the absence of hydrogen atoms bonded to tetravalent carbon atoms. The tertiary C—H absorption is weak.
(8) 2850–2760	C—H stretching of an aldehyde Intensity: weak	Either one or two bands *may* be found in this region for a single aldehyde function in the molecule.

B. *The triple-bond region (2300–2000 cm^{-1}).* Absorption in this region is associated with the stretching vibration of triple bonds.

$\bar{v}$(cm^{-1})	Functional Group	Comments
(1) 2260–2215	C≡N Intensity: strong	Nitriles conjugated with double bonds absorb at lower end of frequency range; nonconjugated nitriles appear at upper end of range.
(2) 2150–2100	C≡C Intensity: strong in *terminal* alkynes; variable in others.	This band will be absent if the alkyne is symmetrical and will be very weak or absent if the alkyne is nearly symmetrical.

C. *The double-bond region (1900–1550 cm^{-1}).* Absorption in this region is *usually* associated with the stretching vibration of carbon-carbon, carbon-oxygen, and carbon-nitrogen double bonds.

$\bar{v}$(cm^{-1})	Functional Group	Comments
(1) 1815–1770	C=O stretching of an acid chloride Intensity: strong	Conjugated and nonconjugated carbonyls absorb at the lower and upper ends, respectively, of the range.
(2) 1870–1800 and 1790–1740	C=O stretching of an acid anhydride Intensity: strong	*Both bands* are present. *Each band* is altered by ring size and conjugation to approximately the same extent noted for ketones [see (4) below].
(3) 1750–1735	C=O stretching of an ester or lactone Intensity: very strong	This band is subject to all of the structural effects discussed in entry (4) below. Thus, a conjugated ester absorbs at *ca.* 1710 cm^{-1} and a γ-lactone absorbs at *ca.* 1780 cm^{-1}.
(4) 1725–1705	C=O stretching of an aldehyde or ketone Intensity: very strong	This value refers to the carbonyl absorption frequency of an acyclic, nonconjugated aldehyde or ketone in which no electronegative groups, *e.g.,* halogens, are near the carbonyl group. Because this frequency is altered in a predictable way by structural alterations, the following generalizations may be drawn.

$\bar{v}(cm^{-1})$	Functional Group	Comments
		I. *Effect of Conjugation:* Conjugation of the carbonyl group with an aryl ring or carbon-carbon double or triple bond lowers the frequency by about 30 cm^{-1}. If the carbonyl group is part of a cross-conjugated system (unsaturation on each side of the carbonyl group), the frequency is lowered by about 50 cm^{-1}.
		II. *Effect of Ring Size:* Carbonyl groups in six-membered and larger rings exhibit approximately the same absorption as acyclic ketones; carbonyl groups contained in rings smaller than six absorb at higher frequencies, *e.g.*, a cyclopentanone absorbs at *ca.* 1745 cm^{-1} and a cyclobutanone absorbs at about 1780 cm^{-1}. The effects of conjugation and ring size are additive, *e.g.*, a 2-cyclopentenone absorbs at *ca.* 1710 cm^{-1}.
		III. *Effect of Electronegative Atoms:* An electronegative atom (especially oxygen or halogen) bonded to the α-carbon atom of an aldehyde or ketone may raise the position of the carbonyl absorption frequency by about 20 cm^{-1}.
(5) 1700	C=O stretching of an acid Intensity: strong	This absorption frequency is lowered by conjugation as noted under entry (4).
(6) 1690–1650	C=O stretching of an amide or lactam Intensity: strong	This band is lowered in frequency by about 20 cm^{-1} by conjugation. The frequency of the band is raised about 35 cm^{-1} in γ-lactams and 70 cm^{-1} in β-lactams.
1660–1600	C=C stretching of an alkene Intensity: variable	Nonconjugated alkenes appear at upper end of range and absorptions are usually weak; conjugated alkenes appear at lower end of range and absorptions are medium to strong. The absorption frequencies of these bands are raised by ring strain but to a lesser extent than noted with carbonyl functions [see (4) above].
1680–1640	C=N stretching Intensity: variable	This band is usually weak and difficult to assign.

D. *The hydrogen bend region (1600–1250 cm⁻¹)*. Absorption in this region is commonly due to bending vibration of hydrogen atoms attached to carbon and to nitrogen. These bands generally do not provide much useful structural information. In the listing below, the bands that are most useful for structural assignment have been marked with an asterisk.

$\bar{v}(cm^{-1})$	Functional Group	Comments
1600	—NH_2 bending Intensity: strong to medium	This band in conjunction with bands in the 3300 cm^{-1} region is often used to characterize primary amines and unsubstituted amides.
1540	—NH— bending Intensity: generally weak	This band in conjunction with bands in the 3300 cm^{-1} region is often used to characterize secondary amines and monosubstituted amines. In the case of secondary amines this band, like the N—H stretching band in the 3300 cm^{-1} region, may be very weak.
* 1520 and 1350	NO_2 coupled stretching bands Intensity: strong	This pair of bands is usually very intense.
1465	—CH_2— bending Intensity: variable	The intensity of this band varies according to the number of methylene groups present; the more such groups there are, the more intense the absorption.
1410	—CH_2—bending of carbonyl-containing component Intensity: variable	This absorption is characteristic of methylene groups adjacent to carbonyl functions; its intensity depends on the number of such groups present in the molecule.
* 1450 and 1375	—CH_3 Intensity: strong	The band of lower frequency (1375 cm^{-1}) is usually used to characterize a methyl group. If two methyl groups are bonded to one carbon atom, a characteristic doublet (1385 and 1365 cm^{-1}) will be present.
1325	| —CH bending Intensity: weak	This band is weak and often unreliable.

E. *The fingerprint region (1250–600 cm⁻¹)*. The fingerprint region of the spectrum is generally rich in detail, with many bands appearing. This region is particularly diagnostic for determining whether an unknown substance is identical with a known substance, the infrared spectrum of which is available. It is not practical to make assignments to all of these bands, because many of them represent combination frequencies and therefore are very sensitive to the total molecular structure; in addition, many single-bond stretching vibrations and a variety of bending vibrations also appear in this region. Suggested structural assignments in this region must be regarded as tentative and generally taken as corroborative evidence in conjunction with assignments of bands at higher frequencies.

$\bar{v}\,(\text{cm}^{-1})$	Functional Group	Comments
1200	⬡—O— Intensity: strong	It is not certain whether these strong bands arise from C—O bending or C—O stretching vibrations. One or more strong bands is found in this region of the spectra of alcohols, ethers, and esters. The relationship indicated between structure and band location is only approximate and any structural assignment based on this relationship must be regarded as tentative. Esters often exhibit one or two strong bands between 1170 and 1270 cm^{-1}.
1150	—C—O— Intensity: strong	
1100	—CH—O— Intensity: strong	
1050	—CH$_2$—O— Intensity: strong	
965	H\C=C/ /\H C—H bending Intensity: strong	This strong band is present in the spectra of *trans*-1,2-disubstituted ethylenes.
985 and 910	H\C=C/H /\H C—H bending Intensity: strong	The lower-frequency band of these two strong bands is used to characterize a terminal vinyl group.
890	\C=CH$_2$ C—H bending Intensity: strong	This strong band, used to characterize a methylene group, may be raised by 20–80 cm^{-1} if the methylene group is bonded to an electronegative group or atom.
810–840	H\C=C/ /\ Intensity: strong	Very unreliable, this band is not always present, and frequently seems to be outside this range as substituents are varied.
700	\C=C/ H/\H Intensity: variable	This band, attributable to a *cis*-1,2-disubstituted ethylene, is unreliable because it is frequently obscured by solvent absorption or other bands.
750 and 690	H—⬡—H (H, H, H, H) C—H bending Intensity: strong	These bands are of limited value because they are frequently obscured by solvent absorption or other bands. Their usefulness will be most important when independent evidence leads to a structural assignment complete except for position of aromatic substituents.

$\bar{v}(cm^{-1})$	Functional Group	Comments
750	C—H bending Intensity: very strong	
780 and 700	and 1, 2, 3 Intensity: very strong	
825	and 1, 2, 4 Intensity: very strong	
1400–1000	C—F Intensity: strong	The position of these bands is quite sensitive to structure. They are, therefore, not particularly useful because the presence of halogen is more easily detected by chemical methods. The bands are invariably strong.
800–600	C—Cl Intensity: strong	
700–500	C—Br Intensity: strong	
600–400	C—I Intensity: strong	

list of nmr
and ir spectra

Compound	NMR (Figure No.)	IR (Figure No.)
p-Acetamidobenzenesulfonyl chloride		19.9
Acetanilide	19.8	19.7
Acetophenone	16.20	16.19
Acetylferrocene	20.4	20.3
Adipic acid		15.5
2-Aminothiazole	19.13	19.12
Aniline	17.4	17.3
Anthracene		13.4
trans-Benzalacetophenone	16.22	16.21
Benzaldehyde	15.7	15.6
Benzamide	17.2	17.1
Benzanilide	17.6	17.5
Benzoic acid	14.2	14.1
Benzoic anhydride		17.7
Benzonitrile		17.8
p-Benzoquinone		10.4
Benzyl alcohol		15.8
Benzyl chloride		16.12
N-Benzyl-m-nitroaniline	16.11	16.10
N-Benzylidene-m-nitroaniline		16.9
4-Bromoacetanilide	19.16	19.15
Bromobenzene		14.5
1-Bromo-3-chloro-5-iodobenzene	19.24	19.23
n-Butyl alcohol	12.11	12.10
n-Butylbenzene		8.1(a)
sec-Butylbenzene		8.1(c)
t-Butylbenzene	5.14	5.13
		8.1(d)
Butylbenzenes, reaction mixture	8.4 (a), (b)	8.3 (a), (b) and (c)
sec-Butyl-p-xylene ($C_{12}H_{18}$)	4.9	4.9

Compound	NMR (Figure No.)	IR (Figure No.)
Carvone	25.3	25.3
2-Chloro-4-bromoacetanilide	19.19	19.17
2-Chloro-4-bromoaniline	19.20	19.19
2-Chloro-4-bromo-6-iodoaniline	19.22	19.21
1-Chlorobutane	5.6	5.5
Chlorocyclohexane		5.8
2-Chloro-2-methylbutane	12.3	12.2
2-Chloronaphthalene		13.3
Cinnamaldehyde	16.2	16.1
N-Cinnamylaniline	16.8	16.7
N-Cinnamylideneaniline		16.6
N-Cinnamylidene-m-nitroaniline		16.3
N-Cinnamyl-m-nitroaniline	16.5	16.4
Citral	24.2	24.1
1,3,5-Cycloheptatriene	20.6	20.5
Cyclohexane		5.7
cis-Cyclohexane-1,2-dicarboxylic acid		6.23
Cyclohexanol	6.15	6.14
Cyclohexanone	11.6	11.5
Cyclohexene	6.17	6.16
4-Cyclohexene-cis-1,2-dicarboxylic acid		6.22
4-Cyclohexene-cis-1,2-dicarboxylic anhydride	10.8	10.7
1,3-Cyclopentadiene	10.2	10.1
1,x-Dichlorobutanes (mixture)	5.4	
Diethyl ethoxymethylenemalonate	18.1	
2,3-Dimethylbutane		4.3(a)
Dimethyl fumarate	6.21	6.20
Dimethyl maleate		6.19
Dimethyl maleate and dimethyl fumarate, mixture	6.18	
trans, trans-1,4-Diphenyl-1,3-butadiene	16.18	16.17
Ethylbenzene	5.10	5.9
Ferrocene	20.2	20.1
2-Furaldehyde	11.4	11.3
Indene	13.2	13.1
Isobutyl alcohol	4.8(b)	4.8(a)
	15.2	15.1
Isobutylbenzene		8.1(b)
Isopropylbenzene	5.12	5.11
Maleic anhydride		10.3
2-Methoxyethyl acetate	25.2	25.2

Compound	NMR (Figure No.)	IR (Figure No.)
Methyl benzoate	14.4	14.3
2-Methylbutane		4.3(b)
2-Methyl-1-butene	12.7	12.6
2-Methyl-2-butene	12.5	12.4
Methylbutenes, reaction mixtures from dehydrohalogenation of 2-chloro-2-methylbutane	12.8, 12.9	
2-Methyl-3-butyn-2-ol	9.2	9.1
3-Methyl-2-cyclohexenone		4.2
2-Methyl-3-heptanol	14.8	14.7
3-Methyl-3-hydroxy-2-butanone	9.4	9.3
4-Methyl-2-pentanol	6.3	6.2
2-Methyl-1-pentene	6.13	6.12
2-Methyl-2-pentene	6.7	6.6
4-Methyl-1-pentene	6.5	6.4
cis-4-Methyl-2-pentene	6.11	6.10
trans-4-Methyl-2-pentene	6.9	6.8
2-Methylpropanal (isobutyraldehyde)	15.4	15.3
Nitrobenzene		19.6
1-Nitropropane	4.5	
t-Pentyl alcohol	12.13	12.12
1-Phenylethanol	19.3	19.2
Piperic acid		24.6
Piperine	24.5	24.4
Polystrene		19.5
cis-Stilbene	16.15	16.13
trans-Stilbene	16.16	16.14
Styrene		19.4
Sulfanilamide	19.11	19.10
Sulfathiazole		19.14
3-Sulfolene (2,5-dihydrothiophene-1,1-dioxide)	10.6	10.5
Triphenylmethane	5.17	5.16
Triphenylmethanol	14.6	
Triptycene	13.5	

index

---

* Numbers in italics refer to entries in
tables of Chapter 25.

EQUIPMENT COMMONLY USED IN THE ORGANIC CHEMISTRY LABORATORY

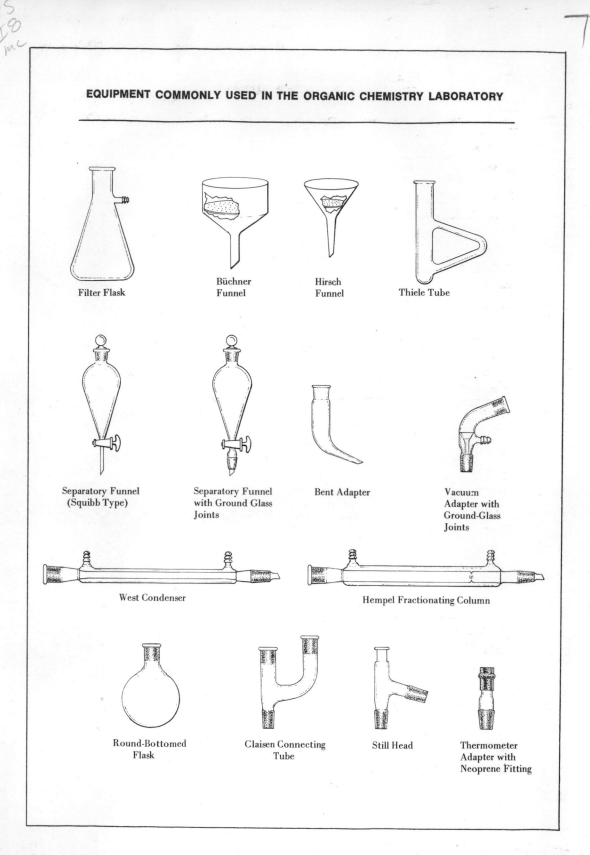

Filter Flask

Büchner Funnel

Hirsch Funnel

Thiele Tube

Separatory Funnel (Squibb Type)

Separatory Funnel with Ground Glass Joints

Bent Adapter

Vacuum Adapter with Ground-Glass Joints

West Condenser

Hempel Fractionating Column

Round-Bottomed Flask

Claisen Connecting Tube

Still Head

Thermometer Adapter with Neoprene Fitting